Physics and Contemporary Needs

―――― Volume 3 ――――

A Continuation Order Plan is available for this series. A continuation order will bring delivery of each new volume immediately upon publication. Volumes are billed only upon actual shipment. For further information please contact the publisher.

Physics and Contemporary Needs

Volume 3

Edited by Riazuddin

Quaid-I-Azam University
Islamabad, Pakistan

Plenum Press . New York and London

The Library of Congress cataloged the first volume of this title as follows:

International Summer College on Physics and Contemporary Needs.
  Physics and contemporary needs. v. 1—
1976—
New York, Plenum Press.
  2 v. ill. 26 cm. annual
  "Proceedings of the International Summer College on Physcis and Contemporary Needs."
  Key title: Physics and contemporary needs, ISSN 0163-2051
  1. Physics—Congresses. 2. Geophysics—Congresses. 3. Technology—Congresses. I. Title.
QC1.I 647a   530   78-647137

Library of Congress Catalog Card Number 78-647137
ISBN 0-306-40282-3

Lectures presented at the Third International Summer College on
Physics and Contemporary Needs, Nathiagali, Pakistan,
June 17—July 5, 1978

©1979 Plenum Press, New York
A Division of Plenum Publishing Corporation
227 West 17th Street, New York, N.Y. 10011

All rights reserved

No part of this book may be reproduced, stored in a retrieval system, or transmitted,
in any form or by any means, electronic, mechanical, photocopying, microfilming,
recording, or otherwise, without written permission from the Publisher

Printed in the United States of America

PREFACE

These proceedings cover the lectures delivered at the Third International Summer College on Physics and Contemporary Needs held from June 17 - July 5, 1978 at Nathiagali, one of the scenic hill resorts in the northern part of Pakistan. The college was organized by The Pakistan Atomic Energy Commission (PAEC) and co-sponsored by the International Centre for Theoretical Physics, Trieste (ICTP). It also received a financial grant by the University Grants Commission for the participation of physicists from various universities of Pakistan. The college was attended by 14 lecturers, 2 invited seminar speakers and 156 participants from 23 countries and consisted of 15 concentrated days of lectures, seminars and informal discussions. These proceedings contain only regular lectures delivered there, but the seminars which were held there are listed in the Appendix.

This year the college put special emphasis on energy, particularly on nuclear energy and its role in the context of energy systems. However the lectures delivered at the college also covered a wide spectrum of physics. The lectures gave an overview of various topics covered at the college and emphasized the inter-disciplinary aspects of physics. Some of the lecturers also indicated the areas where research in developing countries with limited facilities could be carried out. The college had a definite objective of encouraging the physicists, particularly those working at the universities, to apply their knowledge of physics and methodology of research to the needs of modern society. The series of the colleges of which the present college is the third one are an attempt to remove the barrier of isolation for the physicists working in developing countries, far removed from active centres of research. Thus these colleges could help to fill the important gap in communication between the physicists of developing and advanced countries. It is hoped that the colleges would help the cause of science in general and that of physics in particular in the third world countries.

The success of the college is due in large part to the lecturers who gave an excellent presentation of the material covered in their respective lectures, to the participants who took enthusiastic interest in the lectures and discussions, to the local organisation committee who worked very hard and in spite of the remoteness of Nathiagali tried to make the stay of the participants both enjoyable and useful and above all to Mr.Munir Ahmad Khan, Chairman, Pakistan Atomic Energy Commission who took a very keen personal interest in making the college a success. We are deeply grateful to many other persons too numerous to mention who helped us in the organising of this college.

The volume is divided into five parts: Part I covers Physics and Energy with lectures on World Energy Problems, Role of Nuclear Energy in meeting Power Needs, Latest Trends in the Economics of Nuclear Power, Long Term Energy Systems and the Role of Nuclear and Solar Energy and Nuclear Fusion with particular reference to Laser Induced Fusion. This part also contains lectures on various topics in Reactor Physics. Part II covers Physics and Technology with lectures on Amorphous Semi-conductors and Solar Energy Materials. Part III contains lectures on Computational Methods in Physics. Part IV covers Physics and Frontiers of Knowledge with lectures on Observational Traits of Black Holes in the Optical Bands, Fundamental Constituents of Matter and Unification of Weak and Electromagnetic Interactions and Polarisation Phenomena in Microscopic Physics. Finally Part V contains lectures on Science and Development.

In bringing out the proceedings of the college our sincere thanks go to Prof. Fayyazuddin, Dr.A.H.Nayyar, Dr.M.Aslam, Mr.Sajjad Mahmood and Mr.M.M.Ilyas, who helped in various ways. Thanks are also due to Mr.Azhar Hussain for excellent art work and to Mr.S.U.Khan who did a very good job in typing the manuscripts.

Riazuddin

LECTURERS

    Mohammad Ahmed
    P.L.Bernacca
    B.Elbek
    Ugo Farinelli
    D.Faude
    R.Hofstadter
    J.A.Lane
    M.J.Moravscik
    R.L.Murray
    William Paul
    Riazuddin
    K.V.Roberts
    L.Sani
    L.Slack
    M.Yaqub*

*His lectures on Physics of Disordered Materials could not be included in these proceedings.

CONTENTS

PART I : PHYSICS AND ENERGY

Perspective on Pakistan's Energy Problems 3
    Mohammad Ahmad

World Energy Problems 23
    Bent Elbek

Long-Term Energy Systems and the Role of Nuclear and
       Solar Energy 63
    Dieter Faude

Laser Produced Nuclear Fusion 111
    R. Hofstadter

Latest Trends in the Economics of Nuclear Power 149
    James A. Lane

Role of Nuclear Energy with Particular Reference to
       Western Europe 191
    L. Sani

Some Topics in Reactor Physics 229
    Ugo Farinelli

Nuclear Non-Proliferation 271
    Raymond L. Murray

Unified Neutron Transport Theory 285
    Raymond L. Murray

PART II : PHYSICS AND TECHNOLOGY

Amorphous Semiconductors 295
    William Paul

Solar Energy Materials 353
    Lyle H. Slack

PART III : COMPUTATIONAL METHODS IN PHYSICS

Computational Methods in Physics 385
    K.V.Roberts

PART IV : PHYSICS AND FRONTIERS OF KNOWLEDGE

Observational Traits of Black Holes in the Optical Band . . . 475
    P.L.Bernacca, L.Bianchi and R.Turolla

The Role of Polarization in Microscopic Physics 503
    Michael J.Moravcsik

Fundamental Constituents of Matter and Unification of
        Weak and Electromagnetic Interactions 515
    Riazuddin

PART V : SCIENCE AND DEVELOPMENT

Science and Development 547
    Michael J.Moravcsik

Appendix I: List of Invited Seminar Speakers 559

Appendix II: List of Seminars 559

Appendix III: List of Participants 561

PART I

PHYSICS AND ENERGY

PERSPECTIVES ON PAKISTAN'S ENERGY PROBLEMS

Mohammad Ahmad

Pakistan Atomic Energy Commission

Islamabad, Pakistan

1. INTRODUCTION

1.1 Global Energy Picture

The scientific and industrial revolution in the world was only possible with the availability of abundant inanimate energy sources and the development of technology to employ them. In fact energy consumption and economic growth are closely interrelated as clearly demonstrated by the disparity in energy production and consumption figures shown in Table 1 for the developing and the developed regions

Table 1

Region	Population (1974) (10^6)	Energy Production (1974) (10^6 TCE)	Energy Consumption (1974) (10^6 TCE)
Energy Deficient Developing Region	2428 (63.8%)	1259 (14.6%)	1075 (13.6%)
All Developing Regions	2694 (70.9%)	3408 (39.4%)	1247 (15.8%)
Developed Regions	1106 (29.1%)	5233 (60.6%)	6662 (84.2%)
World Total	3800	8641	7909

of the world. There is a serious imbalance in the existing distribution of energy consumption in these two regions, as 86 percent of the world annual energy is used by about 36 percent of the population of the rich countries, while a meagre 14 percent goes to the 64 percent of the people living in the energy-deficient developing countries[1]. As far as production of fuels is concerned, the latter region produces hardly 15 percent of the total energy, the oil-exporting developing countries contributing another 24 percent, which essentially finds its way to the industrialized countries of the world. This pattern, if not modified soon, will seriously undermine the future economic development of the poorer countries. In fact, the world is facing today a serious energy crisis which is deepening year after year and endangering the prosperity of the industrialized countries and threatening to cripple the economic growth of the poorer countries.

In this context, Pakistan's position as regards consumption with respect to the rest of the world, appears to be much more precarious. According to Table 2, Pakistan still lags far behind the African, Asian and the world averages in the use of energy, not to speak of the European or the North American figures.

Table 2

Comparison of energy consumption in Pakistan with rest of the world (1974)

	Elec. Energy		Total Primary Energy	
	KWh/Cap	Ratio with Pakistan	Kg of Coal eq./Cap	Ratio with Pakistan
Pakistan	148	1	180	1
Africa	309	2.1	359	2.0
Asia	376	2.5	625	3.5
South & Central America	739	4.8	1024	5.7
Europe	3776	25.5	4689	26.1
Oceania	4477	30.3	4544	25.2
North America	9580	64.7	11321	62.9
World Average	1613	10.9	2059	11.4

1.2   Energy Consumption in Pakistan

Table 3 summarizes the pattern of growth of population, Gross National Product (GNP), commercial energy, non-commercial energy and electricity consumption in Pakistan for three typical years covering a span of 50 years. Table 4 shows the percentage share of various primary sources in the annual commercial energy consumption over the 1948-77 period whilst Table 5 shows the relative distribution

Table 3

Growth of economy and energy consumption in Pakistan

	1949/50	1976/77	1999/2000
1. Population (million)	33	73	148
Av. Growth Rate (%)	-	3	3
2. Per Capita GNP			
(1959/60 cost, $)	77	116	275
Av. Growth Rate (%)	-	1.5	3.8
3. Commercial Energy Consumption:			
(a) Million TCE	1.9	14.9	97
(b) Kg/Capita	60	200	655
4. Non-Commercial Energy Consumption:			
(a) Million TCE	6.6	12.6	21
(b) Kg/Capita	200	170	142
5. Total Energy Consumption:			
(a) Million TCE	8.5	27.5	118
(b) Kg/Capita	260	370	797
6. Electricity Consumption:			
(a) Million KWh	268	11191	100000
(b) KWh/Capita	8	153	676

Table 4

Historical share of primary energy requirement in Pakistan (percent)

Year	Solid Fuel	Liquid Fuel	Natural Gas	Hydro	Nuclear	Elect. Generation*
1948-49	53	46	-	1	-	6
1951-52	32	67	-	1	-	6
1956-57	28	60	8	4	-	11
1961-62	21	54	18	7	-	17
1966-67	12	49	31	8	-	22
1971-72	7	43	37	13	< 1	28
1976-77	6	40	39	14	1	28

*Part of total primary energy

Table 5

Non-commercial fuel consumption in Pakistan (1973-74)

	(Percent)
- Firewood	40 ± 6
- Dung	32 ± 5
- Cotton sticks, weed and shrubs	13 ± 2
- Bagasse	12 ± 9
- Charcoal	2
- Sawdust	1

WOOD RESOURCES

- Area under forest = 3.7% of land area
    - which is equal to 13% of world average
    - 6% of that in Japan

- Production of timber and firewood 10 cft/acre of productive forest area

- Per capita annual consumption of timber in Pakistan : 1% of USA

---

of non-commercial fuel consumption in Pakistan for the year 1973-74. Table 6 summarizes the present share of energy consumption by fuels, sectorwise. After reviewing the pattern of historical growth given in these tables, the following conclusions can be drawn:

- The per capita energy consumption in Pakistan at its present stage of economic development is extremely low (0.2 TCE for primary energy and 153 kWh for electricity) and would require an accelerated growth to catch up with the world.

- In the commercial energy sector an average annual growth rate of about 8 percent has been maintained since independence. Because of low energy base, at least this much growth rate would be desirable over the next two decades.

- The average consumption of electricity and gas in Pakistan has increased during 1960-77 at rates of about 12 percent and 13 percent per annum respectively. Assuming ample supplies of energy, these rates would slowly decline in future years.

- There is a disparity in the form of growing overdependence on liquid petroleum and natural gas in relation to the reserve base of these fuels. Oil and gas now

Table 6

Present share of energy by fuels and sectors (1976-77)
(percent)

Fuel Type	Domestic and Commercial	Industrial	Agriculture	Power	Transport	Export	Others	Total
Oil	5	3	3	< 1	15	6	8	40
Gas	3	23	-	12	-	-	-	39
Coal	-	6	-	< 1	-	-	-	6
Hydro	-	-	-	14	-	-	-	14
Nuclear	-	-	-	1	-	-	-	1
Sub Total	8	32	3	28	15	6	8	100
Prim. energy for Elec. generation	6	13	5	-	-	4	-	28
Share of total commercial energy								
Total Commercial	14	45* / 37	8 / 16*	-	15	10	8	100
Share of total energy								
Commercial	8	20	9	-	8	5	4	54
Non-Commercial+	31	6	9	-	-	-	-	46
Total Energy	39	26	18	-	8	5	4	100

*Including gas for fertilizer
+Excludes human and animal power

account for about 40 percent each of Pakistan's energy consumption but 7 percent and 65 percent respectively of their known energy resources.

- The generation of electrical energy both by natural gas and hydro power increased at a rate of about 12 percent per annum during 1960-76.

- Non-commercial fuels are an important source of energy in Pakistan. Although no reliable statistics exist for them, their share has been falling e.g. from 63 percent in 1960-61 to 46 percent in 1976-77. After allowing for the non-commercial fuels, the average growth in the overall energy consumption has been about 4.5 percent.

- The respective share of oil, gas, coal, hydro-generation and nuclear in the total commercial energy consumption are 40, 39, 6, 14 and 1 percent respectively.

- The sectoral distribution of the total commercial energy consumed is about 32, 28, 16, 15 and 3 percent for Industry, Power, Domestic (plus Commercial and others), Transport and Agriculture (excluding fertilizer) respectively. Distributing the primary energy used for electricity generation to various sectors, the industrial sector presently takes up as much as 45 percent of total commercial energy consumption (Table 6). Again transferring the gas consumed for fertilizer production into the Agricultural sector would reduce the industrial share to 37 percent and increase Agricultural share to 16 percent.

- Table 6 also shows the sectorwise distribution of non-commercial fuels, (excluding human and animal power mostly used in the Agriculture and Transport sectors), and that of the total primary energy consumption (including both commercial and non-commercial fuels). It is concluded that Domestic plus Commercial sector is the leading consumer of energy at present (39 percent), followed by Industry (26 percent), Agriculture (18 percent), Transport (8 percent), Export (5 percent) and other (4 percent) respectively.

2. ENERGY RESERVES

2.1 Hypothetical Potential

Pakistan's indigenous energy resources consist of hydro-electric potential, coal, natural gas and some oil. Figure 1 shows the existing and potential sites for hydroelectric stations whereas Figure 2 shows the location of oil, gas and coal fields discovered so far.

Table 7 shows the hydro potential in Pakistan, which is mostly based on desk studies. In view of the characteristics and limitations typical of our hydro potential and the fact that some of the sites are mutually exclusive, it is doubtful that, with the present technology, it would ever be possible to economically develop more than 10,000 MWe of hydro resource. This is equivalent to primary energy consumption of about 17 million TCE per year at 50 percent plant factor.

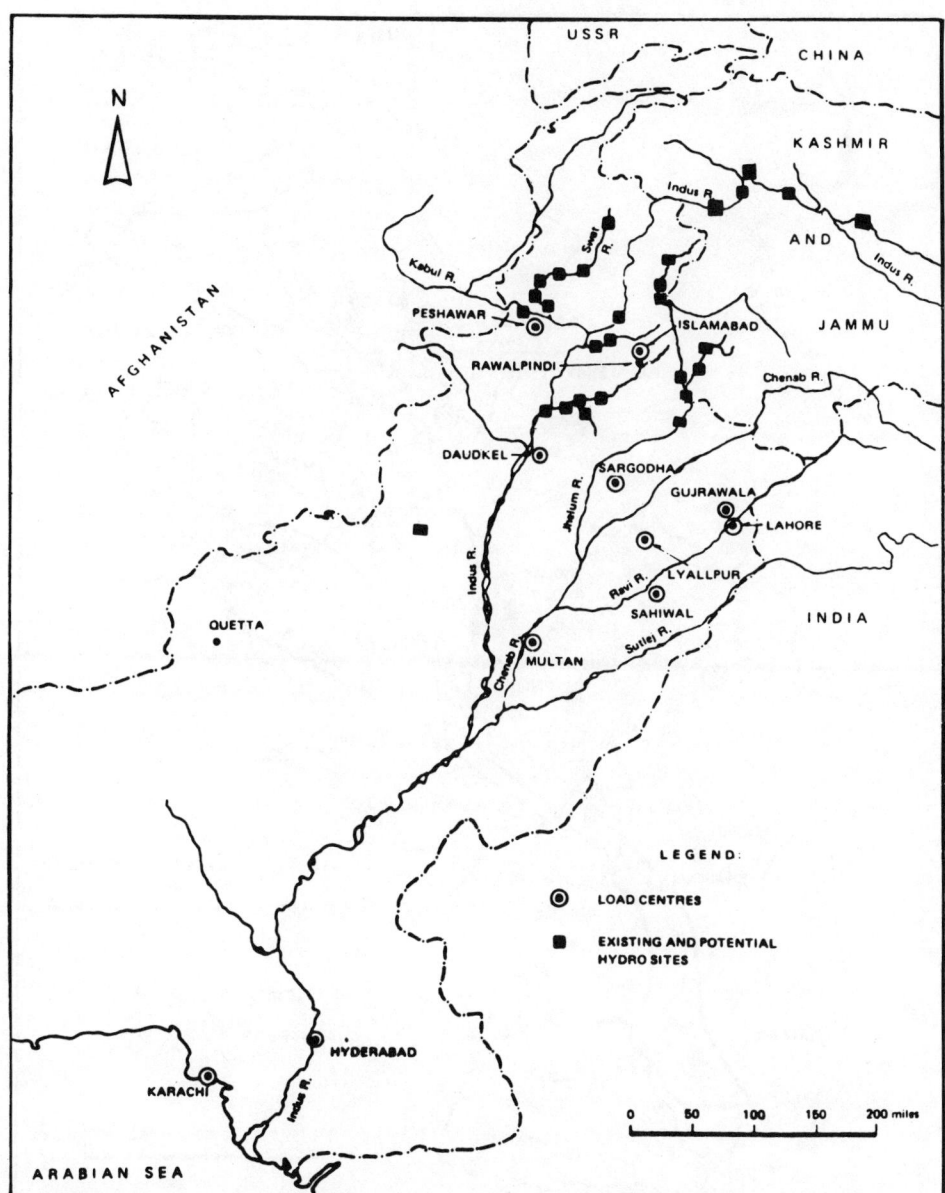

Figure 1. Existing and potential sites for hydroelectric stations.

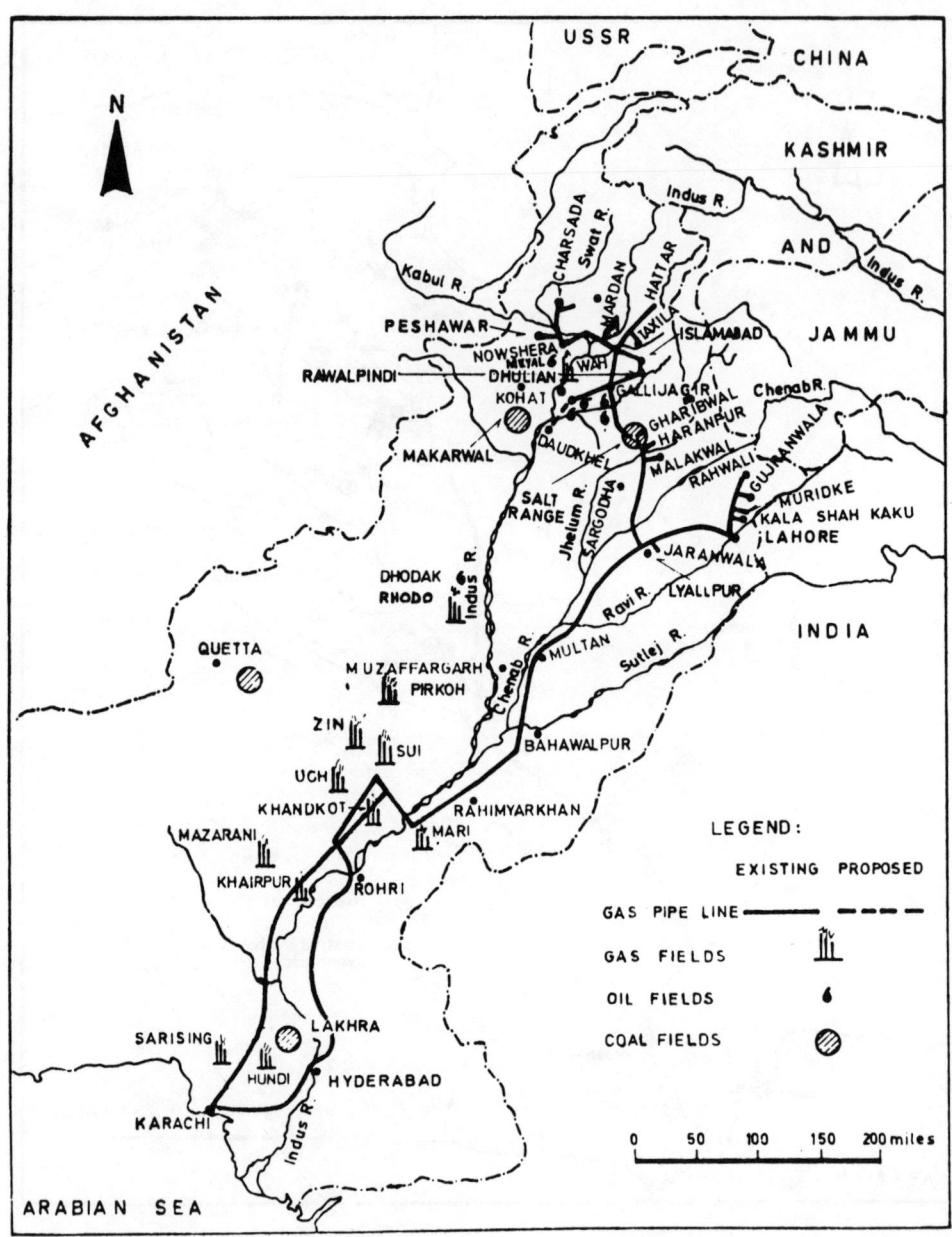

Figure 2. Location of oil, gas and coal fields.

Table 7

Hydro potential in Pakistan

(MWe Range)

1. Upper Indus Storages	10000 - 30000
2. Mid. Indus Storages	3000 - 4000
3. Off Channel Storages	2000 - 4000
4. Jhelum River Basin Storages	3000 - 5000
5. Lower Swat Storages	2000 - 4000

Limitations

- Poor geology of sites
- Inaccessibility and remoteness from Load Centres
- Wide seasonal fluctuations in river flows
- Location in siesmically active areas
- Heavy sediment load
- Require multi purpose development
- Landslides and glacier actions
- Flooding of prosperous valleys, cities and towns.

In the case of oil and natural gas various optimistic views, based on sedimentary rock formation (which is most suitable for oil deposits) have been expressed about the availability of their deposits in Pakistan. After making some reasonable assumptions about the nature of rocks[2] the 'Recoverable Potential Reserves' of the Indus Basin have been estimated to be of the order of 40,000-45,000 million barrels of oil (8000×10^6 TCE) and 105-110 trillion cubic feet of gas (4000×10^6 TCE). The time frame within which this potential would be fully discovered cannot be predicted but as a guess, it has been suggested that a discovery rate of the same level as in the past, viz, one trillion cubic feet per year, of natural gas could be sustained.

As regards coal, no estimates are available for the ultimate potential, since the coal bearing formations have not been fully explored and evaluated yet. The so-called presently known reserves are only estimated for the areas which are currently in production.

2.2   Presently-known Energy Reserves

In terms of presently-known and locally available fuel and energy resources, Pakistan is amongst the poorest countries in the world. This is demonstrated in Table 8 according to which the known deposit of coal, natural gas and oil represent a total of about 1200 million TCE, or a reserve base per capita of less than 16 TCE which is little over one percent of the world average. The annual per capita consumption in the United States is currently 12 TCE, with a reserve

Table 8

Energy reserves in Pakistan

Energy Resource	Reserves	Million TCE
1. **Coal**		
Proven	176×10^6 tons	118
Indicated and Inferred	314×10^6 tons	215
Sub-total	490×10^6 tons	333
2. **Natural Gas**		
Proven	460×10^9 m^3	434
Indicated	296×10^9 m^3	353
Sub-total	756×10^9 m^3	787
3. **Oil**		
Proven and Probable	55×10^6 tons	85
Total Fossil Fuel Reserves		1205
Reserve base/Capita: Pakistan		15.6 TCE
India	>	10 times
World	:	100 times
USA		5000 TCE
Annual consumption in USA	12 TCE	
4. **Hydro (Renewable Source)**		
Economically recoverable	10000 MW	17×10^6 TCE/Yr (at 50% P.F.)

base of 5000 TCE; in the Soviet Union the fossil fuel reserves per capita are about 14000 TCE. In neighbouring India, the per capita energy resources of 160 TCE, are more than ten times those in Pakistan. It is, therefore, quite evident that Pakistan's energy resource base is disturbingly weak and if any country justifiably needs new energy related technologies, it is Pakistan.

3.   FUTURE ENERGY CONSUMPTION

It is evident that Pakistan cannot hope to undertake any large scale programme for economic and industrial development or envisage an increase in its agricultural output without substantially higher inputs of energy. As an example, let us analyse the agricultural sector which remains the largest sector of Pakistan's economy, contributing more than one-third of the Gross National Product and employing more than half of the labour force in the country.

3.1 Agricultural Energy Needs

The present agricultural production in Pakistan is low in terms not only of absolute quantities but also of per acre yields. After a substantial rate of growth - around 5 percent annually in the sixties, agricultural output has stagnated in the present decade, rising only at an average rate of around 1.5 percent per year. Over the long term, Pakistan's ability to sustain faster growth in agricultural production depends essentially on fundamental improvements in the management of land and water resources and use of advanced agricultural techniques which involve increased application of a number of vital inputs. These techniques are highly energy-intensive. In fact, energy makes all the difference between primitive and modern agriculture. The primitive agriculture draws it entirely from manual and animal power, whereas the present day agriculture in advanced countries taps commercial sources of energy, such as petroleum, natural gas and electricity.

In agriculture, power is needed for levelling and preparing the lands, mechanising the cultivation, planting, harvesting, drying, threshing, operating tubewells for subsurface water control and drainage of water-logged areas and pumping water for irrigation. Energy is also needed to process the agricultural produce, manufacture fertilizers and plant protection chemicals and is equally indispensable for the transportation of agricultural produce to the utilization centres.

The main source of power for agriculture in the country has so far been the animal as reported in the FAO Indicative World Food Plan; according to which the total availability of power for agricultural purposes in 1969 was about 0.1 HP per cultivated acre. Nearly 75 percent of this was available through animals and 14 percent was provided by tractors. The FAO report has suggested a minimum of 0.2 HP per acre by 1985, if Pakistan were to achieve its projected agricultural growth rates. It is physically and biologically impossible, at known rates of bovine reproduction, to achieve even a modest increase in agricultural power from this source, by then. In the longer term, a higher population of draught animals would also mean a greater competition with human beings for food and feed. Under these circumstances, mechanical power is the only source which can fill this gap.

Primary energy requirement for agriculture has been estimated in Table 9 for the year 2000 AD, based on the consumption in 1976-77. These estimates do not allow for energy required for transporting agricultural produce to markets and for processing industry etc. However, included in Table 9 are the non-commercial forms of energy which inevitably go into farm production. Thus the energy requirement in agriculture would grow four-fold by the turn of the century.

Table 9

Estimated primary energy requirement in Agriculture
(10^6 TCE)

	1976-77	1999-2000
Petroleum	0.5	2
Electricity	0.8	4
Gas for Fertilizer	1.2	9
Total Commercial	2.5	15
Non-commercial	2.5	5
Total	5	20

The agriculture energy needs can be checked up by yet another method - from the point of view of adequate supply of food requirements on which the well-being of the people depends. The food diet of Pakistan with respect to that of the advanced countries, shows a remarkable disparity in average daily caloric and protein intakes (Pakistan's 2381 kCal and 63.2 grams relative to about 3100 kCal and 90 gm for developed regions). Thus to increase the nutrition level, advanced agricultural techniques will have to be introduced with consequent heavier inputs of energy.

A unit energy input into agriculture gives a much higher output of food energy. The amplification factor depends on the type of the agricultural system. For some crops in a primitive agricultural system, it may exceed as much as ten. But the output per unit area is substantially less. The advanced agricultural systems optimize food output per unit area and not per unit of input energy. It is the solar energy which is absorbed by the plants and gets converted into food energy - the input energy only acts as a lever.

Using the figures given in Reference 3 for the energy required per unit acre for one crop season of cereal grains in the USA as typical of modern agriculture, Table 10 shows an output of 2.82 kCal of input in the form of labour, mechinery, fertilizer, fuel etc. A high protein diet, such as that in North America and West Europe, requires 5300 kCal of plant product per capita per day[3] The advanced agricultural system will thus call for an energy input of 1880 kCal per day per person. For Pakistan assuming a diet somewhat poorer than the above standards even in the year 2000 AD, (say corresponding to 4800 kCal/capita/day of plant product), an improved agricultural system yielding an output-input ratio of 4, and a population then of 150 million, the total energy input for food production would be about 66 trillion kCal. After allowing for non-food crops and farm product losses, the energy requirement for self-

Table 10

Energy inputs for grain production in US (1970)

10^3 kCal/Hectare

Labour	12.1
Machinery	1037.8
Fuel	1969.4
Fertilizers	2609.4
Seed, Insecticides etc.	210.1
Irrigation	84.0
Drying	296.5
Electricity	766.0
Transportation	173.0
Total	7158.0

Output Crop yield (Av) = 20175.2 kCal

Output per kCal input = 2.82

Plant product for high protein diet in USA etc = 5300 kCal

Energy input per capita per day (US) = 5300/2.82
= 1880 kCal

sufficiency in food and fibre, would thus be around 100 trillion kCal. If additional farm products were required for export, say at 25 percent, the overall energy need would rise to about 130 trillion kCal per annum (18.6 million TCE) which agrees reasonably with the estimate given in Table 9.

3.2   Source-Wise Energy Projections

The overall energy consumption in various sectors of economy is expected to grow at a fast rate in order to provide for the growing population as well as for better quality of life. Forecasts of energy requirements have been summarised in Table 11 based on macro-economic frame-work. It has been estimated that by the end of the century, the population of Pakistan will double to about 150 million and even a modest increase in the standard of living would require a three-fold increase in its per capita commercial energy consumption to about 0.7 TCE which will still be roughly 1/5th of the world average by 2000 AD.

According to Table 11, the use of gas, coal and hydro would grow at about 7-8 percent in the next two decades, whereas liquid fuels would probably grow at about 5 percent per annum because of

Table 11

Projected energy requirement in Pakistan (10^6 TCE)

Year	Gas	Oil	Coal	Hyd.	Nuc.	Sub-total Commercial	Non-Commercial	Total
1979-80	9.1	7.2	1.0	3.4	0.1	20.8	13.5	34.3
1984-85	13.8	8.9	1.5	6.0	0.7	30.9	15.0	45.9
1989-90	18.9	11.3	2.6	7.9	4.3	45.0	16.7	61.7
1994-95	27.2	15.4	3.6	10.6	9.2	66.0	19.2	85.2
1999-2000	38.7	20.5	4.1	14.0	19.7	97.0	21.2	118.2
Growth Rate % (20 Yrs.)	7.5	5.4	7.3	7.3	30.0	8.0	2.3	6.4
Percent Shares:								
1979-80	27	21	3	10	1	61	39	100
2000 AD	33	17	4	12	17	82	18	100

their higher price. The role of nuclear energy would be coming from its use in electric power generation. Starting from a very small base in 1980, nuclear energy has been projected to grow at 30 percent per annum.

The non-commercial energy will maintain a growth rate of about 2.3 percent, as such the total energy growth is expected to be around 6.4 percent. The share of non-commercial energy fuels is expected to further decline to 18 percent by the year 2000 AD: and subject to its availability, the natural gas would become the largest source of energy in Pakistan (33 percent). Likewise, depending upon adequate financial resources and project implementation as per present schedule, nuclear energy would take up about 17 percent, the same as that for oil, followed by hydro, 12 percent and coal 4 percent.

4. GROWTH OF ELECTRIC POWER

The consumption of electricity in Pakistan has experienced a manifold increase in the last 27 years. However, the present per capita generation is still 160 kWh. Likewise, the generating capacity has gone up some 28 times to the present level of 3280 MWe. Nevertheless, the installed capacity has often lagged behind the electric demand with practically no reserves in the power system.

Table 12 summarizes the existing power capacity in Pakistan. This capacity is shared by hydro 58 percent, gas/steam 37 percent, gas turbine 11 percent and nuclear/coal 4 percent. The output of hydro power plants, being dependent upon the seasonal flow of rivers and further limited by irrigation requirements, varies considerably during the year. For instance, during the month of March, the firm hydro capability of the existing units would be hardly 57 percent of the installed capacity, thereby reducing the system capability to 73 percent. The present maximum demand is about 1825 MWe (as of Jan 1978) and 440 MWe (June 78) for WAPDA and KESC systems respectively.

Table 12

Existing power system capacity (MW) (January, 1978)

	Capacity			March
	WAPDA	KESC	Total	Capability
Hydro	1567	-	1567	887
Coal	12	-	12	10
G/Steam	703	508	1211	1080
Gas Turbines	353	-	353	320
Nuclear	-	137	137	100
Total:	2635	645	3280	2397

Figure 3 shows the historical and projected growth of electricity generation per capita, taken from Reference 4, covering the span of 50 years. The essential data regarding the growth of population and GNP per capita at constant factor cost are also included in this figure.

It is indeed true that from amongst the conventional energy resources, hydro power alone can play a major role in meeting the anticipated power demand, however, there will still be a large gap left which can be filled only by nuclear energy. Recognising the necessity to find a feasible mode to meet this gap the PAEC, in cooperation with the International Atomic Energy Agency (IAEA), has conducted several nuclear power generation expansion planning studies, as regards the timing the size of plants which could be economically built until the year 2000 AD. These studies have provided the framework for the planning of future nuclear power programme in the country and setting up of the basic infrastructure and facilities to meet these requirements.

Table 13

Peak demand and generation Capability (MW)

Year	Peak Demand	Generation Capability					
		G/Steam	GT	Coal	Nuc	Hyd	Total
1985	4540	1660	700	260	725	1810	5155
1990	7290	1850	600	500	1925	2980	7855

Projected energy generation (GWh)

Year	Energy Demand	Share by various plant types				
		Hydro	Nuclear	Coal	G/Steam	GT
1979-80	15300	8830	240	60	5415	755
1984-85	25850	15000	1800	500	7350	1200
1989-90	42300	19800	11000	2500	7680	1320
1994-95	64000	26000	23500	4000	8600	1400
1999-2000	100000	35500	49800	4000	9300	1400

The projected energy generation by various plant types is shown in Table 13 according to which hydro would provide 47 percent, nuclear about 26 percent, gas about 21 percent and coal about 6 percent in the annual contribution to energy in 1990 AD. Although the number of nuclear reactors and timing of their installation would depend primarily upon the evolution and commissioning of specific hydro power projects, discoveries of additional coal deposits and gas fields at suitable locations and availability of financial

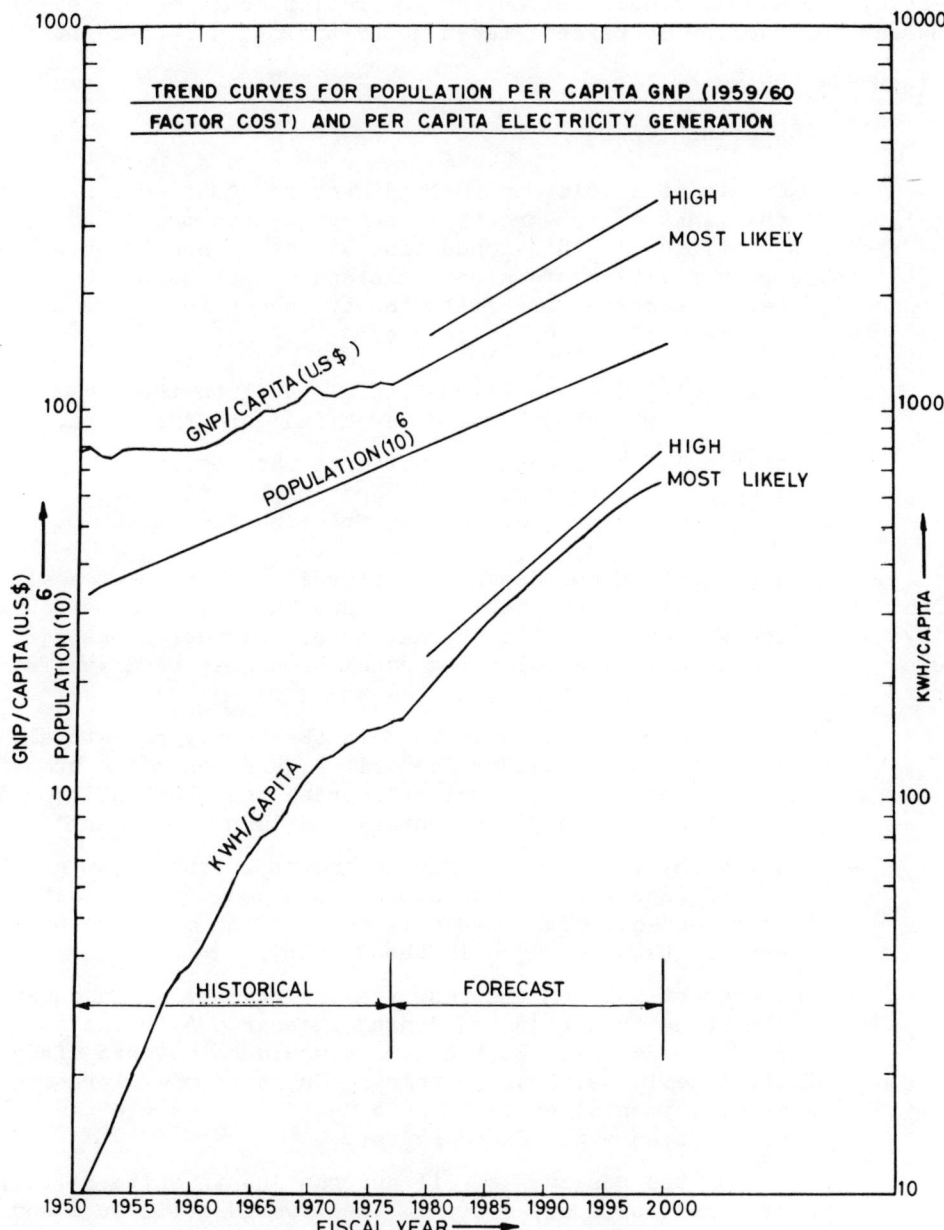

Figure 3. Historical and porjected growth of electricity generation per capita.

resources, it is anticipated that by the year 2000 as much as 50 percent of the annual electrical energy generation could be shared by nuclear followed by 36 percent taken up by hydro and 14 percent from the fossil fuels[4].

5. ENERGY DEFICITS

The rate of development of indigeneous reserves can now be analysed in the light of the growth in energy demand projected in the previous sections. In this connection Tables 14 and 15 show the demand and supply position of natural gas and oil which has been projected after considering the following two scenarios as regards the future discovery of these two sources:

>Scenario 'A' Gas and oil potential equal to the presently discovered/indicated reserves.

>Scenario 'B' 45 percent of ultimate theoretical potential of gas and 16 percent that of oil discovered upto the year 2000 AD.

The production and demand of natural gas have been analysed both with and without the availability of nuclear power for electricity generation whereas for oil, thermal power generation has not been considered at all. The following conclusions can be drawn from this analysis:

- There is going to be a deficit in the supply of natural gas in the 1990's, under Scenario A, and depending upon the contribution from nuclear source, the shortfall would vary between 17 to 39 percent by the turn of century.

- Considering the more optimistic growth in the known gas reserves (Scenario B) the supply position would be satisfactory and deficit would only occur if no more nuclear power plants were added in the country.

- The presently-known oil reserves are obviously inadequate for meeting the projected demand (Scenario A in Table 15). Under Scenario B, the demand would be covered after initial deficits. However, this needs further exploration and developmental activity on a much larger scale than what has been experienced so far.

- Based on the reserves of oil and gas and in spite of full development of nuclear power, there would still be a shortfall in the supplies of oil and gas which would grow from about 27 percent of the primary energy demand in 1979-80 to about 36 percent by the year 2000 AD. The corresponding figure for 2000 AD in the absence of nuclear power generation, would be much higher, 59 percent (there would still be a deficit of about 6 percent with substantial new discoveries in this case).

PERSPECTIVE ON PAKISTAN'S ENERGY PROBLEMS

Table 14

Supply and demand of natural gas (10^6 TCE) in Pakistan with and without nuclear power generation

CASE 'A' (Presently Discovered/Indicated Reserves)

Year	Production	Demand		Surplus/Deficit	
		with NPG	w/o NPG	with NPG	w/o NPG
1979-80	9.6	9.1	9.6	+0.5	–
1984-85	15.5	13.8	14.5	+1.7	+1.0
1989-90	21.6	18.9	23.7	+2.7	-2.1
1994-95	21.6	27.3	37.2	-5.7	-15.6
1999-2000	21.6	38.7	60.4	-17.1	-38.8

CASE 'B' (45% of ultimate potential Discovered by 2000 AD)

Year	Production	with NPG	w/o NPG	with NPG	w/o NPG
1989-90	28.4	18.9	23.7	+9.5	+4.7
1994-95	36.5	27.3	37.2	+9.2	-0.7
1999-2000	43.0	38.7	60.4	+4.3	-17.4

Table 15

Projected supply and demand of oil (10^6 TCE)

CASE 'A' (Presently known oil reserves: 436×10^6 bbl)

Year	Production	Requirement	Surplus/Deficit
1979-80	1.1	7.2	– 6.1
1984-85	2.5	8.9	– 6.4
1989-90	2.5	11.3	– 8.8
1994-95	2.5	15.4	-12.9
1999-2000	2.5	20.5	-18.0

CASE 'B' (New Discoveries: 16% of ultimate theoretical Potential upto the year 2000 AD: 6000 million barrels)

Year	Production	Requirement	Surplus/Deficit
1979-80	1.1	7.2	– 6.1
1984-85	3.0	8.9	– 5.9
1989-90	11.7	11.3	+ 0.4
1994-95	22.0	15.4	+ 6.6
1999-2000	32.0	20.5	+11.5

ACKNOWLEDGEMENT

The author is deeply indebted to M/S S.A.Siddiqi and S.H.Mansoor for their help in preparing this article.

REFERENCES

1. World Energy Supplies 1950-74, Department of Economic and Social Affairs, United Nations, 1976.

2. I.B Kabir, et al, 'Petroleum Potential of West Pakistan' paper presented at PIP Symposium, Lahore, November, 1972.

3. J.K. Dillard, et al.,'Nuclear Plant Delays and the World Food Supply' Westinghouse Electric Corporation, USA, 1977.

4. M. Ahmad, 'Demand for Electric Power and Projected Role of Various Sources of Electricity Generation in Pakistan' paper presented at 8th Convention of the Institute of Electrical Engineers, Lahore, 1978.

WORLD ENERGY PROBLEMS

Bent Elbek

Niels Bohr Institute

University of Copenhagen, Denmark

1. ENERGY DEMAND AND ECONOMIC DEVELOPMENT

1.1 Energy Demand in a Growing World

In this first section I intend to examine the problem: How much energy is the world going to need 30-50 years hence? Now, this of course is a question that can be answered only if we are willing to accept a number of basic assumptions. Such as, that we live in a surprise-free world, that the goals of mankind do not change drastically and that we avoid major catastrophies as war, famine or disastrous climatic changes.

Even when we exclude such occurrences, energy demand is a complex question which requires deep excursions into the economic and social structures of our societies. The oil crisis in 1973 illustrated, that the imprecedented economic and technical developments witnessed in many parts of the world to a very large extent had been driven by a convenient, but severely underpriced fuel, namely oil. Cheap oil gave some of the now industrialized countries the means of raising themselves from a modest standard of living to a state of affluence. Today many developing countries could be at the threshold of such a transition, but they can not count on the advantage of a cheap fuel. What are the consequences of this change for the developing countries and what are the consequences to the world?

The relationship between energy and economy has been much discussed and can be analyzed from different points of view. A simple tabulation (Table 1) of national energy consumption (E/year) and gross national product (GNP/year) clearly indicates some kind of relationship between these two quantities but not a straight propor-

Table 1

Energy-GNP Relation for Major Regions (1975)

Region	Energy EJ/Y	GNP 10^9 \$/Y	Energy/GNP MJ/\$
North America	80.5	1678	48
Latin America	12.5	327	38
Japan	14.2	496	29
W. Europe	50.3	1695	30
USSR, E. Europe	59.6	925	64
China	16.3	315	52
Asia (excl. China)	10.9	210	52
Africa	5.0	163	31
Middle East	4.1	153	27
Oceania	3.1	94	33
World	256.6	6056	42

Based on references 5 and 13.

tionality. The ratio E/GNP varies, even when averages for major regions are considered, almost a factor of two, being highest in North America and USSR and lowest in W. Europe and Japan. The developing countries show considerable fluctuations, but there is no indication that their economies on the average are particularly energy intensive. If any conclusion can be drawn from numbers such as given in Table 1 it is, that local availability of cheap and abundant energy makes the economy energy intensive.

The energy-GNP near proportionality is not of mysterious origin. The GNP is composed of goods and services that all require energy. If we for some typical regions (Table 2) break down the GNP and its associated energy demand into components, several observations can be made:

1) The composition of GNP varies considerably among regions. High-income countries have larger service sectors than low-income countries.

2) The production of goods is more energy intensive than services.

3) Industry is on a world basis the largest energy consumer. Basic industry is more energy intensive than more advanced industry.

Table 2

Composition of GNP and Corresponding Energy Demand (1975)

	Agriculture	Industry	Construction	Trade	Transport	Service
North America						
% of GNP (1678 G$)	3	28	5	17	6	39
% of energy (80.5 EJ)	4	32	1	6	25	6
MJ/$ for sector	64	55	10	17	199	7
Japan						
% of GNP (496 G$)	5	37	7	18	7	31
% of energy (14.2 EJ)	3	53	1	5	15	5
MJ/$ for sector	17	41	4	8	61	5
W.Europe						
% of GNP (1695 G$)	5	38	7	13	6	30
% of energy (50.3 EJ)	2	39	1	5	16	5
MJ/$ for sector	12	30	4	11	78	5
Asia						
% of GNP (210 G$)	44	15	4	10	4	13
% of energy (10.9 EJ)	4	46	1	4	28	4
MJ/$ for sector	5	156	14	21	382	16

NOTES: Derived from energy and GNP totals in Table 1. Disaggregation of energy on basis of WAES Demand Study (reference 1). Disaggregation of GNP from UN Statistical Yearbook (1976). Percentages do not sum to 100 because of omission of duties (GNP) and demand categories (energy).

4) High yield agriculture is energy intensive but in absolute terms energy for agriculture is not of great significance.

5) Transport is a large energy consumer. It should be kept in mind that all transportation energy is derived from oil.

If the observations above are applied to the likely future developments of the world economy, we would expect a steady move towards a less energy intensive situation. More services relative to goods and more refined products in the various industries rather than bulk commodities. This trend is indeed reflected in most projections for future energy demand, which consistently show a decreasing energy intensity. One should however remember, that there in some cases could be a threshold effect. Low-income countries with primarily agrarian economies might enter a phase of increasing energy intensity while agriculture is intensified and basic heavy industries are established.

1.2 Material Content of Gross National Product

One can carry these considerations concerning the structure of present and future economies a step further and look not only into the energy content but also into the 'material' content of the GNP. How much steel, aluminium, paper, cement, chemicals and so on is required for each 1000 $ of the GNP?

Table 3 illustrates this for some selected regions. At first sight it is perhaps surprising that the differences among the

Table 3
Material Content of GNP in Kg/1000 $ (1975)

	N.America	Japan	W.Europe	Asia	USSR	China
Steel	71	205	88	88	217	92
Aluminium	2.6	2.1	1.9	1.0	2.3	0.5
Paper	33	27	20	10	13	19
Cement	43	132	109	199	187	95
Chemicals (A)	60	82	52	33	49	26
Grains (B)	129	35	78	1145	206	667
'Total mass' (C)	339	483	349	1430	674	900
Energy (D)	1147	693	717	1242	1529	1242

Based on references 5 and 9.
(A) Oil equivalent of feedstock. (B) Rice + Maize + Wheat + Barley.
(C) Sum of the 6 entries above. With the exception of Aluminium. These commodities are those produced in greatest quantity (D) Oil equivalent of energy consumption in Kg/1000 $ of GNP (cf.Table 1).

regions after all are not larger. Apart from grain, the amounts of material entering into each unit of the GNP in most cases varies less than a factor of 4. This is in spite of GNP/caput variations of up to a factor of 35. This illustrates a considerable stability of the GNP-composition irrespective of the size of the GNP. This stability is also reflected in the near constancy of the energy/GNP ratios given in Table 1.

Apart from energy, where the numbers express consumption, the data in Table 3 concern production of the different materials. As export and import can affect the picture considerably, the table should be interpreted with some caution as far as level of consumption is concerned. Nevertheless Table 3 indicates certain trends. In general the more mature economies as those of North America and Europe require aluminium, chemicals and paper (!) whereas the developing economies typically are characterized by steel and cement. The extreme steel-intensity of the upsurging economies of Japan and USSR is noteworthy and could be related to the threshold-question mentioned before: Will the developing countries have to enter a phase with heavy industrial development as a necessary step towards more diversified economy?

The materials listed in Table 3 are, with the exception of aluminium, those handled in greatest quantity in all economies. One can get a primitive measure of the 'matter-intensity' of an economy by adding up all the masses given in Table 3. The 'total mass' of each unit of the GNP is seen to be much larger in the developing economies than in the industrial countries. Now, this is to some extent due to the inclusion of food (grain) in the mass. Even without food some of this trend is however conserved, but with the 'steel and concrete' economies of USSR and Japan put in prominence.

Physicists know, that energy and mass are equivalent. It is perhaps doubtful, that we can make use of this fact in dealing with energy in society. If, however, we for the regions mentioned in Table 3 look at the corresponding energy consumptions, we find that the industrialized countries use relatively more energy in order to handle a given mass. Thus, although the 'mass-intensity' is less for industrialized countries, the matter goes through more elaborate processes and the 'energy-intensity' differences among regions are therefore less pronounced than the 'mass-intensity' differences.

The information contained in Table 3 tells us in short, that as an economy develops its 'mass-intensity' (mass per unit GNP) decreases (cf. Figure 1).

This information is however put in a quite different perspective, if we instead look at total mass expressed as mass produced by each individual. This can be seen in Table 4, which shows the

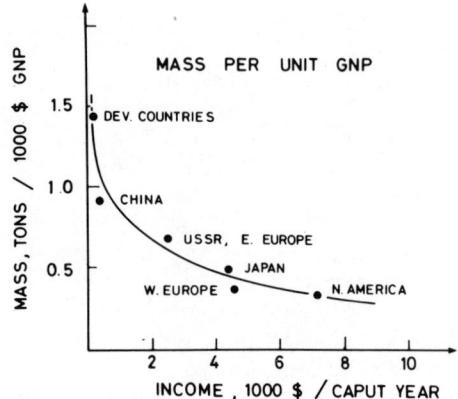

Figure 1. Mass intensity (Mass per unit GNP) of some selected economies. For definition of mass see text.

Table 4

Material Content of GNP in Kg/caput

	N.America	Japan	W.Europe	S.Asia	USSR	China
Steel	504	912	400	9	217	92
Aluminium	18	9	9	0.2	2	0.5
Paper	234	120	98	2	13	19
Cement	305	587	495	41	187	95
Chemicals	426	365	236	7	49	26
Grain	915	155	355	240	525	253
'Total mass' (Kg/cap.y)	2402	2148	1593	299	993	485
GNP ($/cap.y)	7120	4450	4645	210	2550	380
Energy (Kgoe/cap.y)	8144	3035	3211	259	4024	472
Population (10^6)	236	112	365	1002	254	822

See notes for Table 3.

total mass of the six commodities studied expressed per caput in the different regions. Clearly, the total mass is much larger in the developed economies than in the developing economies. Each North American produces about 2.4 t of materials per year whereas the average for a South Asian is 0.3 t per year. So wealth is still to a considerable extent associated with matter (Figure 2).

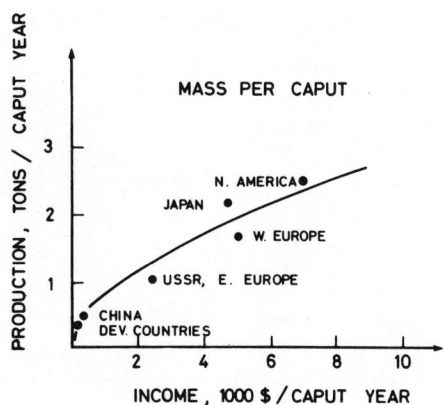

Figure 2. Mass per caput year of some bulk commodities for selected economies.

1.3   The Importance of the GNP for the Energy Demand

I have spent this much time discussing these various aspects of the GNP because the GNP and its growth is of decisive importance for the evaluation of the world's future energy needs. The GNP concept has been much criticised for not being a proper measure of human welfare. Apart from the fact that probably nobody has said that it was, the GNP is still the most comprehensive expression for the total activity level in a society. And, as we have discussed at some length, activity requires materials and it requires energy.

Most countries have for long periods had economic growth but this growth has not been evenly distributed. The trend has been, that the highest growth rates were in the middle-income countries whereas high-income countries and low-income countries had somewhat lower economic growth rates. Some low income countries indeed have and have had substantial economic growth, but the growth per caput has been moderate because of the high population growth rates.

What can we economically expect of the future? Again looking apart from major catastrophies, probably a continuation of the trend mentioned. If we for the world assume a moderate economic growth of 4% p.a. in the period until 2000 a distribution of this growth among the major regions as shown in Table 5 is likely and in agreement with present trends. The overall tendency is a moderate and slowing growth in the mature economies and a high, although declining growth in the developing countries.

Given the economic growth as input assumption one can pro-

Table 5

Economic Growth Rates in % p.a.

	1965-1973 (A)	1972-1985 (B)	1985-2000 (B)
North America	3.6	3.6	3.5
W. Europe	4.8	3.7	3.5
Japan	10.9	6.5	3.0
USSR, E. Europe (C)	4.3	4.0	3.5
China (C)	6.4	6.0	5.0
Dev. World	6.1	6.1	4.5
World	5.0	4.5	3.7

(A) World Bank Atlas 1975.
(B) WAES High Growth Assumptions (C-C1)
(C) Author's estimate.

ceed and calculate the resulting energy demand. As already mentioned, energy and GNP are related to each other, but not proportional. The relative increment in energy demand ($\Delta E/E$) divided by the relative increment in GNP ($\Delta GNP/GNP$) is often called the energy coefficient or the income elasticity of energy demand. Historically this coefficient has been close to unity, but most studies agree, that we in the future can expect much lower energy coefficients than in the period 1950-70. For the industrialized world perhaps this period can be called the welfare transition. Here, indeed the economic growth rates and the energy growth rates were nearly proportional with factors of proportionality close to unity. There are several reasons to expect a lowering of the energy coefficient for the developed countries. These reasons can not always be separated, but some of the more important are:

(1) Structural changes in the economy
(2) Maturing and saturation in developed economies
(3) The impact of higher energy prices
(4) Energy savings through suitable energy policies
(5) Technical developments.

The energy coefficient gives a condensed representation of the combined impact of these factors. Table 6 shows, mostly on the basis of the WAES study[1] how the energy coefficient is assumed to change with time. The impact of economic maturity is particularly evident.

In judging the significance of the energy coefficient as a prognostic tool, it should be stressed, that it is not an input assumption but the result of a long series of detailed evaluations

Table 6
Estimated Energy Coeffieicnts

	1965-1973	1972-1985	1985-2000
North America	1.2	0.5	0.7
W. Europe	1.0	0.8	0.7
Japan	1.0	0.8	0.6
USSR, E. Europe	1.1	1.0	0.9
China	0.8	1.0	1.0
Dev. World	1.1	1.1	1.0
World	1.0	0.8	0.9

for each nation and each sector of the economy akin to the discussions around Tables 1-4 in this lecture and taking into account the points (1) - (5) enumerated above. Thus a considerable insight in the mechanisms behind the growth in material quantities and energy is a prerequisite for a calculation of the energy coefficient.

1.4   Method for Calculation of Energy Demand.

The procedures for calculating future energy demands can be cast in a simple mathematical form. Apart from its simplicity the method described has the advantage of directly showing how demand growth is affected by a number of exogenously given parameters.

The basic assumption behind the calculation of the energy demand E corresponding to a certain economic activity A is

$$E = A \cdot I \qquad (1)$$

where I is the energy intensity of the activity measured e.g., in joule per constant dollar. This equation is valid for each of the economic sectors considered. For a calculation for instance one can use the 8 sectors given in Table 7.

The time development of A is given by the economic growth rate α of the sector:

$$A = A_o e^{\alpha t} \simeq A_o (1 + \alpha)^t. \qquad (2)$$

Likewise the energy intensity is assumed to be time-dependent. The intensity is affected by the price of energy, the energy policy and long term technical developments. We can describe these influences by exponents β, γ and δ respectively so that

$$I = I_o e^{(\beta+\gamma+\delta)t} \simeq I_o (1 + \beta + \gamma + \delta)^t \qquad (3)$$

Table 7
Main Economic Sectors and Energy Demand Coefficients

	α	β	γ	δ
Transport	4.0	-1.0	-0.3	1.0
Industry	4.9	-0.5	-0.5	-2.2
Agriculture	3.0	-1.0	-0.5	4.0
Construction	2.0	0.0	-0.5	-3.6
Commercial	4.1	-1.0	-1.0	-2.7
Public	5.0	-0.5	-0.7	-3.0
Residential	3.7	-0.7	-1.4	-2.8

The coefficients correspond to % p.a. growth rates of fossil fuel induced by economic growth (α), energy price (β), government policy (γ) and inherent trends (δ). The values given apply to Denmark for the period 1972-1985 for a 4.1% GNP-growth, 50% increase in energy price and vigorous government policy.

The economic growth rate α is usually given exogenously (e.g., by politicians) or is obtained from economic models. The other exponents can be estimated as follows:

(1) The price elasticity of energy demand is defined as

$$q = \frac{\Delta E/E}{\Delta P/P} = \frac{d \ln E}{d \ln P} \qquad (4)$$

where P is the price of energy. The price elasticity (4) here is 'pure' in the sense that only the intensity (3) of energy use is affected. Thus the activity levels, because they are exogeneously defined, are considered independent of energy price. Such pure price elasticities are quite small because energy is a necessity for most applications. A value $q = 0.2$ is typical, that is demand is reduced by 0.2% for each 1% increase in energy price.

In the long term one can assume that the price of energy (especially oil) will increase exponentially in real terms. That is

$$P = P_o(1 + r)^t \qquad (5)$$

From (4) one gets $E \propto P^q$ or

$$E \propto (1+r)^{qt} \simeq (1+qr)^t \qquad (6)$$

Comparison with (3) then gives

$$\beta = qr \qquad (7)$$

where r is the rate of price increase. If we assume a 2.5% p.a. increase in energy price and q = 0.2 then β is -0.5% p.a.

(2) Energy policy is hard to quantify. But let us assume that a suitable policy, e.g. by conservation measures over a period t could reduce demand from D_o to D. Then the policy exponent is

$$\gamma = \frac{1}{t} \ln \frac{D}{D_o} \qquad (8)$$

In the next lecture we shall see, that D/D_o typically is from 0.85 to 0.95 for a 10 year period. Thus γ ranges from -0.5% to 1.5% p.a.

(3) Finally the parameter δ describes some changes in energy intensity that neither can be ascribed to price, nor to government policy. These changes are best described as <u>inherent trends</u> and can cover a variety of technical changes as mechanization of agriculture, substitution of diesel trains for steam locomotives or increased automation. One obvious additional example is electrification of all kinds of equipment with a resulting improvement in end use efficiency. Trends can be estimated from historical developments. If energy intensity has changed from I_o to I over t years then

$$\delta \simeq \frac{1}{t} \ln \frac{I}{I_o} . \qquad (9)$$

Historic trends have been of the order +2.0% p.a. for electricity and -1.0% p.a. for other energy.

In the procedure described above the effects of growth price, policy and trends are supposed to be independent, so that their effects can be factored (addition of exponents). This is perhaps not a very good approximation. However, the method allows a very fast estimate of future energy demand once the initial values A_o and I_o have been determined for each sector of the economy. The method is particularly valuable for investigation of many different combinations of growth, price, policy and trends.

The energy coefficient referred to above can be expressed as

$$\frac{d \ln E}{d \ln (GNP)} = \frac{d \ln E}{d \ln A} = \frac{1 + \alpha + \beta + \gamma + \delta}{1 + \alpha} \simeq 1 + \beta + \gamma + \delta \qquad (10)$$

In future situations all exponents β, γ and δ tend to be negative thus giving energy coefficients less than unity. In the absence of energy price increases and government policy the energy coefficient (in this case better called the pure income elasticity) equals 1 + δ.

1.5   World Energy Demand to Year 2000 and Beyond

Application of the method outlined above combined with careful evaluations of all parameters for each region of the world and each sector of the economy makes it possible to calculate the future energy demand for the world. The energy co-efficients shown in Table 6 correspond to one of the cases[1] studied by WAES. It assumes constant oil price until 1985 and then a 50% increase. In addition moderate to strong energy conservation policies in all countries and continuation of present technical trends.

The lower energy coefficients reflect the ability of society to develop under an energy constraint. An absolute ceiling over the use of energy would imply either an energy coefficient of zero or a zero growth economy. None of these extremes are likely or desirable in the near term, but could for the industrial countries become mandatory early next century.

In order to calculate the energy growth rates one must assume a general economic development as e.g. defined in Table 5. Consistent with these aggregated growth rates the growth rate of each country must be determined and subsequently the growth rate for each sector of the economy. As an example the trend towards a more service-oriented economy is depicted by assigning higher growth rates to the service sectors than the industrial sectors. In the long run such structural changes in the economy will have a considerable impact on the energy demand and could eventually lead towards a stabilisation of energy demand.

The resulting energy growth rates obtained by combining the figures in Tables 5 and 6 are given in Table 8. The calculated absolute energy demands are plotted in Figure 3 together with historic data to date.

Table 8

Projected Energy Growth Rates in % p.a.

	1965-1973 (A)	1972-1985 (B)	1985-2000 (B)
North America	4.2	1.9	2.4
W. Europe	5.0	2.8	2.5
Japan	11.1	5.4	1.9
USSR, E. Europe [C]	4.4	4.0	3.5
China [C]	5.0	6.0	5.0
Dev. World	5.9	6.7	4.9

(A) Based on Reference 7.
(B) Corresponding to WAES cases C-C1.
(C) Author's estimate.

In spite of the moderate economic growth assumed and the impact of energy price and energy policy, demand is projected to

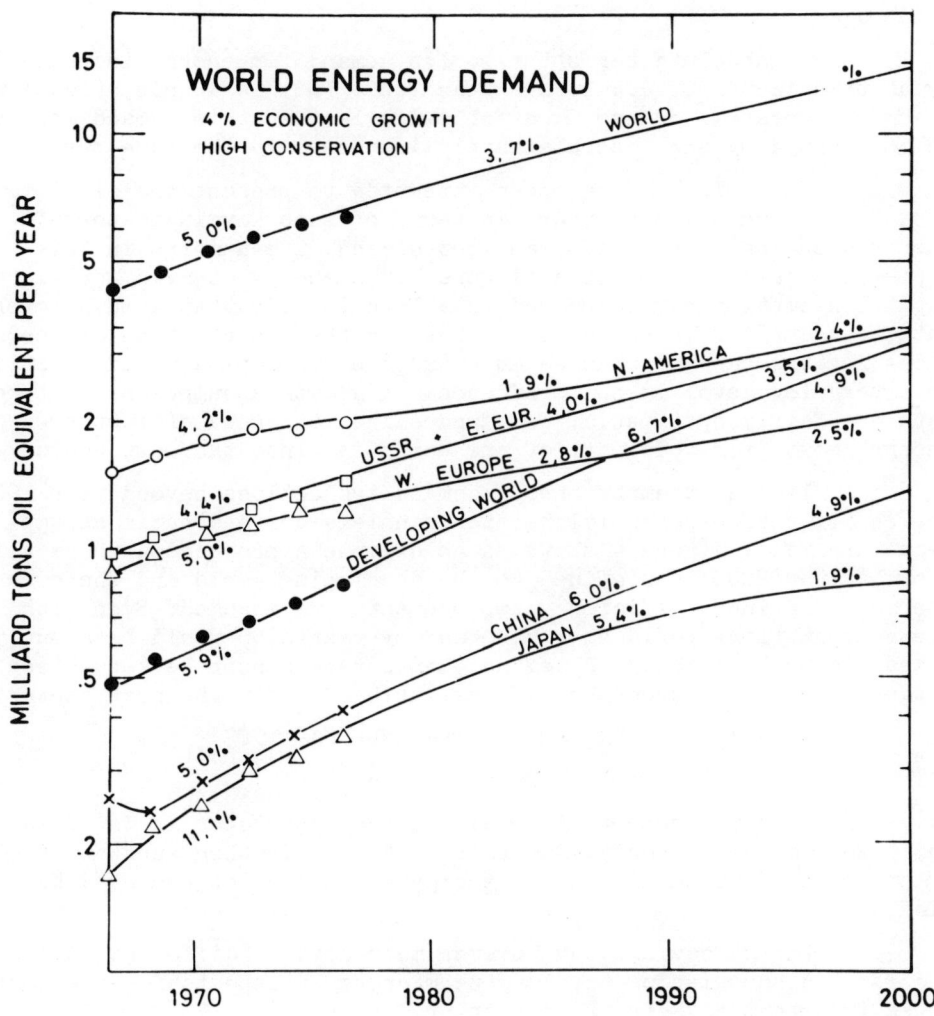

Figure 3. Projected world energy demand to year 2000.

increase in all regions. The demand growth is, compared to history, moderate in the industrialized countries but fast in the developing countries and the socialist world. These tendencies are in good agreement with the historic trends. Japan is an interesting case where, in the period studied, a transition from a fast growing economy to the more sedate development of a mature industrial country is accomplished.

In absolute terms the worlds demand for energy increases by about a factor of 2.5 before year 2000. This is in historic terms a fairly moderate increase. In a following lecture this demand increase is confronted by the possible supply increases in the same period.

In Table 9 it has been attempted to present the situation around the turn of the century in terms of some key indicators for energy assuming that supply can meet demand. The numbers in this table perhaps do not paint a picture of an energy starved world. Per caput consumption has increased in all regions, but at a much lower rate than earlier in the century. However the distribution of energy among regions has not changed much and the developing world is still at a very low level in spite of economic growth assumptions that perhaps are fairly optimistic. This underlines the need of further energy growth for this part of the world far into the next century.

If one attempts energy demand projections beyond year 2000 one easily gets exponential headaches unless quite drastic assumptions are made. Figure 4 shows as an example a projection where all growth after 2000 is confined to the developing world and where furthermore continued efficiency improvements are assumed. Even with these assumptions world energy demand by year 2050 would be about 5 times higher than today. A demand that surely cannot be met unless large new energy sources become available early in the next century.

2. ENERGY CONSERVATION - MYTHS AND REALITIES

2.1 Different Types of Energy Savings

As a response to impending energy shortages it is often said: We must save energy. And this statement is then substantiated by examples of wasteful uses of energy which are not too hard to find.

Energy savings can however mean several different things and for a discussion of the long term energy prospects for the world it is important to keep these meanings separate.

Firstly, there are simple cases of carelessness as leaking hot water taps or gas-pipes, lights being left on where not needed and the like. These could and should be cured by just paying a little more attention to how energy is used - an attention the more easily mobilized as the price of energy goes up. Savings of energy obtained this way are economically very attractive as they cost nothing and have a not insignificant savings potential, especially in countries

Table 9

Energy Indicators for year 2000[A]

	N. America 1972	N. America 2000	W. Europe 1972	W. Europe 2000	Japan 1972	Japan 2000	USSR-E.Eur. 1972	USSR-E.Eur. 2000	China 1972	China 2000	Dev. World 1972	Dev. World 2000
Energy. Toe/cap-y	8.1	9.6	2.9	4.4	3.1	5.7	3.3	6.9	0.4	1.1	0.4	0.8
Electricity, KWH/cap-y	7600	14000	2800	8000	3000	9000	3300	–	140	–	80	300
Electricity % of Gross	27	36	29	36	24	–	–	–	–	–	8	25
% of world Energy[B]	33	23	19	15	5	6	21	22	6	10	12	21
Intensity, Mj/$(1972)	60	41	41	32	38	28	74	70	66	66	48	52
End Uses %												
Transport	25	19	16	14	15	16	–	–	–	–	30	28
Industry	32	38	39	43	53	52	–	–	–	–	43	46
Agriculture	5	3	3	3	4	3	–	–	–	–	5	4
Commercial	11	13	10	10	10	13	–	–	–	–	8	8
Residential	24	22	25	21	10	11	–	–	–	–	9	9
Non Energy	3	5	7	9	8	5	–	–	–	–	5	5

(A) Corresponds to projection in Figure 3 reduced by 6% to ensure supply-demand balance. Economic growth rates as in Table 5, UN Medium Population Projection.
(B) Sum less than 100 because of omission of Oceania and Bunkers.

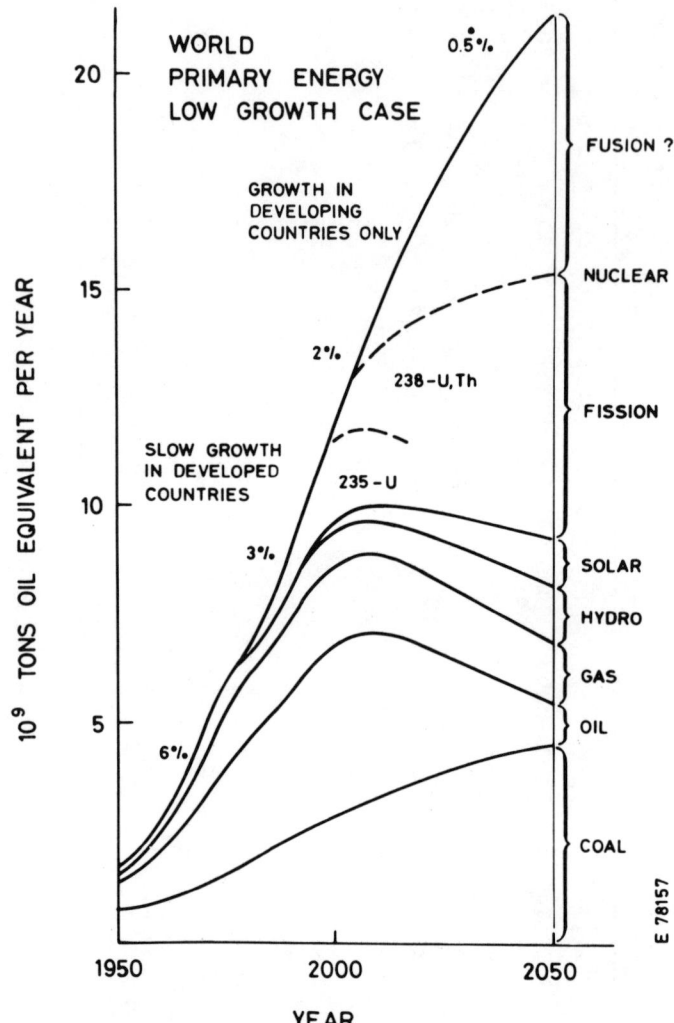

Figure 4. Long term estimate of energy demand and supply. Low growth case with demand saturation in developed economies. Supply estimates for solar, fission and fusion after year 2000 are speculative. Projections for fossil fuels and hydro-power are in agreement with present estimates of ultimately recoverable resources.

where the energy consumption is high. The industrialized countries have since the oil crises improved their energy GNP-ratio by 5-10% (cf. Table 10) an improvement that to a considerable extent can be ascribed to an increasing awareness of the fact that energy should not be wasted.

Table 10

Energy-GNP Ratio Changes 1970-75 (A)

	GNP-growth 1970-75 % p.a.	E-Growth 1971-76 % p.a.	E/GNP change 1971-76 %
N. America	2.6	1.5	-5.2
W. Europe	3.2	1.8	-6.6
Japan	5.3	3.2	-9.6
USSR	4.0	4.5	+2.4
China	7.0	6.2	-3.7

(A) Note GNP growth 1970-75 but energy growth 1971-76. The one year shift in the periods does not change E/GNP ratio significantly.

Secondly, significant improvements in the energy efficiency can be obtained by better energy using equipment. More efficient tractors or motors, better insulation in houses, improved processes in agriculture and industry and in numerous other ways involving new or improved techniques. The necessary changes do however not happen overnight and they cost money. More efficient equipment is generally more expensive and require larger investments. Investments that must pay for themselves through the energy saved over the years.

Thirdly, it is possible to save energy by organizing society in a different way. Energy is saved if automobiles are unneccesary, forbidden or so expensive that nobody can afford them. Energy is also saved if cities are built with efficient mass transport systems or if international conferences are conducted via satellite TV instead of meetings. Common for all such energy savings are, that they are based upon a society different from the one we have or the one we get extrapolating known trends. Thereby energy conservation has become part of the political discussion on how society should develop and therefore opinions are divided. My personal point of view is, that we should structure society to reach our political goals in general and then by technical efforts supply the energy needed - within reason of course.

2.2 Energy Saving Techniques

If we return to energy savings through technical improvements, it in some case has been possible to estimate fairly accurately

Table 11
Investments for Energy Savings

	Potential (A) savings %	Invesement (B) $/Toe	Pay-back period years
1. Insulation, Greenhouses	0.4	360	4
2. Thermostats, Public Buildings	0.3	410	5
3. Heat Recovery, Industry	0.8	460	5
4. Heat Recovery, Public Buildings	0.2	630	7
5. Wall Insulation, Houses	0.8	650	7
6. Roof Insulation, Houses	1.4	970	11
7. Thermostats, New Houses	0.2	1020	11
8. Roof Insulation, Apartments	0.1	1040	12
9. Thermostats, Old Houses	0.9	1370	15
10. Floor Insulation, Houses	0.4	1460	16
11. Windows, Double Glass	0.7	1670	19
12. Building Codes, Houses	0.6	1860	21
13. Floor Insulation, Apartments	0.1	2200	24
14. Wall Insulation, Apartments	0.1	2200	24
15. Heat Recovery, Apartments	0.1	3600	40
16. Diesel Engines, Automobiles	1.0	5100	57
17. Windows, Triple Glass	0.3	5600	62
Total	9.0		

(A) In % of gross national energy demand.
(B) Based on Danish data from 1975 inflated by 30% (reference 2).
    Toe = Ton oil equivalent. Oil price today is about 90 $/Toe.

how much energy can be saved for a given investment (Table 11). This is especially the case for large industrial processes and for residential heating. As an example Figure 5 illustrates the impact of certain saving measures under conditions found in northern Europe. Without going into the details, which are somewhat dependent on locality, it is obvious that energy can be saved - up to 10% of the total consumption - but the cost of saving energy is strongly dependent on how much you save. Of course, by improved insulation one e.g. in principle can get to a point where no heating energy at all is

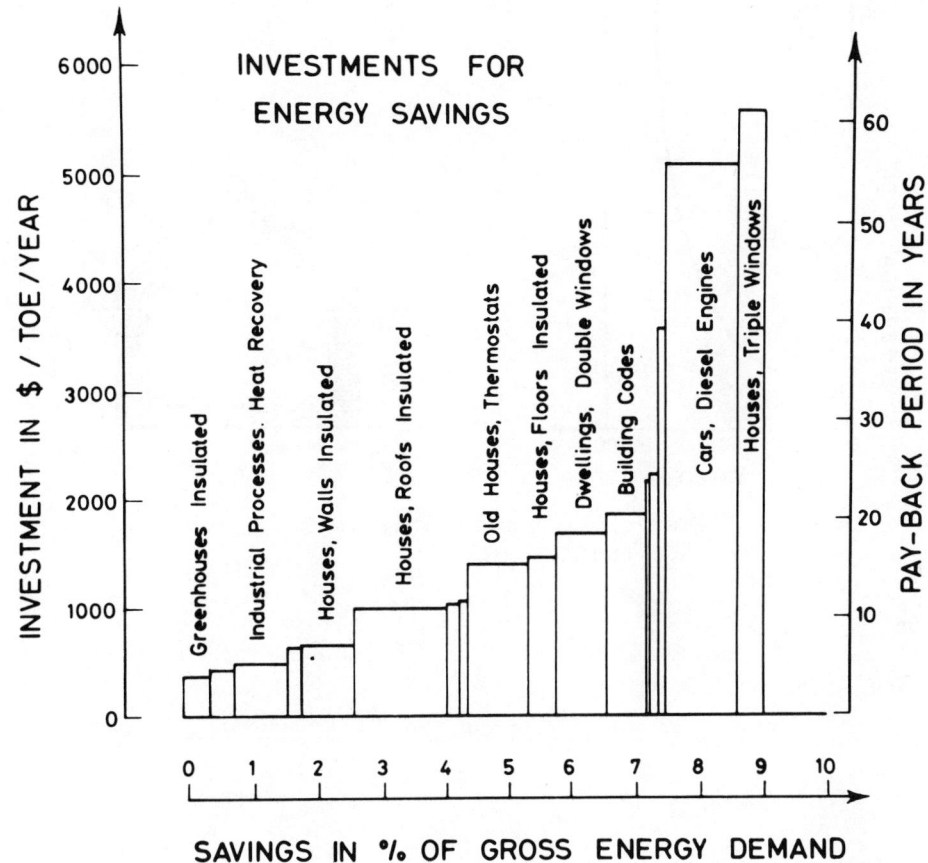

Figure 5. Investments for possible energy saving measures in Denmark. Pay back period calculated for an energy price of 90$/toe (reference 2).

required, but the cost of doing so on a nationwide basis would be astronomical. At any given energy price there are saving measures that are economical in the sense that the cost of the fuel saved can pay for interest and depreciation of the capital invested. Measures beyond that are uneconomical. Indeed, the cost of saving more increases quite steeply with the quantity saved and one soon approaches something like a sonic wall for energy savings.

As an example that can be analyzed in simple physical terms we can consider a building (Figure 6) to be kept at an inside temperature T_i against a (lower) outside temperature T_o. The building has a surface A and is, for the sake of argument, without doors and windows. The original walls are equivalent to a thickness ℓ_o of insula-

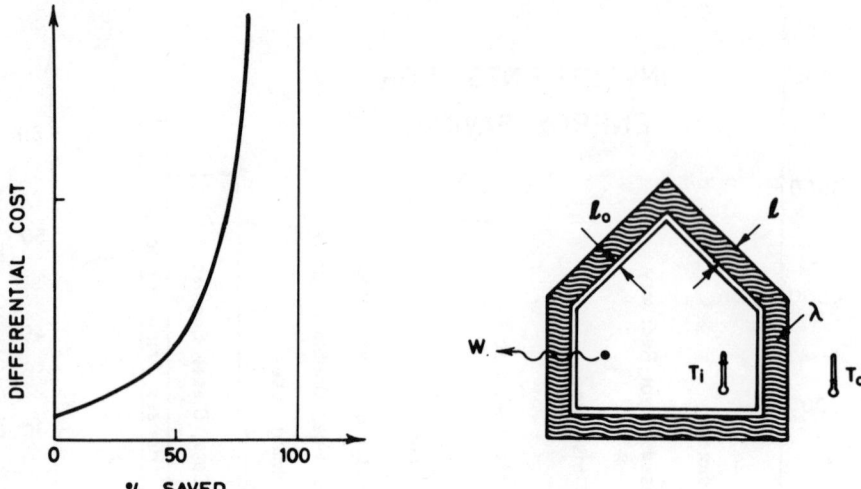

Figure 6. Differential cost of building insulation for model house (see text).

tion material of heat conductivity λ. If a layer of thickness ℓ of this material is added the heat loss is reduced from W_o to W and the energy flow savings amount to

$$S = W_o - W = A \cdot \lambda \cdot (T_i - T_o) \left(\frac{1}{\ell_o} - \frac{1}{\ell_o + \ell} \right) \quad (11)$$

with differential savings

$$\frac{dS}{d\ell} = A \cdot \lambda \cdot (T_i - T_o) \frac{1}{(\ell_o + \ell)^2} \quad (12)$$

The <u>cost</u> C of the insulation is $p \cdot A \cdot d\ell$ where p is the price per volume material. Thus the differential cost per unit energy flow saved is

$$\frac{dC}{dS} = \frac{p \cdot (\ell_o + \ell)^2}{\lambda (T_i - T_o)} \quad (13)$$

If we introduce the 'savings fraction' $F = S/W_o$ then

$$\frac{dC}{dS} = \frac{1}{\lambda(T_i - T_o)} \frac{\ell_o^2}{(1-F)^2} \quad (14)$$

Thus the cost increases steeply with the fraction saved and of course diverges for $F = 1$ (Figure 6).

I cannot prove it, but I suspect that this type of relation is quite general (cf. also Figure 5): The more you want to save, the less profitable it becomes. Therefore, in energy conservation in order to get the most out of the money one should spend them on improving many processes a little instead of improving few a lot.

In the first section we saw, that industry in all regions of the world is by far the largest energy consumer and is projected to increase its share from about 38% in 1972 to 43% in 2000.

Into these projections there has already been built a considerable amount of energy efficiency improvements. It is however difficult to get a clear picture of the changes that are expected, because industrial output is inhomogeneous and often measured in money terms rather than physical units.

However, most of the energy going into industrial production is consumed by a few sectors, namely steel, aluminium, paper, cement and chemicals. Just the high-mass products earlier discussed (see Tables 3 and 4). In North America these sectors are reponsible for about 70% of industrial energy demand, in Europe 65% and in Japan 80%. Accurate figures are not available for countries in the developing world, but also there the heavy productions are known to dominate the industrial energy demand.

As an example the worlds steel industry in 1975 produced 6.4×10^8 t of crude steel which, at an estimated energy requirement of 30 GJ primary energy per ton steel, corresponds to 19 EJ of energy out of a world energy consumption of 260 EJ. Steel thus accounts for 7% of all energy used in the world and about 20% of all industrial energy. It is therefore of considerable interest to investigate whether the energy efficiency of steel production can be improved.

To get some idea of the possibilities one can ask: Are all manufacturers equally efficient today? Dependent on the source one gets somewhat different answers. As an example Table 12 shows the energy required to produce 1 t of steel as derived from official

Table 12

Energy to Make one ton of Steel

	GJ/T
Germany	20.4
France	20.7
Italy	17.8
Belgium	19.6
United Kingdom	22.7
EEC-9	20.8
Japan	20.7
USA	20.9

Based on references 9, 10 and 11.

statistics on energy and steel quantities. The differences are surprisingly small and probably mostly just reflect statistical uncertainties. On the whole one would however say, that all producers operate at nearly the same efficiency.

As an example of more detailed character Figure 7 shows the amount of coke required for production of 1 ton of steel in West Germany and Japan through the last 35 years. Firstly one can notice,

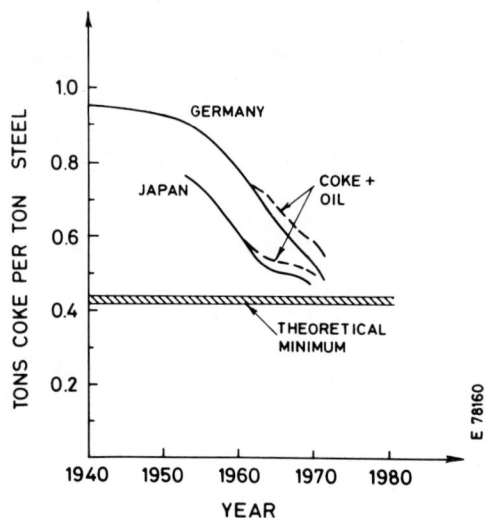

Figure 7. Coke requirement for steel production in Germany and Japan (after reference 12).

that the Japanese production has required only about 80% of the coke required in Germany. Secondly, that the coke requirements have been steadily decreasing by approximately 3% p.a. Thirdly, that the gain in coke efficiency has become less in recent years as the theoretical limit is approached. Reduction of iron ore is a well defined chemical process with the net result.

$$2Fe_2O_3 + 3C \rightarrow 4Fe + 3CO_2$$

and this process cannot proceed without supplying the necessary carbon atoms and the heat for the reaction. In fact, the coke use is now only about 130% of the theoretical minimum which explains why future improvements will be slow.

The average energy efficiency of future steel production is a complicated question of the rate of introduction of new equipment, the efficiency of such equipment, the amount of scrap and the final product mix. Estimates of the combined impact of such factors have been performed by several groups. The estimated improvements in the specific energy intensity of steel production are about 1% p.a. for the next 10-15 years. Similar improvements can be expected in other branches of heavy industry as cement, non-ferrous metals and paper (cf. Table 13).

Table 13

Projected Industrial Efficiency Change 1972-85 in % p.a.

	Germany	Italy	Japan	Holland	USA
Steel	-1.7	-0.7	-0.8	-0.4	-1.8
N.Ferrous Metal	-0.8	-2.2	-0.9	-0.5	-1.7
Paper	-2.0	-3.0	-1.1	-0.4	-1.8
Chemicals	-1.7	-0.4	-0.9	-0.1	+0.3

Based on final energy demand for unit value added.

It is not always possible or even desirable to measure industrial production in physical units as ton or m^3. Indeed, for a large fraction of the industry with a complex product mix the production volume can only be measured in monetary units. Often the value added is a convenient measure of production and the energy intensity is then expressed as energy/value added e.g., in MJ/constant $. Thereby a direct connection is also obtained to the economic projections behind most energy demand forecasts.

A general lowering of the energy needed for each unit of value added is particularly evident in industries where new and more

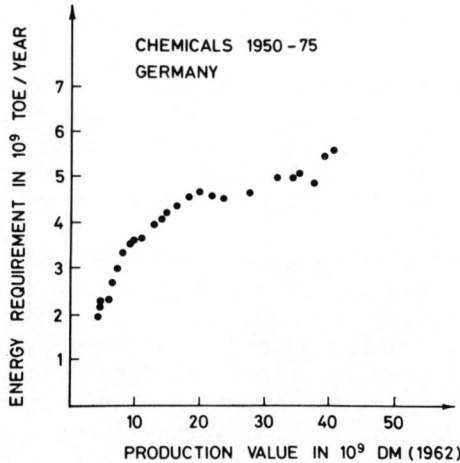

Figure 8. Energy requirement and value of production for German chemical industry 1950-75 (cf. 12).

valuable products are steadily being marketed. As an example Figure 8 shows the relation between energy demand and volume of production for the German Chemical Industry where the non-linear relation between energy and output implies a considerable lowering of the energy intensity.

A somewhat different example is shown in Figure 9 for the aggregate of French industry in the period 1963-75. The average reduction in energy per unit value is 2.2% p.a. It is remarkable, that

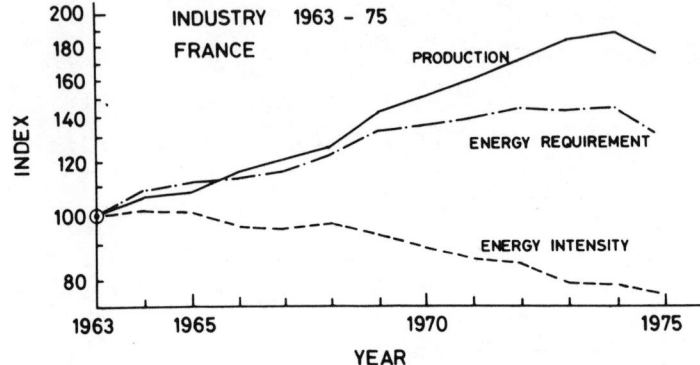

Figure 9. Energy requirement and production volume for total French industry 1963-76 (reference 12).

most of this change took place in a period where energy was cheap. Much of the change therefore must be ascribed to what I in the first section called 'inherent trends'

2.3   Energy and Food Production

In the available statistics concerning energy uses it is quite difficult to isolate energy for agriculture. Agriculture disappears among the really large energy consuming sectors as industry and transport. A rough estimate shows, that agriculture is responsible for only about 2% of the worlds energy consumption.

Such an estimate however, is misleading for several reasons. It refers to the energy directly used in agriculture for the running of tractors, harvesters, irrigation and so on. But in addition agriculture indirectly requires substantial quantities of energy feedstocks for production of fertilizer and other chemicals. Also, one must not forget, that in many countries agriculture is only a first part of an extensive food system. Further treatment and distribution of agricultural products require substantial energy inputs.

In Table 14 it has been attempted to illustrate the similarities and differences among some highly developed agricultural system.

Table 14
Key Numbers for Food Production(A)

		Denmark	Holland	United Kingdom	Japan	USA
Field work	GJ/HA-yr	8.8	7.1	9.4	18.2	7.2
Fertilizer	KGN/HA-yr	104	176	74	140	40
Field work	GJ/CAP-yr	5.1	1.1	2.1	1.1	6.3
Fertilizer	"	4.7	2.1	1.3	0.6	2.7
Food industry	"	5.9	6.0	4.5	2.3	5.9
Household	"	5.1	6.0	3.8	2.3	8.1

(A) Not corrected for import/export. From reference 8.

Although the basic statistical material is somewhat shaky and the categories not entirely consistent there are certain conclusions which seem to hold:

(1) The direct energy for agriculture is only about 25% of all energy for food.

(2) The indirect energy - the energy for the food industry and for the household processes - are each of the same size as direct agricultural energy.

(3) There are considerable differences in the amount of nitrogen fertilizer used.

These conclusions are approximately valid for industrial food systems, but conditions in mainly agrarian economies could be quite different, and points to considerably higher energy efficiencies.

In the discussion of these matters one often speaks of the so-called energy-ratio, R, which gives the ratio of the total agricultural output measured in energy units (nutritional value) and the input of artificial energy

$$R = E_{out} / E_{in} \tag{15}$$

Some values of R are given in Table 15 for some different systems. For the industrialized systems this ratio is considerably less than unity. In contrast primitive agrarian systems with mainly a vegetable production, much manual work and careful husbandry of animal and human wastes can show quite high energy ratios as 20 or higher.

Although the energy ratio in a concentrated form tells us something characteristic of the efficiency of an agricultural system one should not uncritically focus on its size.

Table 15

Energy Ratios for Agriculture

	E_{out} GJ/ha-y	E_{in} GJ/ha-y	R
Denmark	14	31	0.5
England	11	24	0.5
Holland	43	36	1.2
USA	9	13	0.7
Australia	6	2.4	2.8
India (example)	10	0.7	15
China (example)	281	6.8	41

E_{out} vegetable + animal output
E_{in} fossil fuel + human and animal labour
Based on references 6, 7 and 8.

Firstly, that a dimensionless ratio can be derived is due to the fact that agricultural products happen to be measurable in energy units. This is not the case for most other productions. That the ratio can exceed 1 is caused by the sun which generously gives an energy subsidy to agricultural production.

Secondly, the energy ratio does not tell anything about absolute yield. It is clear, that the high crop yields per area are conditional upon a high energy input. Production can be increased by application of energy (work and fertilizer) but as for other growth-factors the law of diminishing returns is also valid for energy. It seems that agricultural yield (plant production) increases approximately as the square root of the energy input.

Finally, the energy ratio does not tell us about the quality of the production. A high protein diet has a higher energy requirement per unit food but also a higher nutritional quality.

One can ask, how much energy is required to produce the vegetable and animal matter sufficient to feed the world population around year 2000? If we as a basis take a minimum daily diet of 10500 kJ of which 15% or 1600 kJ is of animal origin we can estimate the energy requirements. For animal matter we can assume they are 10 times higher than for vegetable matter. It is furthermore assumed that an energy ratio of 4 can uniformly be achieved for plant production.

The world population in the year 2000 can be estimated to be 6.5×10^9 (UN population division). Thus the total yearly energy requirements for food production in terms of vegetable and animal produce would be at the farm gate:

Vegetable: $6.5 \times 10^9 \times 8900 \times 365 \times \frac{1}{4}$ kJ = 5.3 EJ/y

Animal: $6.5 \times 10^9 \times 1600 \times 365 \times \frac{1}{0.4}$ kJ = 9.5 EJ/y

Total: 14.8 EJ/y

This is certainly a considerable amount of energy (about 6% of present world energy use and about 3 times present world agricultural energy). Although agricultural energy mainly would have to be oil and oil products this simple calculation underlines, that energy is not a severely limiting factor in providing an adequate diet for the world population. This in spite of the fact, that agriculture is one area where energy conservation has the wrong sign. In order to step up production we need to use more energy per unit product, not less.

2.4   Transport

Transport is not the world's largest energy consuming activity. But it is by far the largest oil consuming activity (cf. Table 18) responsible for around 40% of the oil demand. If we consider the projections until year 2000 this share could have increased to around 55%.

Transport without oil as a mobile energy source is at present unthinkable. What could happen in the long run is difficult to say, but until year 2000 and probably a good deal longer oil is the only realistic energy source for transport.

It is therefore obvious that improvements in the energy efficiency of transport is of overriding importance. From engineering changes in automobiles, airplanes and ships we can only expect moderate gains, perhaps 10-15% improvement over 20 years. We are here at an example where change in society's attitude to transport is much more decisive than technical improvement.

The different modes of transport are not equally efficient. Table 16 gives some examples of the efficiency of passenger and freight transport. Such efficiencies vary with local conditions and load factors, but those given can be considered as typical averages.

Obviously the automobile is a less efficient mode of transportation than bus or train. Therefore there are considerable savings in energy if the daily traffic in urban areas goes by public transportation, which might have other advantages as well. The advantages of individual transport are however obvious too, and there will be a considerable pressure to keep it in countries where already widespread and to introduce it where it is rare.

Table 16

Average Transport Efficiencies

Passenger Transport	MJ/Pass-Km
Automobile, Europe	2.1
Automobile, USA	4.0
Bus, urban	1.2
Train, diesel	1.0
Train, electric	1.1
Ferry	6.0
Airplane, international	3.3
Airplane, domestic	4.2

Freight	MJ/Ton-Km
Rail	0.7
Truck	3.2
Ferry	3.7
Airplane	25.0

Average load factors assumed throughout.
Based on reference 2.

In the light of a coming scarcity of liquid fuels the near term future of individual transport is dependent upon oil savings in other sectors than transport, considerable efficiency improvements in the automobiles themselves and in the long term on synthetic fuels from coal, tar sand or oil shale.

Air traffic is even more dependent on petroleum fuels, but the present and even the projected fuel uses are much less than for road based traffic.

2.5   Energy Conservation Will Not Do It Alone

If we take the more restricted view of energy conservation and define it as efficiency improvement (reduction in energy required per unit activity) it is clear from numerous analysis that the potential for energy conservation is there, but it is limited to about 10-20% of present day energy use efficiency. Higher energy prices would increase the potential where presently limited for economic reasons, but such increases would be relatively small because of the steepness of the cost-savings curve (cf. Figure 6). In general one should not base expectations on more than the mentioned 10-20% for a twenty year period. Even as the result of the combined action of energy price, energy policy and long term technical development.

WORLD ENERGY PROBLEMS

Nevertheless such efficiency improvements are essential for achieving the reduced energy coefficients and thus the moderate growth rates for energy demand without cutting into the economic growth rates.

In making more efficient use of energy, the developing countries might have an advantage because they start an industrial development where the energy aspect can be taken into account right from the beginning. The whole aspect of technology transfer is presently under discussion and probably nobody knows what the best procedures are. However, the need for a basic industrial development in many countries of the world is great. At the same time the world is in need of something which can be called a third industrial revolution (counting coal and the steam engine as number one and electricity and oil as number two). This third generation technology would call for much closer attention to the closed flow of materials, to minimal use of energy and low environmental impact.

The highly industrialized countries have now reached a level of energy use where stabilization ought to be within reach, although it cannot be expected overnight. In the developing countries energy demand must grow for several generations to come, even with the highest possible efficiency of energy use.

3. THE CONCEPT OF THE ENERGY GAP - SUPPLY AND DEMAND MISMATCHED

3.1 Can Demand be Met?

The former two sections have mostly dealt with the demand for energy. It is now time to ask: Can we supply energy to meet the demand and from what sources?

The demand, discussed earlier in these sections is the desired demand, the demand that is a consequence of the needs, plans and aspirations of our societies. As we saw (Figure 3) even a moderate economic development would imply an energy demand by year 2000 of 2.5 times that of today.

It is of course quite certain, that the world never can use more energy than it can supply. The calculated demand is _desired_ demand, and it might never be met. In a real world, demand and supply will always be in balance. But the consequences of such a forced balance at a level much below the desired demand could be so alarming for the economic and social stability in the world, that one must find them unacceptable.

3.2 Supply Expansion Becoming Difficult

We all know that the most crucial supply is that of oil. Oil today covers 44% of the worlds demand for energy. If we look at market-economies only the share of oil is even greater. All experts and governments agree that the share of oil must be reduced. Typically the forecasts say that by year 2000 oil can not supply more than about 35% of the world demand. That this goal is not easy to reach is illustrated by the fact, that the share of oil for the market economies from 1975 to 1976 grew from 50.9% to 51.7%. This seemingly modest increase corresponds to 6 times the volume of oil produced in the North Sea in 1976.

There have been many estimates of the possible future oil supply. The estimates I shall present here are those[1] of WAES supplemented with data from independent sources.

Oil is a limited resource. The part of the resource we have discovered and can produce economically is called the <u>reserves</u>. Presently known oil-reserves are about 90×10^9 tons (Table <u>17</u>). If we divide this quantity by the present world production of $2.7 \times 10^9 t/y$ we find that the reserves could last 33 years. This is for many

Table 17

Proved Reserves at end 1976

Region	Oil 10^9 T	Gas 10^9 Toe
N. America	5.8	6.6
Latin America	4.7	2.2
W. Europe	3.3	3.4
Middle East	50.0	13.0
Africa	8.1	5.1
USSR, E. Europe	11.1	22.6
China	2.7	0.6
Asia	2.6	2.9
World	88.3	56.4

From reference 5.

reasons an extremely misleading calculation and should at most be used to give an impression of the fairly short period in which we can count on oil. Some of the reasons for rejecting the calculation are:

(1) In spite of intensive prospecting all oil in the world has definitely not been discovered yet. There is a steady addition to reserves which for many years has exceeded the production. We do

not know how large the discoveries will be in the coming years, but somewhere between 1.5 and 3.0×10^9 tons per year is likely for the world outside the socialist countries. Thus, somewhat dependent on luck, new discoveries will correspond to between 60% and 120% of todays production in the same regions. However, demand is expected to increase, and within a few years certainly will exceed any conceivable rate of discovery.

(2) There have been several guesses as to the total quantity of oil on the globe. Such estimates have in recent years been converging (Figure 10) and point to a total resource of about 300×10^9 t. The present reserve is thus about 30% of the total resource. However, it would take a long time to discover all the remaining oil and much of it is porbably located in unaccessible areas. So again, in spite of oil to be discovered, reserves might soon be on the decline.

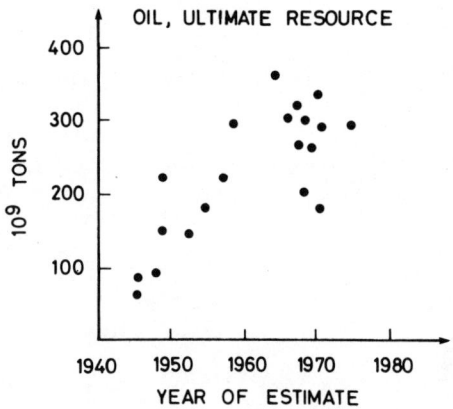

Figure 10. Estimates of ultimately recoverable oil (based on data contained in reference 1).

(3) Production can not proceed at a constant rate until the reserve is completely exhausted. In general one can for technica reasons produce, per year, only about 7-10% of the total world reserve, the lower rate being the most realistic. Thus present reserve could at the maximum produce around 6×10^9 t/y. After the maximum was reached and with no added reserves, production would decline exponentially.

In total, new discoveries and maximum production rates would lead to an expected profile for world oil production as shown in Figure 11. It is a bell-shaped curve starting at zero at some tim in the former century and ending at zero some time in the next century.

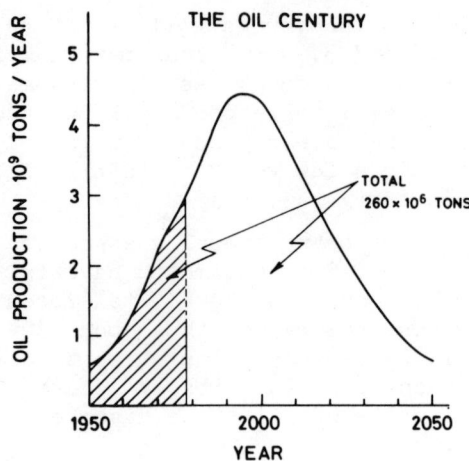

Figure 11. Schematic world oil production profile 1950-2050.

(4) Most importantly, oil production is limited not only for technical or geological reasons, but also for political reasons. The large oil producers in the Middle East can not be expected to produce to full capacity to meet an increasing demand that under all circumstances will have to be curtailed soon after

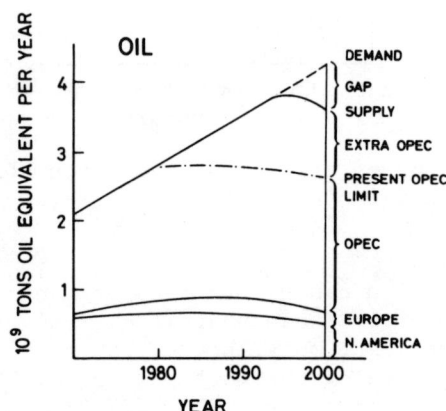

Figure 12. Estimated oil production profile for world outside socialist areas (market economies). The Figures 12-17 are based on data from the WAES study (ref.1-4) for high economic growth and increasing energy prices. The production profiles Fig. 12-16 refer essentially economically and technically feasible developments with disregard of major political constraints and can consequently be regarded as upper limits of production.

WORLD ENERGY PROBLEMS

A combined evaluation of the factors mentioned above led the WAES group to the oil-production profile in this century shown in Figure 12 for the non-socialist part of the world. The development in the socialist countries, at present producing 20% of the global total, is more difficult to estimate. However, large exports can not be envisaged so it is a reasonable approximation to consider the two regions separately.

Figure 12 shows that the OPEC countries are decisive for the development. Production in North America and Europe is small and declining towards the end of the century. However, because of their large reserves, the OPEC countries can meet demand (estimated to increase by 3% p.a., later 2% p.a.) until the early 1990's. Total production in the world would then be declining with desired demand still on the increase.

3.3 The Concept of the Oil Gap

As Figure 12 shows the independent estimates of oil supply and oil demand lead to a widening 'gap' between them. By year 2000 the prospective gap is about 20% or about 700×10^9 t/y.

As already pointed out, supply and demand must always be in balance and an energy gap can not really occur. Nevertheless the concept of the 'energy gap' is a useful pedagogical tool, because it directly measures the distance between what we hope to have and what we are likely to get.

One can react to the energy gap in different ways: We must increase supplies is one reaction. But the supply estimates for oil, if anything, are probably already on the optimistic side and can for technical and political reasons not be increased significantly. We must use something else is another reaction. The possibilities of doing that is discussed in more detail below. Finally: If supply cannot increase, demand must come down. This is an obvious conclusion, but not much of a solution. The demand estimates referred to in the first section were intimately connected to the economic and social development, and as discussed in the second section, the possibilities of increasing the efficiency of energy use beyond what is already assumed in the demand estimates are small. Indeed, it appears doubtful, that all the efficiency improvements will be realized in time.

Therefore, closing the energy gap will almost certainly imply a further reduction of the economic growth rates with a negative feedback to all parts of the world and all sectors of society.

3.4 Substitution of Other Energy For Oil

The oil shortfall which could become a reality as early as 1985 makes it mandatory that other energy sources are rapidly brought into use to supplement the oil supplies. Unfortunately, the sources we in this century can hope to have at our disposal are only few: Natural gas, coal, nuclear energy and hydro-power. Solar energy in its various forms could give useful supplements especially for uses calling for low temperature heat, but large scale solar energy for the generation of electricity or the production of synthetic fuels seems very far off. Also advanced sources of nuclear energy as breeder reactors or fusion reactors hardly will have an impact in this century but they could be crucial in the beginning of the next as indicated in Figure 4. There is therefore every indication, that the oil deficit mainly must be covered by increased reliance upon sources already in use today.

Natural gas is an obvious substitute for oil for many stationary applications. The total resource is of the same order of magnitude as oil. But whereas oil today constitutes about 45% of the total energy supply, gas covers only 18%. The increase in natural gas use has not been able to match that of oil. In the period 1971-76 world oil consumption increased by 2.8% p.a., natural gas consumption by only 1.3%. The reason for this modest increase in a world facing an oil crisis is clear enough: Although gas is relatively plentiful, there is a shortage of gas in the large energy consuming areas, North America and Japan. In other parts of the world, especially South Asia, Africa and Europe gas consumption is increasing, but is still at a low level compared to North America and USSR.

The future of gas is therefore bright in the regions where it is available, but its future as a major world source of energy is dependent on the creation of an international trade with Liquified Natural Gas (LNG). Figure 13 shows the potential gas production in

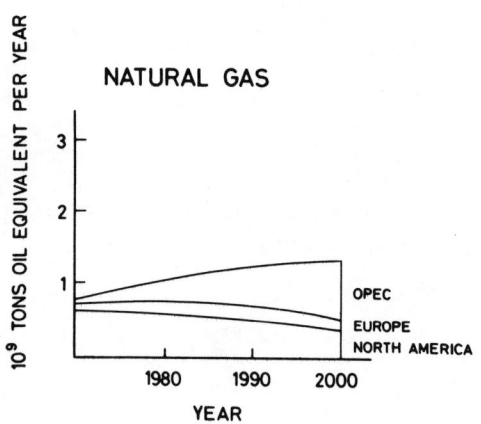

Figure 13. Gas production profile. Realization of the large OPEC production is conditional upon a major expansion of international LNG trade..

the non-socialist world until year 2000 as estimated by WAES. Whether there is any reality to this picture is entirely dependent on the future of LNG.

Coal is another plentiful energy resource which to some extent can be substituted for oil. Coal, however, is not a preferred fuel and its use has in many parts of the world been stationary or even declining. Coal production can certainly be increased (see Figure 14) in many parts of the world rich in coal, but especially in North America, USSR and China. International coal trade has been on a very modest scale. Steam coal trade, based on energy content, has only corresponded to 3% of the oil trade. So the role that coal has to play as a most necessary substitute for oil is again dependent on the creation of a huge system for the international transportation and use of coal.

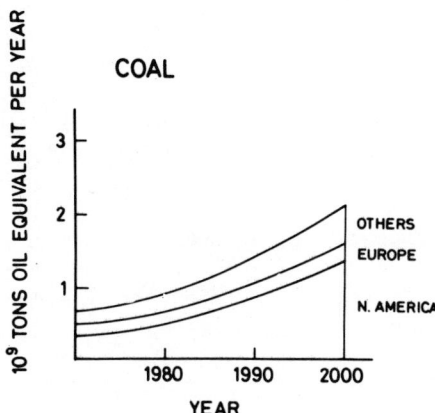

Figure 14. Coal production profile.

Nuclear energy today contributes about 2% of the worlds primary energy. Its share is however rapidly growing in spite of political resistance in several countries. The potential of nuclear energy is considerable because of the very large capacity of each nuclear plant. A primary energy contribution of about the same size as that of coal is technically possible around the turn of the century. This would correspond to about 20% of all primary energy, a contribution so large that it is inconceivable that the world could do without it (Figure 15).

The hydropower resources of the industrialized countries are today almost fully exploited. However, in many developing countries the hydropotential is huge and could be of great value for the expansion of industry and general welfare where it is available. Electri-

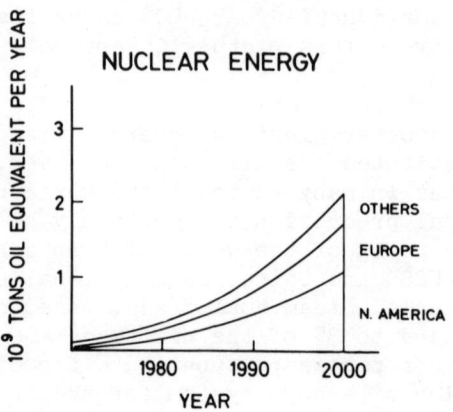

Figure 15. Nuclear production profile (thermal input basis). The profile corresponds to an installed capacity of about 1600 GW(e) by year 2000.

city too offers a possibility of creating a third generation basic industry for the third world without many of the negative aspects that makes basic industry unattractive in its present forms.

Finally solar energy as mentioned can make some contribution either directly as a source of low temperature heat and cooling or indirectly via combustion of biological matter. The contribution

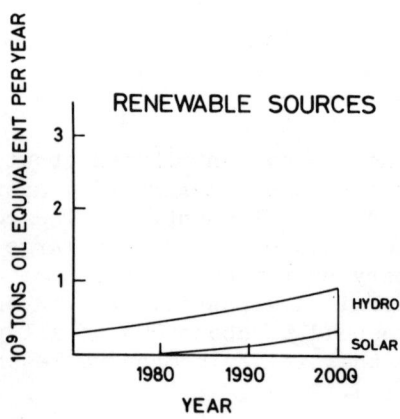

Figure 16. Renewable energy profile (thermal input basis). Solar energy contribution mostly for domestic comfort heat and hot water.

is limited partly by a limited demand for the types of energy that solar energy can supply and partly by adverse economy even at considerably higher prices for competing energy.

The total estimated contribution from renewable energy sources is shown in Figure 16.

3.5   Living With the Energy Gap

We are now in a position where we can compare the total available energy supply to the total energy demand. Figure 17 shows this comparison for the non-socialist world until year 2000. As for oil a gap appears between desired demand and possible supply. This gap is of the order of 10% of total energy and is mostly a reflection of the corresponding gap for oil. In spite of sizable substitutions of coal and nuclear energy for oil it has not been possible to close the gap.

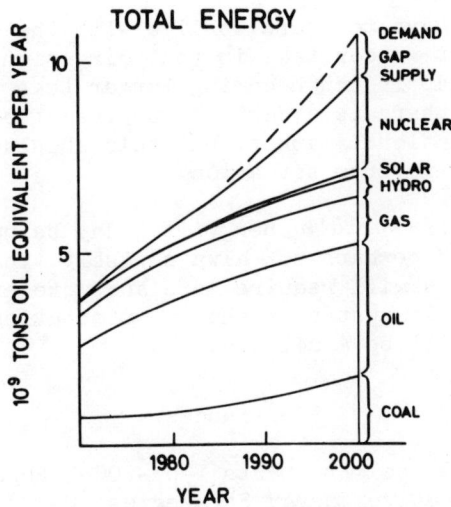

Figure 17. Total energy demand and supply corresponding to market economies demand from Figure 3 and supply from Figures 12-16. This is a high nuclear situation which reduces the coal component. A high coal situation is also feasible with nuclear reduced to about 60% of the level shown here. The demand-supply gap essentially shows a shortfall of non-substitutable oil.

A deficit of 10% in total energy perhaps does not sound too alarming and could be eliminated by a moderate reduction in the world economic growth rates from say 4% p.a. to 3.5% p.a. The appearance of the gap does however point to a tight situation for energy almost irrespective of basic assumptions. Furthermore, the demand

and supply picture in Figure 17 is probably overly optimistic because it is crucially dependent on a surprise-free world. There are assumed no major political restrictions on the free flow of oil and LNG, coal is assumed to be a commodity in large scale international trade it is assumed that the nuclear contribution can be expanded in an orderly manner. If just one of these assumptions does not hold, then the energy gap increases from a perhaps managable 10% to a much larger and unmanagable 20-30%.

Our projections of demand and supply are not better than their underlying assumptions. But I think it highly improbable that the assumptions made here concerning desirable or probable economic development, on the possibilities of increasing the overall efficiency of energy use and of the rate of production of various fuels, should prove far from reality. At least not so far from reality that the energy problems predicted towards the end of this century could simply go away. Prudence tells us to prepare for a situation which is all too likely to occur.

We will have to learn to live with the energy gap. The most important step we can take in that direction is to make sure, that the looming gap is not becoming larger than we already think it is. At present there is a certain surplus of energy, especially oil, on the international market, but this should not make us forget to prepare for the reverse situation.

Most of the world's energy is today being used in the industrial countries and perhaps not always wisely. In the future the developing countries will require more and more energy for their basic development. It must be a shared international responsibility, that such energy will be available.

References

(1) Energy: Global Prospects 1985-2000, Report of the Workshop on Alternative Energy Strategies (WAES), McGraw-Hill Book Company, New York 1977.

(2) Energy Demand Studies: Major Consuming Countries, First Technical Report of WAES, The MIT Press, Cambridge 1976.

(3) Energy Supply to the Year 2000, Second Technical Report of WAES, The MIT Press, Cambridge 1977.

(4) Energy Supply-Demand Integrations to the Year 2000, Third Technical Report of WAES, The MIT Press, Cambridge 1977.

(5) BP Statistical Review of the Oil Industry, British Petroleum Company Limited, London (1976).

(6) R.M.Gifford, Search, $\underline{7}$ (412) 1976.

(7)   G.Leach: Energy and Food Production, London (1976).

(8)   B.Elbek, Nordisk Lantbruksekonomisk Tidskrift, $\underline{2}$, 4 (1977).

(9)   United Nations Statistical Yearbook (1977).

(10) Eurostat, Energy Statistics Yearbook (1976).

(11) An Initial Multi-National Study of Future Energy Systems (IEA). BNL-50641, Jül-1406 (1977).

(12) H.Reents, Berichte der Kernforschungsanlage Jülich, Jül-1452 (1977).

(13) World Bank Atlas, The World Bank, Washington (1977).

LONG-TERM ENERGY SYSTEMS AND THE ROLE OF NUCLEAR AND SOLAR ENERGY

Dieter Faude

Karlsruhe Nuclear Research Centre

Federal Republic of Germany

As never before in human history, energy demand and supply has become a political theme. The problems of the 'limits to growth', i.e. steadily increasing human activities on the one hand, and - in many aspects - limited resources on the other hand, have been spelled out very distinctly in the context of energy consumption.

The following contribution is an attempt to survey the present world energy situation and future aspects of energy demand and supply. First, some general aspects of an energy economy and the question of world-wide increasing energy demand are treated. Then various energy supply options are discussed: the traditional fossil fuel resources of coal, oil and gas; options for a long-term energy supply with special emphasis on nuclear and solar power, but also nuclear fusion and geothermal energy; and some additional energy supply options, such as water and wind power.

1. SOME ASPECTS OF AN ENERGY ECONOMY

The energy consumption of a society or a nation can be judged in the most meaningful way by looking at the consumption of the individial. Figure 1 shows the historic development of the per capita energy consumption since man's origin - i.e. primitive man one million years ago up to technological man in a highly industrialized society.

Primitive man, in East Africa one million years ago without the use of fire, had only the energy of the food he ate (2000 kcal/day). The use of fire is 400,000 to 600,000 years old, and a hunting man in Europe 100,000 years ago had more food, and also burned wood for heat and cooking (5000 Kcal/day). In the first stage of

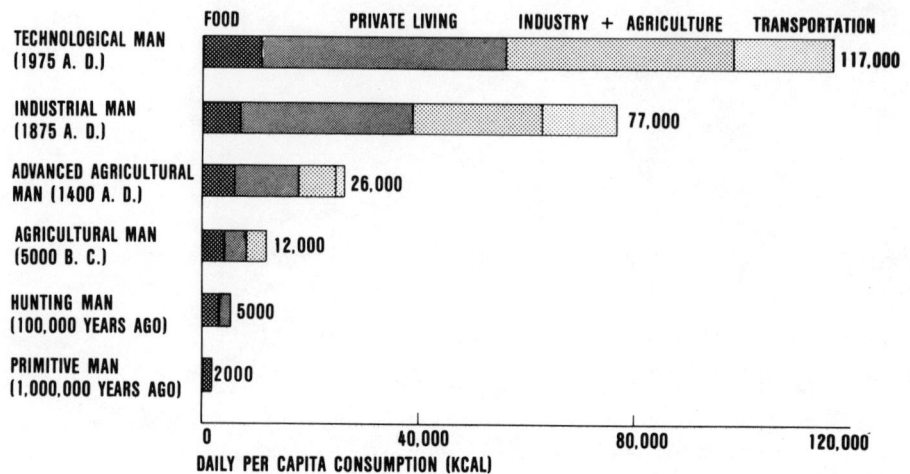

Figure 1. Development of per capita energy consumption (from Scientific American, Reference 1).

agricultural life (5000 B.C.), man was growing crops and he had gained animal energy (12,000 Kcal/day). Advanced agricultural man (e.g. in Europe in 1400) had some wood and coal for heating, as well as some water power, wind power and animal transport (26,000 Kcal/day). Industrial man in England in addition made use of the steam engine which led to the consumption of large amounts of coal (77,000 Kcal/day). It is here where commercial energy economies start, whereas previously all energy consumption - wood for burning, animal energy, wind power for sailing ships or for water pumping, and water power for milling corn, etc. - can be regarded as non-commercial energy consumption. Technological man, in a highly industrialized economy like the F.R.G., in 1975 consumes about 120,000 Kcal per day ($\cong$ 6 tce/year) of commercial energy, the non-commercial fraction is negligible. This value of 6 tce/year is an average for industrialized countries - as we will see later, the consumption is twice as high in the U.S. and half as much in, for example, Japan or Italy.

The difference between the basic energy consumption of an individual and the energy consumption of man in a highly industrialized country is 1 : 60, and this is remarkably high.

A per capita consumption of 6 tce/year requires an extensive and complex energy flux, as is shown, for example, in Figure 2 for the F.R.G. Figure 2 shows the different energy sources that contribute to energy consumption: 50% oil which has to be imported almost completely, 35% hard coal and lignite which are domestic resources, 12% natural gas, and the balance is water power, imported electricity and nuclear energy. In Figure 2 the flow of energy is plotted from the left to the right starting with the primary energy consumption which amounts to about 360 million tce/year net. This is followed by the energy conversion sector (e.g. electricity generation or oil refinement), and further to the right the consumption of end energy is shown (which is divided into the sectors residential + commercial, industry and transportation). The official energy balance stops at this point. It provides an answer to the question: Who consumes how much of each energy source? But it does not tell one anything about the purpose for which the energy is used.

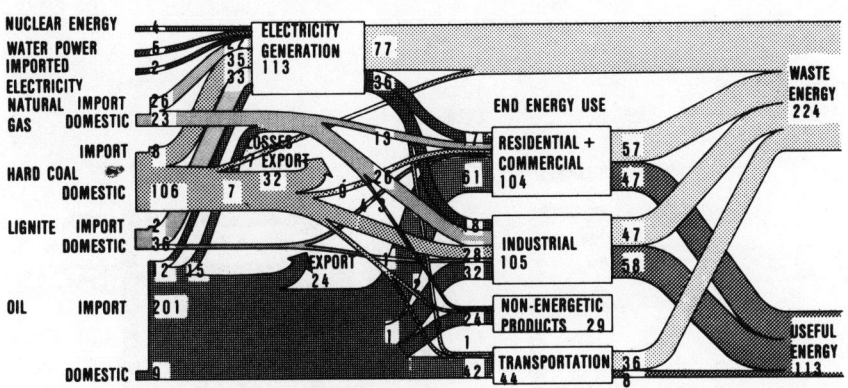

Figure 2. Energy Balance 1974 in the Federal Republic of Germany (million tce) (From Reference 2).

The break-down on the right hand side of Figure 2 into useful energy and waste energy is only a crude estimate, according to the conversion efficiencies of the various conversion processes.

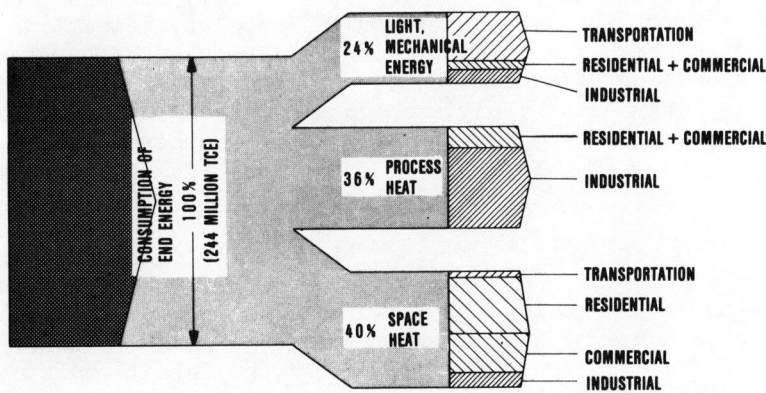

Figure 3. Useful Energy Balance 1974 in the Federal Republic of Germany (after H.Schaefer, F.R.G.)

This leads to the question of a useful energy balance, but such a balance can only be estimated. Figure 3 shows the break-down of the end energy consumption into different categories of useful energy: space heat (low-grade heat), process heat (intermediate range) and light and mechanical energy (high-grade energy). Moreover, one should not start at end energy consumption when talking about useful energy and waste energy, but with the primary energy source taking into consideration the complete energy conversion process.

Figure 4 is an example of the question of how much useful energy can be obtained by water heating using electricity that originates from coal as a primary energy source.

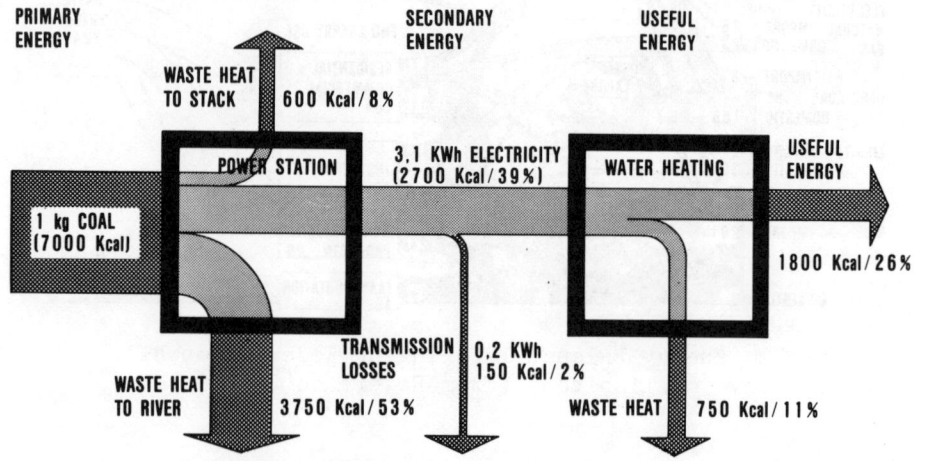

Figure 4. Energy Conversion: Coal - Electricity - Heat.

There is one further remark: Strictly speaking, energy cannot be consumed. Energy consumption of a society actually means the conversion of high-grade forms of energy, such as chemical binding energy of coal or hydrocarbons, or electromagnetic energy into heat. The total energy 'consumed' in an economy is ultimately converted into low-grade heat and emitted into the atmosphere.

2. SOME REMARKS ON THE GROWTH OF ENERGY CONSUMPTION

One very important aspect in today's discussions about energy policy is the question about the further growth of energy consumption - both nationally and globally. The consumption of energy, and thus the growth of energy, is not an end in itself, but is closely linked with general economic growth. This is only mentioned here but will not be discussed in greater detail.

Figure 5 shows the development of world energy consumption over a period of 100 years. Consumption has increased from 0.25

billion tce/year in 1875 to almost 9 billion tce/year in 1975, which means an average annual growth rate of 3.7%. If one considers only the last 25 years, i.e. the period from 1950 until today, the growth rate is even higher: 5%/year on the average, which corresponds to a doubling time of 14 years.

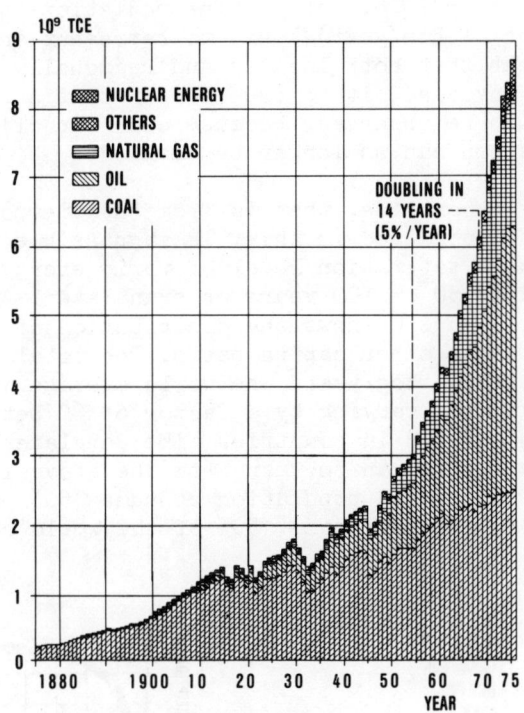

Figure 5. Development of World Energy Consumption.

100 years ago coal was the only energy source in the commercial energy economy. Oil entered the market not before the 1920's, corresponding to the break-through of the automobile, and oil did not reach a dominant market share before the 1960's. Today, the spectrum of primary energy sources is more or less the same in the industrialized countries as in the developing countries. The world energy consumption of 9 billion tce/year today breaks down to 50% oil, 25% coal (both hard coal and lignite), almost 20% natural gas, with the balance being water power, nuclear energy and others.

Let us return to the growth rate in world energy consumption. If - just for argument's sake - one extrapolates the world-wide energy consumption with 5%/year into the future, one arrives at an annual consumption of 100 billion tce/year after 50 years, of 1200 billion tce/year after 100 years, and of 180,000 billion tce/year after 200 years from now, which is in the range of the total

solar radiation incident upon the earth. Since all the energy used
is ultimately converted into heat, and since any disturbance of the
heat balance may influence the climate, the question here is: what
are the limits to global energy consumption if climatic changes on
the globe only on account of heat generation should be completely
avoided? This question cannot be answered accurately. Globally speaking, a value of 1 or 2% of total solar radiation (which is several
thousand billions of tce/year) seems to represent an upper limit,
but it is apparent that both locally and regionally this level will
be exceeded. In any case, it is true to say that a sufficiently large
safety margin must be observed, because once our climate has been
upset, the situation can no longer be amended.

In the other side, that is from the viewpoint of energy
demand, the question raised is this: Is it possible to estimate in
any approximation a saturation level of world energy consumption that
might be reached in 50 or 100 years or even later? As an introduction
to this problem, Figure 6 shows the distribution of our present annual
energy consumption on a per capita basis. The total shaded area corresponds to 9 billion tce/year. The world average is around 2.2
tce/year, the extremes varying by a factor of 60 between 12 in the
U.S. and Canada, and 0.2 in countries like Bangladesh or in Central
Africa, for example. As can be seen from the above integral curve in
Figure 6, 10% of the world population consumes 50% of the total amount
of energy and, on the other hand, 50% of the world's population live
with 10%.

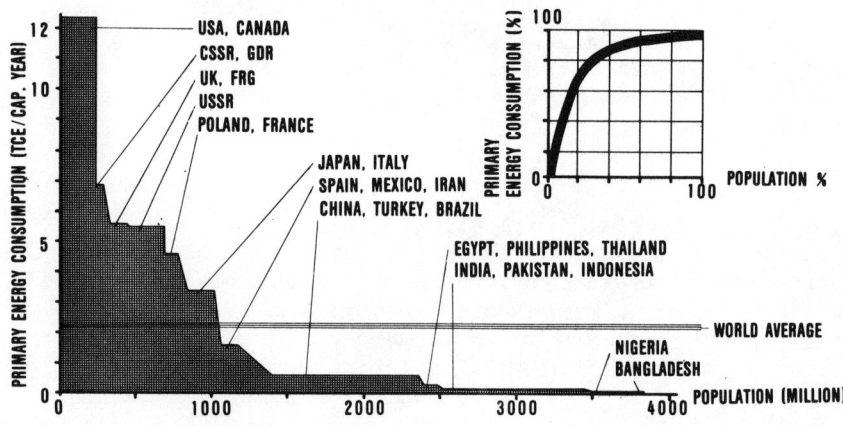

Figure 6. Per Capita Energy Consumption.

It does not require a major scientific systems analysis to
say that this extremely uneven distribution does not represent a
stable situation, and that it is going to change in the future -
peacefully or by violence. This will cause the average to rise, but
it is difficult to say up to what level. In addition the world popu-

lation will continue to increase markedly, from 4 billion at the present time to 8, to 10 or even 12 billion people within the next 100 years. An average per capita energy consumption in the range of 3 to 6 tce/capita and year in 100 years from now should be envisaged as a more or less realistic value and this would lead to a total world energy consumption between 30 and 60 billion tce/year. This is in line with the most important world-wide energy forecasts: The IAEA at its Salzburg Conference[3] in 1977 or the MIT 'Workshop on Alternative Energy Strategies'[4], for example, give forecasts of the world energy consumption for the year 2000 of around 20 billion tce/year, and IIASA has established reference scenarios of world energy consumption ranging between 25 and 40 TW (TW $\cong$ billion tce/year) in the year 2030[5].

On the whole, it should be borne in mind that world-wide energy consumption will increase to several ten billion tce/year within the next 100 years, and one should use this value as a yardstick when looking at long-term energy sources.

3. PRINCIPAL ENERGY SUPPLY OPTIONS

All energy supply options that can be used on the earth can be reduced to two original energy sources: one is the sun or, more specifically, the nuclear fusion process in the sun, and the other is the planet earth itself with its internal heat storage, its minerals and its rotational energy.

The solar radiation energy incident on the earth can be used either directly via thermal conversion or photovoltaic conversion systems, or indirectly via food, animal energy and photosynthesis (e.g. wood). Another kind of solar energy use is with water power (including the use of ocean thermal energy) and wind power (including waves energy), - which are both indirect forms of solar energy. And there is a second kind of indirect solar energy which is used today very extensively with a time delay of millions of years: it is the chemically bound energies of coal, mineral oil and natural gas.

The earth itself offers the use of geothermal heat from the inner parts, the minerals uranium and thorium for nuclear fission processes, heavy water and lithium for nuclear fusion processes, and tidal energy (which is a special form of rotational energy related to the gravitational interaction of the earth and the moon). As will be shown in the next sections, the traditional energy sources oil and gas are fairly limited, and will be exhausted within a few decades. Nuclear fission, nuclear fusion, the use of direct solar radiation and possibly also geothermal heat are, at least in principle, able to supply large amounts of energy over long periods of time (which means to provide practically unlimited amounts). They can be regarded as long-term energy supply options. Coal is somewhere in-between an exhaustable and an 'unlimited' energy source, as will be illustrated. Finally, water and wind power, photosynthesis and tidal energy which often are called 'renewables' are unlimited in time but limited in amount, and should be regarded as additional energy sources.

4. THE TRADITIONAL FOSSIL ENERGY SOURCES COAL, OIL AND GAS

Before entering the details of energy resources, let me add a word of caution. In general, one finds that data about energy reserves or energy resources in the literature vary considerably, depending on the different categories used such as assured or proven reserves, reserves which can be mined with today's technology, probable reserves or total estimated resources. Data also depend on the state of the art assumed for the exploration of new deposits.

Another quite important aspect to keep in mind is economical extraction, i.e. the cost level up to which an economic exploitation of reserves seems feasible.

Figure 7 explains these kinds of problems; it shows various categories of fuel reserves or resources illustrated by the example of coal resources in the U.S., and has been worked out by McKelvey of the U.S. Geological Survey.

	IDENTIFIED			UNDISCOVERED	
	DEMONSTRATED		INFERRED	HYPOTHETICAL (IN KNOWN DISTRICTS)	SPECULATIVE (IN UNDISCOVERED DISTRICTS)
	MEASURED	INDICATED			
ECONOMIC	50	147	NONE	1.482	NONE
SUBECONOMIC	63	238	935		

← INCREASING DEGREE OF GEOLOGICAL ASSURANCE →

INCREASING DEGREE OF ECONOMIC RECOVERY

Figure 7. Classification of Resources - Total U.S. Coal Resources (million tons)(from Mckelvey, Reference 6,7)

The following figures are taken from the World Energy Conference, Istanbul, 1977 (Reference 7).

The most important energy source of today is <u>mineral oil</u>. The economically exploitable and proven oil reserves amount to approximately 100 billion tons of oil, equivalent to 140 billion tce. Another 160 billion tons ($\cong$ 220 billion tce) are estimated as additional reserves which are extractable with today's or with a slightly advanced technology. This makes a total of 260 billion tons ($\cong$ 360 billion tce) of 'ultimately recoverable' or conventional resources. There are other considerable oil resources such as deep offshore oil and oil in polar zones, oil shale, and oil sands, but their exploitation can neither technically nor economically be forecasted today. The recoverable fraction of this 'unconventional' oil is estimated at 200 to 300 billion tons (300 to 400 billion tce).

The annual oil production today amounts to 3 billion tons, equaling 4 billion tce per year. The main oil producing countries are the countries of the Middle East, the Soviet Union, North Africa and the U.S. It is also worth mentioning that approximately 70 billion tce oil have been exploited so far. Depending on further oil discoveries, an exhaustion of mineral oil has to be envisaged at the beginning of the next century.

Natural Gas occurs mostly together with oil, and the resources are in the same order of magnitude. One estimates 80 billion tce proven reserves and 260 billion tce undiscovered resources. Present natural gas production is 1.8 billion tce per year; the increase in production rate is far higher than that of oil, so natural gas will probably play a much shorter role in the history of human energy consumption than oil.

Coal resources are remarkably high: 640 billion tce are technically and economically recoverable reserves and about 10,000 billion tce are the total estimated geological resources (for details see the following Table). On the other hand, there is a very irregu-

Table
Distribution of World Coal Resources (billion tce)
(from Reference 7)

	Technically and economically recoverable reserves		Geological Resources	
	hard coal	lignite	hard coal	lignite
Europe	94	34	540	55
Asia	220	30	5500	855
- U.S.S.R.	83	27	3400	870
- P.R.China	100		1400	
America	127	71	1310	1410
- U.S.A.	113	64	1190	1380
Africa	34	-	170	
Australia	18	10	210	
World Total	493	145	7730	2400

lar distribution of coal: 80% of the total estimated resources of hard coal are distributed over only three countries, i.e. the U.S., the U.S.S.R., and the People's Republic of China. The Soviet Union alone has 45% of the world's hard coal resources. The world consumption of hard coal today is around 2.2 billion of tons per year, which would mean that, with a constant consumption rate, the assured world hard coal reserves of 500 billion tons would last for 220 years, and, for example, for only 60 years if the consumption rate increased by 4% per year. But one should not pay too much attention to these figures - they are rather hypothetical. An increase in coal consumption of 4% per year would mean a consumption rate of 24 billion tons of coal per year in 60 years from now, and this is highly unrealistic. The range of lignite reserves is in the same order of magnitude.

The question is this: How and by how much can the future consumption of coal be increased world-wide? There are two reasons why coal is not likely to attain the same importance in the world market as mineral oil has today: (1) The cumbersome and expensive method of extraction and handling, and the environmental impacts of burning coal greatly hamper an increasing world-wide use of coal as an energy source; (2) Coal resources are located in a few countries that already have a large total energy consumption rate and that do not depend on large-scale coal exports.

In order to improve and to enlarge the handling of coal as an energy source, techniques for gasification or liquefaction of coal are being studied. They may well become more important in the future. The heat input they require may be either autothermal, i.e. by burning a certain fraction of the coal, or allothermal, i.e. by addition of external process heat (for example, from nuclear high temperature reactors or from solar thermal energy conversion systems). Research and development work today mainly concentrate on the coal gasification process, but it is too early to tell at what time gasified coal will enter the energy market on a larger scale, or how economical the gasification process will be. Let me just briefly mention the environmental aspects of the fossil energy sources.

Fossil fuels cause environmental pollution primarily in the combustion process. Its pollutants are released into the atmosphere, especially SO_2 (sulfur dioxide) and NO_2 (nitrogen oxide). In principle, it is possible to reduce or even prevent these pollutant emissions by technical means. In addition, any combustion process generates CO_2 (carbon dioxide). It is well-known that atmospheric carbon dioxide has increased by a measurable amount, which is possibly due to the burning of coal and oil. The increase in the atmospheric CO_2 content will cause the risk of global changes of the climate. This is due to the so-called greenhouse effect, i.e. (short-wave) solar radiation can penetrate the earth atmosphere, but (long-wave) thermal reradiation from the earth is largely absorbed by the carbon dioxide. This could result in a temperature rise of the atmosphere with subsequent climatic changes, which may be much more severe than those due to direct additional waste heat production, resulting from energy consumption.

The CO_2 problem and the related climatic risks could thus limit a large-scale use of coal.

5. FAST BREEDER REACTORS - A LONG-TERM NUCLEAR ENERGY OPTION

5.1 The Present State of Nuclear Energy

In nuclear fission reactors, energy is produced by splitting heavy atomic nuclei by means of neutrons in a controlled chain reaction. If, for instance, a neutron collides with a uranium-235 atomic nucleus, the U-235 nucleus may be fissioned into two or three fragments, the energy being released mainly in the form of kinetic energy of the fission products. At the same time, two or three neutrons are released which are capable of splitting further U-235 atoms; thus a chain reaction of nuclear fission can be realized. With the fission of one uranium atom about 200 MeV of energy are produced; this means that the complete fission of one gram of uranium results in a thermal energy release of one MWd. To gain the same amount of thermal energy from coal almost three tons of coal would have to be burned.

The fuels to be used for this fission process are uranium or thorium, which are utilized either directly (U-235) or indirectly (U-238 via plutonium, or thorium via U-233).

It is the light-water reactor that has achieved a general economic break-through all over the world. But besides, the heavy-water reactor should also be mentioned as an economically working reactor system. Light-water reactors use normal (i.e. light) water in the core as coolant and moderator, and their fuel is uranium oxide, enriched to approximately 3% of U-235 (in natural uranium the fraction of U-235 is only 0.7%).

Enriched uranium permits a very flexible reactor layout with regard to materials selection for cladding, moderator and coolant. This is not so much the case for natural uranium, especially if graphite is used as a moderator. Natural uranium also requires very large reactor cores, as the fissile material is rather diluted. The advantage of enriched uranium, on the other hand, implies dependence on a separation facility for the isotopic enrichment of U-235. Contrary to that, heavy-water reactors use natural uranium, and one of the attractive features is their independence of separation plants.

The availability of the large diffusion separation plants in the U.S. built as part of the Manhattan Project, has greatly stimulated the development of light-water reactors. The first U.S. light-water reactor for electricity generation, the Shippingport nuclear power station, went into operation in 1957. In 1964 light-water reactors reached a commercial break-through: It was for the first time that a utility ordered a nuclear power plant without any governmental financial support. It was the Oyster Creek plant in the U.S. Since that time, LWR have increasingly penetrated electricity production all over the world. The present status of nuclear energy is as follows:

	Main Reactor Type	Nuclear Power Plants	
		In Operation MWe	Under Construction MWe
World Total		96,000	320,000

Countries (with more than 2000 MWe plant capacity):

USA	LWR	45,000	190,000
Japan	LWR	8,900	10,000
USSR	LWR[+]	7,800	13,000
GB	GGR/AGR	7,500	4,300
FRG	LWR	7,000	18,600
France	GGR	4,600	28,400
Canada	HWR[++]	4,000	4,800
Sweden	LWR	3,800	5,700

Reactor Types:

LWR-PWR	52,000	210,000
LWR-BWR	27,000	85,000
HWR	5,000	18,000
GGR/AGR	11,000	4,000
Others	1,000	3,000

[+]Modified version

[++]Besides in Canada, HWR are operated in Argentina, India, Pakistan

LWR = light-water reactor

GGR = gas-cooled graphite reactor (MAGNOX)

PWR = pressurized-water reactor

AGR = advanced gas-cooled reactor

BWR = boiling-water reactor

HWR = heavy-water reactor

5.2 Fuel Utilization - Fuel Resources

Light-water reactors (as well as HWR or GGR) cannot solve the energy problem in the long run, due to incomplete utilization of natural uranium: they use up U-235 and only inadequately convert U-238 into fissile plutonium. The capability of a reactor to convert fertile material into fissile material can be expressed by its neutron economy. The following table shows some very crude figures of the neutron economy of the LWR (in comparison with the FBR):

	LWR	FBR
Fission Process	U-235 Thermal Neutrons	Pu-239 Fast Neutrons
Number of Neutrons Produced per 100 Fissionable Atoms Consumed	220	280
Neutron to Sustain Chain Reaction	100	100
Neutron Losses (Total)	65	65
Neutrons Used to Convert U-238 into Pu-239	55	115
TOTAL	220	280

A light water reactor uses U-235 as fissile material and thermal, i.e. slow, neturons for the fission process. 100 neutrons captured by U-235 result in about 220 neutrons newly produced. 100 of them are needed for continuation of the chain reaction. Then, there are different kinds of neutron losses - some will escape the core region, others are captured in structural materials, etc. The balance of 55 neutrons are captured by U-238 and subsequently converted into plutonium-239, which is also a fissionable material. On the whole, in the case of an LWR there is a ratio of 55 fissionable atoms newly produced for every 100 consumed.

This results in a fuel consumption which is shown on the left side of Figure 8. A light-water reactor requires 160 t of natural uranium per 1000 MWe and year. After passing the enrichment plant one gets 30 t of light-water reactor fuel, 130 t are depleted uranium that mainly consists of U-238, and that can no longer be used in light-water reactors. After leaving the reactor, the spent fuel elements break down into 28.7 t of uranium (with about 0.85% of U-235), around 300 kg of newly produced plutonium, and about one ton of fission products. In this case, i.e. without reprocessing and recycling, fuel utilization is 1:160 or 0.66%.

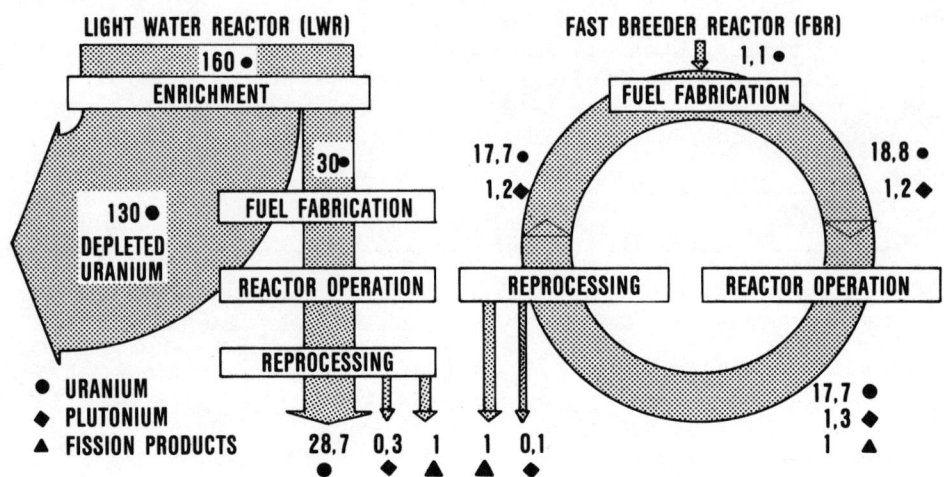

Figure 8. Annual Fuel Throughputs of LWR and FBR
(tons per year for a 1000 MWe plant at 80% loead factor)

In the case of reprocessing, both plutonium and the residual uranium can be extracted and recycled; complete recycling results in fuel savings of around 60 t of natural uranium. This means that in the case of reprocessing and recycling of the spent reactor fuel, the <u>net</u> fuel consumption of a light-water reactor is 100 t of natural uranium, which results in a natural uranium fuel utilization of 1:100 or 1%.

This leads to the question of natural uranium reserves on the earth and their energy content. On the one hand, uranium is rather abundant; the average uranium content of the earth crust is measured to be 4 ppm, which would add up to a theoretical total of many thousand billions of tons. The uranium of the sea is comparatively more easily accessible and amounts to some four billion tons. On the other hand, the uranium reserves estimated to be available for economic utilization in today's thermal reactors (that is in the cost category of up to 50 \$/lb U_3O_8 extraction costs) at present amount to only 4 million tons (for further details, see for example reference 8).

Four million tons of natural uranium correspond to an energy equivalent of 80 and 120 billion tce without and with reprocessing, respectively. Thus, it can be stated that the energy content of uranium used in today's LWR technology is on the same order of magnitude as are oil or natural gas reserves.

With increasing ore extraction costs, however, uranium reserves will increase, but as yet there are only crude estimates available (around 100 million tons of uranium up to ore extraction costs of 250 to 300 $/lb U_3O_8).

5.3 The Principle of Breeding

The important question is: How can utilization of uranium reserves at more than 1% be obtained? Since uranium utilization corresponds to the conversion of fertile U-238 into fissile plutonium, the next question is: How can the conversion ratio be increased? Is it possible to design a reactor in which the conversion ratio is increased in such a way that it produces more fissile atoms than it consumes? Indeed, it is with the breeder reactor. The neutron economy of the LWR shows two, at least principally, different ways increasing the conversion ratio: by reduction of the neutron losses of the reactor core, and by provision a better neutron yield in the fission process. The first possibility yields only marginal improvements. The second makes use of the fact that the number of neutrons released in the fission process varies with the fissile material and with the energy of the initial neutron initiating the fission. This is shown in Figure 9, which illustrates the neutron yield for different fissile materials as a function of the neutron energy. U-235 with a thermal neutron spectrum has a neutron yield of about 2.2, as already mentioned. The neutron yield can be improved in the thermal spectrum if U-233 is used as fissile material. On the other hand, plutonium in the thermal spectrum decreases the neutron yield.

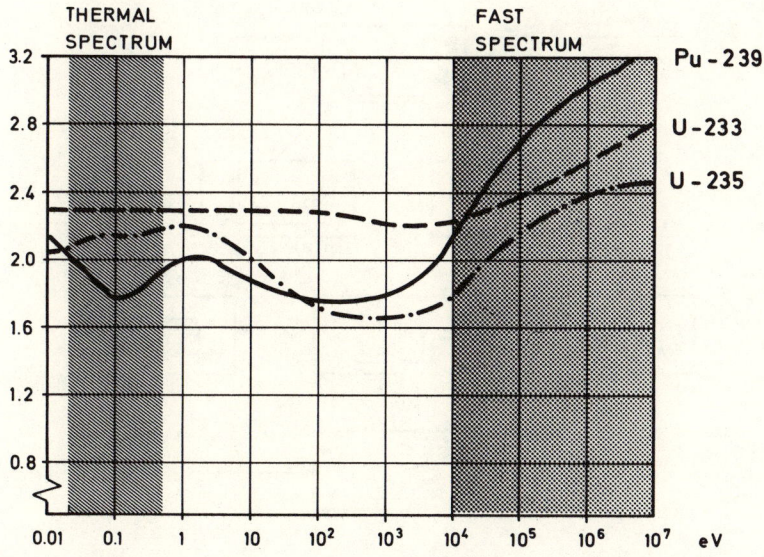

Figure 9. Neutron Yield for Different Fissile Materials

A breeding process asks for a neutron yield considerably higher than two, which is principally possible with a thermal neutron spectrum, but very difficult in practice. A much better solution is to use a fast neutron spectrum. The most favorable neutron yield takes place with fast neutrons and with plutonium as fissile material; this leads to the concept of a Pu-fueled fast breeder reactor (FBR).

Such fast breeder reactor, for example, has a neutron yield of 280 neutrons produced per 100 fissionable atoms consumed (see preceding table). Again, 100 neutrons are required to sustain the chain reaction, and the neutron losses are again on the order of 65. But in the case of a fast breeder reactor, the balance that can be used for converting U-238 into plutonium is 115. This means that not only all fissioned atoms have been replaced, but what is even better that there is a surplus of 15 fissionable atoms.

The right side of Figure 8 shows the practical realization of a breeding process: a fast breeder reactor requires an amount of almost 20 t of uranium plus about 1.2 t of plutonium per 1000 MWe and year (the figures are set as an example - they may vary with the layout of the reactor core).

After the fuel has passed the reactor, the amount of plutonium has increased, the amount of uranium-238 has decreased, and about one ton of fissile material has been burned into fission products.

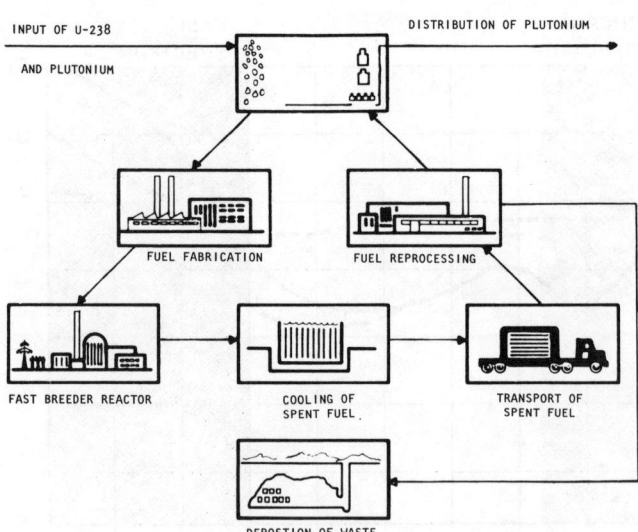

Figure 10. Closed Fuel Cycle of a Fast Breeder Technology

Two essential points should be mentioned here:

(1) Fast breeders require a closed fuel cycle, i.e. reprocessing of the spent reactor fuel and refabrication are <u>integral</u> parts of fast breeder technology, in order to replace the plutonium consumed by the plutonium newly produced (see Figure 10).

(2) A fast breeder reactor needs plutonium for the first-core inventory to be provided from outside; in practice it is from light-water reactors. Again, this means that light-water reactor reprocessing technology is an unalterable prerequisite for the introduction of fast breeder reactors.

The fuel ultimately consumed in a fast breeder is uranium-238; the amount used up is on the order of 1.1 t per 1000 MWe and year (including fuel cycle losses, this would amount to 1.5 t).

The general interdependence between fuel utilization and conversion or breeding ratio is shown in Figure 11.

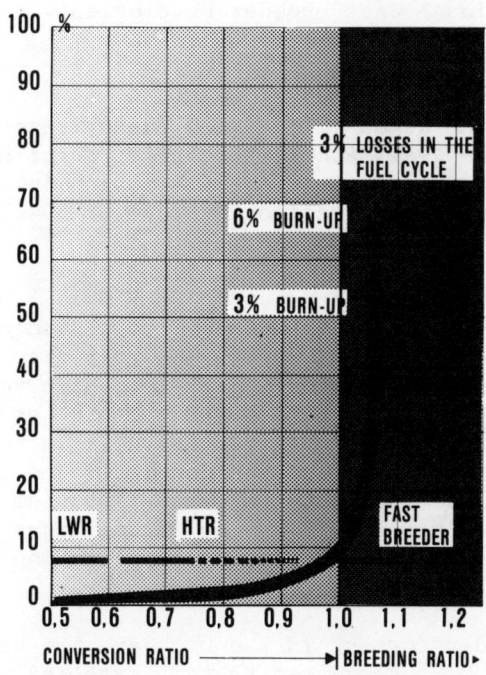

Figure 11. Utilization of Natural Uranium

In practice, fast breeder reactors are expected to attain a fuel utilization of natural uranium about 60 times greater than

that of of a LWR - depending on the fuel cycle losses and on the
burn-up of the reactor fuel. This results in two consequences: On
the one hand, much more efficient use is made of the cheap uranium
reserves (if breeders are used in addition to LWR), and, on the
other hand, economically reasonable uranium costs may rise by just
this factor. This will allow uranium resources to be exploited for
use in the FBR that would not be accessible otherwise.

Four million tons of natural uranium utilized in fast breeders correspond to around 8000 billion tce, which is almost as high as the total estimated coal resources, and up to several 100,000 billion tce can be made available for uranium ore extraction costs up to about 300 \$/lb U_3O_8. On the whole, the technology of fast breeder reactors permits the use of a practically unlimited source of energy. This was spelled out very clearly by A.Weinberg when he spoke of 'burning the rocks' with fast breeder reactor technology[9]. (At the same time Weinberg spoke of 'burning the sea' with respect to nuclear fusion.)

5.4 International Fast Breeder Development

The principle of breeding has been recognized from the very beginning of nuclear reactor development. As early as in 1944, Enrico Fermi and Walter Zinn began to design a faster reactor, this was little more than one year later after Enrico Fermi had achieved the first self-sustaining controlled chain reaction in a nuclear reactor. This historical event took place on 2 December 1942, with the Chicago Pile 1 (CP-1).

The first fast test reactor 'CLEMENTINE' was constructed in the years 1945-46, it was also the first plutonium-fueled reactor, with a thermal power output of 25 KW. The next step in the U.S. was the Experimental Breeder Reactor I (EBR-I), which went into operation in 1951. EBR-I was the first nuclear reactor in the world producing electricity from nuclear energy. This is the more remarkable, as reactor development at that time was completely military-oriented: The aims of reactor development then were plutonium production for nuclear weapons and propulsion systems for military application, especially for submarines. The peaceful application of nuclear energy did not start before 1954-55.

Further development in the U.S. and the beginning of development in the U.S.S.R. and Great Britain have led to fast reactors that are often referred to as fast reactors of the first generation. Besides the EBR-I, there are the EBR-II and the Enrico Fermi Reactor in the U.S., the Russian BR-5 and the British DFR.

Consistent with the general approach to reactor technology of those early years, the principal fuel was metal, i.e. U-235 metal.

Na was chosen as coolant, because the cores were small but the specific power densities were high. With respect to long-term reactor strategies, most attention was paid to the breeding ratio and to the doubling time; core inventory and especially fuel cycle costs were not so much in the focus of interest.

Around 1960, economic considerations developed that caused attention to shift to the nuclear fuel cycle as a whole. In particular, it became clear that the burn-up of the fuel must be increased in order to achieve economic feasibility of fast breeder reactors. Metallic fuels only allow for a modest burn-up in the range of 5000 to 10000 MWd(th)/t, which is prohibitive for an economic reactor operation.

A light-water reactor of today, for instance, requires a burn-up of 25,000 to 35,000 MWd(th)/t in order to burn effectively all original fissionable atoms. Fast reactors inherently require high enrichment, and burn-ups of around 100,000 MWd(th)/t are necessary in order to keep the number of passes of an individual fissile atom through the whole fuel cycle tolerably low.

This led to the so-called second generation of fast reactors with UO_2/PuO_2 as reactor fuel, whose development started around 1960.

Since 1960, extensive fast breeder development programs of the second generation have been underway in the U.S., U.S.S.R., Great Britain, France, the F.R.G. together with Belgium and the Netherlands, and Japan.

All programs had broad basic research and development activities in the sixties in the following three major areas:

- large physics experiments and reactor safety investigations,

- Fuel development and fuel irradiation experiments,

- Coolant investigations and complete reactor design studies.

Complete reactor design studies in the sixties showed 1000 MWe power output to be a target size for a commercial fast reactor. At the same time, it turned out that 300 MWe is a reasonable size for a prototype plant as a first step or an interim stage towards large commercial breeder reactors.

This led to the class of prototype or demonstration reactor plants, the development and construction of which have been the main objective in the late sixties and in the seventies. The following table summarizes the international fast breeder development:

	Installation	Country	Power (MWe)	Startup
	EBR-I / EBR-II	USA	0.2 / 20	1951/1965
	DFR	UK	15	1962
	RAPSODIE (adv.)	France	(40 th)	(1967)/1970
Experimental	BR-5/BOR 60	USSR	(5 th/(60th)	1960/1969
Reactors	(KNK-I)/KNK-II	FRG	20	(1973)/1977
	JOYO	Japan	(50 th)	1977
	FFTF	USA	(400 th)	1979
	BN 350	USSR	350 eq.[+]	1973
	Phénix	France	233	1974
	PFR	UK	254	1974
Power	BN 600	USSR	600	1979
Reactors	SNR 300	FRG	295	1983
	Superphénix	France	1200	1982
	(CRBR	USA	350	1984)
	Monju	Japan	300	1985

[+]150 MWe + 120,000 m³/day seawater desalination

    The first fast breeder prototype plant to go into operation was the Russian BN 350, which was built at Shevchenko on the Caspian Sea. The dual purpose plant has been designed for 150 MWe electricity generation plus the production of 120,000 m³/day of desalinated seawater (which corresponds to a total of 350 MWe equivalent). BN 350 became critical in November 1972. The next one was the French fast breeder reactor PHENIX of 250 MWe power output, which was built on the Rhone River in Southern France, PHENIX reached its first criticality in August 1973. The British prototype fast breeder PFR of 250 MWe became critical in 1974. Construction of the 300 MWe German/Belgian/Dutch prototype reactor SNR 300 at Kalkar on the Lower Rhine started in 1973. Full power operation is scheduled for 1983-84.

    In the U.S., there is the 350 MWe Clinch River Breeder Reactor project which has become a major point of discussion within the U.S. breeder development. A final decision has not yet been made. Besides this project, a very broad fast reactor research and development program is being pursued; one major part of this program is the Fast Flux Test Facility (FFTF) at Hanford, Washington, a large test reactor of 400 MW (thermal) power output. Construction of the FFTF is almost complete; it is scheduled to go into operation in 1979.

The Japanese prototype fast reactor MONJU should also be mentioned. A site has been envisaged for MONJU, construction may start in 1979, and first criticality is scheduled for 1985. As a preceding step Japan has built JOYO, a 50 MW (thermal) test reactor, which has reached criticality in April 1977 and which reached full power in June 1978.

The next step towards commercial feasibility will be the construction of commercial-sized fast breeder reactors in the range of 1200 to 1500 MWe electrical power output. The Russian BN 600 (600 MWe) is a first step in this direction. BN 600 is in an advanced state of construction and will be completed in 1979; besides this, a 1600 MWe plant is under design in Russia.

In France, construction of SUPERPHENIX, a 1200 MWe plant, started early in 1977. There are also designs for 1300 MWe fast breeder reactor plants in Great Britain and in Germany, which have not as yet materialized.

5.5 The Technical Concept of Fast Breeder Reactors

A fast reactor has some principal design features in common with a thermal reactor, for instance a light-water reactor. A fast reactor, however, has no neutron moderator, and fast neutrons exclude the use of moderating materials such as liquid water.

The nuclear heat generation takes place in the reactor core. The fission zone of the reactor core contains the fuel pins of about 5 to 8 mm diameter, about 1 meter active length, and some 3 to 5 meters total length. These fuel pins contain mixed Pu/U oxide pellets as fissile and fertile materials, with an enrichment of the fissile Pu between 15 and 30%. Around 160 to 320 fuel pins per fuel element are assembled in a hexagonal geometry. A 300 MWe plant like the SNR 300 contains around 200 fuel elements with 169 fuel pins each, which amounts to more than 30,000 fuel pins. The fission zone is completely surrounded by the UO_2 blanket. The breeding process takes place both in the fission and in the blanket zones.

The fission gas produced in the fuel during irradiation is collected in a fission gas plenum at the end of each pin. Finally the core contains the absorber elements made of B_4C rods. Boron carbide is capable of absorbing neutrons, and thus allows the control or shutdown of the reactor power by moving these absorber elements into or out of the core.

The heat generated is transferred from the core by means of liquid sodium; the LMFBR concept is mainly determined by the choice of sodium: sodium has a high melting point (100°C) and a high boiling point (900-1000°C) which allows for high coolant temperature conditions under very low system pressures (6 to 10 bars). Furthermore, Na has excellent heat transfer properties, such as a high specific heat capacity and a very good thermal conductivity. Its disadvantages are due to the fact that it reacts chemically with water and air, and that it becomes radioactive under neutron irradiation within the core. This leads to a plant design with a primary coolant circuit which contains the radioactive sodium heated up in the core, a secondary coolant circuit coupled with the primary one by intermediate heat exchangers, and a tertiary water circuit producing steam for electricity generation by means of the turbine-generator system. Figure 12 shows the heat transfer system of a dodium cooled fast breeder reactor.

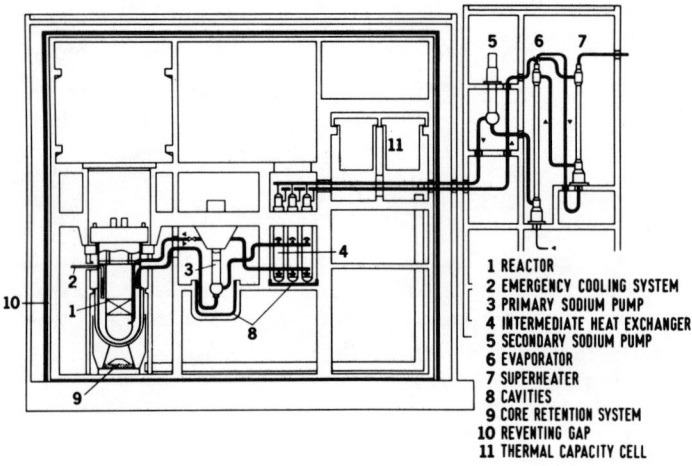

Figure 12. Heat Transfer System of SNR 300.

5.6   Safety Considerations

The most important hazard potential of any nuclear fission reactor - and thus also of the fast breeder reactor - is connected with the radioactive fission products arising during reactor operation. The main important safety requirements, therefore, is to prevent uncontrolled release of radioactive substances into the environment, both during normal operation and in the case of off-normal operating conditions.

To prevent radioactive leakage during normal operation, there is the principle of subsequent barriers: fuel, fuel cladding, reactor vessel and primary coolant system, and one or more containments.

In the case of an incident, the safety system is designed in such a way that preventive measures and containment measures will work together.

Preventive measures are active measures to prevent that deviations in normal reactor operation may lead to a major incident. This leads to two major conditions of fast reactor safety: (1) to shut down the reactor safely at any time, and (2) to guarantee the decay-heat removal from radioactive fission products after reactor shut-down at any time.

Containment measures preferrably are passive measures to mitigate the consequences of reactor incidents, and to limit them to the reactor plant itself.

As already mentioned, sodium in a fast reactor has some specific safety-relevant properties:

(a) Sodium reacts chemically with oxygen or water and becomes radioactive in the reactor core. To avoid interaction of radioactive sodium with steam in the case of leakage in the steam generator, fast breeders use an intermediate (non-radioactive) sodium cirtuit.

(b) Voiding of sodium in the reactor core may cause an increase in reactivity, i.e., in neutron density, and thus a power increase (e.g. through integral sodium boiling). But integral sodium boiling is extremely unlikely due to the high temperature difference (400°C) between operation and boiling temperature.

Only if all preventive measures fail independently, sodium boiling might occur, which could be the beginning of a power exursion. This leads into the consideration of hypothetical accidents.

Fast breeder reactors are controlled through the same mechanism as thermal reactors, i.e. through the fraction of delayed neutrons. Not all neutrons originating from the fission process will be released immediately or promptly, i.e. in 10^{-14} seconds, a small fraction of neutrons is released by further decay of fission products, with half lives between 0.2 and 54 sec after the fissioning of the original atom; these are the so-called 'delayed' neutrons. Delayed neutrons are fundamental for reactor control. A nuclear reactor is operated in such a way that the reactor remains subcritical with only the prompt neutrons; the reactor reaches criticality with the delayed neutrons, and thus can be controlled without any time problems due to the life-time of the delayed neutrons.

The fraction of delayed neutrons in a PWR, for example, is 0.5 to 0.7%. It is only 0.4% in the case of a plutonium-fueled fast breeder reactor, but the important point is that the delay times are the same in both cases, therefore both light-water reactors and fast breeder reactors have similar characteristics in the control and shut-off systems (for example, the speed of the control rod or shut-off rod movements is almost the same).

In order to shut down the reactor safely at any time, the German-Belgian-Dutch prototype fast reactor SNR 300 has been equipped with two independent and different shut-down systems. The neutron absorber of the first shut-down system is located above the core and is of a rigid guided type, it may drop into the core even with gravitational forces. The second shut-down system is arranged underneath the core; it is of a flexible guided type, which means that it can be drawn into the reactor core if the fuel elements have already been deformed. On the whole, the principle of <u>redundance</u> and <u>diversity</u> has been realized.

In order to guarantee a safe decay-heat removal after reactor shut-down, the SNR 300 is being equipped with three independent and different decay-heat removal systems; and again, the principle of redundancy and diversity has been realized.

There remains the consideration of hypothetical, i.e. extremely improbable, accidents.

In the case of the fast breeder reactor, one can think of such a hypothetical accident if the complete core cooling devices and both shut-down systems fail. This would lead to a temperature increase within the core, and, as a consequence, to sodium boiling in the core, and as a result, to a power increase. This power increase would be limited by the neutronic Doppler effect, but consequently the core would be disrupted because of the high fuel temperature. The reactor vessel of the SNR 300 is designed to withstand the mechanical energy releases of such an event. This means that the vessel would remain intact, and that decay-heat could be removed. But the safety consideration goes one step further in the case of SNR 300. An external fuel retention system, a so-called core catcher, will be installed below the reactor vessel. This additional cooling device must be understood as another barrier of defense, if molten fuel should find its way from the core region through the core support plate and the bottom of the reactor vessel. It would then fall into the core catcher and could be cooled there over a long period of time. Furthermore, a special geometrical arrangement of the core catcher would avoid a critical fuel configuration. On the whole, the consequences of such an accident would be limited within the reactor containment.

5.7 Some Remarks on the Risk of Proliferation

Fast breeder reactors can only operate in a closed fuel cycle with reprocessing of the spent fuel element and refabrication of the fuel. This means that large amounts of plutonium will have to be handled. The first core Pu inventory is taken from LWR fuel reprocessing. In principle, fast breeder reactors can also start with U-235, and they can also work with the thorium-U-233 cycle, but they will have their best performance in the U-Pu-cycle. At present, a major controversy is going on, mainly in the U.S., about the question whether or not fast breeder reactors have the potential to considerably increase the risk of diverting fissile material (in this case plutonium) from the peaceful application of nuclear energy to the production of nuclear weapon devices.

A major input in the discussion in the U.S. was the report on nuclear energy of the Ford Foundation 'Nuclear Power: Issues and Choices', which was published[10] early in 1977. The main results of this report are as follows:

- The U.S. has large reserves of coal and uranium. In many parts of the U.S. coal will be economically attractive compared to nuclear energy;
- The risks and environmental pollution of today's LWR are believed to be acceptable to the public;
- The ultimate storage of radioactive waste is considered as sufficiently safe;
- Because of the large coal and uranium reserves in the U.S. it was recommended to slow down fastbreeder development, to embargo the export of sensitive nuclear technologies, and to renounce civilian reprocessing facilities for a long time to come.

The Ford Foundation Report has been the background of the Statement of U.S. President Carter of 7 April 1977, in which he announced his new nuclear power policy.

The main decisions of this Statement are as follows:

(1) to defer indefinitely the commercial reprocessing and recycling of the plutonium produced in the U.S. nuclear power programs;

(2) to restructure the U.S. Breeder Reactor Program giving greater priority to alternative designs of the breeder;

(3) to redirect funding of U.S. nuclear research and development programs to accelerate research into alternative nuclear fuel cycles that do not involve direct access to materials usable in nuclear weapons;

(4) to continue to embargo the export of equipment or technology that would permit uranium enrichment and chemical reprocessing.

To set a mile-stone for the new policy of the Carter Administration, the Clinch River Breeder Reactor project is to be dropped. Discussion about the new nuclear power policy is still going on in the U.S. One of the most important points is that the U.S. is a country rich in energy sources, and it may possibly be able to do without both breeder technology and LWR reprocessing technology for some time, but that there are many other countries which cannot afford to renounce these technologies because of their limited energy sources.

In order to investigate in more detail the risk of proliferation within an extensive U/Pu cycle, and in order to look at alternate nuclear fuel cycles which perhaps are more resistant against proliferation, the 'International Nuclear Fuel Cycle Evaluation Program' (INFCE) was established in October 1977, and many countries of the world are working together in this program. For a final judgement of the whole problem one has to wait for results that are expected to be available after the two year program.

Preliminary results show that various alternate fuel cycles differ only quantitatively with respect to proliferation, and one has to make a trade-off between advantages for a better proliferation resistance or possible disadvantages in the technological process of an alternate fuel cycle.

The preliminary conclusion up to now is that all fissile materials (U-233, U-235 and Pu) bear the risk of a potential misuse as weapons material. Furthermore, many experts believe that a civilian nuclear power reactor program is the most expensive and time-consuming way for a state to obtain fissile material for nuclear weapons. On the whole, it can be stated that the risk of proliferation is not only typical of fast breeder technology but is a problem which is inherently connected to the development of nuclear energy in general. It may turn out from INFCE that preventing misuse of fissile nuclear materials requires political and institutional solutions in the first order, whereas technical fixes are of secondary importance. In any case the proliferation problem has to be solved.

5.8   Market Introduction

As already mentioned, the utilization of uranium-238 in fast breeder reactors requires two conditions:

> (1) A fast breeder needs plutonium for the first-core inventory. This plutonium of the first-core inventory may be supplied from light-water reactors, and this leads to a temporary symbiosis of light-water reactors and fast breeder reactors, where LWR start a nuclear economy and FBR are installed according to the available plutonium, thus taking over the nuclear energy demand;

(2) Fast breeders require a closed fuel cycle i.e. reprocessing of the spent fuel is an <u>integral</u> part of fast breeder technology, in order to replace the consumed plutonium by the newly produced plutonium.

The development of a new major technology such as the fast breeder technology follows a scheme that distinguishes three thresholds of feasibility:

(a) scientific feasibility,

(b) engineering feasibility,

(c) commercial feasibility.

Fast breeder reactors have already passed scientific feasibility and they are in the midst of passing engineering feasibility. They have not yet passed commercial feasibility.

Commercial introduction of fast breeder reactors is expected not before the years 1995 to 2000; besides the availability of plutonium, there is the question of producing electricity economically which is of major concern; and, with some time-delay, the fuel cycle has to be closed on a commercial basis.

The introduction rate of fast breeders then depends on some 'environmental' aspects of a country, such as energy demand, available reserves of natural uranium, electricity generating costs, independence of energy supply and alternative energy sources. In this sense, the fast breeder reactor today is an option for a long-term energy supply system.

6. SOLAR ENERGY - A LONG-TERM ENERGY OPTION

6.1 Introduction

By far the largest power source that can provide energy to the globe is 150 million km away from the earth. It is a huge spherical ball of 1.4 million km in diameter, and it has a power output of 400,000 billion TW (thermal) ($\cong$ 400,000 billion × billion tce/year). But only a very small fraction, i.e. 180,000 TW, arrives at the earth, an amount of energy that corresponds to 20,000 times the present world-wide energy consumption. This power plant is fueled with hydrogen, and energy is generated by means of a complicated nuclear fusion process. The plant has been in operation for approximately 5 billion years, and it will operate for billions and billions of years to come; the fuel resources are estimated to amount to 6×10^{27} TW·years. The energy generated is transmitted to the earth by means of electromagnetic waves within 8 minutes and 18 seconds time.

I am talking of the sun, of course. In fact, the sun is an immense energy resource, and, in terms of the human time horizon, it is really unlimited.

Figure 13 shows the energy balance of the atmosphere of the earth that is maintained by the sun. As already mentioned, 180,000 TW of solar radiation are received by the earth. 35% of the incident

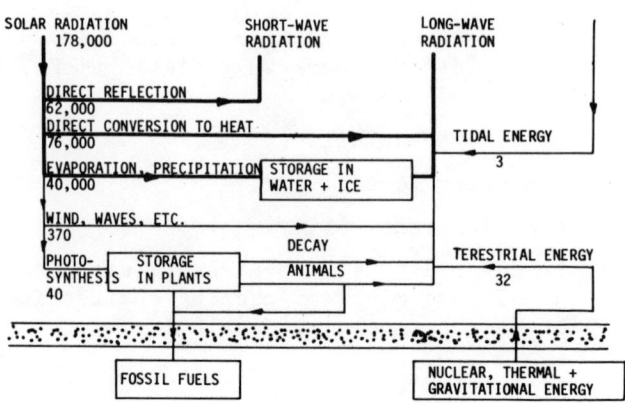

Figure 13. Flow of Energy to and from the Earth (in TW) (after Scientific American, Reference 1).

solar energy (62,000 TW) is directly reflected and scattered back into space in the same way as it arrives, that is, as short-wave radiation. Another 43% (76,000 TW) is absorbed by the atmosphere, the land surface and the oceans, converted directly into heat at the ambient surface temperature, and radiated back to space as heat, i.e. as long-wave radiation. Another 22% (40,000 TW) is consumed in the evaporation, convection, precipitation and surface run-off of the water cycle of the atmosphere. A small fraction, about 370 TW, is transformed into wind and waves energy and oceanic circulations, and is also dissipated into heat by friction. Finally, an even smaller fraction - only about 40 TW - is used for photosynthetic processes. Regarding the overall energy balance of the earth, one should also mention the storage of solar energy in fossil fuels - and additional energy sources, e.g. tidal energy or terrestrial energy such as geothermal heat, gravitational energy and radioactive decay heat, but these contribute very little to the energy balance.

On the whole, the energy balance of the earth is a complicated equilibrium, with one third of the incident solar radiation being directly reflected and two thirds being reradiated to space as long-wave heat.

If one looks at the total amount of solar radiation incident upon the earth, i.e. 180,000 TW, one should expect no further problems of energy supply to occur. However, one should not be deceived by this figure. There are three principal problems that more or less hamper an extensive utilization of solar radiation energy:

- energy density,
- energy storage,
- transportation of energy.

A fourth problem to be mentioned in this context is solar energy costs.

Energy Density: Outside the atmosphere the solar radiation power density is approximately 1400 W/m^2 (the so-called solar constant). At the earth surface, the power density is limited by the properties of the atmosphere, a medium that scatters and absorbs radiation. The solar radiation at the surface is a mixture of direct and diffuse radiation which together make up the so-called global radiation. The ratio of direct and diffuse radiation depends on the state of the atmosphere. Under best atmospheric conditions, the maximum global radiation can be as high as 1000 W/m^2, 90% of which is direct radiation.

Figure 14 gives the geographic distribution of solar radiation on the earth. The following key figures should be borne in mind:

	W/m^2	KWh/m^2·year
Solar constant	1400	12,000
Maximum earth surface value	1000	(8,800)
Average for sunny arid regions	200 - 250	2,000
Average for central Europe and Northern U.S.A.	100 - 120	1,000

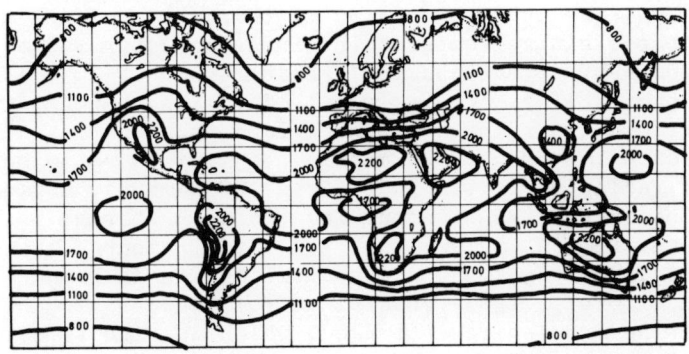

Figure 14. Geographic Distribution of Solar Radiation on the Earth (in KWh/m^2·year).

These figures characterize the problem of land requirement in the case of solar energy utilization.

Energy Storage: Besides the geographic distribution of solar radiation, there is the daily and the seasonal distribution. It is a well-known fact that solar energy can be harvested only during 50% of a 24-hour day. This means that energy from solar radiation has to be stored during daytime in order to be used during the night (short-term energy storage).

An example of the seasonal distribution is Figure 15 giving the annual distributions of solar radiation of Washington (38°N.L.) and Hamburg (54°N.L.). The maximum of solar radiation for each is between May and August and the minimum between November and February. An overall solar energy utilization system would have to have adequate, i.e. long-term, storage facilities in order to bridge the gap between high summer and low winter radiation.

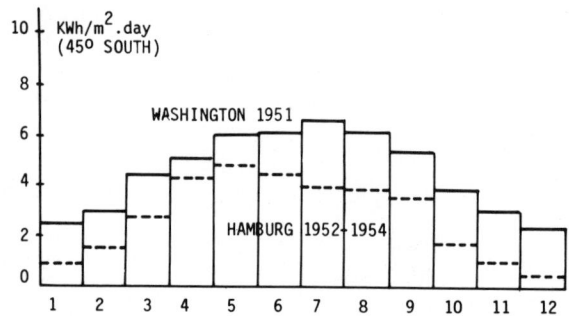

Figure 15. Annual Distribution of Solar Radiation (from Collmann, F.R.G.).

Now, the question raised is this: How do land and storage needs for these cities compare? The integral solar radiation for Hamburg is 1000 KWh/m$^2 \cdot$year and for Washington about 1650 KWh/m$^2 \cdot$year, and the ratio is 1.65. A solar plant of a given power output in Hamburg, therefore, would need 65% more land than one in Washington. As regards seasonal storage, the maximum to minimum ratio between summer and winter radiation is ten in the case of Hamburg, and only around three in the case of Washington. The seasonal storage problem then is much more severe - by a factor of 3.3 - for Hamburg than it is for Washington. In short, solar energy utilization in Hamburg would require an effort of 1.65 (resulting from land requirements) times 3.3 (resulting from energy storage), which is five times the effort necessary for Washington.

Energy Transportation: Solar energy cannot be used in its original or primary energy form, in every case it must be converted into a secondary energy carrier. Together with this conversion process the question of energy transportation comes up. For example, with solar energy for space-heating in solar houses, solar radiation is converted into low-grade heat that is used at the same place. With this type of conversion process, almost no transportation problems will arise. A large-scale solar energy scenario, on the other hand, in which large solar plants are located in sunny regions, faces the problem of transmission of large amounts of energy. Electricity could become one secondary energy carrier, but perhaps not the best one, all the more as electricity cannot directly be stored. It may turn our, for example, that hydrogen is much more suitable as a secondary energy carrier over long distances, for several reasons: hydrogen energy could easily be stored; hydrogen energy could be transported in pipelines or tankers without major energy transmission losses; and hydrogen could be used with only minor technological changes in already existing energy system.

I will now briefly describe the various solar energy conversion systems that are used or under consideration at the present time, either as small-scale ('soft' solar) or large-scale ('hard' solar) technologies. These conversion systems can be subdivided into four different categories:

(1) Decentralized solar thermal conversion, e.g. for water distillation or in solar houses; the end energy is low-grade heat;

(2) Centralized solar thermal conversion in solar power plants; the end energy is high-grade heat, electricity, hydrogen;

(3) Photovoltaic or photo-electric solar energy conversion in solar cells, the end energy is electricity;

(4) Photosynthesis or bioconversion; the energy is chemically stored and has to be further converted.

6.2   Decentralized Solar Thermal Conversion - Water Distillation

Water distillation by means of solar radiation is probably the oldest and simplest application of solar energy. The first large solar distillation plant was built 1872 in the mountains of Chile, supplying miners with drinking water. It was in operation for 40 years and had a water output of 22 m^3 per day. At that time the brine itself was used as a collector absorbing solar radiation. Today's collectors are generally more efficient.

A solar distillation plant is being developed in the Federal Republic of Germany especially for use in developing countries. This plant uses a collector together with a so-called heat pipe. Solar radiation is absorbed by the double-glazed absorber area of the flat collector, and converted into heat. The heat is transported

to the distillation equipment above the collector by means of the
heat-pipe, which is a closed inclined tube filled with a liquid with
a low boiling point. The liquid evaporates through the solar heat
source, and the evaporated medium carries the heat to the distillation tank. Here it undergoes condensation by conducting the heat to
the brine, and, by means of gravity, flows back to the absorber area
of the collector. The advantage of a heat pipe is that it works automatically with no moving parts and without maintenance. The seawater
or brine is continuously supplied to the distillation tank where it
evaporates and it recondensates at the cooling surface of the tank
by means of cold seawater. A prototype plant is now in operation in
Jordan near Aqaba on the Red Sea; it consists of 15 integrated modules,
each with heat-pipe collector and distillation tank. Each module has
an area of 25 m^2, which makes a total of 375 m^2 of collector area;
the overall plant is designed for 4 m^3/day of fresh water. The total
auxiliary power for plant operation is 200 We for the pumping equipment. In my opinion, the technology of seawater desalination with
solar energy by means of heat pipe collectors could well attain
importance for a decentralized application in developing countries.

6.3   Decentralized Solar Thermal Conversion - The Solar House

In a 'solar house' solar radiation energy is converted into
low grade heat and directly used for space heating or water heating.
The technical principle of a solar house is shown in Figure 16. The

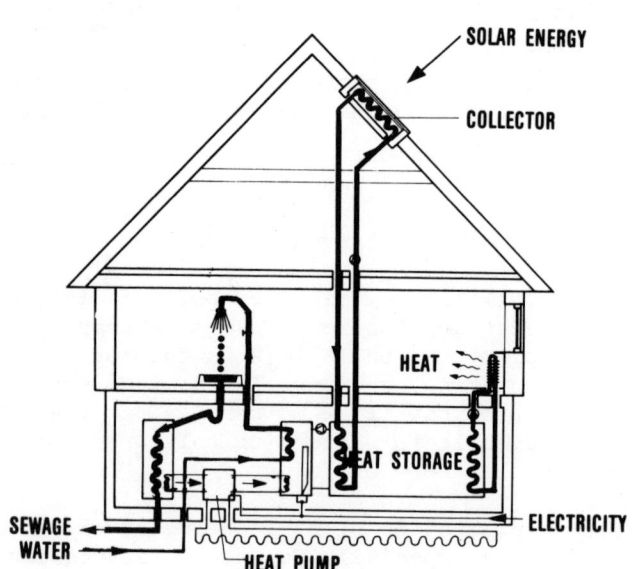

Figure 16. Principle of Decentralized Solar Thermal
Conversion ('Solar House') (After Philips, F.R.G.).

incident solar radiation upon the roof of a house is converted into heat by highly efficient heat-absorbing layers of the solar collectors, and transported to a heat storage system via a heat transfer medium which normally is water. The temperatures attained are on the order of 60 to 90°C and sometimes even more, depending on the solar radiation conditions and the technical sophistication of the absorbing collector. The heat storage system is able to supply the energy necessary for water heating or space heating, and can also be used to bridge periods of less or no solar radiation; for example, the day and night cycle, or short bad weather periods. If its capacity is insufficient a conventional auxiliary heating system has to be added. A heat pump may also be installed to reduce special energy losses, for example, that of warm sewage.

Quite a number of solar houses are already installed all over the world, especially in the United States, although most of them can not meet the total heat requirement of a private household over a year. But, as will be shown later on, the target of a solar house cannot be to meet the total heat requirement, at least not in areas like Central Europe.

There is no doubt that solar heating for decentralized residential application is of interest in the sunny regions, that is, in regions where the average solar radiation incident is in the range of 200 W/m^2 or 2000 $KWh/m^2 \cdot year$, especially for water heating, possibly also for cooling devices and, if necessary, also for space heating. These solar houses might become economically attractive much earlier than, for example, in Central Europe. The question is whether or not solar heating may at all become economically attractive in regions like Central Europe or northern U.S.A., that is, in regions with an average of 100 W/m^2 or 1000 $KWh/m^2 \cdot year$. Let me explain this, for example, by some rough calculations for the Federal Republic of Germany.

A one-family house, the type of house most suitable for the installation of a solar heating system, may have an annual consumption of heating oil of 3000 l or 20 barrels, which corresponds to roughly 30,000 KWh (thermal) per year. On the roof of a one-family house, there is room for approximately 20 m^2 of solar collector area. With an average solar radiation incident of 110 W/m^2 in the Federal Republic of Germany, this amounts to roughly 20,000 KWh of incident solar energy per year. Depending on the technological sophistication of the absorber system, the solar radiation energy will be converted into useful heat with an efficiency ranging between 0.2 and 0.5, resulting in useful thermal solar energy between 4,000 and 10,000 KWh (thermal) per year. In every case, there is a remarkable gap between heat requirement and heat supply.

Considering first the possibility for a reduction of the heat requirement by an advanced insulation technique, it is estimated that the total heat requirement could be reduced to about 20,000 KWh (thermal) per year, and to even 10,000 KWh (thermal) per year if a maximum possible insulation technique is applied.

The second and more important aspect is the investment costs of solar heating systems. The following estimates for investment costs of solar heating systems are taken from IIASA publications[11]:

Application	System Costs DM	Thermal Efficiency	Useful End Energy KWh	Fuel Oil ltr	Savings DM
A) Water Heating	5,000–10,000	0.30	6,000	600	180
B) Space and Water Heating	15,000–20,000	0.43	8,600	860	260
C) Space + Water Heating Incl. Heat Pump, etc.	30,000–40,000	0.50	10,000	1,000	300

For each of the three cases the investment cost have to be weighed against the annual savings; as a result solar heating technology in the F.R.G. does not appear economically attractive at the present time. The savings are even smaller than the annual interest rates. This means that solar heating systems can only be introduced into the market with substantial support from the government or by enthusiastic solar fans. Another important conclusion from the figures above is that the economic attractiveness decreases with the increasing fraction of solar energy use. For the time being, the economic optimum of solar energy utilization in a solar house is at a fraction of 30 to 40%. With greater solar share the respective investment costs will increase in an overproportional manner, thus leading away from the economic optimum.

The question remains how to bridge the difference between today's heating cost based on mineral oil and today's solar heating costs. The answer is twofold: Oil prices, on the one hand, will doubtless increase in the future. On the other hand, as the production of solar heating systems will increase in the future, a reduction in fabrication costs may have to be expected. This could lead to a break-even of solar heat energy with energy from oil heating in 20 or 30 years time from now.

On the whole, decentralized solar thermal conversion systems will be introduced into the energy market in the future, and they may well be suited for partial substitution of oil. To what extent it may contribute to the energy supply in developing countries is subject to specific evaluations.

But one should note that decentralized thermal energy conversion systems cannot solve the energy problem on a world-wide basis.

6.4   Centralized Solar Thermal Conversion - Solar Power Plants

Large-scale utilization of solar energy asks for conversion systems that can attain higher temperatures than decentralized solar heating, and secondary energy carriers other than heat only. Higher temperatures mean concentration of solar radiation, and as far as secondary energy carriers are concerned, electricity is given most attention at the present time. Activities in solar thermal electricity technology in most countries (such as U.S.A., France, the F.R.G. or Italy) are basically directed at the development of the central receiver or solar tower configuration. Figure 17 illustrates the technical principle of a solar tower plant. The central receiver

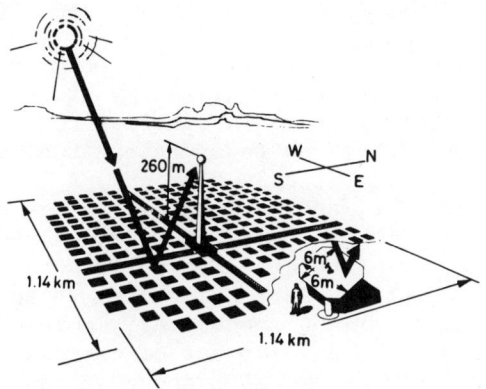

Figure 17. Solar Thermal Electricity Conversion - Solar Tower (from IIASA, Reference 12).

system consists of an array of reflectors steerable at two axes (called heliostats) which focus sunlight onto a receiver at the top of a tower. The receiver produces either superheated steam (560°C), for operation of a steam turbine of conventional design, or very hot air or other gases (e.g. helium), for operation of a gas turbine cycle. The numbers in Figure 17 are typical of a plant of 100 MWe. Current heliostat development efforts suggest a reflecting surface per heliostat of approximately 40 m^2. The tower would be approximately 260 m in height. The ground covered by the reflecting surface is about 40 to 60% of the total area needed in order to minimize blocking and shading of the heliostats.

For solar thermal electricity conversion systems (STEC), only the direct fraction of the global radiation is focusable and usable. Approximately 85 to 90% of the direct radiation incident onto the reflecting surfaces is reflected to the receiver system; the diffuse fraction cannot be used for thermal conversion. About 95% of the direct radiation reflected to the receiver is converted to steam or hot gases. The conversion into electricity takes place at a net efficiency of 25 to 30%, depending on the engineering design details. This results in an overall efficiency of converting solar radiation energy incident on the heliostats to electricity at the busbar of the plant of approximately 20%.

Here are some typical numbers for a solar tower plant:

Plant capacity: 100 MWE → 500 MWth

Day-time solar radiation in sunny regions:

750 - 800 W/m^2, → total heliostat area 0.64 km^2

→ 16,000 heliostats of 40 m^2 each; ground cover ratio 0.5

→ Total required land area 1.3 km^2.

In the case of an electric solar power plant which is operated within an interconnected grid, energy storage is much more crucial than for decentralized solar thermal conversion. The solar power plant described above can be operated at a 100 MWe power level only for about 8 hours within a 24-hour day. A 100 MWe power output plant of at least 300 MWe, and additional storage equipment. One way of storing energy is with hydrogen, as I have already mentioned. This could be done either by means of electrolysis or by means of thermochemical water dissociation. It remains open which of the two processes will be technically and economically more attractive in the future. In every case, the development of commercially interesting electrochemical and thermochemical hydrogen production technologies ranks among the greatest technical challenges in connection with large-scale solar energy utilization.

Solar thermal electricity conversion systems with central receivers or solar towers are still in an early stage of design and development. Only one prototype plant in the one MW range is in operation or is just about to be operated in the U.S., but is not yet designed for electricity production. Other projects are under design in Spain, Italy and the U.S.A. On the whole, it is very difficult to assess the introduction of this kind of energy technology into the energy market.

Another kind of solar thermal electric conversion system is the 'solar farm', which is shown in Figure 18. In this context, one should mention A.B. and M.P. Meinel in the U.S., who have been highly engaged in the development of solar energy. A solar farm is

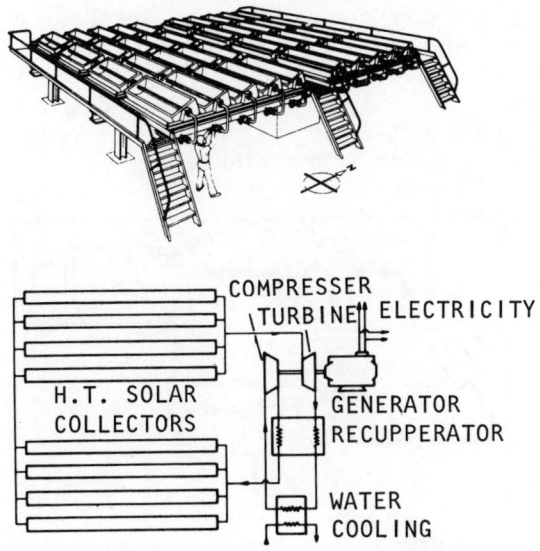

Figure 18. Solar Thermal Electricity Conversion - Solar Farm (from reference 13).

more favorable than the solar tower concept for smaller power outputs in the 1 to 10 MWe range. In this concept, solar radiation energy is harvested by means of high-temperature solar collectors that concentrate sunlight in parabolic reflectors of linear shape onto tubes carrying the heat transfer medium. Figure 18 shows a plant design of 10 KWe power output with a mirror area of 180 m^2. Heat is transformed into electricity by means of a closed gas turbine cycle with argon. Solar farms attain lower efficiency rates of energy conversion, i.e. approximately 10%.

A solar thermal conversion system with special application is the 'solar furnace', a prototype of which is located in Odeillo in the Pyrenees on the French-Spanish border. This special research facility is designed to achieve very high temperatures of approximately 3000°C. Solar radiation energy is reflected from the heliostats upon a hill (2800 m^2 of reflecting area) to a parabolic mirror of 40 m height and 50 m width, from which solar radiation is focussed by the mirror in 18 m distance. With a solar radiation incident of 600 W/m^2, the power output of the furnace is 1000 KW (thermal). The plant has been in operation since 1972 and is used for high-purity melting processes with temperatures up to 3000°C. Some solar furnaces in the U.S. can attain even higher temperatures on the order of 4000°C.

Small solar power plants in the range of 10 to 100 KWe power output are being developed in the F.R.G. They are especially designed to supply small communities in developing countries with electricity in a decentralized and autonomous way. Autonomous here means that the plant needs no energy from the outside and no continuous maintenance, the only provision from the outside is cooling water for the condenser; Figure 19 shows the schematic arrangement of such a prototype solar power plant.

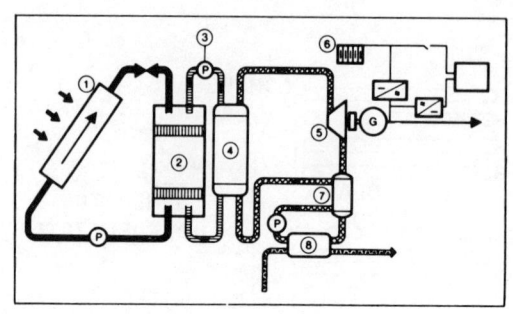

1 SOLAR COLLECTOR	2 THERMAL STORAGE	3 PUMP
4 EVAPORATOR	5 TURBINE	6 BATTERY
7 RECUPERATOR	8 CONDENSER	

Figure 19. Small Solar Power Plant (10 KWe) (from Messerschmidt-Bölkow-Blohm, Reference 14).

The layout of the plant is 10 KW of net electricity during eight hours during day-time, and additional 2.5 KW of electricity during six hours in the evening and in the morning giving a total energy output of almost 100 KWh electricity per day.

Solar energy is converted into heat through double-glazed flat collectors with 350 m^2 of total collector area. Heat is transported by water to a thermal energy storage with a capacity of 30 m^2 and an upper storage temperature of 95°C. Electricity is generated from the stored heat via a thermodynamic circuit with freon as working medium.

A prototype of this plant has already undergone extensive testing in Germany and is now being tested in Jordan.

6.5   Photovoltaic Conversion

Photovoltaic conversion, that is direct conversion of solar radiation into direct current electricity in solar cells, for example silicium cells, is already a well-established technology for small-scale application in space missiles. Thereby, a thermal efficiency in the range of 5 to 10% can be reached. For the time being, an extreme bottleneck for a broader application of solar cells also on the earth is the very high investment costs. Advanced production methods and a widening of the market may reduce the cost of solar cells.

It is not very likely that a large-scale use of photovoltaic conversion will be made within the next decades, e.g. for a solar satellite power station (SSPS) in which the energy generated is sent to earth by means of microwaves and is further distributed by regional power supply systems.

On the whole, the economic utilization of solar cells will probably be in smaller units in the Watt and Kilo-Watt range.

6.6 Photosynthesis

By harvesting solar radiation through photosynthesis, a variety of biological processes can be used for the production of synthetic fuels. Photosynthesis can be used directly through the growth, harvesting, preparation and combustion of various plants. An energy production system via photosynthesis would consist of large-scale forms of high-yielding plants, including a thermal power plant for the processing and combustion of the materials. Sugarcane, for example, may become attractive for its high conversion efficiency of 2 to 3% of solar radiation. A 100 MWe plant requires about 50 to 80 km^2 of harvesting plant area.

Other - indirect - processes include the microbiological conversion of plant materials or of algae to useful fuels such as methane, methanol or hydrogen. The question remains open as to what share of the future commercial energy market energy conversion through photosynthesis may obtain.

7. NULCEAR FUSION - A LONG-TERM ENERGY OPTION

Nuclear fusion is another long-term energy supply option. Here, energy is produced by fusioning heavy hydrogen nuclei. The reaction almost certain to serve as the principal energy source in first generation fusion reactors is that of deuterium and tritium:

$$D + T \rightarrow He^4 \ (3.5 \ MeV) + n \ (14.1 \ MeV)$$

The energy released amounts to 4 MWd (th) per gram of reacting nuclei. Deuterium is a natural isotope, and tritium which is almost nonexistent in nature must be bred. The main breeding reactions involve the neutron-induced fissioning of lithium:

$$Li^6 + n \rightarrow He^4 + T + 4.8 \ MeV,$$
$$\text{or} \quad Li^7 + n \rightarrow He^4 + T + n - 2.5 \ MeV.$$

Therefore, the effective overall reactions of D-T fusion are:

$$D + Li6 \rightarrow 2\ He4 + 22.4\ MeV\ \big(3\ MWd(th)/g\ reactants\big),$$

or $\quad D + Li7 \rightarrow 2\ He4 + n + 15.1\ MeV\ \big(2\ MWd(th)/g\ reactants\big).$

There is an additional fusion reaction of deuterium itself which could be used in advanced fusion reactor systems:

$$D + D \rightarrow He3 + n + 3.3\ MeV$$

$$D + D \rightarrow T + H + 4\ MeV.$$

Each of these two D-D reactions occurs with roughly 50% probability.

In the case of the D-D reaction the amount of deuterium on the earth determines the energy resources available as fuel for this process; in the case of the D-T reaction the lithium resources will be the limiting factor because lithium is less abundant than deuterium.

Deuterium is contained in normal water at 16.7 ppm which means that it is available in the seas in practically unlimited quantities. Lithium is also abundant, both in the earthcrust and in the seas. Very little lithium prospection has been carried out so far, hence the reserves that are easy to extract are not entirely known. The following list is meant to give an impression of the orders of magnitude of energy resources available in nuclear fusion:

D - T process		Billion tce
lithium in the earth crust:	assured	4,000
	total estimated	250,000
Amount of lithium in the sea (theoretically)		400 million
D - D process		
amount of deuterium in the sea (theoretically)		500 billion

The figures show that nuclear fusion is a technology of energy production in which the fuel supply is not a problem at all.

For nuclear fusion to occur, the reacting nuclei must be heated up to more than 100 million °C. Only at such high temperatures will they be able to overcome their repellent forces generated by their identical positive charges, so that the nuclear fusion process can be initiated. At these high temperatures the material is in the plasma state, i.e. the nuclei and the electron shells are separated from each other. A plasma of this type can no longer be enclosed by a material wall in the ordinary way; at these high temperatures the plasma is held together either by magnetic confinement (which is realized in the TOKAMAK machine) or by inertial confinement (e.g. the laser principle). Up to now, research has shown that the D-T process will be the first to be implemented, because it will work at lower temperatures than the D-D process; hence reactors with the D-T process are considered to represent the first generation of nuclear fusion reactors (fusion breeder reactors), with D-D reactors constituting a possible second generation of fusion reactors.

In order to heat the plasma, first of all energy has to be supplied to the system, and only when certain reaction conditions such as temperature, plasma density and the confinement time are met, the energy balance will become positive. This is the so-called Lawson criterion. Thus far the Lawson criterion has not yet been attained under laboratory conditions, i.e. no controlled nuclear fusion has been brought about in which the energy generated was at least as high as the energy input. Extensive fusion research programs are being carried out particularly in the U.S. and the Soviet Union, with the prospect that by the early 80ies controlled nuclear fusion on D-T basis can be demonstrated with an energy balance that equals unity.

Figure 20 shows the various international efforts on the way to reach the Lawson criterion.

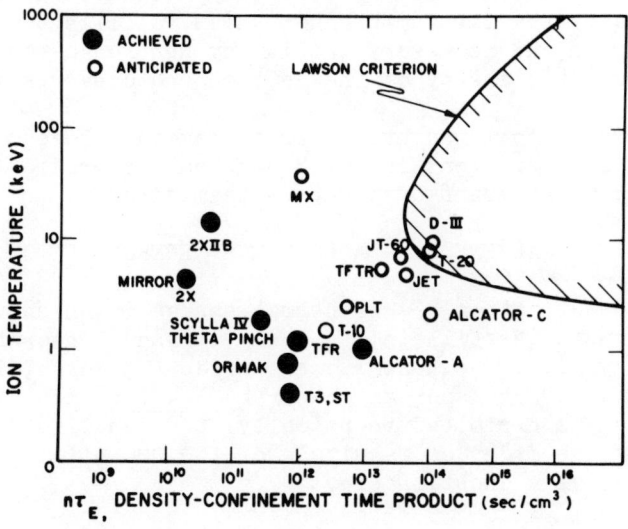

Figure 20. Break-Even Plasma Conditions for Fusion Power (from IIASA, Reference 15).

The following time-table can then be derived for an economic use of nuclear fusion (the data are taken from the U.S. fusion development program):

around 1950:	Beginning of fusion research (or plasma research)
around 1980:	Successful controlled nuclear fusion with the energy balance equal to unity (Lawson criterion). This would mean scientific feasibility;
1990:	A physical fusion test reactor may be in operation;
2000 - 2010:	Small experimental power reactors may be in operation. This would mean engineering feasibility;
2020 - 2030:	Economical electricity generation from nuclear fusion reactors. This would mean commercial feasibility.

The TOKAMAK system is technically more sophisticated than the laser principle, but by now it has reached a higher development level.

In a TOKAMAK, the plasma is magnetically confined in a torus, i.e. an annular arrangement where the nuclear fusion processes occur. As with nuclear fission the energy produced is first released as kinetic energy of the reactants. The energy of the He nucleus is used to maintain the temperature in the plasma, whereas the neutrons generated can penetrate the first wall and thus escape from the plasma zone because they are electrically neutral. The first wall is surrounded by the lithium blanket which fulfills two functions: (1) It generates the necessary tritium by neutron capture. (2) It absorbs the kinetic energy of the neutron and converts it into heat.

This is followed by a thermodynamical process, typical of any thermal power station, in which the heat generated is converted into electricity by means of a water-steam circuit.

8. GEOTHERMAL ENERGY - A LONG-TERM ENERGY OPTION

The extraction of geothermal energy is an energy option whose development is still in its early infancy. Theoretically geothermal energy, i.e. the heat stored inside the earth, is one of the largest natural energy sources on earth. However, nobody knows how much of it can be exploited technically. One can find data in the literature on the order of magnitude of 100,000 billion tce, but these data are very questionable.

At present, geothermal energy is utilized only in areas with geothermal anomalies, that is in volcanic regions where the heat inside the earth comes very close to the surface. Iceland, for example, has already installed a district heating system with hot geothermal water for its capital Reykjavik.

There are a few power plants in the world using geothermal energy for electricity generation: at Lardarello, Italy (370 MWe); the Geysers plants, USA (500 MWe) and at Wairakai, New Zealand (290 MWe). The total geothermal electricity generation capacity in the world amounts to about 1400 MWe. All these plants use natural hot water or steam. Estimations show that this so-called wet geothermal energy will not exceed 0.1 billion tce per year in the future, as it is limited to volcanic areas.

A large-scale utilization of geothermal energy is only possible with the so-called hot-dry-rock technique, but the hot-dry-rock method has not yet been applied, it is still in its design phase.

On the whole, the question is still open as to what extent geothermal energy will be applied on a large scale and world-wide.

9. ADDITIONAL ENERGY SUPPLY POSSIBILITIES

In addition to the large energy supply options, there are various supplementary supply options which, because of their limited resource potential, cannot solve the energy problem on a world-wide basis, but which can contribute towards easing the situation locally or regionally.

First, <u>water power</u> should be mentioned for electricity generation. Water power is an indirect form of solar energy - and had been in use for a long time before electricity generation, for instance, through water wheels for milling corn.

The total theoretical potential of water power on the earth is estimated to be approximately 5 billion tce/year; the technically and economically extractable potential is assumed to be on the order of 1.5 billion tce/year, which corresponds to almost 3.000 GWe of plant capacity. Only about 10% of this is now being exploited, mainly in the highly industrialized areas in North America, Wester Europe and the Soviet Union. Large unused reserves are still available in Africa, Asia and South America. The technique of electricity generation of the use of water power is well at hand and is also economically attractive. There are two kinds of water power utilization which are of principal difference: one is the utilization of water in large rivers - here we find large amounts of water and only small water heads. The other kind is utilization of the glacier water with smaller amounts of water that can be stored in large dams and with very high water heads.

There is another point worth mentioning: water power plants

can very easily and immediately be switched on to and off the grid, which means that they can very easily follow the load curve of electricity demand meeting the peak load demand. They also can serve as pumpstorage facilities. On the whole, water power is a clean type of energy supply, but available only locally and on a limited scale.

Wind energy is also derived from solar energy and can be used for electricity generation. Utilization of wind power is even older than that of water power. In 3000 B.C., wind power already served to drive sailing ships. Since 700 A.D. wind mills have been used in Persia and since 1200 A.D. in Europe. Wind power has almost the same energy potential as water power. About 1.5 to 2.0% of total solar radiation are converted into kinetic energy of the atmosphere and about 1 to 3 billion tce per year are assumed to be technically convertable. An exact evaluation of the wind energy potential is to be based on the measurement of local wind velocity distributions. Economic utilization requires a specific altitude and a certain constancy in wind speed. If wind energy were to be used on a larger scale either interconnected operation and/or energy storage systems would be necessary in order to harmonize the availability of wind energy and the energy requirement.

Wind power plants have thus been built and operated only in smaller experimental units in the 100 KWe range. Economic electricity generation from wind power requires plant sizes up to 3 MWe for a single plant, and 100 to 300 MWe for an integrated plant system.

On a smaller scale, wind energy will probably be able to contribute to the supply of energy. Large-scale projects, such as the 'Off-shore wind power system' in the Gulf of Maine north of New York, will hardly stand a chance for implementation (this project aims at 15,000 floating towers with a total of 500,000 wind turbines covering an area of 60,000 square kilometers, resulting in a power output of 20,000 MWe).

The utilization of tidal energy requires a sufficiently high level of tides and a closed bay or river estuary as a natural storage basin. Only a few sites or regions would fit these requirements. The only plant already in operation is La Rance in Britanny (France) with a capacity of 250 MWe.

In exploiting ocean thermal energy, use is made of the temperature difference between a warm surface current of about 30°C and a cold current below of, for instance, 4°C at 600 to 900 m depth. Between these two temperature levels, a thermodynamic cyclying may be installed and, thus, electricity be generated with an efficiency of approximately 3%. Power plants making use of ocean thermal energy can only be installed in the warm equatorial zones. Economic utilization on a larger scale cannot be forecasted today.

<u>Wave energy</u> and its utilization already exists on a small scale for the supply of electricity to buoys, light-houses, etc. There are no prospects as yet for large-scale technical utilization.

10. CONCLUSIONS

- The spectrum of energy consumption is determined world-wide, both in industrialized countries and developing countries, mainly by the fossil fuels coal, mineral oil and natural gas. At the present time, oil is the most important source of energy.

- There is no doubt that energy consumption will continue to increase in the future, more in the developing countries, but to some extent also in the industrialized countries. For 100 years from now, a world energy consumption of several ten billion tce/year (30 to 60 billion tce/year) has to be envisaged, and it is uncertain if this level of consumption will then have already reached a saturation point.

- Mineral oil and gas are not able to ensure the supply of energy over long periods of time. From the point of view of resources, coal would be able to provide this assurance on a certain scale, but this is met by major problems in extracting and handling, and by environmental problems.

- In principle, there are four possibilities known today how to provide energy on a large scale and over long periods of time:

  (1) nuclear fission with breeder reactors,
  (2) solar energy,
  (3) nuclear fusion,
  (4) geothermal energy.

- Nuclear energy is being introduced into the energy market. Today, nuclear energy is produced economically mainly in nuclear power plants equipped with light-water reactors which, however, achieve only an insufficient utilization of natural uranium. Fast breeder reactors are able to solve the uranium problem on a long-term basis. Their commercial introduction into the market may be expected around the turn of the century. In this case, practically unlimited energy resources will become available.

- Solar energy is used on a small scale today (soft solar). Technologies of this kind may become important in local or regional energy systems in the future. Introduction of solar energy on a large scale (hard solar) requires huge areas of land; there are the problems of energy storage and of energy transport. On the whole, major technical, structural and economic problems still have to be solved. No definite time schedule can as yet be given for the introduction of large-scale solar energy conversion.

- Nuclear fusion has not been experimentally proven to date. In nuclear fusion power plants, major technical problems still have to be solved. A commercial application cannot be anticipated before the year 2020, but, once this has been achieved, unlimited energy resources would be available.

- The use of geothermal energy is still in its infancy. To what extent it will ever be able to largely contribute to the supply of energy remains open.

- In addition, there are energy supply options of a more supplementary character (called 'renewable' energy sources) which may well attain local or regional importance. Water power is the most prominent among these renewable sources, but also wind power, tidal energy and others should be mentioned.

- Energy systems in 100 years from now may turn out quite different from what they are today. Generally speaking, the problems of a transition from today's energy systems of exhaustable fossil fuels to an unlimited nonfossil fuel economy have to be solved. Whereas the question of a temporary or local fuel shortage dominates today's energy policy, an energy system in 100 years would have to meet other challenges. Fuel shortage, then, would no more be a problem, but there would be constraints relating to the energy conversion process, e.g. land requirement, materials requirements, and - the most severe constraint - capital requirement.

- There remains the question of how nations should decide in realizing an advanced energy system. The answer here can only be very general. Globally speaking, mankind has to make use of every energy source available, it may be fossil, nuclear or solar, and it may be both small-scale (soft) and large-scale (hard) energy technologies. Only quite a broad energy-mix may protect nations from severe energy shortages. Detailed answers require more detailed analysis.

I would like to finish with quoting the famous German philosopher Immanuel Kant of the 18th century:

> "Die Notwendigkeit der Entscheidung ist immer größer als die Möglichkeit der Erkenntnis".

> The urgency for a decision is always greater than the ability to obtain the relevant knowledge.

This is true, I think, also for energy policy decisions.

REFERENCES

1. Energy and Power - A Scientific American Book (September 1971 issue of Scientific American) W.H.Freeman and Co., San Francisco (1971).

see especially:

M.King Hubbert, The Energy Resources of the Earth,
and:
Earl Cook, The Flow of Energy in an Industrial Society.

2. Energiebilanzen der Bundesrepublik Deutschland (Energy Balance of the Federal Republic of Germany) Arbeitsgemeinschaft Energiebilanzen, Frankfurt, F.R.G.

3. Nuclear Power and its Fuel Cycle, Vol. I: Nuclear Power Prospects and Plans, Proceedings of an International Conference, Salzburg, May 1977, International Atomic Energy Agency, Vienna (1977).

4. Energy: Global Prospects 1985 - 2000, Report of the Workshop on Alternative Energy Strategies (WAES), McGraw-Hill Book Company, New York (1977).

5. V.C.Chant, P.S.Basile (ed.), Global and World-regional Energy to the Year 2030, International Institute for Applied Systems Analysis, Laxenburg, Austria, to be published.

6. See for example: M.Grenon, Coal: Resources and Constraints, in: Second Status Report of the IIASA Project on Energy Systems 1976, International Institute for Applied Systems Analysis, Laxenburg, Austria, IIASA-RR-76-1 (1976).

7. World Energy Resources 1985 - 2020, Executive Summaries of Reports on Resources, Conservation and Demand to the Conservation Commission of the World Energy Conference, Istanbul, 1977, IPC Science and Technology Press, Guildford and New York (1978).

8. Uranium - Resources, Production and Demand, Nuclear Energy Agency and International Atomic Energy Agency, OECD, Paris (1977).

9. A.M.Weinberg, Energy as an Ultimate Raw Material, or - Problems of Burning the Sea and Burning the Rocks Physics Today, November (1959), p. 18-25.

10. Ford Foundation, Nuclear Power: Issues and Choices, Report of the Nuclear Energy Policy Study Group, Ballinger, Cambridge, Mass. (1977).

11. Data presented at the: Third Energy Program Status Report, International Institute for Applied Systems Analysis, Laxenburg, Austria, January (1978).

12. J.M.Weingart, Systems Aspects of Large-scale Solar Energy Conversion, IIASA-RM-77-23 (1977).

13. Energiequellen für Morgen? - Nichtnukleare, nichtfossile Primärenergiequellen (Energy Sources for Tomorrow? Non-Nuclear, Non-fossil Primary Energy Sources), Umschau-Verlag, Frankfurt (1976).

14. Sonnenenergie II (Solar Energy II), Umschau-Verlag, Frankfurt (1977).

15. W.Häfele et al., Fussion and Fast Breeder Reactors, IIASA-RR-77-8 (1977).

LASER PRODUCED NUCLEAR FUSION*

R. Hofstadter

Max H. Stein Professor of Physics

Stanford University, Stanford, California 94305, USA

1.   INTRODUCTION

In these lectures I will try to give a coordinated view of the overall field of fusion with special emphasis on laser produced fusion, with which I have been associated on a part time basis, since 1972.

There are three primary inexhaustible energy sources:
1. Solar Energy
2. Fission Breeder
3. Fusion, and perhaps a fusion breeder, i.e., a kind of hybrid in which fusion neutrons make a fissionable fuel.

The main advantages of fusion power, as listed on a Livermore chart, are as follows:

1. Effectively infinite fuel supply at low cost (<< 1 mill/kwh)
2. Inherent safety -- no runaway
3. No chemical combustion products
4. Relatively low radioactivity and attendant hazards
5. No emergency core cooling problem
6. No use of weapons grade materials and hence no diversion possibilities
7. Flexibility to site plants near load centres, possibly even in urban areas.

---

*Notes taken by Dr. Masud Ahmad of Pakistan Atomic Energy Commission.

The main reactions which one can consider for fusion purposes are listed below:

	Total Energy (MeV)	Charged Particles Energy (MeV)
$D + T \to {}^4He\ (3.52\ MeV) + n(14.06\ MeV)$	17.58	3.52
$D + D \to {}^3He\ (0.82\ MeV) + n(2.45\ MeV)$	3.27	0.82
$D + D \to T\ (1.01\ MeV) + p(3.03\ MeV)$	4.04	4.04
$D + {}^4He \to {}^4He(3.67\ MeV) + p(14.67\ MeV)$	18.34	18.34
$T + T \to {}^4He + n + n$	11.32	–

The most promising reaction at present seems to be the D + T reaction. The second reaction D + D is harder to achieve, although it has the advantage that it produces low energy neutrons which mean lesser radioactivity. All other reactions listed above are even more difficult to achieve. It may be interesting to remark that most, if not all of the magnetic confinement work has been confined so far to the D + D reaction, whereas the laser fusion work has been done exclusively on the D + T reaction. The main characteristics of the plasmas produced in magnetically confined devices are quite different. The magnetically confined plasmas are characterized by low pressure, long time of confinement, and high temperature. On the other hand, the characteristics of inertially confined plasmas are very high pressures (millions of atmospheres), very short confinement time (nanosecond-picosecond), and temperatures relatively lower than those required in the case of magnetic confinement.

The magnetically confined fusion systems can be grouped into 'closed systems' (Tokamaks) and open systems (mirror or mirror tandem devices). Similarly in the inertially confined fusion systems, the drivers can be different. For instance the driver may be a beam of a laser, a beam of electrons, high energy heavy ions or low energy ions. The main advantages and disadvantages of the various drivers are discussed below.

In the case of magnetic confinement the main advantage is that it is fairly straightforward and no great new technologies need to be developed. Another great advantage is that the subject is completely unclassified and there is international collaboration on a world-wide basis. The possible disadvantage is the presence of large current-carrying coils which impair accessibility to neutrons. Probably one will require large superconducting coils, which may be difficult to make and maintain.

In laser fusion, one of the advantages is the easy accessibility of neutrons and the relatively small size of the apparatus. Moreover, the laser fusion device will just work like an automobile engine which one can start or stop, unlike the fission reactor. On the side of disadvantages one can list the requirement of developing very high quality and high powered lasers which are recyclible in very short times. There is also the problem of protection of the 'last mirror' which may present some engineering problems. Moreover, unlike magnetic confinement, laser fusion work is still a partly classified field. Unlike its magnetic counterpart there is less international collaboration in this case.

The idea of using heavy ion beams for fusion is a relatively recent idea and it mainly involves using beams of say lead or uranium ions with a charge of 2 or 3 at an energy of 5-6 GeV. In this case the heavy ions beams are required to deliver energy of the order of a few megajoules in a nano-second at a repetition rate of, say, 1 per second. This seems to be attainable without the development of new accelerator technology. Once the energetic beam is produced, it will deposit its energy without failure or diversion. The ion is stopped as it loses energy in the target. The disadvantages of the approach are that one requires very long machines for production of heavy ion beams and complicated storage rings for accumulation of ions. Also the technique is very expensive and needs a large pellet.

The advantages of electron beam fusion are that beams are simple and relatively easy to produce with large currents. The disadvantages are that relativistic electrons have minimum energy loss. And, moreover, one needs various tricks, for instance, in producing a magnetic field in the plasma to confine electron paths to multiple passes and this is quite a difficult thing to achieve.

Going back to different fusion reactions, the $D+T$ reaction seems to be the most favoured reaction. This one can see from the energy vs fusion cross-section plot given in Figure (1.1). As one can see in the case of $D+T$ reactions the peak in the cross-section occurs at about 100 KeV energy whereas the corresponding number for $D+D$ reaction is about 125 KeV. However, what really counts is not the cross-section alone but the velocity averaged cross-section. As one can see, from Figure (1.2) at about 10 KeV ion temperature the DT reaction is favoured by about a factor of 100 over the DD reaction.

In order to compare the relative sizes of the magnetically confined and inertially confined systems, Figure (1.3) shows a recent sketch of a 100 MW Tokamak-like device designed by a group at Oak Ridge. One may be struck by the large size of the machine. However, the size of the machine seems considerably smaller, if one imagines the conceptual designs of Tokamaks of a few years ago, when they could have been thought of being as big as the pyramids of Egypt. This great reduction in the size of the Tokamak machines has become possible by using the concept of an evacuated building, which has proved very useful.

In Figure (1.4) we show the corresponding laser fusion

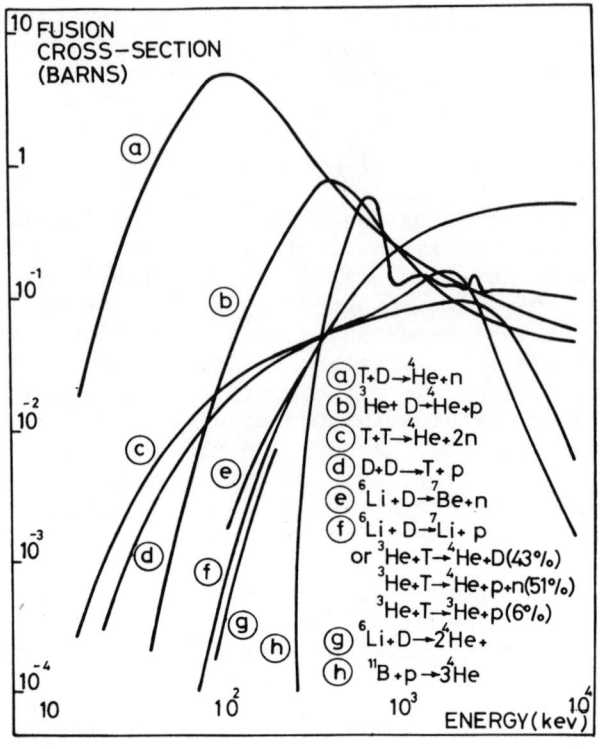

Figure 1.1. Energy vs fusion cross-section curve for different fusion reactions.

system which has been designed by a group at the University of Wisconsin. This device has been designed for producing 1000 MW (electric) and makes use of 6 or 8 laser beams for deposition of energy and uses LiO_2 powder for breeding to produce tritium.

On the technological side also various problems have to be solved. First of all some techniques have to be developed for handling large quantities of tritium. Lots of work in this connection is going on in the United States especially in the Los Alamos Laboratory. One has also to study the effect of 14 MeV neutrons on materials. If these neutrons damage or break up the reactor materials, then one has to change the liner in the reactor quite often, which may pose a considerable problem in actual practice. Thus, one has to develop good stainless steel or some other materials which can withstand the bombardment of neutrons. Then there is the pellet manufacturing technology. These pellets have to be made inexpensively in great numbers. Work at present is going on not only the simple pellets but also pellets with several layers inside them.

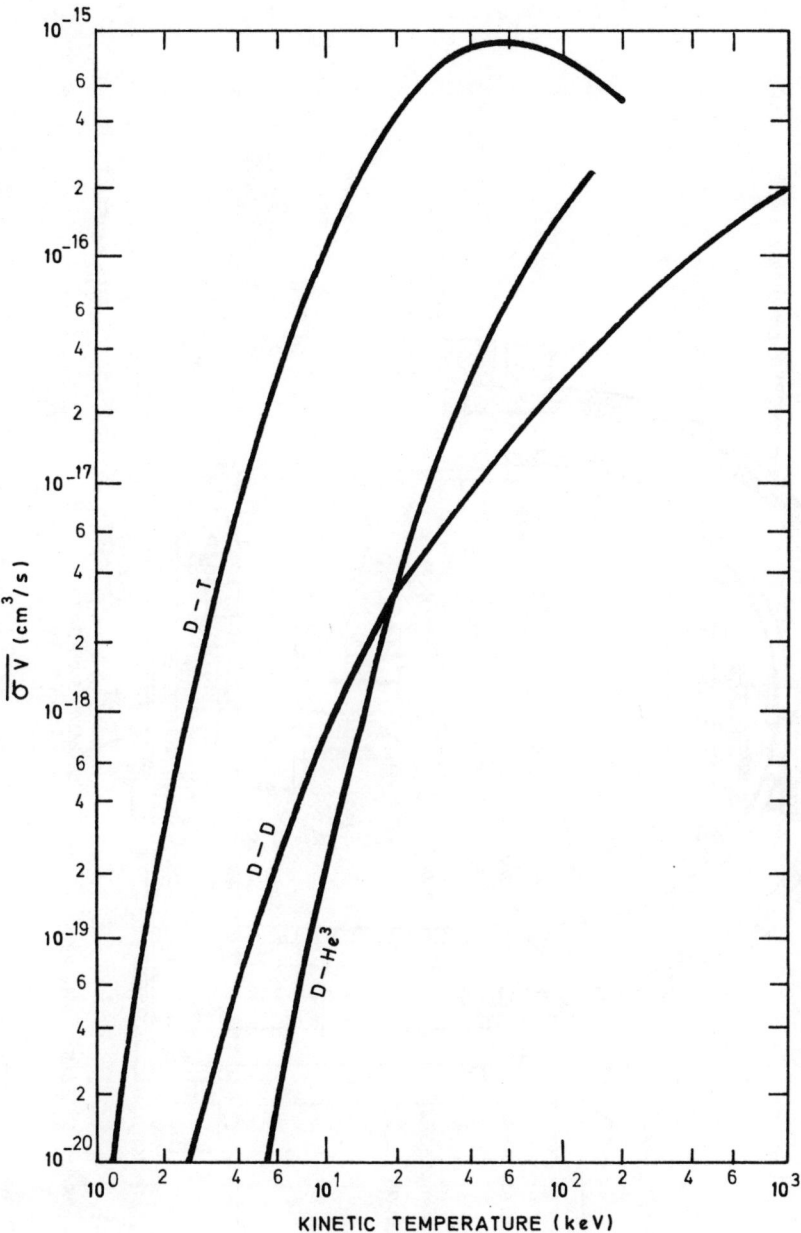

Figure 1.2. Velocity averaged cross-section vs kinetic temperature for different fusion reactions.

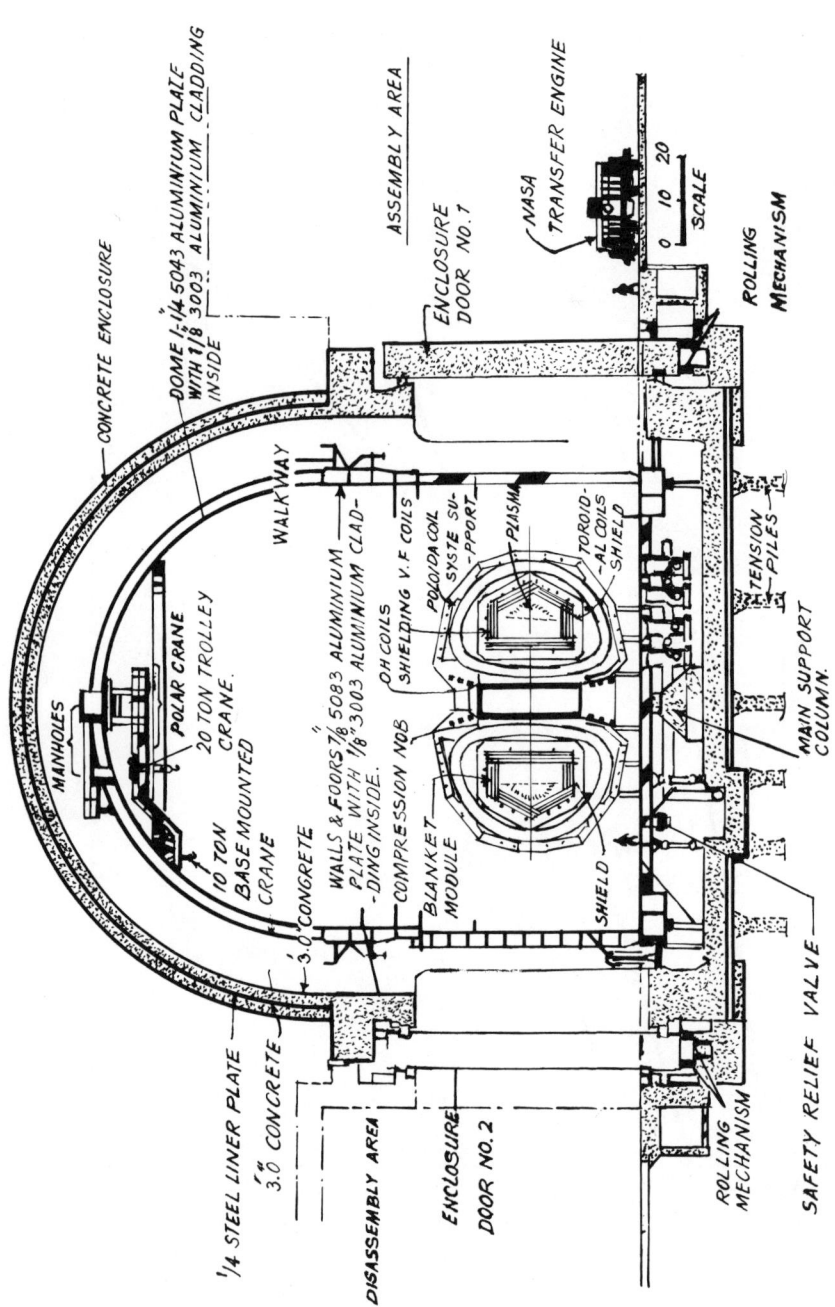

Figure 1.3. The ORNL reactor design superimposed on a cross-section of the NASA Plum Brook vacuum facility at Sandusky, Ohio.

LASER-PRODUCED NUCLEAR FUSION

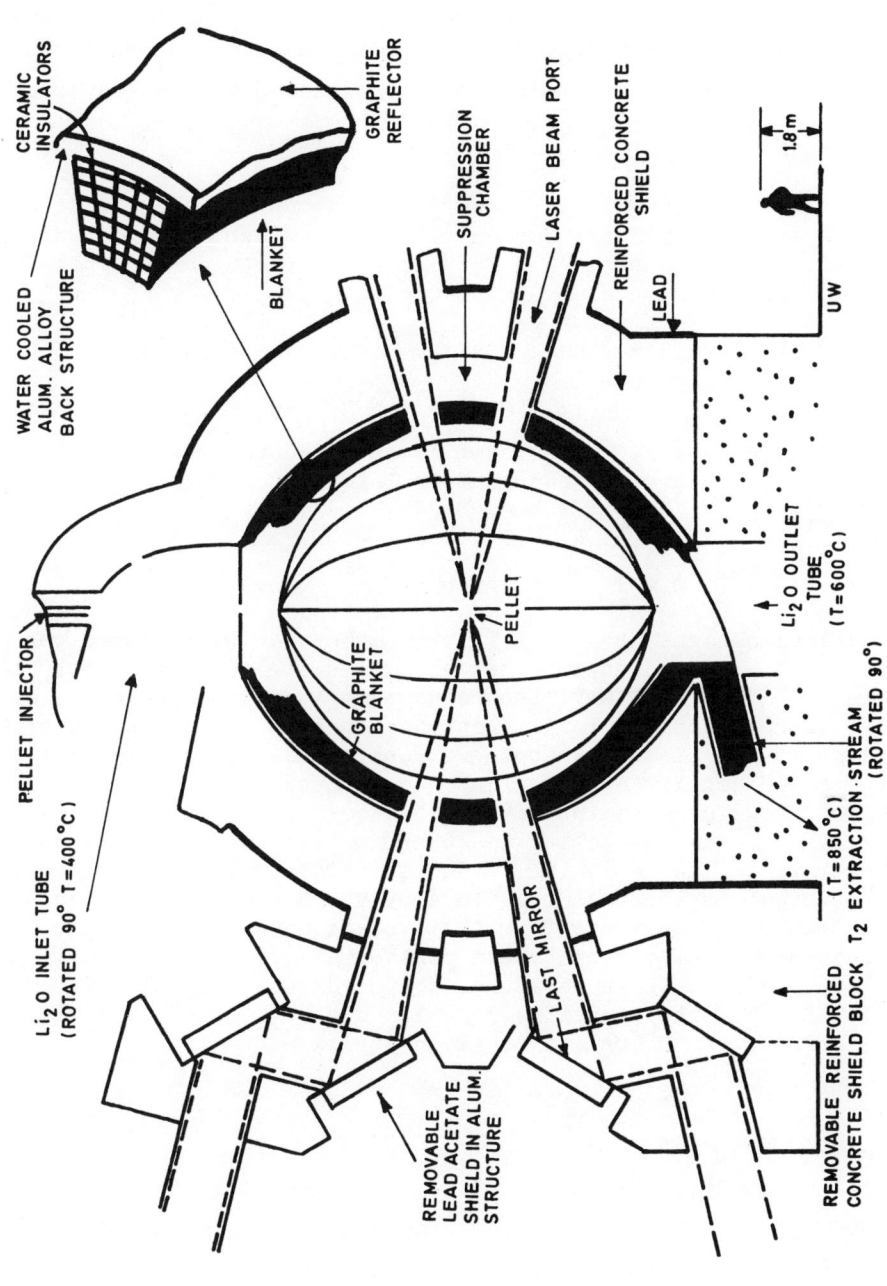

Figure 1.4. The Solase Laser Fusion Reactor System.

As far as the production of electricity from fusion is concerned, it is generally considered to be done in the conventional way by using heat exchangers etc, steam, and electricity. However, this is not the only way, the second way is the production of hydrogen gas by radiolysis of H_2O by neutron irradiation. As hydrogen is a perfect fuel, it will not add to CO_2 pollution of the atmosphere.

2. LASER FUSION

After these general observations about fusion, I will now discuss specifically laser fusion. If one considers a pellet of an equal mixture of Deuterons and Tritons with a total density of N atoms/cm^3, the thermonuclear power per cm^3 is

$$P = (N/2)^2 <\sigma V> Q_T$$

where $Q_T = ME_n + E_\alpha + M (14.1 \text{ MeV}) + 3.5 \text{ MeV}$ and M is the blanket multiplication factor. The blanket is designed to contain Lithium, which produces Tritium by the neutron interactions:

$$^6Li + n \rightarrow {}^4He + T + 4.80 \text{ MeV}$$
$$^7Li + n \rightarrow {}^4He + T + n - 2.47 \text{ MeV}$$

In practical designs about 1.1 Tritium atoms can be produced per Tritium atom consumed and the net energy multiplication factor can be about 1.3 or Q = 22 MeV. Let us now consider the conditions under which fusion can occur. I will reproduce here the argument given by R.Perkins of Los Alamos. Suppose we consider a 1 mg pellet of frozen DT at a solid density of 0.2 g/cm^3. The number density is 5×10^{22} atoms/cm^3, so 1 mg contains 2.4×10^{20} atoms in a sphere of 1 mm radius. If we heat the pellet to 10^8 K° or 8.6 KeV instantaneously we must invest 3 nkT = 1×10^6 J. The inertial confinement time $T_e = R/4C_s$, that is the radius divided by sound speed in the hot plasma (the factor of 4 arises from the average mass in the sphere during its disassembly). The sound speed or thermal velocity of the ions is 10^8 cm/sec at this temperature, so the disassembly time $T_e \sim 2.5 \times 10^{-10}$ S. The reaction time $T_r = 1/n<\sigma V> \sim 4 \times 10^{-7}$ S. The fractional burn up of the pellet is given approximately by

$$f \sim \frac{\tau_e}{\tau_r} = \frac{<\sigma V>}{4m_i C_s} \rho R = \frac{1}{1600}$$

The energy output, assuming Q = 17.5 MeV is therefore E_{out} = 326 MJ × 1mg × $\frac{1}{1600}$ = 0.2 MJ, only 1/5th of what was invested within the pellet. Obviously this process is very inefficient. Higher gains are possible if we compress the pellets to high densities. This can be seen by noting that the fractional burnup is proportional to the product ρR. The other factors are only a function of the temperature at which the reaction occurs. The product $\rho R \sim 1/R^2$, so if we can raise the density of the pellet, we will increase the fraction burned, even though the disassembly proceeds more quickly. Suppose we are able to compress the pellet radius by a factor of 10 i.e. 0.1 mm radius. Then the density increase would be 1000 fold and

$$\tau_e \sim 2.5 \times 10^{-11} \text{ s}$$
$$\tau_r \sim 4 \times 10^{-10} \text{ s}$$

and
$$f \sim 0.06$$

In this case E_{out} = 20 MJ, so we have obtained 20 × the incident power. This is not a very considerable gain, because one has also to consider the inefficiency of converting heat to electricity, ~ 0.35, of producing laser energy from electricity (0.05) and finally losses due due to lack of perfect absorption of laser energy by pellet 0.30. The product of all these factors is about 1/200, so one is removed from the breakeven point by a factor of 10. However, at these conditions, the range of α particles is only 0.3 g/cm^2, whereas the absorption length in a pellet radius is ρR = 2 g/cm^2. Twenty percent of the nuclear energy is redeposited in the pellet, providing a bootstrap mechanism which reduces the incident laser power requirement. In fact one needs to ignite only the centre of the pellet to create a propagating burn wave front from the α particles which can then ignite and burn a substantial portion of the pellet.

In the subsequent section I will try to illustrate the different aspects of laser fusion, using some simple arguments. For this purpose I shall be considering very small pellets - very small compared to those which will be ultimately used for reactor purposes. The laser fusion program can be divided into three phases. The first phase corresponds to ignition in which the fusionable material can be compressed to ignition temperature in which most of α particles produced during fusion remain inside the material and contribute most of their heat energy to the material and ignite it. This stage corresponds to a level of $10^{11} - 10^{12}$ neutrons per event. The second stage is that of scientific breakeven in which the output energy, in the form of neutron energy, equals the laser input energy. This will roughly correspond to a neutron level of $\sim 5 \times 10^{15}$ neutrons/event which is a factor of 5×10^5 from the present experiments. The Lawrence Livermore Laboratory has already achieved a yield of 10^{10} neutrons compared to the first yield of 10^4 neutrons obtained at KMS Fusion in 1974. And probably we can expect to gain another factor of a million in the next four years, in which case, we would be very close to the scientific breakeven point. 'Engineering breakeven' will probably require an additional factor of 10^4 or so.

The simplest way of producing laser fusion is to focus laser light with the help of two lenses on a glass shell (a microbaloon) filled with DT gas as shown in Figure (2.1).

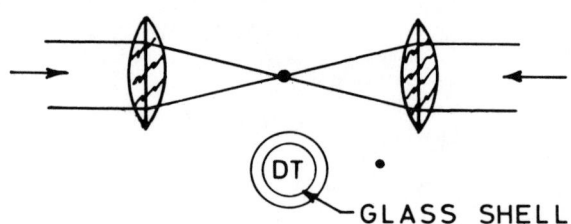

Figure 2.1. A simple system of producing laser fusion.

If the radius of the microbaloon is say 50×10^{-4} cm, then the scale of the time of collapse of the pellet is

$$\frac{R}{c_s} = \frac{50 \times 10^{-4} \text{ cm}}{5 \times 10^7 \text{ cm/sec}} = 100 \text{ ps}$$

and the power level is $1 \text{ TW} = 10^{12}$ W.

At KMS Fusion we have achieved a power of 1 TW only once. Most of our experiments are conducted at ½ TW or even less. The Shiva laser at Livermore, working at the same wavelength as ours (1.05 µm) has attained a power level of 26 TW, whereas the CO_2 laser developed by Los Alamos attained 15 TW at a much longer wavelength.

In order to calculate the number of neutrons produced in laser fusion in one event, we use the following expression derived by MAYER and others on the assumption of adiabatic compression of a glass shell filled with DT gas

$$N_{DT} = 3 \times 10^{-28} n_i^2 R^4 c^{2/3} \theta_o^6$$

where N_{DT} = number of neutrons produced per event, n_i is the Deuterons, Tritons density, R is pellet radius, and C is the compression. Using typical numbers, we get

$$n_i = 1.5 \times 10^{21}/\text{cm}^3, \; R = 35 \text{ µm}, \; c = 100$$

$$\theta_o = 700 \text{ ev}, \; N_{DT} = 3 \times 10^5$$

$$\theta_o = 1.5 \text{ KeV}, \; C = 1000, \; N_{DT} = 2 \times 10^7$$

3. LASER - PLASMA COUPLING

The coupling i.e. how the laser light interacts with the

plasma produced at the glass surface, is a very complex phenomenon which involves a number of mechanisms. The most common process considered is the classical inverse Bremstrahlung mechanism in which a photon passes by an atom and gives energy to electrons which in turn transfer energy to other electrons and ions. However, classical inverse Bremstrahlung is not the only process for this purpose because otherwise the energy absorbed would be much less than observed experimentally. The other processes which take part in this connection are Ion Accoustic Turbulence, Resonance Absorption etc. However, I don't want to go into details of these processes, because I do not understand them fully myself. The subject is undergoing a lot of investgation at present. The essential point is that after the electrons pick up energy by one process or the other, by conduction they impart energy to ions which produce conduction effects and there is also the ion-electron energy exchange process. After the energy has thus been invested, the hydrodynamics takes over. There are various possible ways of describing the hydrodynamics. But the essential idea is to compress the fusionable material adiabatically. The various requirements on the lasers and targets for fusion purposes can be listed as follows:

   1. The lasers should have adequate power and the beam profile should be good.

   2. The temporal behaviour of the laser pulse should be correct and matched to the pellet.

   3. Target illumination should be uniform and orthogonal and targeting and viewing should be good.

   4. Pellets need both fabrication and characterization.

   5. Diagnostics.

The laser coupling occurs probably through some mechanism sketched schematically in Figure (3.1). In this figure the logarithm of the density of the pellet is plotted against radius/(velocity of sound × time). The curve represents a moving front going outwards. It is thought that there is some sort of steepening effect at the critical density region where the light is absorbed. The Los Alamos people think that there is a very important discontinuity or near-discontinuity at which the light is absorbed. They believe that it does not matter whether one uses long wavelength or short wavelength. The steepening that occurs in the profile will allow the long wavelength to be absorbed at the same part of the pellet corona as the short wavelength. People had previously thought that the light would be absorbed farther out and the scale length would be larger with longer wavelength and the coupling would not be so good. However, this subject is still not very well understood and has to be looked into more carefully.

   In Figure (3.2) are shown some computer calculations for

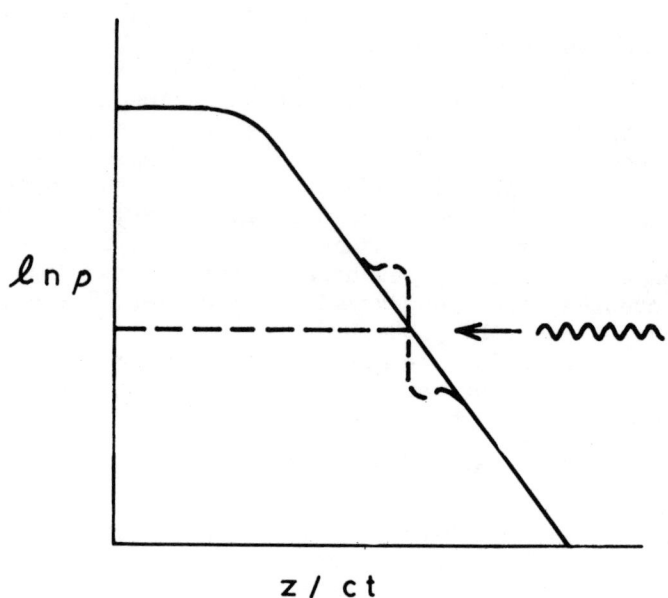

Figure 3.1. A schematic sketch of laser coupling mechanism. The logarithum of the density of the pellet is plotted against radius/(velocity of sound × time).

the density distribution of the pellet at different times. The initial pellet consists of gas at 10 atmospheres. One can see the subsequent increase in the density with time. These are one-dimensional calculations. The one dimensional model gives satisfactory results for the experiments at the present level.

Figure (3.3) shows the KMS Fusion laser configuration. Figure (3.4) shows the optical system developed at KMS Fusion by Sandy Thomas. Light comes through the two thick lenses, and by reflection from ellipsoidal mirrors, is focussed on the pellet. The system gives uniform and nearly orthogonal illumination.

Figure (3.5) gives a typical early pulse distribution. It shows four pulses spaced 60 ps apart. One can make individual pulses

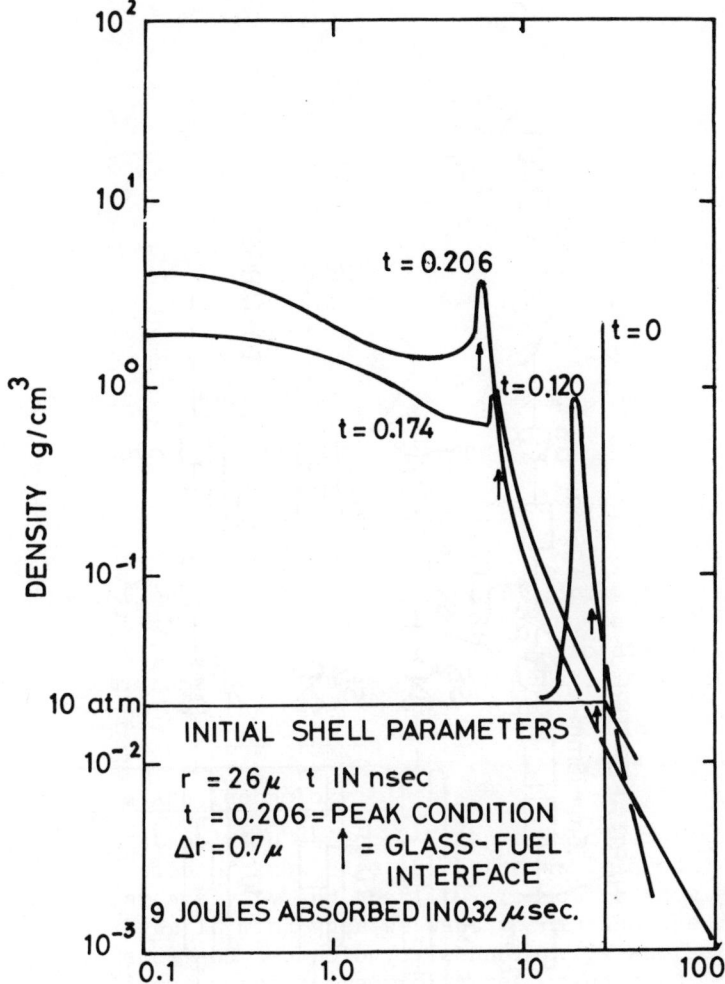

Figure 3.2. Computer calculations for the density distribution of the pellet at different times.

of any size and shape with a 'pulse stacker'. In small pellets only 3 or 4 pulses are important but in a reactor type of pellet a longer pulse compound of 100 sub-pulses may be important for properly shaping the entire input.

Figure (3.6) shows the amount of reflected light versus the average intensity. Lot of data have been taken around an average intensity of 10^{15} watts/cm^2. This intensity is already sufficient for the reactor regime. As can be seen about 70-80% light is reflected, leading to absorption of only 20-30% which is very small. One

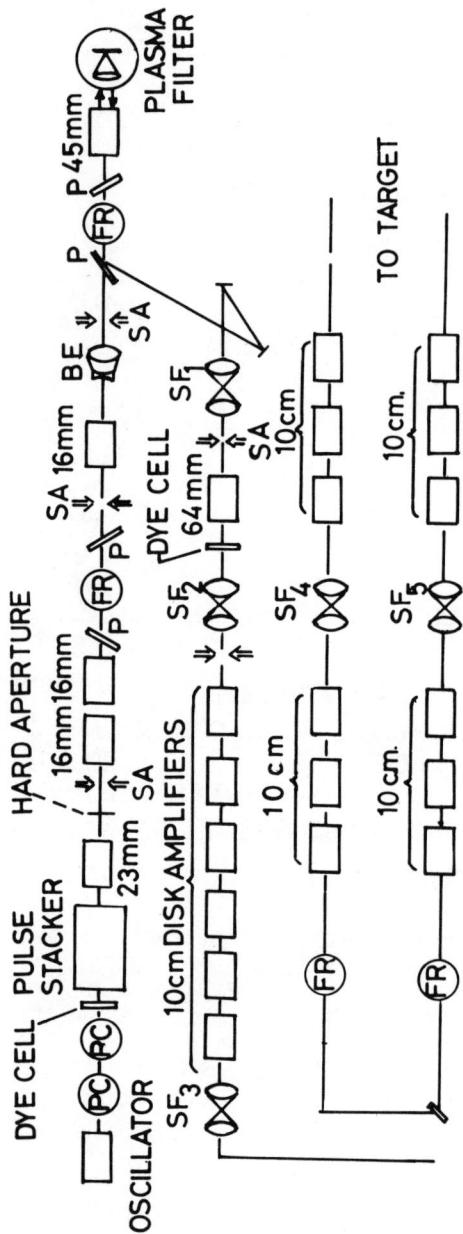

Figure 3.3. KMS Fusion laser configuration.

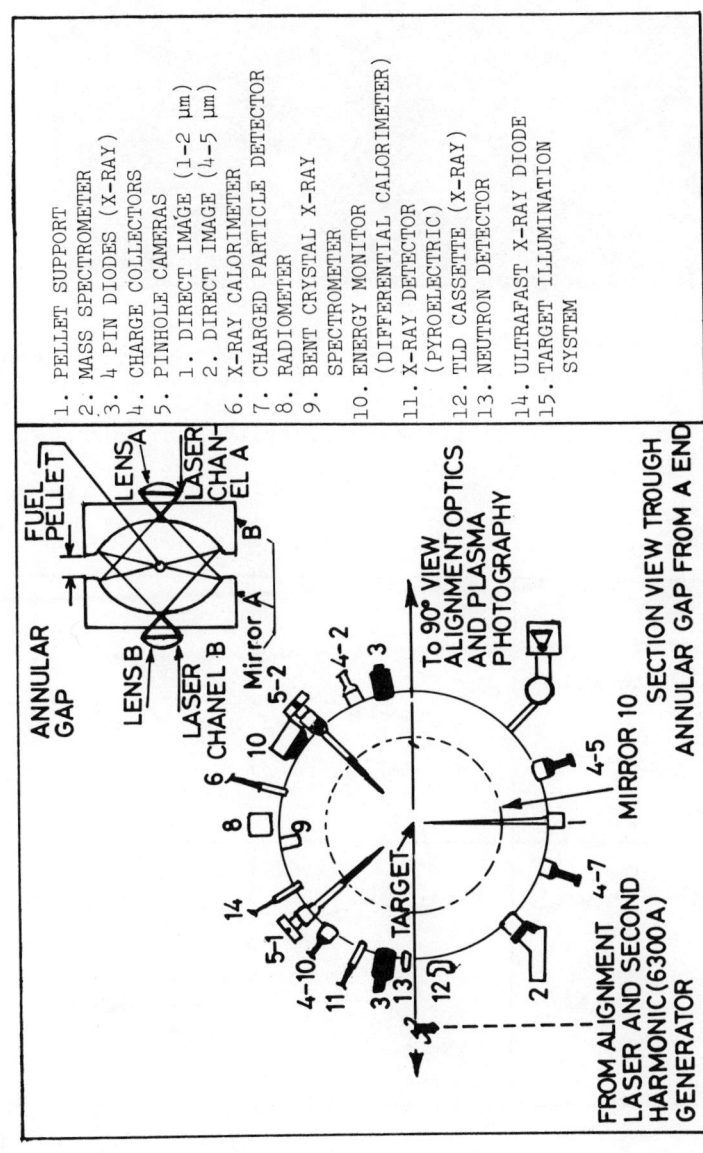

Figure 3.4. The optical system developed at KMS Fusion by Sandy Thomas.

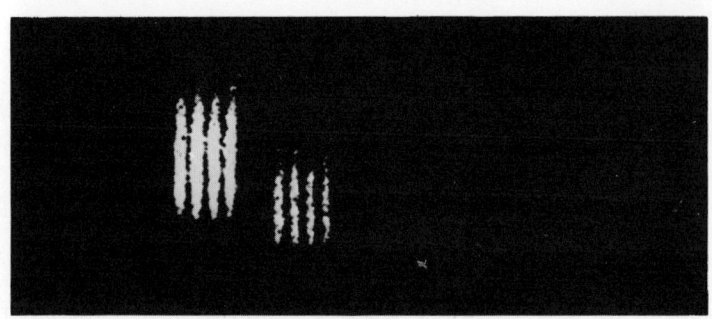

4 pulses spaced 60 psec
successive images 320 psec

Figure 3.5. Typical early pulse distribution. Four pulses spaced 60 psec, successive images 320 psec.

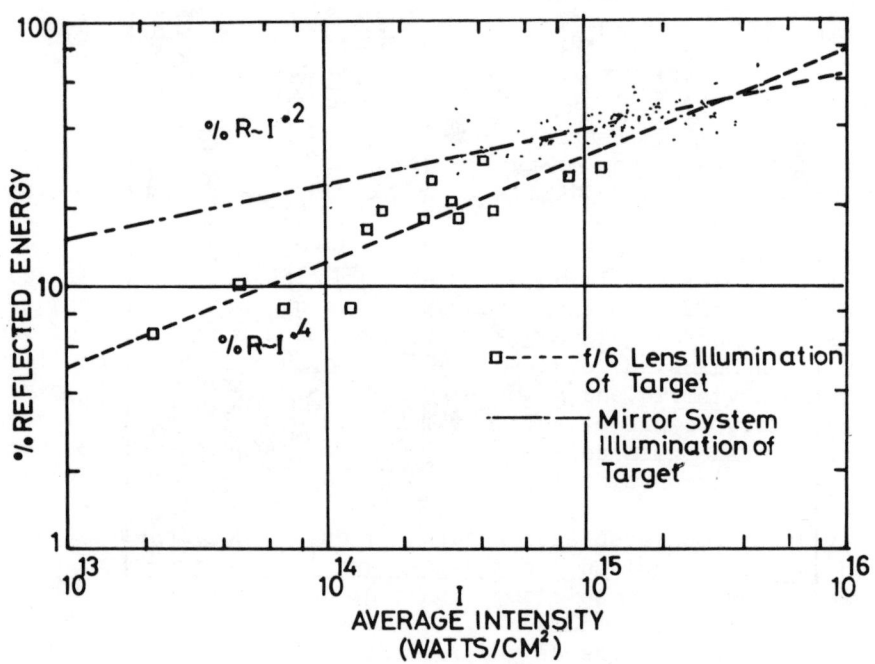

Figure 3.6. Percentage reflected energy (horizontally polarized) for both the lens and mirror illumination systems.

of the important problems at the present time is how to improve the coupling so that more light is absorbed.

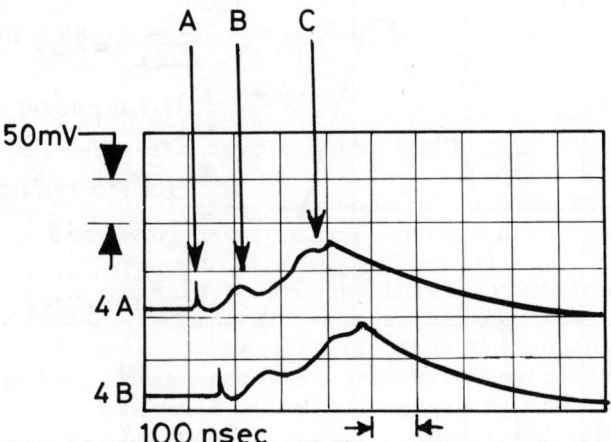

Figure 3.7. Response of the two Faraday cups or charge collectors to the ions coming out of the imploding targets. The horizontal axis is the time of flight and the vertical scale is the ion current.

Figures (3.7, 3.8) show the response of the two Faraday cups, or charge collectors, to the ions coming out of the imploding targets. The horizontal axis is the time of flight and the vertical scale is the ion current. The two cups are displaced in time from each other but the curves are essentially similar in nature. There are two groups of ions - the fast ion group (small time) and slow group (large time). One can see a little bump in the fast ion group which was previously not expected. The main features of the data are illustrated in Figure (3.8). The dotted curve is the energy associated with the ions and the other curve represents the mass associated with the ions. One can see that the mass associated with these fast ions is very small, but due to the great velocity with which they move, the total energy associated with the fast ions is roughly half of the total energy. It would be nice to understand the cause of the fast ions.

Figure (3.9) shows the α particle spectra obtained with the help of a magnetic spectrograph. The width of this curve is important because the α-particle momentum will be affected by the thermal energy appearing in the pellet core and the α curve will be broadened, the higher the temperature. By measuring the half width of the curve, one can obtain the approximate temperature of the ions. Thus one can find whether the implosion corresponds to a thermonuclear process or not.

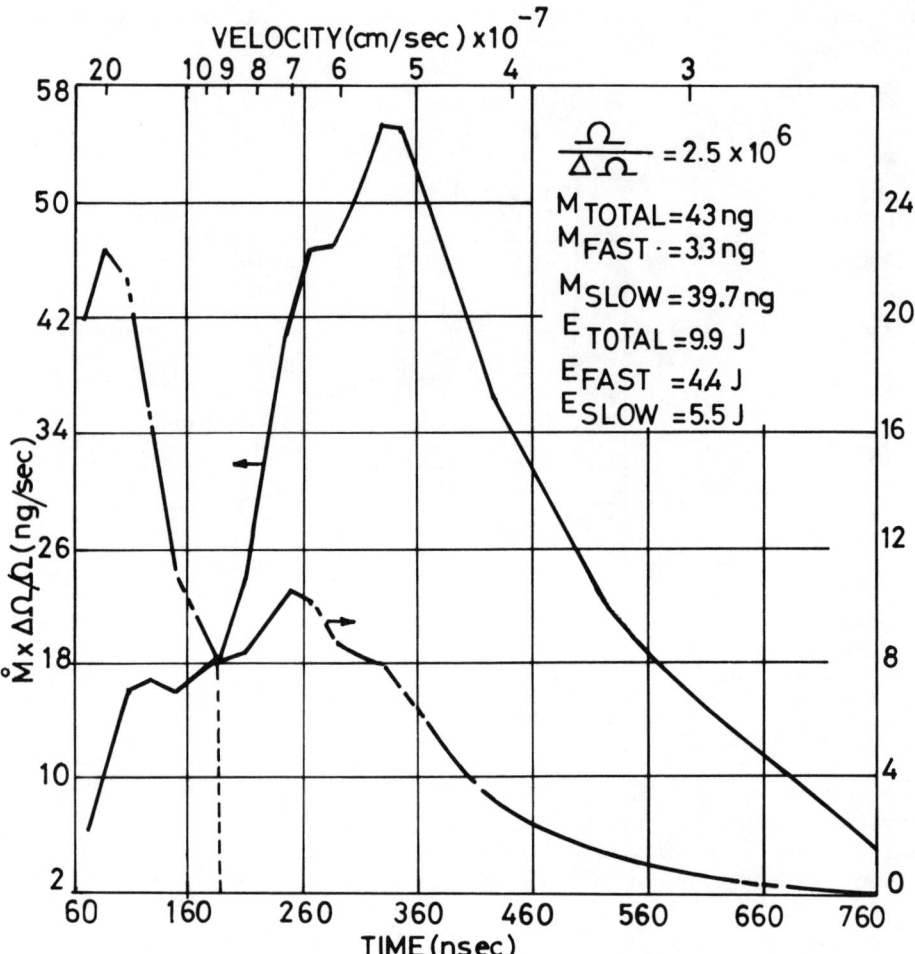

Figure 3.8. The dotted curve is the energy associated with the ions and the solid curve represents the mass associated with the ions.

Assuming that all broadening is due to the thermal distribution we have

$$\Delta E_\alpha = 177 \, \theta_i^{1/2}$$

For $\Delta E_\alpha = 430$ KeV, $\theta_i = 5.9$ KeV.

Figure (3.10) shows the spectrum of the more energetic fast ions coming out of the implosion. It is possible to make a simple theory of the corona with the help of these fast ions. If one assumes that there is an isothermal expansion, the spectrum tells us that the corona is far away from the inside of the pellet, about 200-300 micron away from the centre of the implosion.

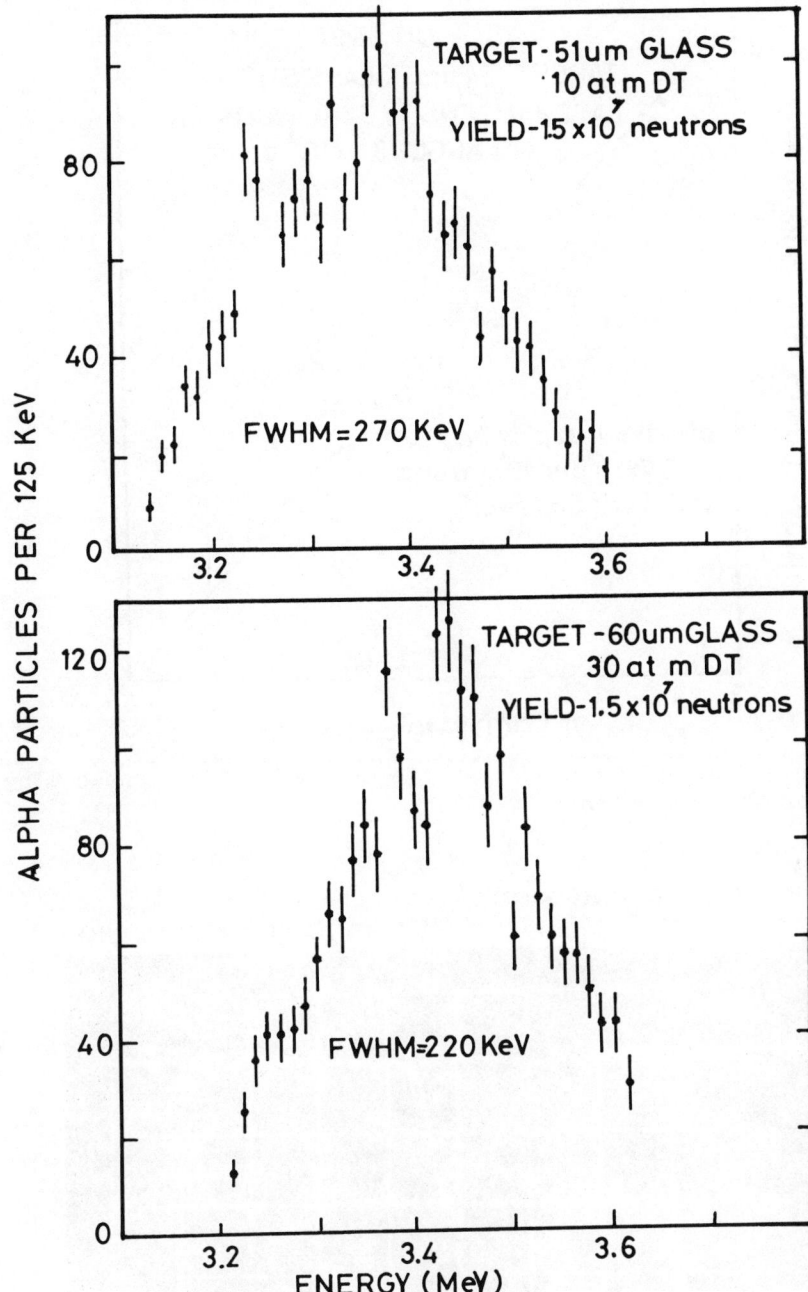

Figure 3.9. The α particle spectra obtained with the help of a magnetic spectrograph.

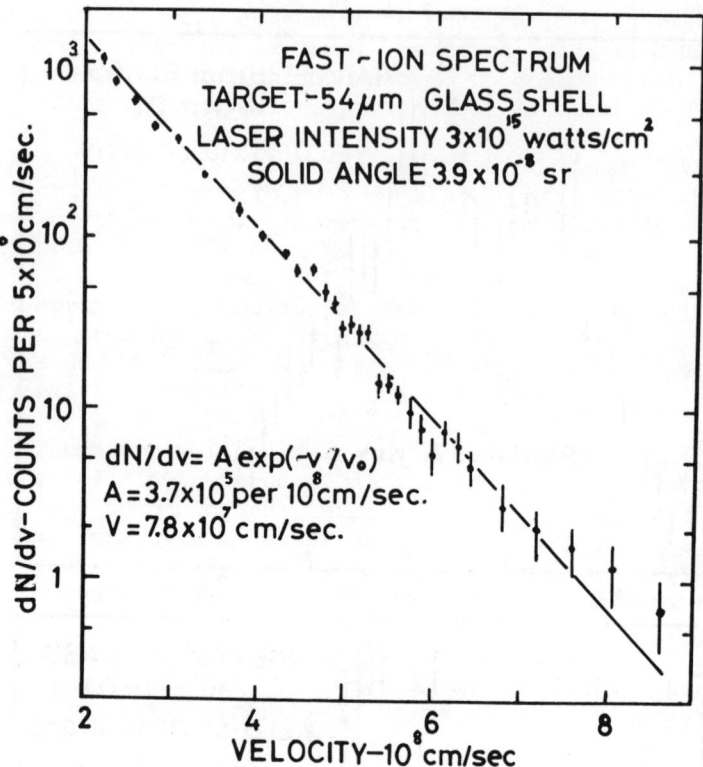

Figure 3.10. Spectrum of the more energetic fast ions coming out of the implosion.

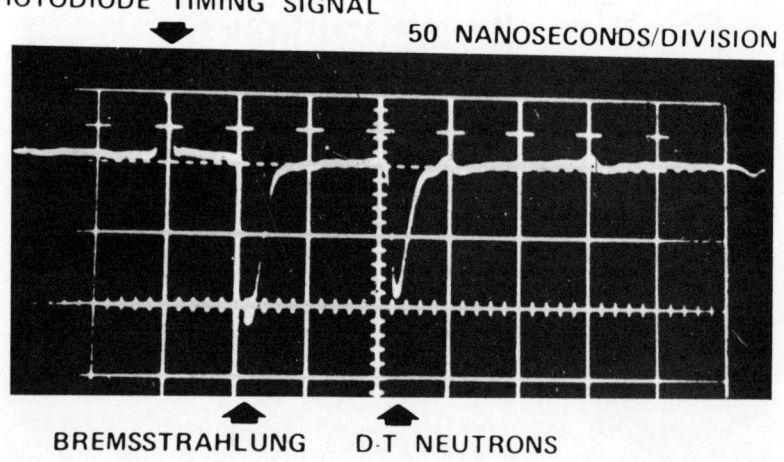

Figure 3.11. The neutron pulse and the preceding x-ray flash. Horizontal axis is the time of light.

Figure (3.11) shows the neutron response. It shows the neutron pulse and the preceding x-ray flash. Horizontal axis is the time of flight. When the number of neutrons coming out becomes large, $\sim 10^9$-10^{10}, one starts seeing structure in the forward edge and the time of flight technique will give the temperature. This has been done at Livermore.

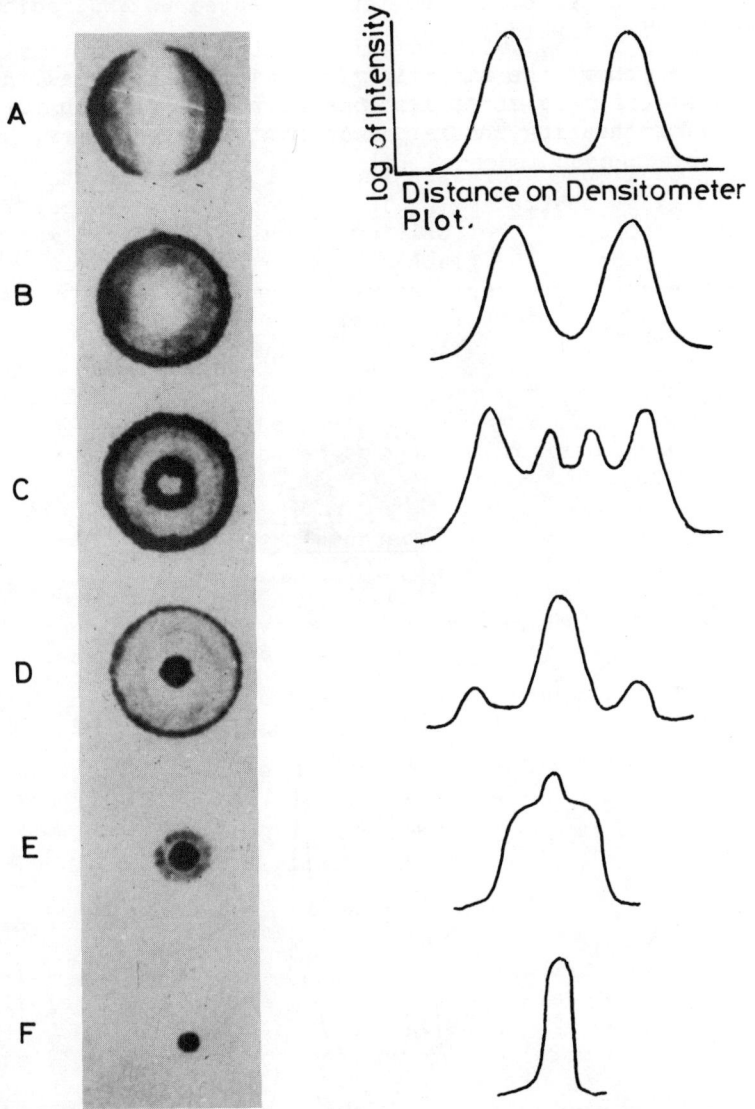

Figure 3.12. A schematic idea of what one might expect theoretically when one looks at x-ray pin hole pictures of imploding targets corresponding to different manners of illumination of targets.

Figure (3.12) gives a schematic idea of what one might expect theoretically when one looks at x-ray pin hole pictures of imploding targets corresponding to different manners of illuminating the targets. The different situations are described as follows:

A. When the two-lens system is used for illuminating the pellet, it leads to heating of both sides of the pellet, but not the equator. The sides of the pellet get heated up and radiate x-rays without moving inwards.

B. shows the situation when mirrors are used instead of two lenses. As can be seen it leads to more uniform illumination and more uniform heating and x-radiation at the periphery.

C. If one increases the input energy into the mirrors, so that the glass starts moving in, we get the two rings. The outer ring corresponds to the position in which the glass was originally radiating. Most of the glass ablates but about 10% of it moves in, thus compressing the gas inside it. The process stops when the kinetic energy of the glass outside equals the potential energy of the gas inside. This point is called the stagnation point.

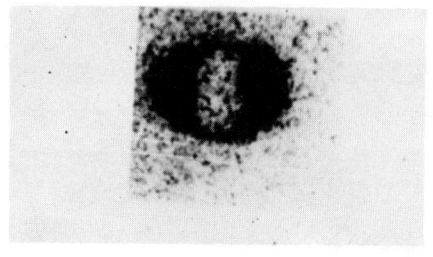

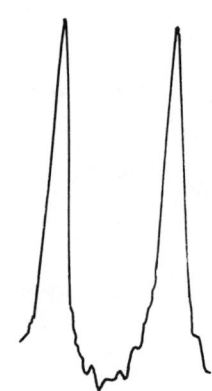

Figure 3.13. A pin hole photograph of the two lens illumination situation.

LASER-PRODUCED NUCLEAR FUSION

If one still further increases the input energy, one starts seeing less and less of the outer ring (situation D and E), until in a very powerful implosion it is reduced to a single bright spot. We can now show some experimental results illustrating these conceptional expectations.

Figure (3.13) shows a pin hole photograph of the two lens illumination situation (case A).

Figure (3.14) shows the case of uniform illumination. The pellet stem is cold and appears at the bottom.

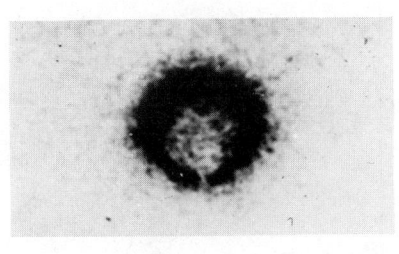

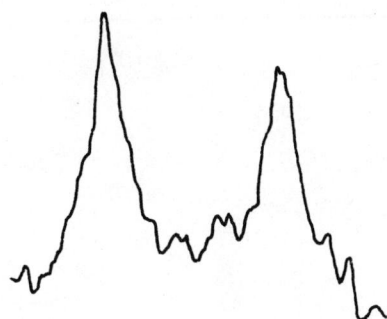

Figure 3.14. Case of uniform illumination.

Figure (3.15) shows the two-ring case and it corresponds to a compression of about 60. One sees the inner stagnation ring.

Figure (3.16) shows a pellet with the radius of 198 μm compressed to a radius of 9.8 μm. This is the case of a pellet which did not contain Deuterium-Tritium gas by accident. It contained only

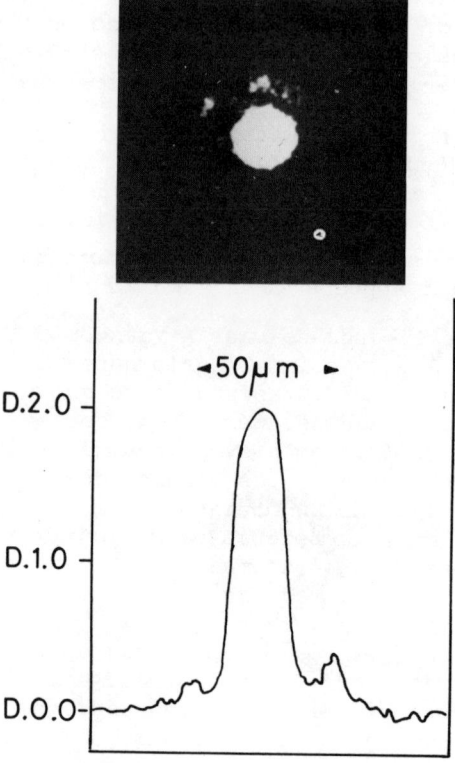

Figure 3.15. The two ring case.

Nitrogen and CO_2 at a pressure of 0.1 - 0.2 atomsphere. It was compressed by a factor of 7000.

In Figure (3.17) one can barely see the outside of the pellet.

In Figure (3.18) shows significant compression of about 500 to 1000.

Typical laser implosion parameters and the energy balance are given in Table 1.

The most frequently used wavelength for fusion purposes is 1.06 μm and is characteristic of the light from the Neodymium doped - Glass laser. It has been predicted that absorption will be higher at a wavelength shorter than 1.06 μm. To get shorter wavelength, we used crystals of potassium dihydrogen phosphate (KDP), a material that has a property of doubling the frequency of intense light passing through it, so that 1.06 μm light from laser will emerge

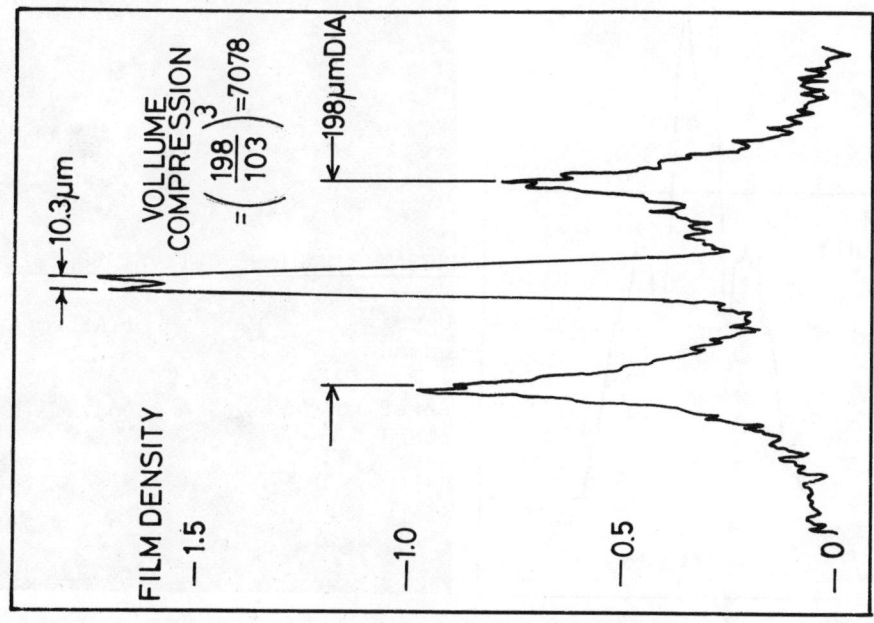
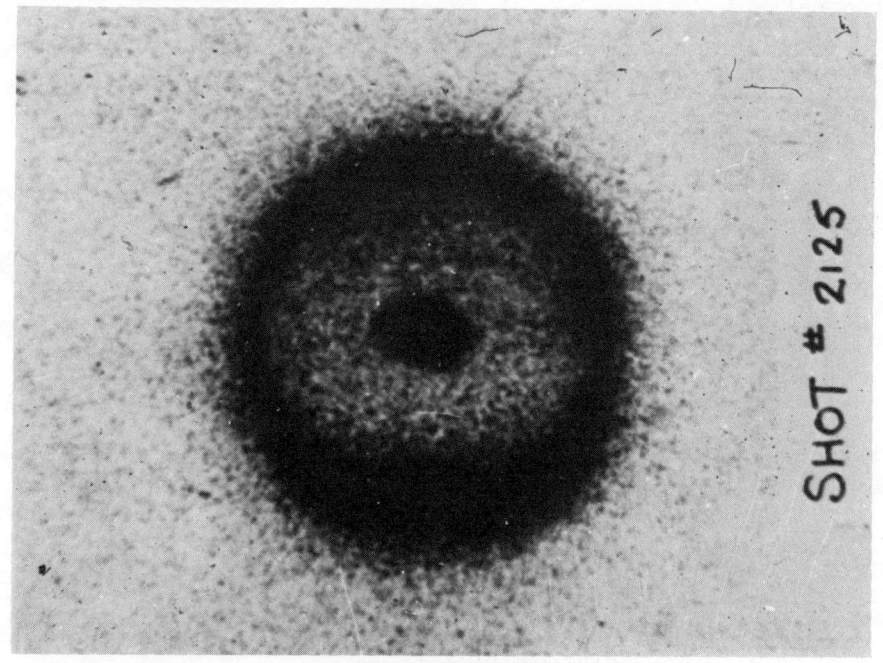

Figure 3.16. Case of a pellet compressed by a factor of 7000 accidentally.

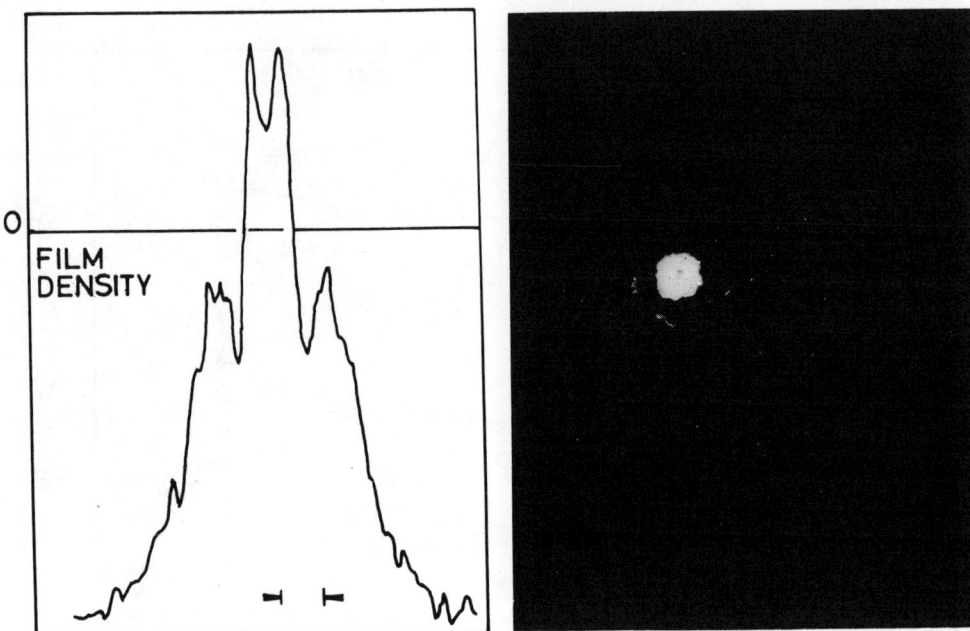

Figure 3.17. Case where one can barely see the outside of the pellet.

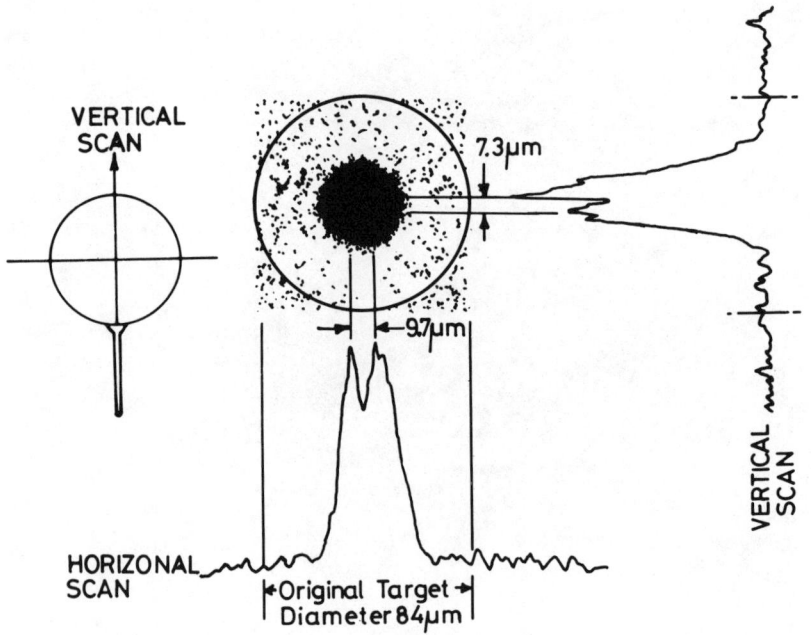

Figure 3.18. Case corresponding to a compression of 500-1000.

Table 1
Typical laser-diven implosion parameters

Shell Diameter	55 μm
Shell Thickness	0.7 μm
Shell Mass	15 nanograms
DT Gas pressure	10 atmospheres
Power on Target	0.4 terawatt
Energy on Target	18 joules
Pulse length	45 picoseconds
Reflected/Refracted Laser Energy	10.7 joules
Scattered Laser Energy into Mirror Gap	2.5 joules
Absorbed Energy	3.4 joules
Kinetle Energy	2.8 joules
Fraction of Kinetic Energy $\geq$ 5 keV/nucleon	0.6 joules
Ionization Energy	0.2 joule
X-Ray Energy	0.4 joule
Low Temperature Component	0.75 KeV
High Temperature Component	11 KeV
Energy in High Temperature Component	2.5 millijoules
Neutron Yield	2×10^7
Alpha Particle Energy Loss	$\sim$ 5 percent
DT Fuel Ion Temperature	$\sim$ 2.5 KeV
DT Fuel Compression	$\sim 10^3$

from a KDP crystal with its wavelength shortened by half to 0.53 μm - the green light, see Figure (3.19). Table 2 lists the various green light experiments that have been done till now. Figure (3.20) shows a plot of laser power on targets vs the neutron yield. In general, the green light experiments show better results than the Neodymium experiments. However, due to recent doubts on the calibration of the green light, it is difficult to say conclusively that green light results are better than the Neodymium results.

Figure (3.21) shows the laser development program of Lawrence Livermore Laboratory. Shown also in Figure (3.21) are the expected target gains (LASNEX predictions) for each system, with actual target parameters to-date.

Table 2

Target shot #	Diameter (μm)	Wall (μm)	Gas pressure (ATM)	Laser power (TW)	Pulse width (PSEC)	Neutron yield
CRYOGENIC TARGETS						
January 28, 1977 3043	68	0.7	81.0	0.54	176	6.9×10^7
January 28, 1977 3046	68	0.7	81.0	0.62	138	3.1×10^7
February 7, 1977 3077	55	0.7	81.0	0.29	200	4.3×10^7
GLASS TARGETS						
August 23, 1977 3224	79	0.9	10.0	0.79	84	1.03×10^8
August 25, 1977 3235	84	1.0	8.7	0.65	78	7.6×10^7
August 25, 1977 3237	82	0.9	8.7	0.74	76	7.7×10^7
PLASMA FILTER EXPERIMENTS						
September 30, 1977 3293	84	0.9	8.7	0.58	75	5.3×10^7
October 11, 1977 3317	80	1.0	8.7	0.31	96	4.6×10^7
October 14, 1977 3330	83	1.0	10.0	0.41	90	4.5×10^7
POLYMERIC TARGETS						
October 26, 1977 3352	99	2.6	10.9	0.45	94	2.3×10^7
October 26, 1977 3353	102	2.1	10.9	0.59	103	2.7×10^7
November 2, 1977 3371	86	2.1	9.0	0.47	141	3.9×10^7
GREEN EXPERIMENTS (GLASS)						
December 12, 1977 3425	85	0.9	8.7	0.076	342	1.28×10^6
December 14, 1977 3431	85	1.2	8.7	0.084	346	1.66×10^6
December 19, 1977 3444	51	0.6	8.9	~ 0.12	~ 100	1.41×10^6
GREEN EXPERIMENT (PVA)						
December 21, 1977 3452	82	1.4	9.8	0.097	296	7.7×10^5
December 22, 1977 3456	61	1.6	10.1	0.089	358	1.01×10^6
December 22, 1977 3457	75	1.5	10.1	0.083	399	5.5×10^5

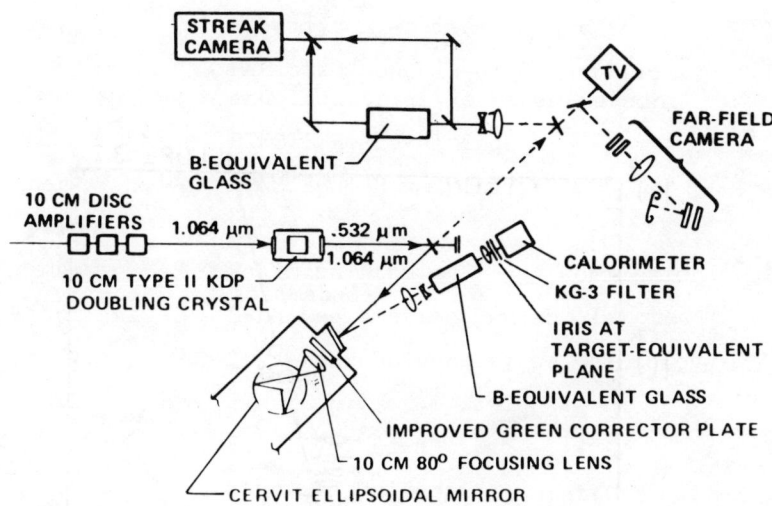

Figure 3.19. Configuration of a Green experiment.

Janus, a small two-beam laser system, produced a 0.4 TW of power. With Janus, laser driven Deuterium-Tritium implosions yields 10^7 neutrons and allowed proof of the thermonuclear nature of the reaction. Cyclops was the developmental test bed for Janus, Argus and Shiva. It was the first laser to deliver 1 TW. Argus II, designed as a 3 TW system, has actually delivered upto 4.6 TW. Neutron yields with Argus II exceeded 10^9 permitting for the first time, confirmation of thermonuclear nature of DT reaction with neutron time of flight spectroscopy. Argus IV was not built because of the funding limitations. The Shiva system at 20 - 30 TW is being completed in 1977 and is expected to achieve significant thermonuclear burn within the next two years. Shiva-Nova has been designed to provide the power needed to drive high energy gain microfusion events. It will produce 10-15 time the power and 20-30 time the energy of Shiva (see Figures 3.22-3.26).

Figure (3.27) gives the driver source option now being considered and the key fusion physics milestones leading to scientific feasibility. Because Neodymium Glass lasers are the most advanced in terms of high peak power, the inertial confinement program has been structured to achieve the fusion physics milestones with them, culminating in the proof of scientific feasibility with Shiva-Nova. After proof of scientific feasibility Shiva-Nova will fill the dual role of a reactor pellet test feasibility and a military applications facility. The development of other driver sources proceeds in para-

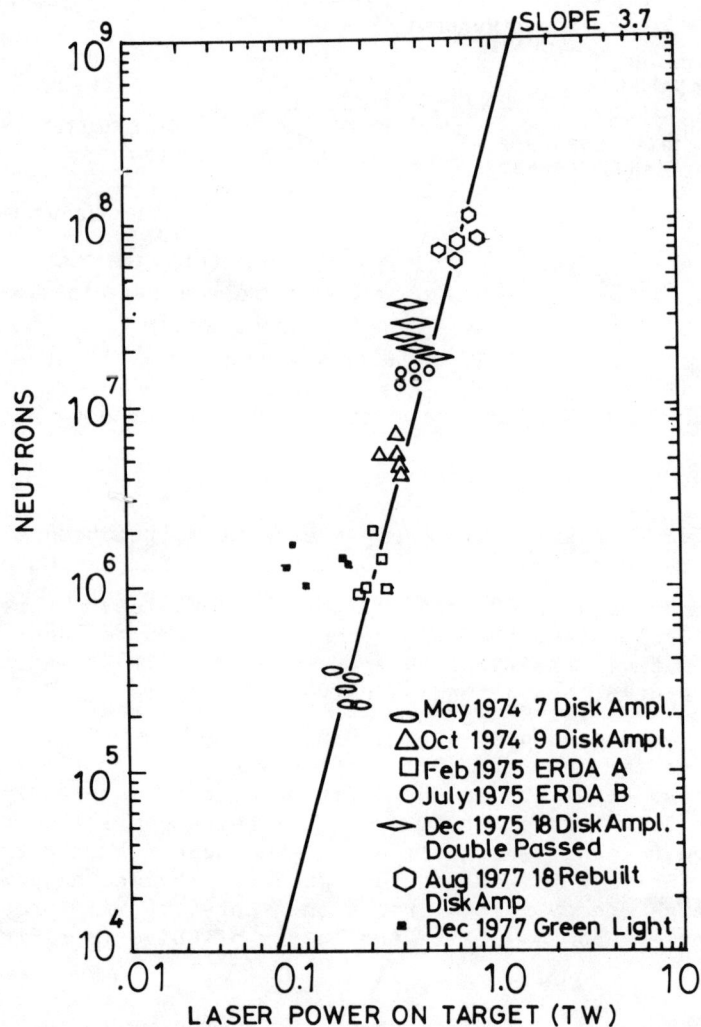

Figure 3.20. Laser power on target vs the neutron yield.

LASER-PRODUCED NUCLEAR FUSION

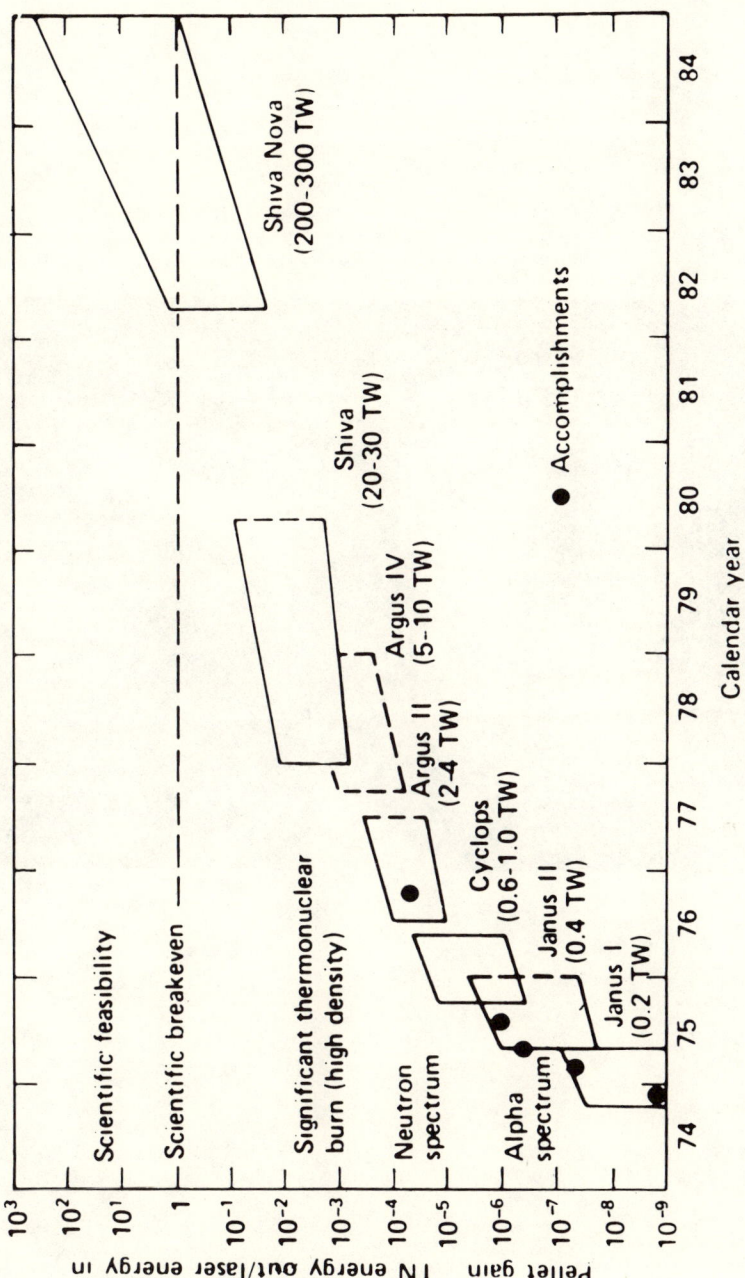

Figure 3.21. Laser fusion energy yield projections.

Figure 3.22. Shiva Model.

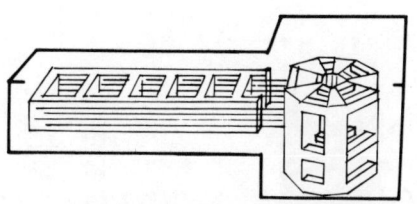
Shiva (20TW; 1977; $25M)

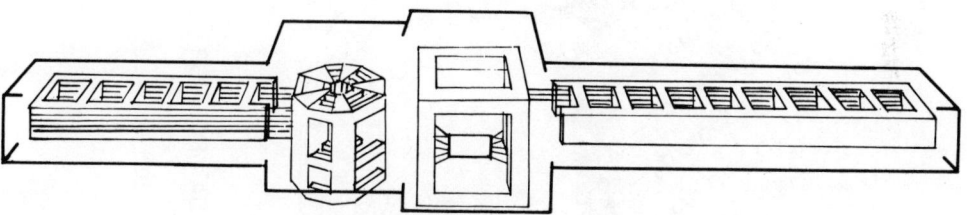
Shiva (20TW) and Shiva Nova 1 (100 TW. 1981. $94M)

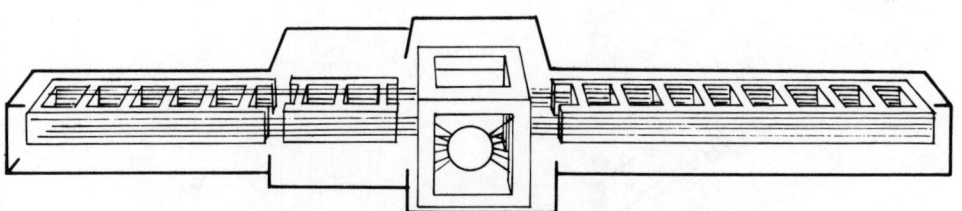
Shiva Nova II (200-300 TW; 1983; $94M + $91M = $185M)

Figure 3.23. Shiva/Shiva Nova configuration.

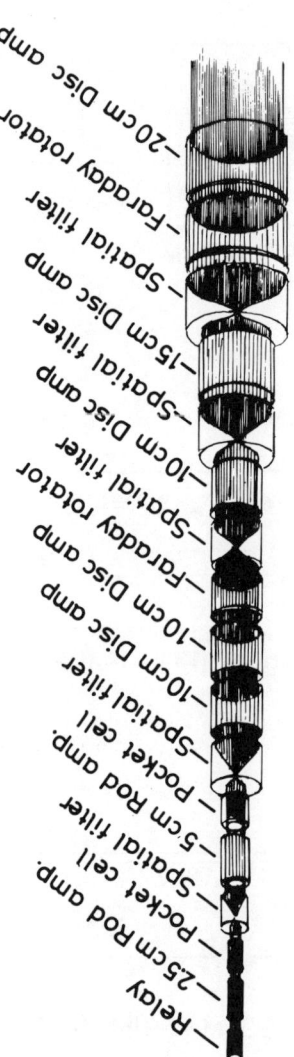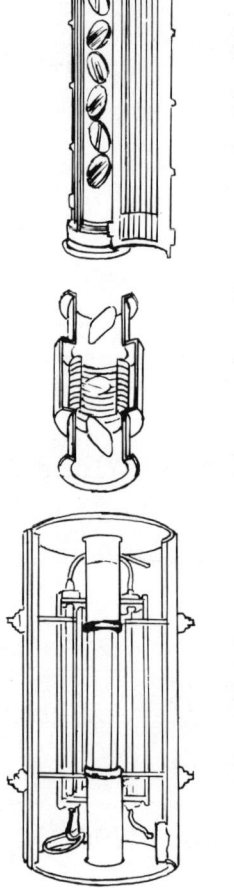

Figure 3.24. Shiva Laser Chain.

LASER-PRODUCED NUCLEAR FUSION 145

Figure 3.25. Automatic alignment system of Shiva.

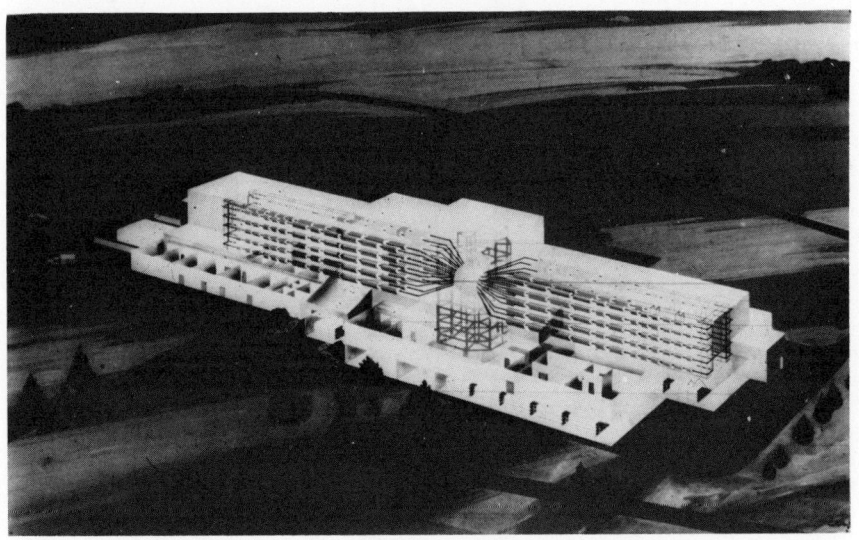

Figure 3.26. Shiva Nova Laboratory.

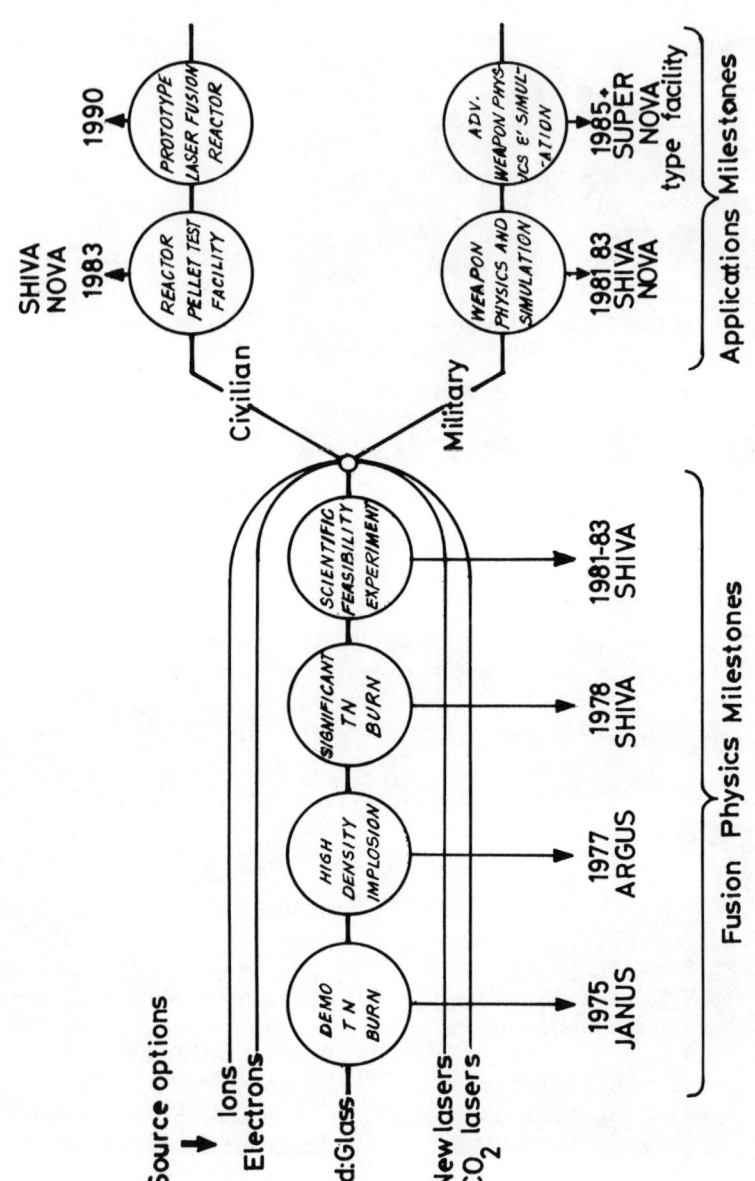

Figure 3.27. Inertial confinement fusion development logic.

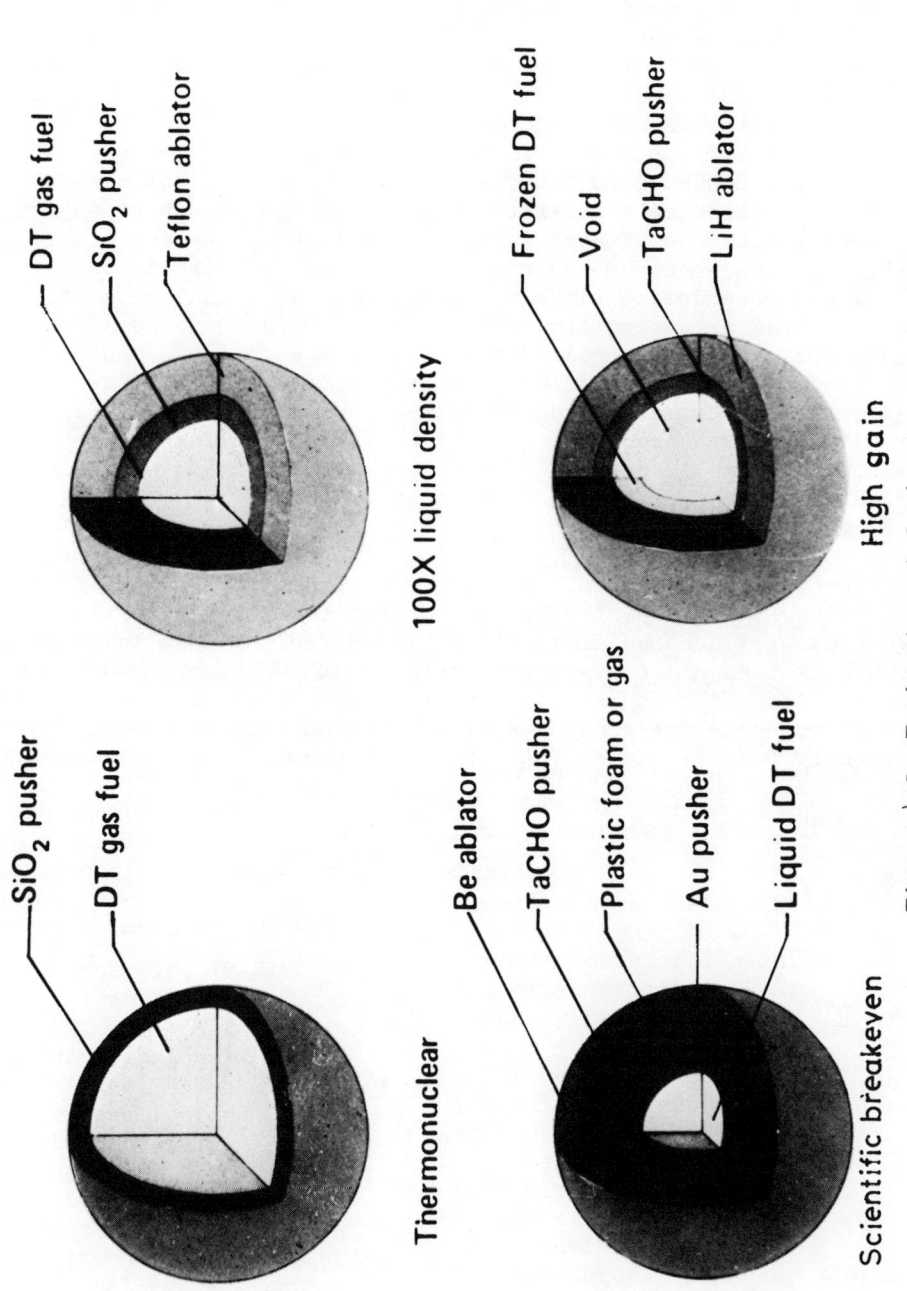

Figure 4.1. Fusion target designs.

llel, to offer the widest choice of optimum drivers for the various applications of inertial confinement fusion.

4. TARGET DESIGNS

The LASNEX computational code of Livermore has been used to design an increasingly sophisticated set of fusion experiments that match well with the planned increases in laser power and energy capability. For each set of experiments at a given laser capability, LASNEX has been used to design targets that will achieve a well defined step towards reactor level performance. Figure (4.1) shows a sequence of target designs progressing from exploding - pusher design used in early laser fusion experiments to high density, isentropic, high gain targets for fusion reactors. These targets are successively more difficult to produce and implode. Increasing laser energies and power, more complex pulse shapes and more complex and more nearly perfect targets fabrications are required.

So this is, briefly speaking, the present state of art in the laser produced fusion. At this stage I am not in a position to state clearly which of the two approaches to fusion (inertial or magnetic) will succeed first but I have a feeling that laser fusion has a slight edge over magnetic fusion. But if laser fusion does succeed first, the magnetic fusion will be closely behind. Irrespective of the relative prospects of the various approaches to fusion, I strongly believe the fusion power will play a vital role for the good of all mankind.

LATEST TRENDS IN THE ECONOMICS OF NUCLEAR POWER

James A. Lane

International Atomic Energy Agency

1. COST TRENDS IN THE UNITED STATES

1.1 Introduction

Although the prospects for nuclear power are almost always assessed on the basis of cost comparisons, the economics of nuclear power is a very inexact science because of ever changing costs and the unpredictability of future events. An examination of past trends reveals that nuclear and conventional plant power cost projections have rarely been correct relative to actual costs. Nevertheless, it is of interest to review these trends with the hope that some insight might be gained to improve the accuracy of future cost estimates. These two lectures have this goal as their main objective.

1.1.1 Review of More than Three Decades of Nuclear Power Cost Comparisons:

Although the title of my two lectures is 'Latest Trends' etc., I would like to go all the way back into history and trace the evolution of nuclear power economics from its very beginning.

To do this, I will have to rely solely on U.S. experience because I am not aware of comparable studies in other countries.

The very first study of the technical feasibility and costs of a central station nuclear plant was carried out in Oak Ridge in September 1944 by myself and another engineer, J.T.Weill. The conclusions were that a helium-cooled graphite moderated power reactor was technically feasible but that fuel costs would be too high because of the high cost of enriched uranium. At that time, the only source of enriched uranium was the Oak Ridge Electromagnetic Enrichment plant producing U-235 at $1000/gr. That report was classified secret and never published.

During the late 1940's, numerous design and economic studies of nuclear plants were carried out by the USAEC and others but most involved the use of highly enriched U-235 and could not compete with the then prevailing price of U.S. coal. In September 1952, W.H. Zinn, the director of Argonne National Laboratory published a paper[1] showing that even at $20 per gram U-235, reactors fuelled with highly enriched uranium could not compete with coal-fired plants.

In June 1954, I published a paper on nuclear power economics[2] and presented the cost data given in Table 1. Here it is seen

Table 1

Estimated Power Costs (1954)

Investment, $/kW	Coal	Nuclear
Boiler or Reactor	50	100
Turbogenerator	100	120
	150	220
Power Costs, m/kWh		
Capital	3.2	4.7
Fuel	2.8	1.0
Operating + Maintenance	1.0	1.0
Total	7.0	6.7

Basis

$7/ton coal, $35/lb. nat. U
15% fixed charges, 80% plant factor
500 MWe nat. U -D_2O reactor

that nuclear power just edges out coal at $7 per ton. Shortly after this paper came out, the USAEC released a study[3] which analyzed the cost of electricity in 1952 from post war steam generating plants in the United States. These ranged from 2-12 mills/kwh; however, 70% of the electricity was generated at costs from 4-7 mills/kwh. From these results, it was concluded that the potential U.S. market for nuclear power during the 1955-1975 period would increase from 2 GWe at 10 mills to 22 GWe at 7 mills and 140 GWe at 4 mills/kwh[4].

LATEST TRENDS IN THE ECONOMICS OF NUCLEAR POWER

As a follow up of these results, we carried out a study in Oak Ridge of the conditions under which one might achieve nuclear power costs at 4, 7 and 10 mills/kwh respectively. The findings were presented at the First Geneva Conference in August 1955[55]. Table 2 illustrates how one might reach power costs of 4 mills/kwh by 1975 in a 500 MWe light water reactor. These costs were actually achieved by nuclear plants operating in 1975 (after correcting for inflation from 1955); however, the conditions were quite different from those postulated.

Table 2

How To Achieve 4 Mill Power (1955)

Item	Unit Cost	Power Cost
Reactor Investment	$60/kW	1.1
Turbogenerator Investment	$120/kW	2.2
Operating and Maintenance		0.6
Fuel		
Natural UF_6	$40/lb U	1.04
Enrichment	$25/g. U-235	0.32
UF_6 to U metal	$5/kg U	0.07
Metal Machining	$9/kg U	0.13
Zirconium Clad	$7/kg U	0.10
Reprocessing	$14/kg U	0.20
Fuel Inventory	$4%/yr	0.24
Gross Fuel Cycle Cost:		2.10
Pu-239 Credit	$25/g.	-2.00
Net Fuel Cycle cost		0.10
Total Power Cost		4.0

Basis

1% enriched LWR, 10,000 MWD/T irradiation

14% fixed charges, 90% plant factor

Table 3 gives a 1958 nuclear power cost estimate by K. Cohen of the General Electric Co.[6]. This seemed to indicate that

4 mill power was not yet in sight. But then came what might be called "the optimistic trend"[7]. During the period 1958 to 1966, both capital and fuel cycle costs in nuclear plants appeared to be dropping.

Table 3

Reactor Supplier Estimate (1958)

	mills/kwh
Capital	4.2
Fuel	3.3
Operating and Maintenance	0.6
Total:	8.1

Basis

580 MWe BWR
12% fixed charge rate, 80% plant factor

The resulting trend is shown in Table 4. It is no wonder why buyers as well as suppliers of nuclear plants were optimistic about the economic competitiveness of nuclear power in the U.S. when reported costs of nuclear plants ordered in 1965 and 1966 were in the range of $106/kwe to $143/kwe as shown in Table 5. It was believed at that

Table 4

The Optimistic Trend (1958 - 1966)

Year of Estimate	Estimator	Station Type	Station Capacity/MWe	Total Power Costs, m/kWh
1960	USAEC	LWR	300	7.6
1962	USAEC	"1966"LWR	500	5.6
1962	USAEC	"1970"LWR	500	4.8
1962	USAEC	"1975"LWR	500	4.2
1964	Jersey Central	BWR	620	3.5
1966	P. Sporn	BWR	800	4.4
1966	USAEC	"1970"LWR	800	4.0
1966	USAEC	"1975"LWR	1000	3.5

time, moreover, that with further economies of scale and plant capacity factors approaching 90%, total power costs in investor-owned nuclear plants might reach the order of 2.3 to 2.6 mills/kwh. Such estimates are now termed "unbridled optimism" but they seemed very real at the time.

Table 5

Costs of Nuclear Plants Ordered in 1965-1966

Station Name	Reactor Type	Vendor	Plant Capacity, MWe	Unit Cost, $/kW
Oyster Creek	BWR	G.E.	515	129
Dresden 2	BWR	G.E.	715	106
Brookwood	PWR	West.	420	120
Millstone Pt.	BWR	G.E.	550	118
Indian Pt. 2.	PWR	West.	870	123
Turkey Pt. 3,4	PWR	West.	690	143
Palisades	PWR	Comb.	710	141

Let us now jump ahead and look at power costs from plants which might be oredred in the United States in 1978. These are shown in Table 6. One can rightfully ask, "What happened, what went wrong?"

Table 6

Present Power Costs (Jan. 1978)
(In the United States)

Plant Type	Unit Cost, $/kW	Power Costs, m/kWh			
		Investment	Fuel	O+M	Total
Nuclear LWR	980	24.1	7.3	2.0	33.4
High Sulfur Coal-Fired	630	15.5	11.3	2.2	29.0
Low Sulfur Coal-Fired	540	13.3	15.8	2.0	31.1
High Sulfur Oil-Fired	500	12.3	15.8	1.4	29.5
Low Sulfur Oil-Fired	435	10.7	24.8	1.2	36.7

Basis

900 MWe plant capacity
14% fixed charges, 65% plant factor
Fuel prices based on average cost to U.S. utilities, Feb. 1978

Investment Costs from Reference 14.

1.1.2 Factors influencing historical cost trends

Without going into any detail at this time on the answer to this question, I believe that there are three reasons for the drastic turn-about in the nuclear power cost trend. The first of these was the unusually high inflation rate which prevailed in the United States during the last decade. During this period, inflation proceeded at a 6% per year rate compared to 1 to 2% in the previous decade and there is no sign that the high rate will not continue. The second reason was the sudden increase in world oil prices in December 1973 which brought to an end the era of low cost energy. Finally, the pressure of the environmental/antinuclear movement has led to excessive regulation of all kinds of power plants, thereby adding significantly to their complexity and cost. This point will be covered in more detail when I talk about capital cost trends.

1.2 Discussion of Fossil Fuel Costs

1.2.1 Basis for estimating costs

Any discussion of the economics of nuclear power on a world wide basis must necessarily involve the question of fuel oil prices. Unfortunately, there is no such thing as 'a fuel oil price' because there is no specific relationship between the price of crude oil at its production point and the price of fuel oil consumed by utilities. The crude oil price in the main producing areas of the world is set somewhat arbitrarily by the producers themsleves, while the price of fuel oil is set by refineries and is mainly dependent on supply and demand. One can make the assumption, however, that the price of fuel oil will fall somewhere between an upper and a lower limit imposed by the availability of substitutes. The upper limit, of course, is the price of crude oil itself, since this can and has been used as a fuel. With regard to the lower limit, a historical study of the relationship between fuel and crude oils of similar characteristics shows that the differential between them has seldom exceeded 10%[8]. Thus 90% of the price of crude oil can be considered to represent a lower limit to the price of fuel oil. To simplify matters for a general discussion such as this, one can assume that this minimum fuel oil price represents a reasonable basis for estimating power generating costs in oil-fired plants. The addition of transportation costs from the Persian Gulf would being the figure approximately back to the original posted price of crude oil.

1.2.2 World crude oil price trends

Figure 1 shows the historical trend in the posted price of light Arabian crude. The prices are given in $US/barrel where 7.2 bbl. = 1 MT crude. It is seen that the current posted price for this type of curde is $12.70/bbl. ($91/ton).

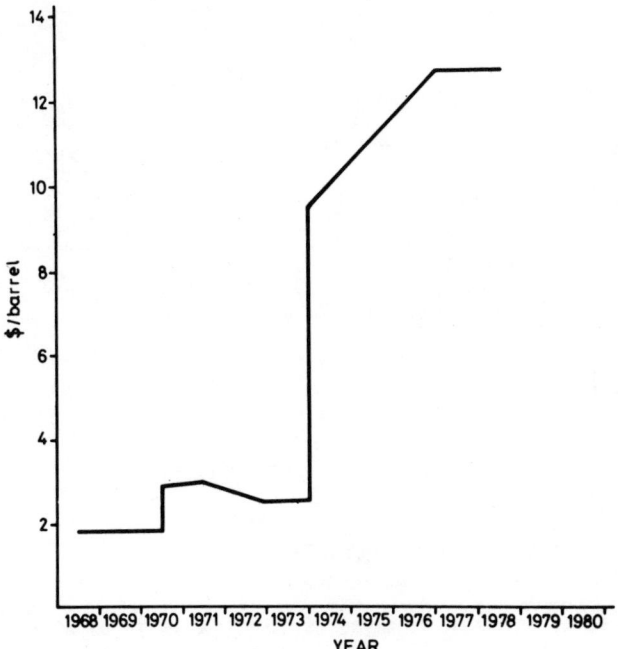

Figure 1. Crude oil price trends

1.2.3   Other fuel cost trends

The sudden rise in OPEC prices in December 1973/January 1974 and subsequent increases since that time has an influence on alternative fuels as shown in Figure 2[9]. Although these data are for the United States, similar trends occurred elsewhere in the world.

1.3   Discussion of Nuclear Fuel Cycle Costs

1.3.1   Description of fuel cycle

In contrast to fossil fuel costs, the cost of nuclear raw material (natural U_3O_8) accounts for only a portion of the total nuclear fuel cycle costs, the remainder being made up of the various operations comprising the fuel cycle. In a light water reactor these are, the conversion of U_3O_8 to UF_6, U-235 enrichment, fuel element fabrication, shipping of spent fuel and reprocessing for recovery of unburned U-235 and fissile plutonium. As an alternative, one can store the spent fuel at the reactor site, eliminating the need for shipping and reprocessing. With present costs for these two steps in the fuel cycle, the economic penalty for this latter course of action is quite small as will be seen later.

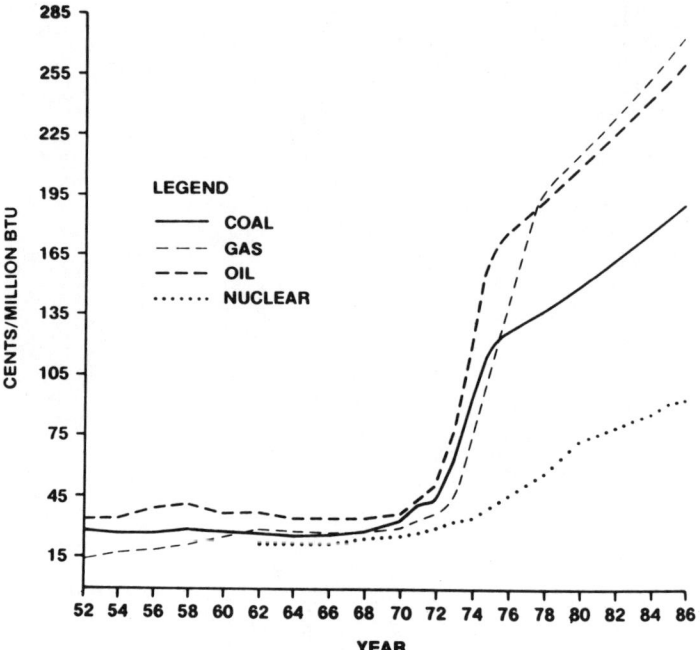

Figure 2. Average annual cost of fuels burned for electrical generation in the United States 1952-1986.

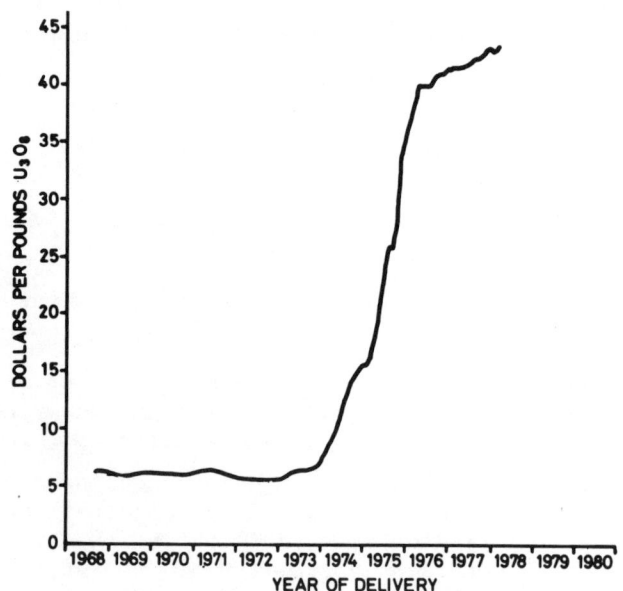

Figure 3. Historical U_3O_8 exchange value

1.3.2 Natural uranium price trend

It is axiomatic in the present energy situation that costs of alternative energy sources tend to seek their highest permissible level. This is apparent in the case of the U_3O_8 price trend shown in Figure 3 which closely paralleled that of crude oil. It is seen in this figure that U_3O_8 prices as measured by U_3O_8 exchange values[10] increased from \$7/lb U_3O_8 at the end of 1973 to \$43/lb in early 1978. Despite this 6-fold increase, however, the effect on fuel costs was considerably smaller than in the case of oil as shown in Figure 4.

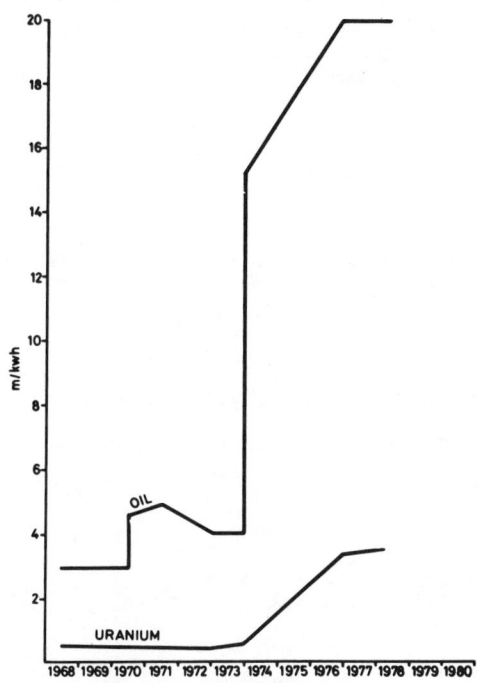

Figure 4. Influence of price trends on fuel costs

Whereas, in late 1973 the price differential between uranium and oil corresponded to a difference of 3.5 mills/kwh, by January 1978, this difference increased to about 16.5 mills/kwh. One might have expected that the result of such an increase would have improved the competitive position of nuclear power. This was not the case, however, because of other cost changes which occured.

1.3.3 Unit fuel cycle costs

Table 7 shows how costs of the various components of the nuclear fuel cycle have changed over the past five years[11]. Here the current price of yellow cake has been taken to be $40 per lb U_3O_8 rather than the march 1978 spot price of $43.25 per lb. It is seen from this table that there have also been major increases in costs for almost every other component of the fuel cycle. The credits for U-235 and fissile plutonium from spent fuel are also higher, however, with the result that the overall net fuel cycle cost has increased a factor of 4.4.

Table 7

Nuclear Fuel Cycle Cost Trends

Item		1973	1978
Natural U_3O_8,	$/lb	7	40
Conversion to UF_6,	$/kg	2.6	4
Enrichment,	$/S.W.U.	32	100
Fabrication,	$/kg	80	170
Shipping,	$/kg	4	50
Reprocessing,	$/kg	36	250
Uranium Credit,	$/kg	-35	-150
Plutonium Credit,	$/g.	-10	-20
Net Fuel Cycle Costs, m/kWh		1.68	7.30

1.3.4   Sensitivity to generating costs

The sensitivity of changes in the unit fuel cycle cost to generating costs is shown in Table 8 based on fuel cycle costs for

Table 8

Sensitivity of Unit Fuel Cycle Costs to Generating Cost

Item	Change Equal to 0.1 m/kWh
Uranium	$ 1/lb. U_3O_8
Conversion	$ 3/kg U
Enrichment	$ 4/kg. S.W.U.
Fabrication	$18/kg U
Shipping	$30/kg Heavy Metal
Reprocessing	$32/kg Heavy Metal
U Credit	- $25/kg U
Pu Credit	- $ 5/g Fissile Pu

LATEST TRENDS IN THE ECONOMICS OF NUCLEAR POWER   159

a 600 MWe PWR. If one combines the data in this table with the cost
increase shown in Table 7, it can be shown that most of the increase
in total nuclear fuel cycle costs can be related to the increase in
uranium and enrichment costs.

1.3.5   FUELCASH Computer code

The nuclear fuel cycle costs just presented were calculated
using the FUELCASH II computer code. This code which is described in
detail in Reference 12 is available in the IAEA for use of its Member
States. It was developed several years ago jointly by CFE of Mexico
and NUS of the U.S.A. especially for planning studies and nuclear
fuel evaluation. Costs are calculated for each fuel batch over the
economic life of a 626 MWe PWR and the present worth value of each
cost item is determined to find the levelized cost.

1.3.6   Illustrative example

To illustrate how FUELCASH works, a sample run was made
using the unit fuel cycle data for 1978 given in Table 7. Technical
and economic data used as input to the code are given in Table 9 and
the results summarized in Table 10. The indirect costs (fuel cycle
working capital) were based on the lead and lag times shown in Table
9 and were obtained directly from the computer output by taking the
difference between the total (levellized) cost and the direct cost
for each item.

Table 9

Basis for Fuel Cycle Cost Calculation
(Equilibrium Core)

Initial Enrichment	3.4% U-235
Final Enrichment	0.95% U-235
Enrichment Plant Tails	0.25% U-235
Burnup	33,000 MWD/MT
Heat Rate	10,242 Btu/kWh
Present Worth Factor	0.10
Natural U Lead Time[1]	2.5 years
Enriching Lead Time[1]	2.0 years
Fabrication Lead Time[1]	1.5 years
Shipping Lag Time[2]	0.75 years
Reprocessing Lag Time[2]	1.00 years

(1) Prior to charge date
(2) From discharge date

Table 10

Current Nuclear Fuel Cycle Costs (626 MWe PWR)

(from FUELCASH)

Item	m/kWh Direct	Indirect	Total
Uranium	2.85	0.99	3.84
Conversion to UF_6	0.11	0.03	0.14
Enrichment	1.83	0.61	2.44
Fabrication	0.70	0.22	0.92
Shipping	0.20	− 0.04	0.16
Reprocessing	1.00	− 0.20	0.80
U Credit	− 0.74	0.15	− 0.59
Pu Credit	− 0.53	0.12	− 0.41
Net Fuel Cycle	5.42	1.88	7.30

1.3.7   Fuel costs versus plant size

Unfortunately, the FUELCASH code is based on a 626 MWe PWR and does not compute costs for other sizes or types of reactors. To find fuel cycle costs as a function of plant sizes, hand calculations were carried out based on the assumption that the only change would be the slightly lower initial fuel enrichment for larger size reactor cores. The results, shown in Table 11, indicate that fuel cycle costs are not very sensitive to plant size.

Table 11

Fuel Cycle Costs Versus Nuclear Plant Size

Nuclear Plant Size, MWe	m/kWh Direct	Indirect	Total
600	5.43	1.88	7.31
900	5.30	1.83	7.13
1000	5.25	1.81	7.06
1200	5.17	1.79	6.96

1.3.8   Fuel costs for alternative reactor types

Turning now to fuel cycle costs in other types of reactor, L.F.C. Reichle of Ebasco Services in the USA gave a paper at the

Salzburg Conference[13] which compared fuel costs in current reactor types (PWR, HTGR, CANDU, LWBR-Prebreeder) with those for breeder reactors. In making this comparison, Reichle used a set of fuel cycle cost assumptions postulated for 1980 conditions. These shown in Table 12 are seen to be quite different than the IAEA cost assumptions given in Tables 7 and 9. The results for the PWR, however, appear to be comparable as shown in Table 13. This table indicates that nuclear fuel cycle costs for non-breeder all lie in the range of 6.5 - 7.5 mills/kwh, whereas projected costs for most of the breeders are in the range of 2.5 - 4.0 mills/kwh. The LWBR is an exception with costs of the order of 14 mills/kwh.

Table 12
Nuclear Fuel Cycle Cost Assumptions
Basic Assumptions

Plant Size	- 1,000 MWe	Net Cost of UF_6	- \$136/Kg U
Capacity Factor	- 75%	Enrichment	- \$120/SWU
Fixed Charges	- 15%	Tails Assay	- 0.3% U-235
Date of Prices	- 1980	Net Cost U-235	- \$59/gm
Yellowcake	- \$50/lb U_3O_8	Equiv. Value U-233	- \$68/gm
Conversion to UF_6	- \$5.75/Kg U; 0.5% process loss	Indifference Value Pu	- \$42/gm
D_2O	- \$65/lb	Thorium	- \$25/Kg

Fabrication and Recovery Cost Assumptions

	Fabrication	Recovery (Trans. and Reprocessing)
LWR	\$115/Kg U	\$220/Kg U
LMFBR or GCFR	\$650/Kg HM	\$340/Kg HM
LMFBR (Adv. Des.)	\$520/Kg HM	-
LWBR	\$130/Kg U	\$220/Kg U
HTGR	\$3280/Element	\$2985/Element

The estimated duel cycle costs for the CANDU reactor in Table 13 appear to be somewhat high because of the assumed high fixed charge rate (15%) and the fact that allowance was not made for a higher capacity factor possible with on-line refuelling. Table 14 gives my own estimate of CANDU fuel costs of about 6 mills/kwh. It should be pointed out that this figure is considerably above the 1.4 mills/kwh reported for the CANDU-Pickering station in 1975; however, the low costs for Pickering reflect Canadian, not world wide conditions.

Table 13
Nuclear Fuel Cycle Costs for Alternative Reactor Types

Reactor Types	MWD kg	Conv. Ratio	Fissile Inventory kg/MWe	Fuel Cycle Costs, m/kWh		
				Direct	Indirect	Total
PWR	34	0.55	–	5.8	1.8	7.6
HTGR	100	0.65	2.8	2.6	3.8	6.4
CANDU	7.5	0.71	–	3.5	3.9	7.5
LWBR (Pre-Breeder)	51	0.52	–	3.9	2.7	6.6
LWBR (Breeder)	16	1.01	4.3	7.1	6.7	13.8
LMFBR (Early Oxide)	110	1.15	4.0	0.2	3.7	3.9
LMFBR (Adv. Oxide)	110	1.25	3.1	-0.6	2.9	2.3
GCFR	100	1.45	5.0	-1.3	4.4	3.1
MSR (Pu Converter)	–	0.86	2.1	1.1	1.5	2.6
MSBR (U-233)	–	1.06	1.4	0.6	2.6	3.2

Table 14
Fuel Cycle Cost in CANDU Reactor (600 MW)

Item	Basis	Fuel Cost, m/kWh	
		Direct	Indirect[1]
Natural Uranium	$40/lb U_3O_8	2.0	0.7
Fabrication	$50/kg U	1.0	0.2
Spent Fuel Storage	$10/kg U	–	0.2
Heavy Water	$140/kg D_2O	0.2[2]	1.5
		3.4 6.0	2.6

(1) 10% interest, 80% plant factor
(2) Annual Make Up.

LATEST TRENDS IN THE ECONOMICS OF NUCLEAR POWER

1.4 Discussion of Capital Costs

For many years now, the Agency has been keeping abreast of the trend in capital costs of nuclear and conventional (coal-fired) plants in the United States through the assistance of United Engineers and Constructors Inc. of Philadelphia. Since 1967 this company has carried out a series of comprehensive studies on capital costs, first for the USAEC and more recently for the Nuclear Regulatory Commission[14]. These studies serve as a good basis for reviewing the capital cost trends which have taken place over the past decade.

1.4.1 Capital cost trends[15]

Figure 5 compares trends in cost estimates with cost actually experienced. Although these latter costs are quite scattered,

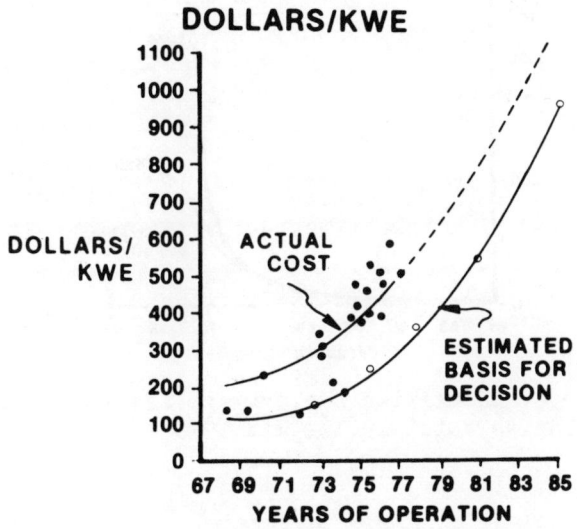

Figure 5. Trends of nuclear power plant costs

they do indicate that there has been about $200/kwe differential between estimated costs and actual costs. This differential used to seem to be quite significant; however, with both estimated and projected actual costs rising to the range of $1000-1200 per kwe it no longer seems important. What is important is the fact that unit capital costs of LWR plants appear to have been multiplied by a factor of six over a span of about eight years. The principal reasons for this increase in order of importance appear to be

    i) Regulatory requirements because of safety and environmental considerations;

ii) Inflation and interest during construction;

iii) Commercial effects.

(i) Regulatory Impact

Starting first with the regulatory impact, the introduction of sweeping new regulations, criteria, codes and guides has been exponential in nature as shown in Figure 6. Here it is seen that the

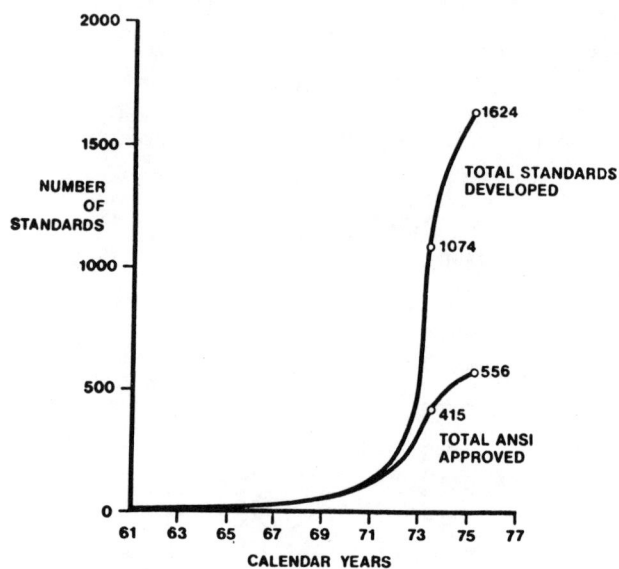

Figure 6. Identified nuclear application standards: Cummulative total 1961 - mid 1975

number of standards applicable to the design and construction of a nuclear plant in the USA grew from about 100 in 1971 to more than 1600 in 1975. In this same period, the number of people in the U.S. working on standards increased from about 1800 to about 8000. The effect of this growth and application of nuclear related standards on nuclear plant costs is shown by the shaded area in Figure 7. It is seen that more than 50% of the cost of a nuclear plant coming on line in 1985 can be traced to the impact of regulations.

To illustrate how safety related standards can influence design, I would like to show two figures provided by Mr. Crowley of United Engineers[15]. The first (Figure 8) shows the design for installing an ordinary light. The second (Figure 9) shows how one must design to provide light under any conceivable circumstances. I believe the figure speaks for itself.

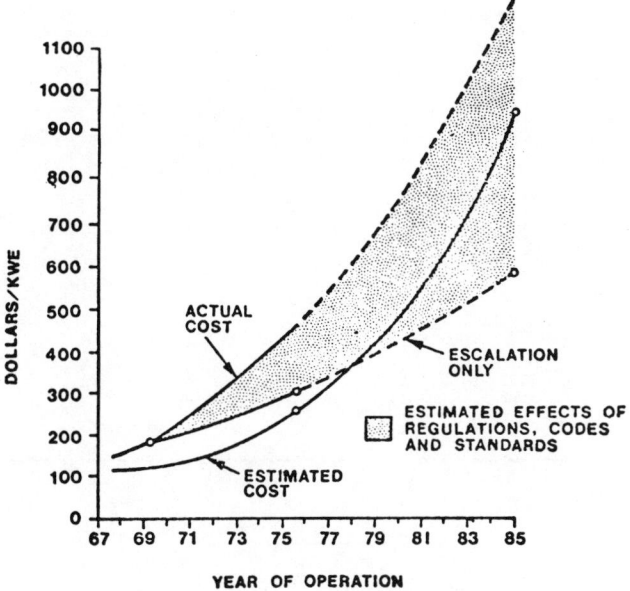

Figure 7. Effect of regulations, codes and standards on power plant costs

Fortunately, not all portions of a nuclear plant require such extreme measures. However, safety and environmental protection measures have led to considerable increases in the scope of supply as shown in Table 15. Here it is seen that the amounts of some of

Table 15

Trends in Scope of Supply and Capital Costs (1971-1976)

(United States Conditions)

	Pressurized Water Reactor		Low Sulfur Coal-Fired	
	Jan. 1971	July 1976	Jan. 1971	July 1976
Plant Size, MWe	1031	1139	1000	1243
Concrete, kg/kW	154	250	134	115
Rebar, kg/kW	9.7	16.7	3.4	4.3
Structural Steel, kg/kW	3.9	8.5	14.6	17.2
Craft Labor, hrs/kW	6.0	9.5	5.6	5.8
Capital Cost, $/kW[1]	205	499	174	324

(1) Base construction cost excluding contingencies, spare parts, owners costs, fuel, escalation and interest during construction.

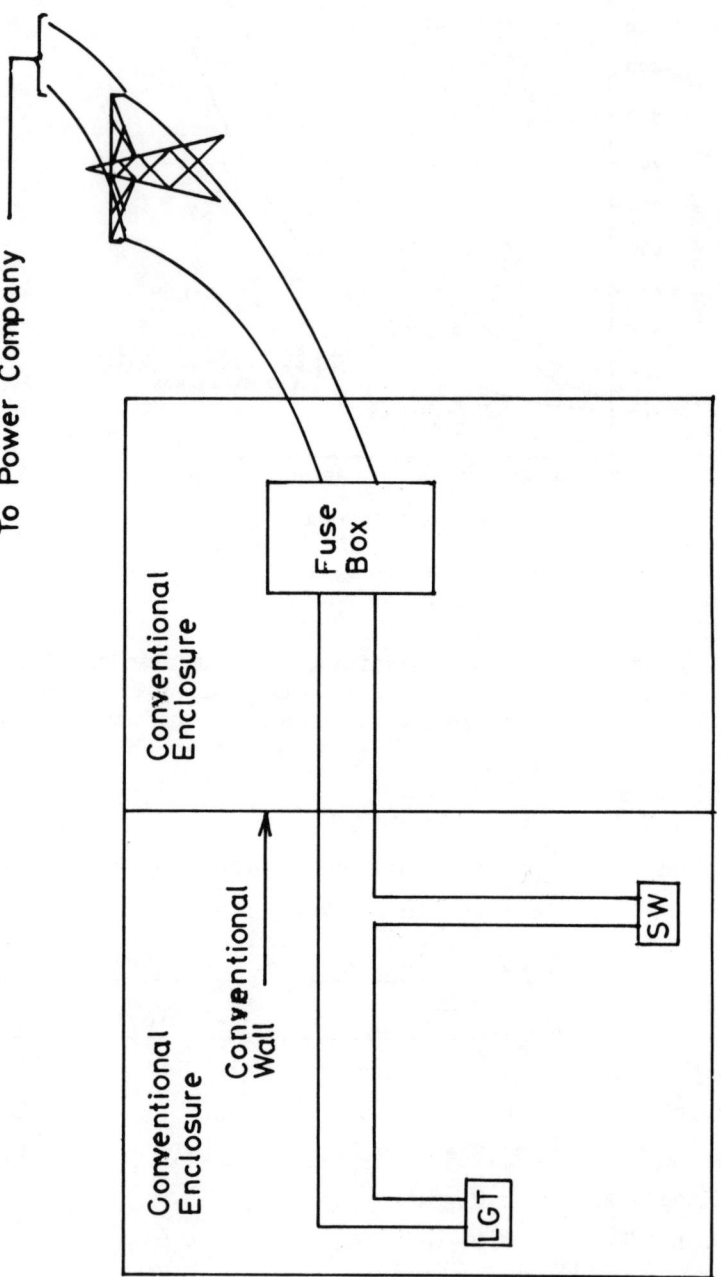

Figure 8. Specification for a conventional system.

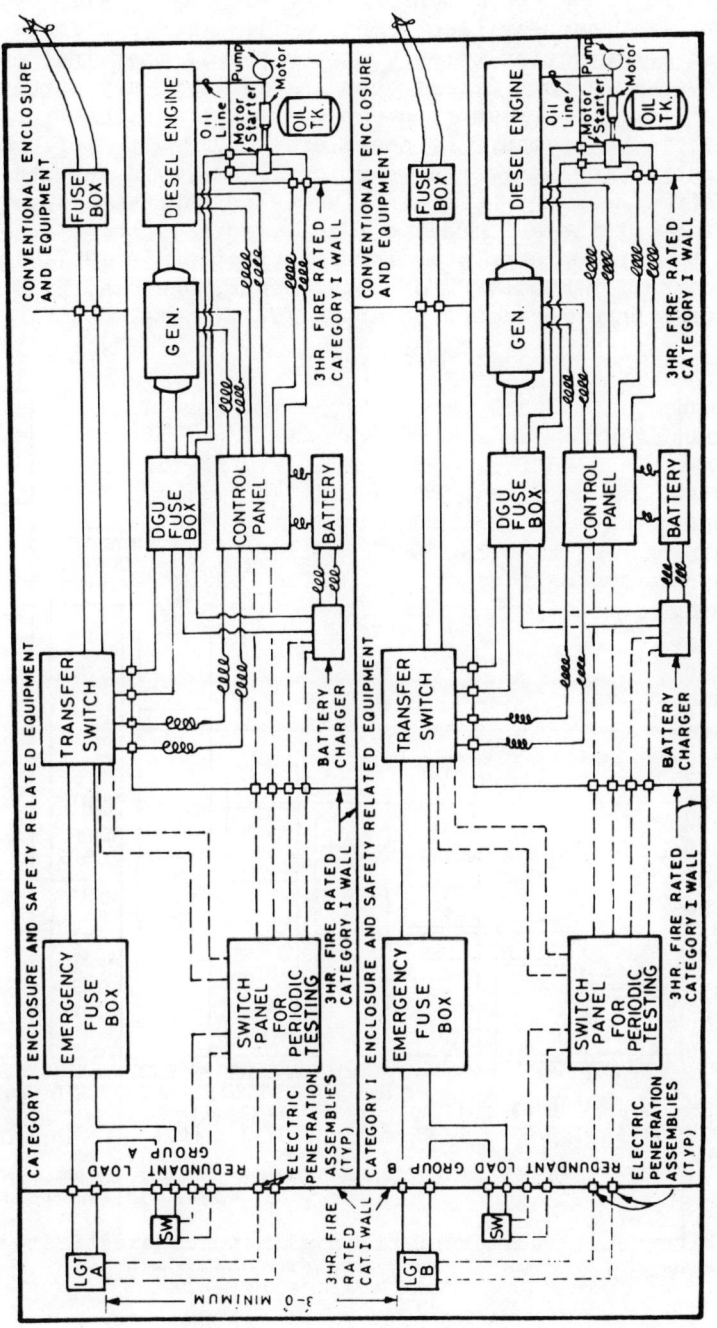

Figure 9. Specification for safety related system.

the basic commodities required per installed kw practically doubled since 1971. The amount of man hours of craft labour also increased significantly for nuclear plant construction. These effects have led to increased construction times with the result that indirect costs increased even more than direct costs. More temporary structures are required to store, label and protect equipment and construction materials. About twice as many engineers are required to perform engineering and construction management services. Quality assurance and quality control measures have also increased substantially. The overall effect of these increased requirements is shown in Figure 10. It is seen that for plants starting construction in 1967, indirect costs accounted for about 30% of total costs, whereas, by mid 1976 indirect costs have increased to about 70% of total costs.

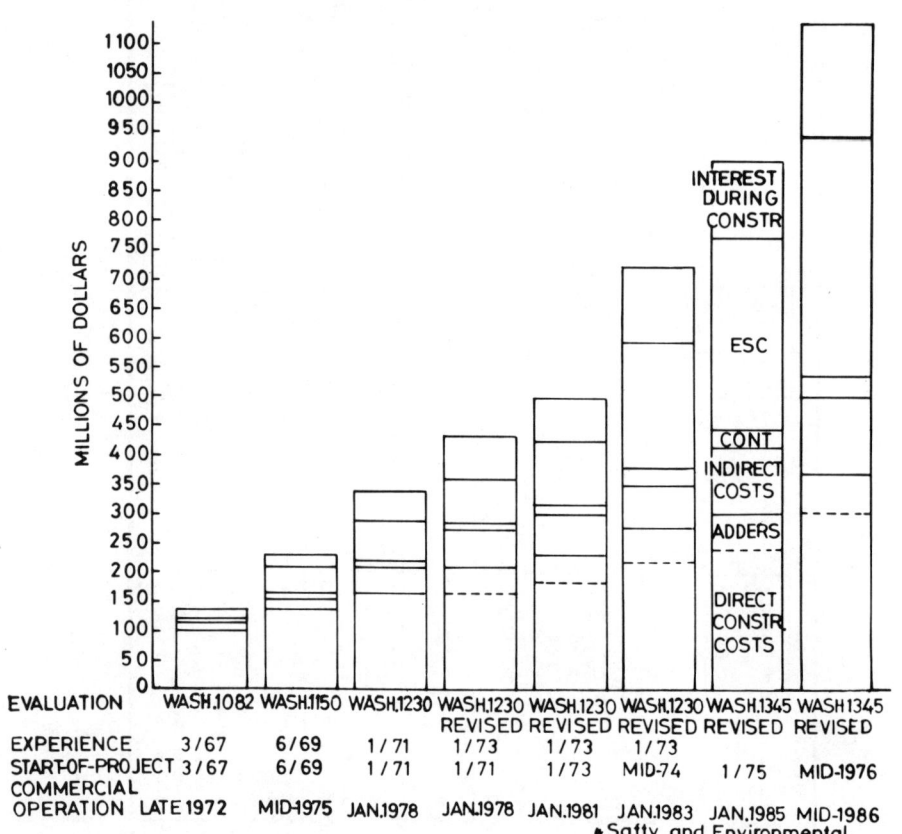

Figure 10. Nuclear plant investment cost estimates: single 1000 MWe water moderated reactor plant

Figure 11 shows that a similar trend existed for coal fired plants in the United States.

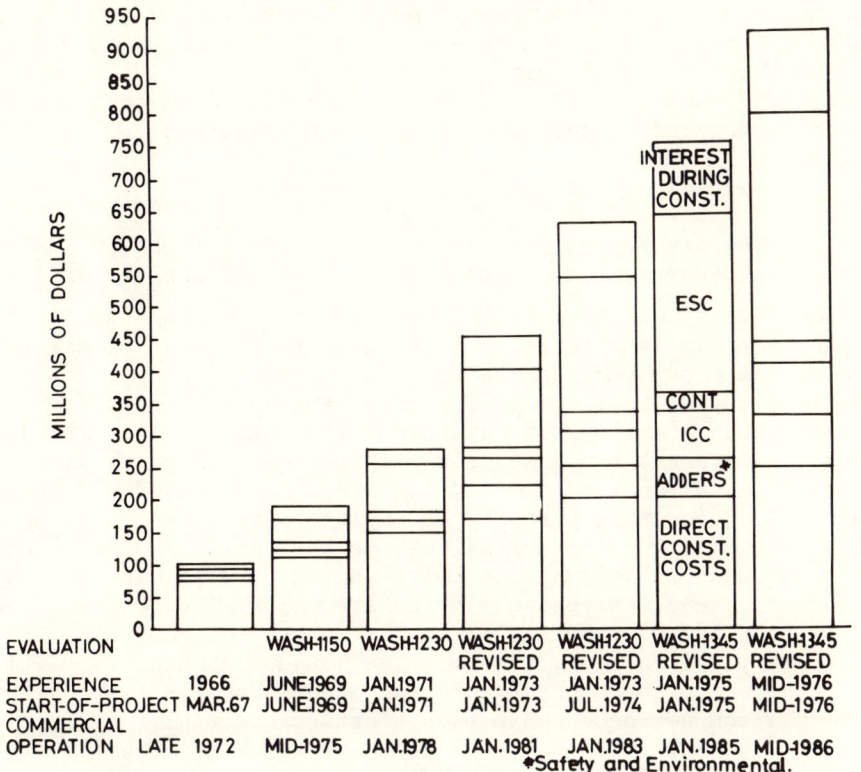

Figure 11. Plant investment cost estimates: single 1000 MWe coal fired plant

(ii) Inflation and interest during construction

Annual inflation rates in the United States and other industrialized countries increased considerably since the early years of nuclear power. They have also led to higher nominal interest rates. Together with extended design and construction periods, this means that both the absolute and relative importance of inflation and interest during construction have grown enormously. Figures 10 and 11 both show that these two items now account for about 50% of total plant costs.

(iii) Commercial effects

Before 1970, reactor manufacturers and architect engineers were willing to undergo substantial commercial risks to enter what appeared to be a very promising new market. A number of low priced contracts were let which reportedly led to substantial financial losses for the vendors. After the quadrupling of oil prices in late

1973, these vendors found themselves in a position to price equipment at a level suitable to cover all of their usual commercial risks. It is unlikely that this situation will reverse itself in spite of the recent downward revision of nuclear programmes of most countries and the subsequent reduction of new nuclear plant order.

1.4.2   Methodology of capital cost comparisons

The development of capital cost data for the evaluation of alternative power system expansion plans is considerably different from the development of costs for a specific project[16]. For planning purposes, capital costs of both nulcear and conventional plants are based on a hypothetical set of conditions. This approach has been found useful at the IAEA for long range economic studies of a country's electrical system expansion. In these studies, the relative merits of different types and sizes of nuclear and conventional power plants for an expanding power system are compared over a given planning period. For this purpose the capital investment costs of electricity generating units are defined as the total of direct and indirect costs of the complete power unit including owners costs, contingencies and interest during construction. The costs are expressed in constant U.S. dollars of a reference year. Costs of the main transformer, switch yard, transmission lines and other facilities outside of the plant boundary are excluded as well as taxes, duties and escalation during construction. Costs of initial fuel loading and other special materials such as heavy water are treated separately.

1.4.3   The ORCOST computer code

In order to facilitate the estimation of capital costs for a large number of alternative generating units, the IAEA uses a computer code called ORCOST developed jointly by United Engineers and the Oak Ridge National Laboratory[17]. This computer program is very easy to use and is available from the IAEA for release to Member States. Costs of PWR, CANDU, High and Low Sulfur Coal and Oil-Fired Plants and Gas-Fired Plants can be estimated for a wide range of economic parameters and other conditions. These include, year of construction start, rate of interest during construction, rate of escalation before and during construction. length of work week, equipment, materials and labour cost indices relative to hypothetical site (Middletown, USA), costs and type of heat removal system (once through cooling or cooling towers). ORCOST also permits one to extrapolate costs from the hypothetical site to a developing country. This point will be discussed in more detail in Section 2.

1.5   Power Generation Costs in the U.S.

Figure 12 shows an estimate by United Engineers and Constructors Inc. of power generation costs in 1985 for various types of

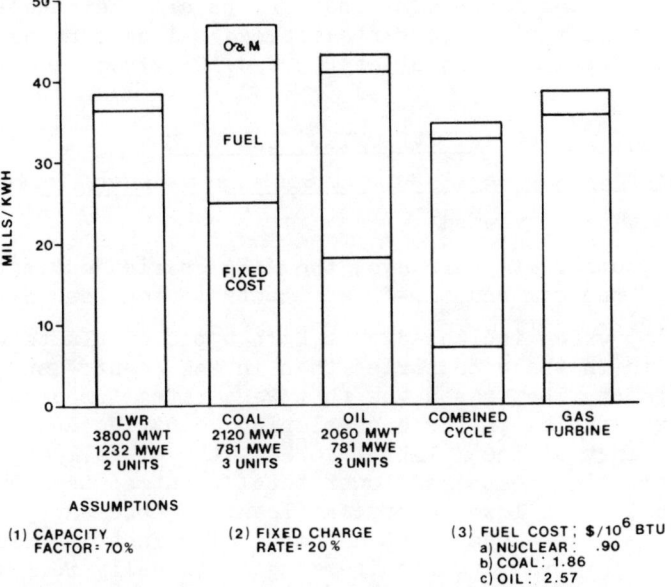

Figure 12. Total generation cost of power plants: rated output - mechanical wet towers - cost, January 1985

generating units[15]. It is seen that under the assumed conditions, a 2 × 1232 MWe LWR station competes favourably with 3 × 781 MWe coal or oil-fired station. Combined cycle and gas turbine plants appear to have generating costs equal to or slightly lower than the nuclear station.

Figure 13 shows the past and projected trends in power generating costs of nuclear and coal-fired plants in the U.S. also

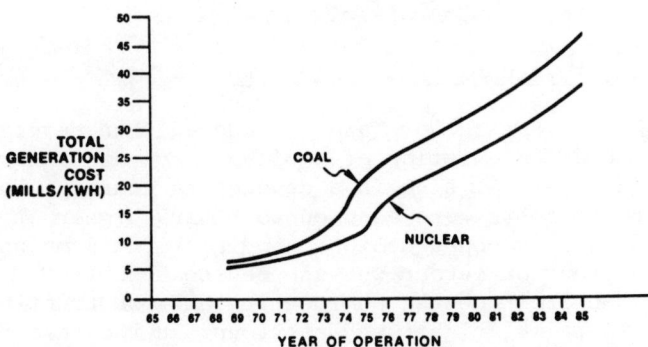

Figure 13. Comparative projections of power cost

estimated by United Engineers. Indications are that nuclear will continue to hold a slight advantage over coal as long as the fuel price differential remains at about 90 ¢/10^6 Btu.

2. COST TRENDS IN DEVELOPING COUNTRIES

2.1 Extrapolation of Capital Cost Data to Developing Countries

2.1.1 Methodology used

A number of years ago, the IAEA carried out estimates of the costs of nuclear and oil-fired plants in fourteen developing countries[18] which indicated that both types of plants would cost less to build in these countries than in the country supplying the plant equipment. The reason was that wage rates for construction labour were very low in these developing countries and despite the lower efficiency of local labour forces, the overall result was lower construction costs and lower total plant costs. Experience show however, that these lower costs, at least for nuclear plants, have never been realized because of a number of offsetting factors. These include, special training requirements for locally recruited workers, a lack of industrial infrastructure, high costs of resident foreign supervisory and technical staff, wide cost variations of locally supplied construction materials and cost adders because of difficult site conditions. In addition, nuclear plant suppliers usually includ an extra contingency factor in their bid in order to cover the increased uncertainty of doing business in a foreign country.

Thus in most developing countries the capital cost of nuclear power plants of a given size are at least as high and, in several cases, higher than those in industrial nations. In addition, there is another element which affects the competitive status of nuclear versus fossil fired stations. Due to less stringent regulations, fossil fired plants in developing countries are often planned with much less sophisticated environmental control systems. As a result, their capital investment costs will be smaller than those in industrialized countries. Similar cost reduction are not possible for nuclear units since these are designed to meet the licensing requirements of the country of origin.

2.1.2 Middletown USA Versus Intertown

All of these factors are considered when developing input data for an ORCOST calculation of capital costs at 'Intertown', a hypothetical quasi-ideal site in a developing country. For the nucle plants, equipment costs were assumed to be 10% higher at 'Intertown' than at 'Middletown', construction materials, the same costs at both sites and construction labour was assumed to be 60% of Middletown labour costs. Indirect costs (engineering and construction managemen were assumed to be 14% higher at Intertown. In the case of fossil fired units, however, costs of equipment, materials and engineering services were assumed to be the same at both sites while construction labour costs at Intertown were taken as 40% of Middletown costs.

Using these assumptions as a basis, costs of nuclear and oil-fired plants at Middletown, USA and Intertown were estimated by ORCOST as shown in Table 16. Here it is seen that in the case of the PWR the lower assumed construction labour costs tend to offset the higher assumed costs of equipment and engineering services with the result that unit capital costs of a plant located at Intertown are only slightly higher than at Middletown. The lower labour costs at Intertown assumed for oil-fired plants, on the other hand, result in unit capital costs about 80% of those for a similar plant built in the USA.

Table 16

Extrapolation of costs to developing countries

Plant Type	Capacity MWe (net)	Unit Cost, $/kW[1] Middletown, USA	Intertown
PWR	600	1260	1285
	900	980	1000
	1200	820	845
Low S. Oil	600	485	395
	900	435	355
High S. Oil	600	550	450
	900	500	405

(1) Excluding Fuel and Escalation during Construction

2.2   Sensitivity of Capital Costs to Plant Size

Savings of scale are much more significant for nuclear than for fossil-fired stations. The cost of a 1200 MWe PWR, for example, is only about 30% higher than that for a 600 MWe PWR. For oil-fired stations, on the other hand, this same percentage increase is realized when the plant size goes only from 600 MWe to 900 MWe. Stated somewhat differently, the scaling exponent for nuclear plants of 600 MWe or larger is of the order of 0.4 compared to 0.75 for oil-fired plants. Hence, the competitive position of nuclear versus oil-fired plants improves with size and there is a strong economic incentive to choose the largest nuclear plant which can be incorporated in a given electrical network. This has led the manufacturers of nuclear plants in industrial countries to concentrate on the maximum sizes acceptable to utilities in their countries. As a result, subject to a few exceptions, 600 MWe is the smallest unit size of proven nuclear power plant currently available commercially. Unfortunately, the trend in nuclear economics has been such that it is difficult at the present time to make a good economic case for choosing even a 600 MWe unit.

2.3   Power Cost Trends in Developing Countries

The trends in estimated power costs in a developing country from January 1974 to January 1978 are shown in Table 17. It is seen

Table 17

Power Costs from Nuclear and Oil-Fired Plants in Developing Countries, m/kWh

	Light Water Reactor		Low Sulfur Oil-Fired	
Plant Capacity, MWe	600	900	600	900
Jan. 1974 Basis				
Capital at 12% FCR, 65% PF	9.0	7.7	4.6	4.1
Fuel	3.0	2.8	14.1	14.2
Operating & Maintenance	0.7	0.5	0.4	0.3
Total	12.7	11.0	19.1	18.6
Jan. 1978 Basis				
Capital at 12% FCR, 65% PF	27.0	21.1	8.3	7.5
Fuel	7.3	7.1	20.8	20.8
Operating & Maintenance	2.2	2.0	1.4	1.2
Total	36.5	30.2	30.5	29.5
Reduction at 10% FCR, 70% PF	-6.1	-4.8	-1.9	-1.7
Net Power Costs	30.4	25.4	28.6	27.8

in this table that in January 1974, 600 MWe nuclear plant generating costs were estimated to be only two-thirds of the power cost from an oil-fired plant of the same size. By January 1978, however, because

LATEST TRENDS IN THE ECONOMICS OF NUCLEAR POWER 175

of higher capital, fuel and operating costs, the economic advantage of 600 MWe nuclear LWR's has completely disappeared. Even 900 MWe nuclear plants, moreover, appear to have a difficult time competing with oil-fired plants. As shown in Table 17, with somewhat lower interest or fixed charge rates and higher plant factors, the competitive position of nuclear power is slightly improved.

2.4 Power Costs Versus Plant Type

A recent study of the cost of a 600 MWe CANDU pressurized heavy water reactor plant (under the sponsorship of the IAEA) was carried out by the CANATOM engineering firm with the assistance of Atomic Energy of Canada Ltd. and United Engineers and Constructors[19] The results of this study indicate that under Canadian licensing conditions, CANDU capital costs, excluding heavy water, might be about 10% lower than costs for a PWR of the same size. With the more stringent USA licensing conditions, on the other hand, the cost of a CANDU plant are estimated to be about 13% higher than a PWR. What this means, however, is that a country interested in buying a CANDU nuclear plant designed to Canadian specifications might find it cheaper than a light water nuclear plant. Such a plant, moreover, might have slightly lower fuel cycle costs than a PWR as shown previously in Table 14. In addition, a higher availability factor possible with on-line refuelling would result in reducing the capital cost component of generating costs. As shown in Table 18 these combined effects bring estimated costs of a 600 MWe CANDU plant below those for an oil-fired plant of the same capacity.

Table 18

Power Costs Versus Plant Type (600 MWe)

(Developing Country, Jan. 1978 Cost Data)

	PWR	CANDU	Oil-Fired
Capital Cost, $/kW	1284	1109	395
Power Costs, m/kWh			
Capital	20.9[1]	15.8[2]	6.4[1]
Fuel	7.3	6.0	20.8
O + M	2.2	2.2	1.4
Total	30.4	24.0	28.6

(1) At 10% interest rate, 70% plant factor
(2) At 10% interest rate, 80% plant factor (higher because of on-line refuelling)

2.4 The System Analysis Approach to Power Cost Comparison

2.4.1 Description of methodology

The traditional method of comparing the economics of nuclear and conventional plants has been to calculate generating costs for each type of plant using suitable capital, operating, and fuel cost data along with an assumed plant factor and cost of money. This approach was adequate until recent years because the choice of generating equipment available to an electric utility was fairly limited. In many cases, one had only to compare nuclear and conventional-fired units of a given size which was selected intuitively depending on the size of the system. Hand calculations were usually adequate for such an approach.

This method of power system planning now appears to be inadequate for a number of reasons. First, the choice of generating units is much wider and includes gas turbines, hydro plants, pumped-storage units, various types of fossil-fired units and even various types of nuclear plants. Second, because of the very high investments involved, the choice of a optimum size of unit becomes quite important. Third, the position of a plant in the loading order influences its capacity factor. Thus different plant types have different lifetime capacity factors and comparisons of alternatives using the same capacity factor may not be valid. Finally, because of the high costs of fossil fuels, particularly imported gas and oil, it is necessary from an economic standpoint to minimize total costs of the system taking into consideration not only existing plants but also plants which might be added over the longer term.

2.4.2 The WASP code[20]

In order to take all of these factors into consideration when evaluating the economics of nuclear power, the IAEA now uses a computer programme called WASP (Wien Automatic System Planning). This programme has been distributed to 32 Member States (including Pakistan) and 5 international organizations. It is also being used by more than 50 utilities and other groups in the United States for power system expansion planning studies.

The input data model for WASP is shown diagramatically in Figure 14. It is evident from this diagram that an extensive amount of information concerning the system being considered is needed in order to use WASP. This includes a forecast of future load demand, a thorough knowledge of the technical characteristics of each power plant in an electrical system that is in operation, under construction or committed, an estimate of the technical and economics of plants which are to be considered as expansion alternatives and a knowledge of system reliability criteria and operating practices. All of these data are fed into one or more of the first six modules shown in Figure 14. The computer then finds the 'optimum' power

LATEST TRENDS IN THE ECONOMICS OF NUCLEAR POWER

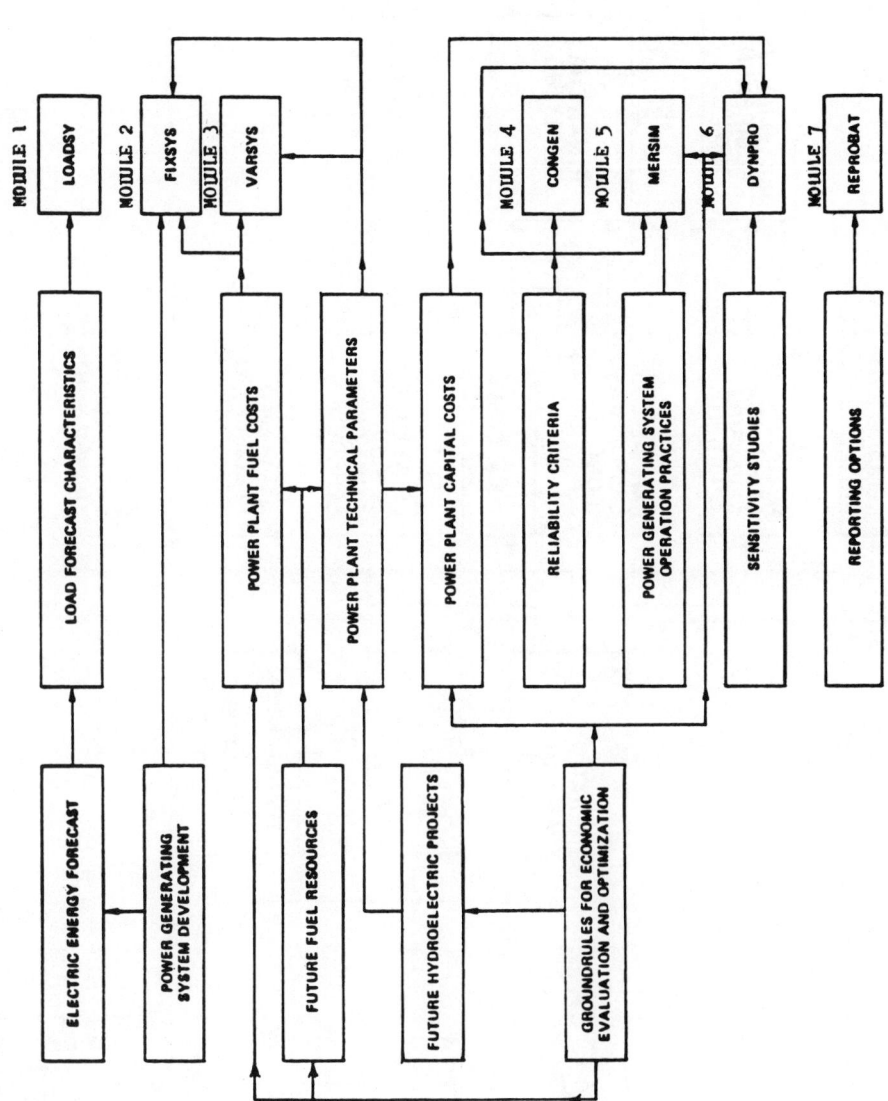

Figure 14. WASP input Data Model.

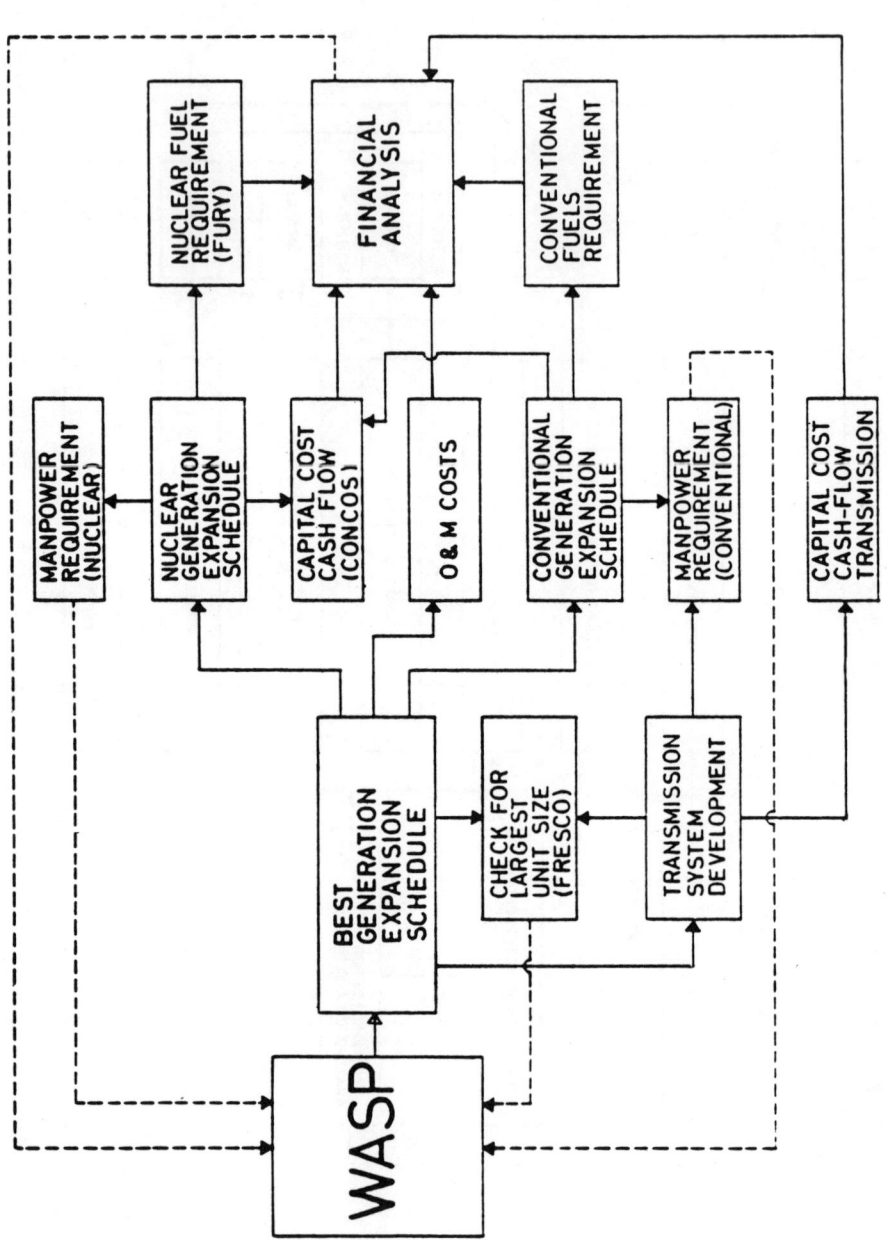

Figure 15. Analysis of WASP output.

system expansion plan within constraints established by the user using the dynamic programme approach. By optimum is meant that the discounted cash flow (capital and operating expenses) is minimized over a given period with provision made to reduce effects of uncertainties beyond that period.

The final module, REPROBAT, develops cash flow information and prints a report of the 'optimum' solution.

The user must then analyze the results of each WASP run as shown in Figure 15 and determine whether the economic optimum expansion schedule is also a feasible schedule from the standpoint of the system characteristics and the country's economy.

2.4.3 Recent nuclear power planning studies

In order to make the WASP package available to as many developing Member States as possible, the IAEA, in cooperation with the U.S. Government, carries out training courses in electric system expansion planning using WASP. The first such course was held at Argonne, Illinois during the period 14 February - 14 April 1978. Because of the lack of space, only five countries (Chile, Egypt, Indonesia, Peru and Yugoslavia) could be accommodated at this first course; however, the course will be repeated in January 1979 and will be larger. Each of the five countries sent to the course teams of from two to four engineers who conducted power system expansion studies while learning to use WASP programme. The results of these studies all tended to confirm that 600 MWe light water nuclear plants are not competitive with plants based on imported oil.

2.5 Outlook for Small and Medium Size Nuclear Plants

With the hope of expanding the potential market for nuclear power in developing countries, the IAEA has focused attention on small and medium size power reactors for the past decade. During June 24-28, 1968, exactly 10 years ago, an IAEA panel meeting on the the subject was held in Vienna[21] and in October 1970 a larger symposium, sponsored by the IAEA was held in Oslo, Norway[22]. The conclusions reached by the participants of both of these meetings were that although there appeared to be a potential demand for small nuclear plants no firm market existed for such plants. Nuclear plant suppliers, moreover, were even then concentrating on plants in the 800-1000 MWe range and were simply not interested in plants of 500 MWe or smaller.

2.5.1 The 1973 market survey of nuclear power for developing countries

The situation on small and medium reactors remained static until October 1971 when the Agency convened a Working Group on Nuclear

Power Plants of Interest to Developing Countries. This group recommended that the IAEA should carry out a survey of the market for nuclear plants in developing countries. This survey was initiated in November 1971 and completed in August 1973 (see Reference 19). One of the objectives of the survey was to identify the specific market for small and medium power reactors. Fourteen countries participated in the survey as shown in Table 19 which also gives the results. As seen in this table, the nuclear power market in the participating countries comprises 58% of the total estimated electric plant market.

Table 19

Results of IAEA 1973 Market Survey of Nuclear Plants for Developing Countries

	Total Electric Plant Market(2) (GWe)	Nuclear Plant Market (GWe)	Percent Nuclear
Market Survey Countries(1)	106	62	58
Other Developing Countries	236	118	50
Total	342	180	53

(1) Argentina, Bangladesh, Chile, Egypt, Greece, Jamaica, Korea, Mexico, Pakistan, Philippines, Singapore, Thailand, Turkey, Yugoslavia.
(2) During 1980-1989 period.

An additional benefit derived from the IAEA Market Survey was that the methodologies developed for such things as load forecasting, capital and fuel cost estimation, reserve margin requirements and permissible sizes of generating units could be used to estimate the nuclear power market in countries not covered by the survey. Such an estimate was carried out by the Agency during the period August to November 1973, the results of which are also shown in Table 19. Here it is seen that 50% of the total electric market in countries not covered by the Survey might be nuclear. These results, of course, were based on conditions which prevailed in early 1973 before the sudden rise in oil prices in December 1973 and the oil crisis which followed.

2.5.2   The 1974 updated survey

The rise in world oil prices quickly made the results of the 1973 Market Survey obsolete. For this reason, the results of the Survey and its extension to other developing countries were adjusted

LATEST TRENDS IN THE ECONOMICS OF NUCLEAR POWER 181

to the economic conditions which prevailed in early 1974[23]. The results of this 1974 updating, shown in Table 20, indicate that the potential nuclear share of the 1981-1990 total electric market (advanced one year from the 1973 Survey) increased to 61% and during the period 1991-2000 that share could be as much as 82%.

Table 20

Jan. 1974 Updated Market Survey of Nuclear Plants for Developing Countries, GWe

Period	Fossil-Fired	Nuclear	Hydro	Total	Percent Nuclear
1981-1990	47	221	93	361	61
1991-2000	38	500	71	609	82
Total in year 2000	307	746	258	1310	57

2.5.3 The influence of the oil price rise on the market for small nuclear plants

The oil price rise not only increased the total potential market but it also, as was believed in 1974, considerably improved the potential outlook for nuclear plants of 400 MWe or less as shown in Table 21. Here it is seen that the potential market for small nuclear plants went up approximately 4-fold. In fact, plants as small as 100 MWe appeared to be competitive as of mid-1974[24].

Table 21

Market for Nuclear Plants 400 MWe or Less

	1973 Survey		1974 Update	
	GWe	Number of units	GWe	Number of units
Market Survey Countries	3.5	10	9.7	34
Other Developing Countries	6.5	18	28.3	106
Total	10.0	28	38.0	140

2.5.4 The 1975 comparison of small plants

Since the capital, fuel and operating costs of nuclear plants continued to increase during 1974 and 1975, the economic competitiveness of small and medium size nuclear plants was re-evaluated in 1975[25]. The results are shown in Table 22. Here a nuclear plant as small as 100 MWe no longer appeared to compete with oil-fired plants, however, the outlook for nuclear plants in the 300-600 MWe size range still appeared promising.

Table 22

Competitiveness of Small and Medium Size Nuclear Plants (1975)

	Light Water Reactor			Oil-Fired		
Plant Size, MWe	100	300	600	100	300	600
Capital Costs, $/kW	1120	725	530	424	302	240
Power Costs, m/kWh						
Capital	23.6	15.3	11.2	8.9	6.4	5.1
Fuel	3.8	3.7	3.4	17.9	17.6	17.5
O + M	5.0	2.1	1.2	4.1	1.8	1.0
Total	32.4	21.1	15.8	30.9	25.8	23.6

Basis: 12% interest rate, 65% plant factor
constant 1975 US $ (no escalation)

2.5.5   The 1978 situation on small reactors

The cost data which have just been presented, when viewed in the light of the current nuclear power cost data (see Table 17) again provide an example of premature optimism. Not only are small nuclear plants uneconomical but also 600 MWe units are too costly[26] On this basis, the outlook for small and medium size nuclear plants is not very encouraging.

There are two exceptions to this dismal picture, both lying outside of the United States and Western Europe. One is the USSR which is successfully marketing 420 MWe (net) PWR's in CMEA countries[27]. The other is India who is building economical heavy water reactors in the 200 MWe size range. Neither of these countries, however, is offerring these plants on a world wide basis.

As for the future, the Babcock and Wilcox Co. of the U.S. is still working on the design of a 400 MWe pressurized water reactor known as the CNSG (Consolidated Nuclear Steam Generator) which they believe can be economically competitive with larger units. Their claim is that two 400 MWe units can be built for the same price as a single 800 MWe unit. Because of reduced site construction requirements, moreover, construction times should be lower than for larger nuclear plants. Although numerous U.S. utilities are expressing interest in the CNSG no one has placed an order yet. This again may be another example of premature optimism.

2.6   Long Term Economics of Nuclear Power

2.6.1   Trends in uranium costs

There are two opposing schools of thought concerning the long term trend in uranium costs. The first, as typified by the Ford Foundation sponsored study of nuclear energy policy[28] speculates that more intensive and extensive exploration should produce enough uranium both in the United States and abroad, to supply the U.S. and foreign requirements of light water reactors (LWR's) well into the next century. The second group, consisting of breeder advocates and others, claims that one should only include reasonable assured resources as a basis for prudent energy planning. These comprise about 50% of total estimated resources of 4.2 million metric tons of uranium[29] minable at prices up to $ 50/lb. U_3O_8 ($130/kg U). Elimination of the 'estimated additional' portion of uranium resources would reduce the amount of 'low-cost' uranium available outside of CMEA countries and China to 2.2 million tons of uranium. This amount of uranium is only sufficient to fuel about 550 GWe of LWR capacity for their 30 year lifetime without plutonium recycle. The most recent estimates of nuclear power growth[29] indicate that market economy countries will reach this level of installed nuclear capacity some time between 1988 and 1991. Under such conditions, a good case can be made for the introduction of breeders as soon thereafter as possible.

An assessment of the validity of either of these schools of thought is very difficult because of the great uncertainty surrounding estimates of future nuclear power growth and the total amount of U_3O_8 that will be mined at costs less than $50/lb. U_3O_8. Let us suppose that this latter figure is 10 million tons of uranium (5 times the present 'reasonably assured' resources). Even with the most conservative nuclear growth estimate, this amount of uranium if used in LWR's would be committed by the year 2025. Thereafter, the nuclear industry would have to depend on much lower grades of uranium such as the Chattanoga Shales which are available in very large quantities but at costs of the order of $800/kg U.

2.6.2   Current reactor types versus breeders

Such costs would increase LWR nuclear fuel cycle costs to about 25 mills/kwh (=$15/bbl oil).

This same high cost ore, if used in breeders would add only 0.5 mills/kwh to nuclear fuel cycle costs. The early introduction of breeders, therefore, would not only remove the uncertainty about estimated resources of low cost uranium but also the economic uncertainty associated with the use of high cost uranium.

2.6.3 Nuclear versus alternative new energy sources

The opponents of nuclear power generally claim that the development of renewable energy sources will obviate the need for nuclear plants. They seem to overlook the fact that such renewable energy sources will have to compete economically with either light water reactors or breeders if they are to replace them. Since renewable energy sources have no fuel costs, their capital investment is of paramount importance. Figure 16, taken from Reference 15, compares capital investment costs of the renewable energy sources with coal nuclear fission and fusion. The bases for these costs are listed in Table 23. To put these data in their proper perspective, each $1000/kw corresponds to a generating cost of about 20 mills/kwh. On this basis, solar energy costs would lie in the range of 60 to 100 mills per kwh and will have a difficult time competing with the other sources of electricity shown in the figure.

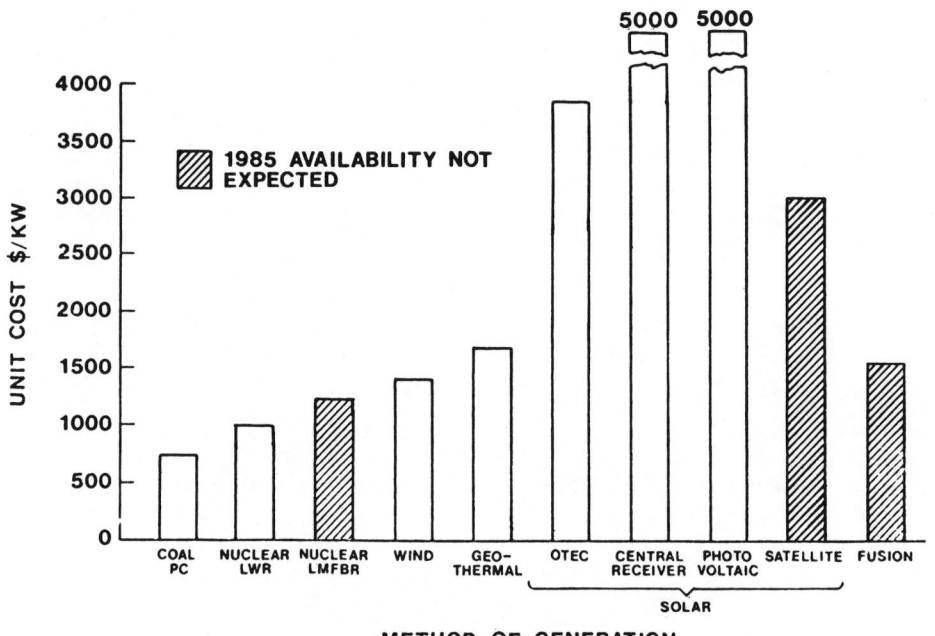

Figure 16. 1985 investment cost estimates.

Table 23

1985 Investment Cost Estimate*

Coal PC Fired 2-781 MWe units with SO_2 Removal	UE&C ERDA Study	$774/KWe
Nuclear LWR 2-1232 MWe units	UE&C ERDA Study	$1000/KWe
Nuclear LMFBR 1-1500 MWe unit	UE&C Projection	$1140/KWe
Wind Generator 1-1.5 MWe	UE&C Proj. From Kaman Info.	$1455/KWe
Geothermal 50-300 MWe Includes est. cost of wells	UE&C Projection - Various Sources	$1730/KWe
SOLAR { Central Receiver 200 MWe without storage	Honeywell/Black & Veath Studies Esc.	$5000/KWe
Ocean Thermal Energy 100 MWe	UE&C SSP Study Esc. Lower Cost is SSP Proj. Esc.	$3900/KWe $2000/KWe
Photovoltaic 1-4 MWe	UE&C Projection	$5000/KWe
Satellite 5,000 MWe	UE&C Projections (Est. operation after 2000)	$3000/KWe
Fusion Tokamak 2,000 MWe	UE&C Esc. Princeton Ref. Design (Est. operation after 2000)	$1575/KWe

*1975 costs escalated 8% per year to 1985.

2.6.4 Economics of nuclear desalination plants[26]

Not many people realize how rapidly the market for sea water desalination plants has grown during the last decade. Whereas during the 1963-1965 period, only 12 desalting plants with capacities greater than 1 million gallons per day were sold world-wide, the number jumped to 95 such plants in 1966-1968 and to 145 plants in 1972-1974. By 1975, the total output of more than 1000 plants scattered throughout the world amounted to about 550 million gallons per day (2.1 million cu.m/day) of desalted water. The desalination project of Saudi Arabia, which is just now being started, will double this output.

Because of the high cost of desalted water, 80% of the present world desalination capacity serves urban areas, the remainder being mainly for industrial use. Only one plant, the 120,000 cu.m/day Shevchenko plant in the USSR, uses nuclear heat. The reason for this is that before the abrupt increase in oil prices in 1973, nuclear heat was not competitive with oil. Now that oil is so expensive, it is worth while taking another look at the economics of nuclear desalation.

The main problem of using nuclear heat for the production of desalted water is that even a single purpose 1900 MWth (600 MWe equivalent) nuclear power reactor would produce from 700,000 to 800,000 cu.m/day of distilled water. The desalination plants that are presently available are not large enough to handle all of this capacity. For this reason, either small single purpose plants or larger dual purpose (water and electricity production) nuclear plants must be considered. Such plants were studied by the Oak Ridge National Laboratory in 1976[27]. Table 24 compares cost of nuclear and oil-

Table 24

Economics of Nuclear Desalination

Heat Source	Heat Cost US ¢/10^6 Btu	Distilled Water Cost, US ¢/cu.m.
Oil-Fired Boiler[1]	306	65
Single Purpose Nuclear Plant (313 MWth)	300	63
Dual Purpose Nuclear Plant (2 x 1875 MWth, prime steam)	180	38
Dual Purpose Nuclear Plant (2 x 1875 MWth, exhaust steam)	118	25

(1) Based on residual oil at $ 12.70 per barrel.

fired desalting plants based on this study. These costs are expressed in early 1976 US $. It is seen that costs of desalted water from an oil-fired plant and a small (313 MWth) single purpose nuclear plant are comparable, while costs of water from a larger dual purpose nuclear plant using either prime or exhaust steam appear to be lower. These costs, however, are only based on estimates and therefore must be viewed with caution.

3. CONCLUDING REMARKS

Only a few years ago it was possible to present an optimistic picture of nuclear power based on economic data that were then believed to be reliable. The latest trends in the economics of nuclear power, however, indicate that it is no longer possible to view nuclear power with any degree of optimism. The reasons can be traced directly to the environmental anti-nuclear movement whose aim has been to stop nuclear power by making it too expensive. Fortunately, they have not been able to stop nuclear power, but they have been able to make it as expensive as possible because of the assistance of various regulatory organizations and the courts. As a result of the higher cost, nuclear power will play a far smaller role in meeting future energy needs than was envisaged even as recent as 2 ½ years ago. In December 1975 the OECD/IAEA forecast of the nuclear capacity of the non-CMEA countries in the year 2000 ranged from 2000 GWe to 2500 GWe. In December 1977, the range was down to 1000 GWe to 1900 GWe. Even if the lower figure prevails, the anti-nuclear movement will cost these countries about $500 billion (constant 1978 U.S. dollars) in extra nuclear plant costs between now and the year 2000.

Will the increasing cost trend ever reverse itself? I believe the answer is no because of the fear of nuclear power now instilled into the mind of the general public. The minimum design basis for protection of a nuclear plant against earthquakes, for example, will far more likely increase rather than decrease. The same is true for protection against radiation. Thus one must assume that the high cost of nuclear power is here to stay. In fact, a few years from now, costs will probably be higher! The world will have to learn to live with these high costs because, as far as is now known, there are no lower cost alternatives generally available to everyone.

References

1. W.H.Zinn, "Basic Problems in Central Station Nuclear Power", Nucleonics 10 No. 9, September 1952.

2. J.A.Lane, "Growth Potential of U.S. Nuclear Power Industry", Nucleonics 12 No. 6, June 1954.

3. "Power Costs: Nuclear vs. Coal", Nucleonics 12 No.8, August 1954.

4. K.M.Mayer, Presented at the National Industrial Conference Board Meeting, October 1954.

5. J.A.Lane, "The Economics of Nuclear Power" Proceedings of the First International Conference on Peaceful Uses of Atomic Energy, Geneva, Switzerland. Vol. 1, August 1955.

6. K.Cohen, "Charting a Course for Nuclear Power Development", Nucleonics 16 No. 1, 1958.

7. J.A.Lane, "The Economics of Nuclear Power", Annual Review of Nuclear Science 16, 345, 1966.

8. R.Krymm, "Future Fossil Fuel Prices", IAEA Market Survey, General Report, Appendix I, September 1973.

9. J.H.Crowley, "Information Presented at the Hearings before the New York Public Service Commission", December 1976.

10. Nuexco (Nuclear Exchange Corporation), Monthly Report to the Nuclear Industry, March 31, 1978.

11. IAEA Staff Estimate, January 1978.

12. "Economic Evaluation of Bids for Nuclear Power Plants", Technical Reports Series No. 175, Appendix C, IAEA, Vienna, October 1976.

13. L.F.C.Reichle, "MSR (Pu Converters) and MSBR's in Commercial Nuclear Power Stations", Proceedings of the Salzburg Conference - Nuclear Power and its Fuel Cycle, Vol. 1, 531, September 1977.

14. United Engineers and Constructors Inc., "Capital Cost: Pressurized Water Reactor", NUREG 0241, "Boiling Water Reactor", NUREG 0242, "High and Low Sulfur Coal Plants - 1200 MWe", NUREG 0243, "Low and High Sulfur Coal Plants - 800 MWe", NUREG 0244, June 1977.

15. J.H.Crowley, Information Presented at the IAEA Consultants Meeting on "Extrapolation of Capital Cost Experience to Developing Countries", Vienna, June 1977.

16. G.Woite, "Capital Investment Costs of Nuclear Power Plants", IAEA Bulletin 20 No. 1, February 1978.

17. "ORCOST II - A Computer Code for Estimating the Cost of Power from Steam Electrical Plants", ERDA 76-38, 1976.

18. "Market Survey for Nuclear Power in Developing Countries", General Report, IAEA, Vienna, September 1973.

19. Canatom Ltd., "CANDU 600 MWe Pressurized Heavy Water Reactor Plant - Summary, Investment Cost Study and Plant Description", December 1976.

20. A.J.Covarrubias and J.A.Lane, "Assessment of Power Program Alternatives using the WASP Package", IAEA Training Course, Argonne, Ill., October 1977.

21. "Small and Medium Size Power Reactors", Proceedings of a Panel, IAEA, Vienna, 24-28 June 1968.

22. "Small and Medium Power Reactors", Proceedings of a Symposium, Oslo, 12-16 October, 1970, IAEA, February 1971.

23. "Market Survey for Nuclear Power in Developing Countries", 1974 Edition, October 1974.

24. J.A.Lane, "The Impact of Oil Price Increase in the Market for Nuclear Power in Developing Countries", IAEA Bulletin 16 No. 1/2, 1974.

25. J.A.Lane, "Nuclear Power for the Developing World", Presented at the IAEA General Conference - Scientific Afternoon, September 1975.

26. R.Krymm, J.A.Lane and I.S.Zheludev, "Future Trends in Nuclear Power", IAEA Bulletin, Twentieth Anniversary Issue, Vol. 19, No. 4, August 1977.

27. J.A.Lane et al, "Nuclear Power in Developing Countries", Proceedings of Salzburg Conference - Nuclear Power and its Fuel Cycle, Vol. 1, 233, Sep. 1977.

28. Nuclear Energy Policy Study Group, "Nuclear Power Issues and Choices", Ballinger Publishing Company Cambridge, Mass. 1976.

29. OECD/IAEA, "Uranium - Resources, Production and Demand", OECD Publication, December 1977.

30. N.Raisić,"Desalination of Sea Water Using Nuclear Heat", IAEA Bulletin 19, No. 1, February 1977.

31. O.H.Klepper, "Small Nuclear Reactors for Industrial Energy", Presented at the Industrial Power Conference, Memphis, 1976.

ROLE OF NUCLEAR ENERGY WITH PARTICULAR REFERENCES TO WESTERN EUROPE

L. Sani

Ente Nazionale Per L'Energia Elettrica

Rome, Italy

INTRODUCTION

The role played by nuclear energy in meeting the electric energy requirements of a country or of a group of countries such as Western Europe is to be considered with reference to the overall energy situation. Western Europe consists mainly of countries that have achieved a high degree of industrialization based prevailingly on conversion processes and therefore characterized by high energy consumption; on the other hand, the resources of primary energy sources are somewhat lacking, as in most of the European countries.

In this connection it should be pointed out that in the world energy situation Europe, as a whole, represents an industrial power almost comparable to that of the United States and the communist countries, as shown in Table 1 where the work primary energy consumptions for the period 1960-1976 are given. This position is even more remarkable when one considers that the European population is only one tenth of the world population.

In the past, electric energy played a rather important role in the European industrial development, through the hydraulic source wherever possible, and by exploiting fossil fuels, coal and oil derivatives. Even if in the future the international economic context should be remarkably different from that of the recent past -- because of its intrinsic characteristics -- electric energy will continue to play a very important role in the social-economic development of Europe, as it is an economically competitive and technically reliable means for the utilization of nuclear energy.

Table 1[1,2]

World Energy Requirements, by Region, 1960-1976

(Million t.o.e.)*

	1960	1965	1970	1972	1974	1976
Canada	96.0 (3.0%)	118.4	154.9	180.5	189.5	196.5 (2.8%)
U.S.A.	1014.2 (31.8%)	1225.5	1570.3	1691.1	1704.3	1743.9 (25.3%)
Western Europe	628.3 (19.7%)	800.9	1038.5	1108.9	1164.0	1180.5 (17.2%)
Japan	94.7 (2.9%)	151.4	284.0	311.8	337.4	349.3 (5.0%)
Sino-Soviet Bloc	986.4 (30.9%)	1144.8	1476.0	1612.8	2076.0	2380.8 (34.6%)
Rest of World	364.8 (11.7%)	494.4	691.2	724.8	902.4	1032.0 (15.1%)
TOTAL:	3184.4 (100%)	3935.4	5214.9	5629.9	6373.6	6883.0 (100%)

* 10^6 tep = $0,042 \times 10^{18}$ Joules = 10^7 Gigacal = $0,0013$ TWy

THE ROLE OF NUCLEAR ENERGY

1. ## THE ENERGY SITUATION IN WESTERN EUROPE

The European energy situation has so far been characterized by a heavy and ever increasing deficit between the total energy consumptions and the internal resources used to cover them. Table 2 shows the increase in the net oil imports to Europe from 30% of the total primary energy requirements in 1960 to 60% in 1974; only the recent economic crisis has slightly lowered this tendency.

The fact that the great majority of the energy resources exploited so far are not renewable and that since the well-known crisis in the autumn of 1973 oil imports bear heavily on the balance of payments leads to conclude that the European countries cannot stand such a situation much longer, without severe consequences in the national and international economy.

On the other hand, the assumption of the so-called 'zero growth' cannot be accepted a priori because social progress within each country and the gradual elimination of the differences among the various regions of a same country are bound to bring an increase in the per-capita energy consumptions. Figure 1 gives an indication

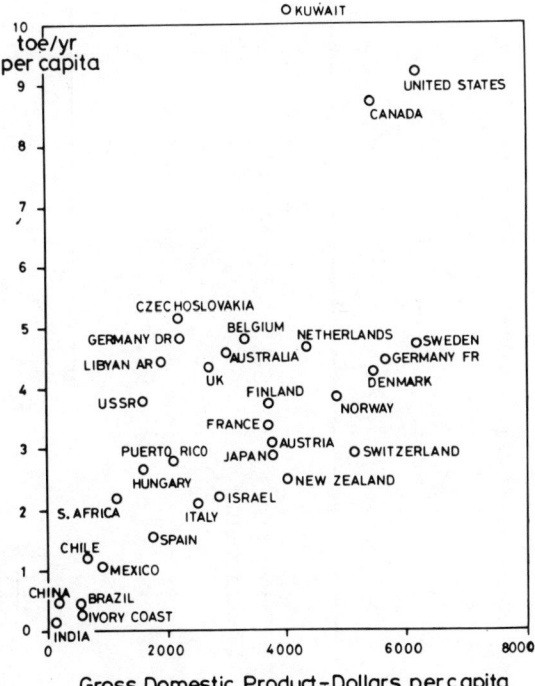

Figure 1. National per capita energy consumption vs GDP (1973)

Table 2[(2)]

Energy Requirements and Net Fuel Oil Imports in the European OECD Countries
(million t.o.e.)

	1960	1962	1964	1966	1968	1970	1972	1974	1976
Primary energy requirements (A)	628.3	690.7	764.7	825.3	904.6	1038.5	1108.9	1164.0	1180.5
Imports (B)	193.0	253.2	333.5	414.6	498.1	618.7	680.6	707.2	666.3
$\frac{(B)}{(A)}$ %	30.71	36.66	43.61	50.23	55.06	59.57	57.63	60.75	56.44

THE ROLE OF NUCLEAR ENERGY

of the per-capita energy consumptions referred to the individual income for various countries. These data evidence not only the difference between a few of the more industrialized European Countries and the United States, but implicitly also the wide gap that today separates the countries under development and those having a medium-to-high income.

In evaluating the assumption of 'zero growth' of the European and world energy requirements, two other factors are to be taken into consideration: the population increase, which occurs in every country of the world, though at different rates, and the progressive exhaustion of raw materials and thus the necessity of recycling waste as much as possible. The latter need arises especially in the industrialized countries and especially in the European countries that lack raw materials. In general, recycling processes absorb more energy than extraction processes at parity of finished usable product.

Finally, it should be borne in mind that the initial stages of industrialization require more energy for an equal increase in the Gross Domestic Product -- and this is the case of developing countries.

From the foregoing, it appears evident that if the governments intend to pursue economic development, raise living standards and employment levels, that is, to create the requisites necessary for social and civil growth in every country, more and more energy must be available.

As for the electric energy, Table 3 shows how in the same

Table 3[2]

Growth of the Total Primary Energy Demand and of the Electric Energy Production in the European OECD Countries

Year	Primary energy demand 10^6 t.o.e	Index (base:1960=100)	Electricity Production 10^9 kWh	Index (base:1960=100)
1960	628.3	100	567.7	100
1962	690.7	109.9	662.9	116.7
1964	764.7	121.7	769.4	135.5
1966	825.3	131.3	871.8	153.6
1968	904.6	143.9	992.3	174.8
1970	1038.5	165.3	1145.5	201.8
1972	1108.9	176.5	1290.6	227.3
1974	1164.0	185.2	1423.1	250.6
1976	1180.5	187.9	1524.1	268.5

period of time the output grew much more rapidly than the total demand of primary energy sources; in fact, referred to the 1960 figures assumed equal to 100, in 1976 the total energy demand increased 1.88 times whereas the electric energy output increased 2.88 times, that is, roughly following the 10-year doubling law.

The economic recession in 1975-1976 due to the oil crisis did not alter the aforesaid rate; in fact, in that period the yearly world growth rate of the electric energy output was 3.5% whereas the primary energy demand rose by 0.72%.

Therefore, it appears evident that the ultimate users of energy have shown an increasing preference for electricity over other available forms of energy. Table 4 illustrates the trend in the EEC countries, which, however, reflects quite closely the energy trend of the whole OECD countries in Europe.

Table 4

Total Primary Energy Requirements (Including Bunkering) and Portion Converted into Electricity Within the European Economic Communities

Years	Requirements (million t.o.e.)	Energy converted to electricity	
		In million t.o.e.	In percent
1969	798	182	22.8
1970	848	193	22.8
1971	856	203	23.7
1972	899	216	24.0
1973	952	230	24.2
1974	934	235	25.2
1975	886	225	25.4
1976	925	248	26.8

As concerns the consumption of primary energy and electric energy, Table 5 shows their percental breakdown by sectors of use in the European OECD countries for the year 1976.

Table 6 shows how the various sources of hydraulic, geothermal, nuclear and thermal production contributed to meet the electric energy requirements in the European OECD countries. It can be noted that hydroelectric energy, though continuing to increase in absolute terms, decreased in percental incidence from over 40% in 1960 to about 22% in 1976; this trend is destined to continue in

THE ROLE OF NUCLEAR ENERGY

Table 5⁽²⁾

Percental Incidence of Primary Energy and Electric Energy Consumptions in the European OECD Countries for the Various Sectors of Utilization (year 1976)

	Percental consumption of primary energy	Percental consumption of electric energy
Industry	40.7	50.8
Transport	20.6	2.6
Other Sectors	38.7	46.6

the future because of the almost complete exploitation of the European water resources. Thermal energy has reached and remains above the 60% mark, and only in the past few years has nuclear energy acquired a significant incidence that keeps going up.

In Europe as a whole geothermal energy plays a negligible role and only in Italy has it reached a certain importance with a yearly production of about 1.5 billion kWh.

To conclude the description of the present energy situation in Europe, a few of the statements that UNIPEDE (International Union of Producers and Distributors of Electric Energy) made at the 17th Vienna Congress in May 1976 are transcribed below:

"(a) Energy especially in the form of electricity is one of mainstays of modern life and it is therefore vitally important to ensure that adequate supplies are available on the most economic terms possible to meet not only the needs of today but those of the future. Europe, then, must develop sources of energy which give it the best opportunity of improving the productivity of its industry, the living standards of its people, the quality of their life and the quality of their environment. Furthermore, Europe needs to be strong in energy and prosperous if it is to give help on an effective scale to the Third World.

(b) The scope for increased supply of water power is severely limited (except for the purpose of pumped storage). Oil has not only become increasingly expensive but its reserves are being steadily depleted; coal is becoming

Table 6(2)

Electric Energy Production in the European OECD Countries, Broken Down by Sources

Year	Geothermal and Hydraulic		Nuclear		Thermoelectric		Total
	10^9 kWh	%	10^9 kWh	%	10^9 kWh	%	10^9 kWh
1960	227.3	40.04	2.6	0.46	337.8	59.50	567.7
1962	233.4	35.20	4.7	0.71	424.8	64.09	662.9
1964	250.0	32.49	12.0	1.56	507.4	65.95	769.4
1966	305.6	35.05	27.0	3.09	539.2	61.86	871.8
1968	317.5	31.99	36.3	3.66	638.5	64.35	992.3
1970	323.3	28.22	44.1	3.85	778.1	67.93	1145.5
1972	336.3	26.06	63.5	4.92	890.8	69.02	1290.6
1974	370.2	26.03	83.9	5.90	968.0	68.07	1422.1
1976	347.7	22.81	125.5	8.23	1050.9	68.96	1524.1

increasingly difficult to mine in Europe. Moreover, events in recent times have shown that security of fuel supplies cannot be taken for granted. Suppliers of primary fuels to Europe increasingly and understandably see oil and gas resources as assets that should be deployed to the best advantage of their own peoples in terms of economic and social development. They may not wish to deplete their precious reserves too fast, but instead seek to increase unit prices as far as possible and to develop the use of oil and gas for premium purposes such as chemical feedstocks. As for coal, the cost of mining it in advanced countries must be expected to rise. Other factors particularly in the case of oil and gas include the entry as buyers into the general market of some previously self-sufficient regions, such as North America, and the likelihood that some Third World countries will continue to develop industrially, producing a rapid rise in their energy demand.

(c) In summary, Europe can no longer rely on fossil fuels as a basis for her further development: they are likely to become more costly and less secure. The techniques of employing renewable sources of energy, such as solar, tidal, wind and wave power are not yet developed other than in minor applications, so that these sources of energy are unlikely to make a material contribution to Europe's needs for electricity in the next few decades. Nuclear power is already providing a significant supply of electricity reliably and safely under economic conditions, and will soon be able to make a major contribution to Europe's energy needs and thus to her prosperity and security.

Effectively the only route at present and in the near future for the utilization of nuclear power is electricity, and through this medium Europe's need for more energy will be met without damage to the environment, not only in the traditional areas of lighting, motive power, electro-metallurgy, electro-chemistry and electric traction, but also be reason of its flexibility and its high utilization efficiency in new thermal and other applications which will be developed for the common benefit as new technologies are perfected. Heat produced jointly with electricity may also be based on nuclear power.

Over recent years, studies in various countries have shown that, from all points of view, nuclear-based electricity provides the principal, the most economic and the best method of improving the supply of Europe's energy needs in the future. While the speed of nuclear development will vary from country to country according to variations in local supplies of fossil fuel, the adoption of nuclear power is only a matter of time."

2. FORECASTS ON THE DEVELOPMENT OF ENERGY REQUIREMENTS IN THE MEDIUM AND LONG TERMS AND CONTRIBUTION OF NUCLEAR ENERGY IN WESTERN EUROPE

The correlation between the parameters characterizing on one side the social-economic development of a country or of a group of countries and on the other side the evolution of energy consumption in general and of electric energy in particular, is quite evident in that the energy consumption is an almost univocal consequence of the economic development, as illustrated previously.

For the forecasts of the total energy requirements until the end of the century, reference is made to the year 1980, since this date is now near and the predictions of energy consumption made by various organizations for that year are in fairly good agreement.

For the following years the predictions of economic development in terms of Gross Domestic Product (GDP) for the OECD European countries can be obtained by assuming a population growth and an increase in the individual income that reflect the present trend of economy (Table 7). This assumption is generally defined 'Present Trend', in opposition to an assumption of 'Accelerated Development', which events in recent years make rather unlikely.

By applying to the GDP growth rate elasticity coefficients taken as the ratio between the yearly growth rate of the energy consumption and the yearly growth rate of the GDP, the growth rates of the energy demand from 1980 can be obtained, as shown in Table 8, and hence the relating primary energy requirements for the European countries.

Against these requirements, the energy reserves of the OECD countries at the end of 1976 were:

- 2,400 million tons of oil   (3.3% of the world reserves)
- 4,600 billion of m^3 of natural gas   (7%   "   "   "   "   )
- 37,000 million tons of coal   (9%   "   "   "   "   )
- 28,000 million tons of lignite   (20% "   "   "   "   )
- 106,000 tons of uranium (ascertained and estimated) (3.5% "   "   "   "   )

An examination of the European reserves versus the world reserves evidences the rather weak position of Europe in respect of

Table 7[(3)]

Social-Economic Development of the European OECD Countries

(Present Trend)

	1980	1980-85*	1985	1985-90*	1990	1990-95*	1995	1995-2000*	2000
Population (millions)	397	0.8	413	0.8	429	0.7	445	0.7	461
Per-capita income (1970 dollar value)	2574	1.7	2799	1.7	3049	1.8	3326	1.8	3633
GDP (10^9 in 1970 dols)	1022	2.5	1156	2.5	1308	2.5	1480	2.5	1675

*Yearly growth rates in the period under consideration.

Table 8[2,3]

Yearly Growth Rates for the GDP and Energy Requirements in the European OECD Countries - Primary Energy Requirements

	1980-1985	1985-1990	1990-1995	1995-2000
Yearly GDP growth rate in the five years under consideration	2.5%	2.5%	2.5%	2.5%
Elasticity index $\frac{\Delta E}{\Delta GDP}$	1.40	1.20	0.80	0.80
Yearly growth rate for the primary energy requirements in the five years under consideration	3.5%	3%	2%	2%
Yearly primary energy requirements at the end of the five-year period (10^6 t.o.e.)*	1648	1910	2109	2328

*The definition of the primary energy requirements was started from the value for the year 1980 recently evaluated by the OECD (February 1978), that is, 1,388 million t.o.e.

other more favoured geographic areas with regard to coverage of the future energy requirements.

As concerns the portion of energy requirements to be covered by electric energy, valid estimates have been made by various international bodies such as OECD, the European Economic Community and governmental bodies of the USA. At present, electric energy covers almost 30% of the total requirements in the OECD countries (about 20% in the other countries); this value falls to about 28% when all the countries in the world are considered. It is expected that in the European countries this percentage may reach 32% in 1985, 35% in 1990 and 38% in 2000, as shown in Figure 2. On the basis of these electric energy requirements, the curve of the installed electric capacity for a constant load factor of 50% on the European network is obtained. The two quantities are tabulated on the first two lines of Table 9.

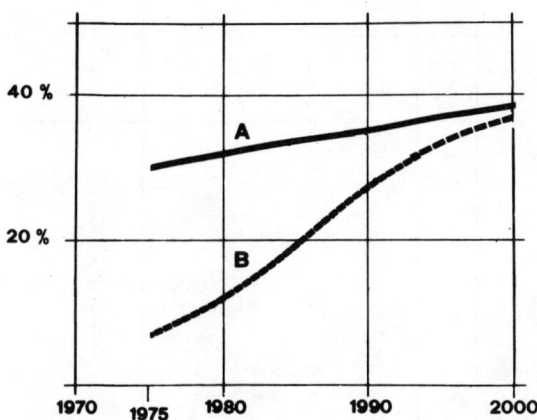

Figure 2. A-Percentage of total energy requirement converted into electricity.
B-Percentage of total electric capacity covered by nuclear power.

Finally, with regard to the portion of the total installed nuclear capacity required, the assumptions are given in Figure 2 and reflect the present trend in the OECD countries. Actually, up to a few years ago the most qualified predictions assumed that nuclear power would have a greater penetration rate than is expected to be likely today. In the United States nuclear power--which in 1970 covered already 10% of the total installed capacity-- should slowly rise to 13% in 1980, to 23% in 1985 and should be somewhere between 32% and 39% in 2000. By the end of the century this percentage in Europe should be about 36%, whereas in the developing countries where nuclear energy still plays a rather modest role, the growth rate should be higher.

Table 9

Forecast of the Development of Electricity Consumption, and of the Electric & Nuclear Capacities to be Installed in the European OECD Countries

	1977	1980	1985	1990	1995	2000
Electricity consumption (10⁹ kWh)	1577	1854	2405	2945	3438	3748
Installed electric capacity (GWe)	359	423	528	672	785	855
Installed nuclear capacity (GWe)	26	60	107	195	260	310
$\dfrac{\text{Nuclear capacity}}{\text{Total capacity}}$ %	7	14	20	29	33	36
Yearly growth rate for electricity consumption (%)		5.6	5.3	4.1	3.1	2.
Yearly growth rate for the installed electric capacity (%)		5.5	4.5	4.9	3.2	2.
Rate of the yearly increase in the installed nuclear capacity (%)		23.9	13.0	12.6	6.5	3.6

THE ROLE OF NUCLEAR ENERGY

For the European OECD countries, Table 9 shows the nuclear capacity to be installed expressed in GWe and as the percentage of the total installed electric capacity. This table gives the yearly average growth rates for electric energy consumptions, total installed electric capacity and nuclear capacity.

A comparison of the growth rates of electric energy consumption and total energy requirements in Table 8 shows how the use of electricity will continue to increase, albeit at a slackened pace, and tend to equilibrium.

3. <u>NUCLEAR ENERGY IN EUROPE TODAY</u>

The present situation of nuclear energy in Europe is described below through an analysis of three characteristic parameters, that is, capacities of the operating nuclear stations, the output of these stations, and stations under construction or ordered:

(a) Nuclear Installed Capacity

As at 31 January 1978 in the European OECD countries seventy-four nuclear power reactors totaling 28,000 MW were in operation:

- Great Britain   31 reactors   8,101 MW
- France   12   "   4,928 MW
- German Federal Republic 10   "   6,413 MW
- Sweden   6   "   3,882 MW
- Belgium   3   "   1,774 MW
- Italy   3   "   590 MW
- Spain   3   "   1,117 MW
- Switzerland   3   "   1,064 MW
- Holland   2   "   522 MW
- Finland   1   "   440 MW

The Reactor types already in operation are:

- 35 gas-graphite and natural-uranium reactors
- 4 gas-graphite and enriched-uranium reactors (AGR)
- 19 pressurized-water reactor (PWR)
- 12 boiling-water reactors (BWR)
- 4 reactors of various types.

The reactor capacity has been increasing with time from 60 MWe of the first units at Calder Hall (GB), which started operation in 1956, to over 1000 MWe at the most recent stations in the German Federal Republic. If we group 80% of the aforesaid 74 operating reactors by their capacity and date of beginning of operation, we obtain the following table where their increasing capacity is evidenced:

Capacity MW	Beginning of operation	Number of reactors	Average Unit capacity MW
0-100	1955-1960	10	56.6
101-250	1961-1965	9	179
251-400	1966-1970	16	308
401-800	1971-1975	16	538
> 801	1976-1977	8	975

The highest-rating unit now in operation is Unit 2 of the Biblis Station in the German Federal Republic with a net electric capacity of 1,240 MW.

(b) Nuclear Energy Production

By 31 January 1978 in the European OECD countries nuclear stations have generated over 912 billion kWh, broken down as follows:

- Great Britain   403 billion kWh
- German Federal Republic 139   "   "
- France   126   "   "
- Sweden   56   "   "
- Switzerland   47   "   "
- Italy   46   "   "
- Spain   45   "   "
- Belgium   39   "   "
- Holland   17   "   "
- Finland   3   "   "

Figure 3 shows the yearly growth of electricity generation in each country.

Limitedly to the years from 1974 onwards, it is interesting to note the percental incidence of the nuclear source on the total electricity output. Table 10 shows this incidence for all the European OECD countries that produce nuclear energy. In this connection it should be noted that the aforesaid percentages, which

THE ROLE OF NUCLEAR ENERGY

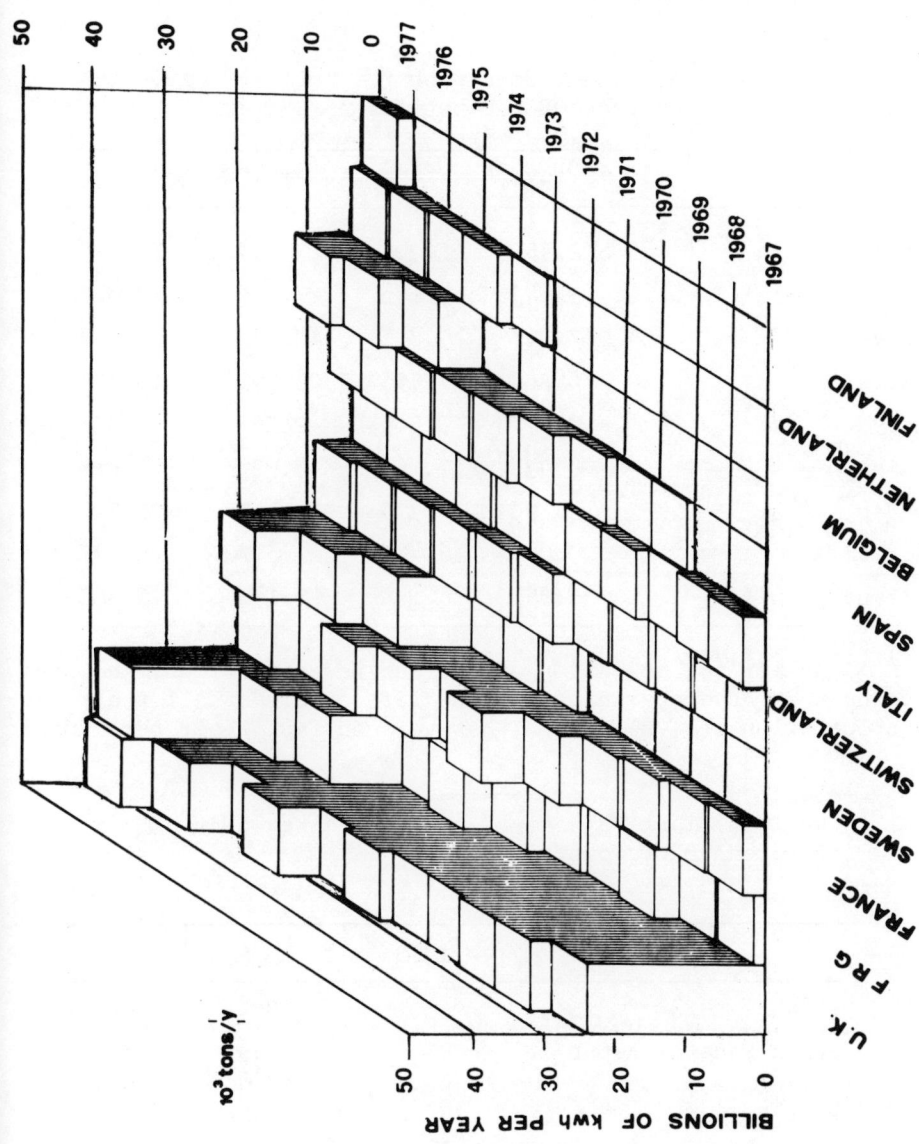

Figure 3. Annual nuclear generation.

are already remarkable in Belgium, Switzerland and Sweden, rise further when referred only to the winter period, that is, when the hydroelectric output reaches its yearly minimum level and the nuclear stations meet most of the base load.

Table 10[2]

Incidence of the Nuclear Source on the Total Electricity output in the European OECD countries

	1974	1975	1976	1977
Great Britain	12.45%	11.20%	13.28%	13.43%
German Federal Republic	3.64%	7.18%	7.35%	10.68%
France	7.81%	9.84%	10.05%	8.53%
Switzerland	18.36%	17.59%	21.29%	18.18%
Italy	2.47%	2.58%	2.32%	2.03%
Spain	8.93%	9.16%	8.32%	7.82%
Sweden	2.19%	14.92%	18.44%	21.72%
Belgium	0.28%	16.47%	21.14%	25.17%
Holland	6.05%	6.14%	7.13%	6.36%

As for the operation of nuclear power stations, the ups and downs that have characterized the life of almost all the reactor now in operation are well known; however, so far, their performance

Table 11[4]

Nuclear Plant Capacity Factors in a Few of the European OECD Countries

(%)

	1975	1976	1977
France	68.1	58.7	63.3
German Federal Republic	71.6	49.9	67.2
Italy	67.8	68.8	64.3
Great Britain	56.8	54.5	81.8
Spain	77.1	25.9	66.7
Switzerland	83.6	85.6	80.2
Sweden	55.0	55.0	59.3
Holland	71.7	79.0	79.7
Belgium	74.9	65.4	77.3
Average plant capacity factor	65.2	52.9	70.2

THE ROLE OF NUCLEAR ENERGY

has been more than satisfactory with an average plant capacity factor of about 60%, as shown in Table 11.

(c) Nuclear Power Stations Under Construction and Ordered

As at 30 June 1977 in the European OECD countries ninety reactors totaling 87,000 MWe had been ordered, that is:

- France   27 units for 27,000 MW
- German Federal Republic   19   "   " 21,400 "
- Spain   14   "   " 13,300 "
- Great Britain   6   "   "   4,000 "
- Italy   6   "   "   4,900 "
- Sweden   5   "   "   4,600 "
- Switzerland   4   "   "   4,100 "
- Belgium   4   "   "   3,900 "
- Finland   3   "   "   1,800 "
- Luxemburg   2   "   "   1,300 "
- Austria   1   "   "   700 "

Their unit capacity averages about 960 MW, which is in line with the average capacity of the stations recently in operation. However, it should be noted that the aggregate of the units considered includes also prototype station such as the SNR-300, HTR and CIRENE, whose capacity is well below the average.

As for the type of reactor adopted, there is a marked predominance of pressurized-water reactors over boiling-water reactors (60 units versus 20); the six AGR now under completion in Great Britain and the prototype reactors are to be added to the light-water reactors. Of the prototype reactors, worthy of mention is the Creys-Malville (France) reactor, which is the first prototype of a commercial station equipped with a 1200-MWe fast reactor. It is being built by NERSA, a company formed by EdF (France), ENEL (Italy) and SBG (Germany).

The stations ordered from 1970, which represent over 80% of the total capacity under construction and ordered as at 30 June 1977, are listed in Table 12 by year and by country. In this connection it should be noted that since 1973 nuclear energy has been receiving a great impulse in France and, on a smaller scale, also in the German Federal Republic and Spain, even though in 1976 and 1977 it marked time, as the utilities considered the advisability of placing new orders for nuclear stations. One of the main causes of this slackening was stagnation of the electric energy consumption due to the recent economic crisis in Europe.

Table 12[4]

Orders for Nuclear Power Stations in the European OECD Countries

(MWe)

	1970	1971	1972	1973	1974	1975	1976
France	850	1874	890	2915	12025	-	3700
German Fedral Rep.	1634	5400	2775	1300	2498	3348	-
Spain	-	2790	1860	975	-	1200	970
Great Britain	1320	-	-	-	-	-	-
Italy	-	-	-	1934	2000	-	-
Sweden	-	1800	1480	-	-	-	-
Switzerland	850	-	940	906	1140	-	-
Belgium	-	-	-	-	3900	-	-
Finland	-	400	660	-	1540	-	-
Austria	-	700	-	-	-	-	-
Luxemburg	-	-	-	-	-	-	-
Total European OECD countries	4654	13004	8605	8030	23103	5795	4670

4. **PROBLEMS RELATING TO THE CONSTRUCTION OF NUCLEAR POWER PLANTS**

Implementation of the European nuclear programs required an adequate development of the technical, productive and organizational capabilities of the industries involved, together with the definition of criteria for nuclear power plant siting and of safety criteria to be adopted in the design, construction and operation phases.

It is deemed advisable to dwell on these subjects because the experience acquired or being acquired in Europe may be useful to the countries that are now beginning to face the problems posed by the implementation of a nuclear program.

4.1 Industrial Aspects

Any industry that intends to engage in a job characterized by high technological requirements, high investment and strict compliance with severe regulations, must first of all be 'qualified'. A few of the technical requisites that that industry has to posses are know-how and high-quality capability. 'Quality' here is intended as the aggregate of characteristics of the product that required performance.

As regards nuclear plants, quality, and the means used to obtain it, bear directly or indirectly on all the main plant requirements, that is, safety, compatibility with the environment, reliability and availability. Therefore, there is a direct connection between the quality and cost of the energy produced in the sense that the price paid for quality is compensated by a better stations performance.

On the other hand, from the very beginning the development of the nuclear sector had to cope with the necessity of ensuring very high safety levels for plant operation and for protection of the station personnel and the surrounding population. Therefore, the industries engaged in this sector had to reach remarkably higher qualifications than those required by other production sectors. In this connection, it should be noted that the large diffusion of American light-water reactors all over the world has actually entailed the acceptance by the industry of regulations, codes and standards developed in the USA.

However, the European industry not only adopted the American standards and procedures, but inserted them in the industrial and legislative context of the respective countries. In fact, a nuclear plant is to comply with the laws and regulations in force in the country where it is being built and with the requirements set forth by the regulatory agencies of that country.

From the standpoint of industrial organization, the construction of a nuclear plant requires the participation of large, medium and small industries: The large industries, in limited number, act as main suppliers, the medium and smaller industries as sub-suppliers at the same level or in cascade. As concerns the degree of a country's industrialization, the recourse to foreign supplies can be reduced only if the main supplier can rely on medium and small industries that are qualified or qualifiable in the nuclear field, thus limiting currency outlays and forming the basis for a gradual and future development of the national industry. Major industries will have to promote the qualification of its sub-suppliers by giving them all the necessary instructions and technical assistance on the strength of its experience. Obviously, this process is not only gradual, but it also branches out in several directions in the sense that qualification of an industry is not to be an isolated fact dictated by temporary interests, but one phase of a more complex process aiming at a continuous technological improvement of the country's industrial texture.

Another aspect of the industrial problems associated with

the construction of nuclear power plants is the personnel effort
required for station design, construction and operation, both in
qualitative and quantitative terms. In fact, transformation of
nuclear energy into electric energy is characterized by a high added
value in terms of work, a great deal of which is of a high technological content. This contribution of human effort in civil, mechanical, electric and electronic engineering is required not only for
construction, but also for plant operation and fuel cycle activities.

According to a recent conservative evaluation, the design
and construction of a 1000-MW nuclear station and its operation for
twenty years require 70 million man-hours, that is, twice as much
as a conventional station of equal rating, and the figure does not
take into account the work of the scientists and nuclear technicians
that work at the development of the various types of reactors. Thus,
it appears evident that, unlike other highly technological inovations, the utilization of nuclear energy for peaceful uses is a rare
example of technical-economic development based on a maximum contribution of human capabilities and on a limited recourse to the primar
sources as compared with the conventional process for electricity
production. In conclusion, because of the higher number of man-hours
required to obtain the same output, the nuclear source differs from
the conventional source not only as concerns economy and diversification of primary energy sources, but also in that it is a valid
means to overcome such current problems as unemployment and under-
employment also at the most qualified levels.

4.2   Nuclear Power Station Siting

The experience acquired in recent years has shown that the
construction of large power stations--especially nuclear stations--
is linked more and more to territorial planning and ecological
policies. To ensure a balanced and ordered development of electric
energy it is necessary to adopt adequate measures aiming at incorporating nuclear power station siting in territorial planning and
in environmental protection policies. In fact, one of the main
problems to be solved now for the development of electric energy
is the selection of adequate sites for new stations. The selection
is made on the basis of several considerations of various nature,
which in any case take into account the many interactions between
site and plant. These interactions can be schematically divided into
two categories: (a) influence of the site on the plant and (b)
influence of the plant on the site. The first category refers mainly
if not exclusively, to plant safety; the second category refers
mainly to population health and environment protection.

The possible influence of the environment on plant safety
depends on the station siting and can be ascribed to natural events
(seisms, floods, exceptional weather conditions) or to accidental
events connected with the presence of man-made structures or human

activities. This influence can be overcome with an adequate design, as described in the section of this report dealing with nuclear safety.

The influence of the station on the environment has 'conventional' aspects that are common to all industrial settlements of the same importance and to major engineering works of various types.

This environmental impact occurs mainly during construction, which entails excavation and earthwork, alteration of the landscape, occupation of a larger area than that covered by the plant, continuous passage of transport means and materials, etc. The greatest alterations to the natural environment occur in this phase; therefore, great care should be taken in selecting the sites, avoiding those subject to environmental restraints.

During normal operation the nuclear power stations release a certain amount of liquid and gaseous radioactive products to the environment. The technical means now available for liquid and gaseous effluent treatment allow these releases to be reduced to very low levels, thus limiting remarkably the amount of environmental receptivity taken up.

Normal station operation entails also disposal of a remarkable amount of heat. This may pose a certain limitation on the power concentration at the same site or at several sites located on the same watercourse. For instance, the construction of high-rating units along large watercourses, such as the Rhine, requires the use of cooling towers because of the thermal saturation reached by the rivers. In this connection it should be pointed out that in the future a good environment protection policy will require that every effort be made to use waste heat as much as possible in order to reduce the thermal impact of nuclear and conventional stations. The development and implementation of techniques for the recovery and utilization of this heat are highly desirable.

4.3   Safety Problems in Nuclear Power Stations

In Europe, light-water reactors were generally preferred. The first designs of these plants were therefore based mainly on the safety criteria in force in the USA. Subsequently, the evolution in nuclear safety led to an increased capability of the protection systems in new stations, even though the criteria adopted in Europe have not been standardized yet. A few instances of this evolution are given below.

(a) Redundancy of the safety related systems:

The design criteria of the early plant safety systems required the installation of at least two separate, independent and fully redundant circuits to face the various assumed accidents of the plant even with one of the safety systems out of service (single failure).

This philosophy, called redundancy criterion N-1 (safety of the station is ensured even with one of the safety systems out of service), was expanded in some cases with the adoption of the redundancy criterion N-2 in which it is assumed that, in addition to the single failure, maintenance is being performed on a major component of another safety system.

(b) Containment systems:

A double-containment system is now used, instead of the single container around the nuclear steam generator. Thus, two aims are reached:

- the reduction in the radioactive material released to the atmosphere in the case of an internal failure, and the possibility of filtering and checking the primary containment leaks;
- protection of the systems inside the container from possible external accidents either natural or caused by man.

(c) Protection against aircraft crash:

Because of the very remote probability of a civil aircraft crash on a nuclear plant, a few European countries adopt protection only against military aircraft crashes, independently of the location of the station. The main parameter is the aircraft fall speed which, on the basis of the impacting mass distribution, gives the diagram of the forcing function. Remarkable efforts have been made in the development of methods of analysis whose results led to increase the 1-m thickness of the concrete structures designed to withstand earthquakes, to 2-metre thickness in those areas of the plant that may have to withstand air crashes.

(d) Protection against external explosions:

Europe considers advisable to adopt protections against deterministic (and not probabilistic) external explosions, whereas in the USA the pressure waves caused by these explosions are covered by the provisions against tornadoes. In Europe, studies have been carried out not only with conventional explosives (TNT), but especially with potentially explosive gas clouds that may float towards the station and deflagrate or detonate upon reaching the plant structures.

The lack of knowledge of these phenomena sometimes led to oversize some of the structures and thus to render them more cumbersome without improving their actual safety. Only recently, has a tendency set in, whereby external explosions are considered less conditioning than an aircraft crash and sabotage.

5. PROBLEMS RELATING TO THE FUEL CYCLE

Since it was first applied to peaceful uses, nuclear energy presented several aspects of the fuel cycle phases that have called for increasing attention, mainly at an international level.

A short review of the technical, economic and political aspects of greater interest at present calls for general considerations on a few problems such as those relating to the cycle costs, the source material procurement and enrichment services, and irradiated fuel reprocessing.

5.1 Economic Considerations:

One of the first aspects that should be pointed out is the total amount of the expenses associated with the nuclear fuel cycle, particularly with its main components. In this connection, it should be noted that in the twenty-year lifetime of a nuclear power station the total amount of expenses for the nuclear fuel cycle (inclusive of reprocessing), evaluated at constant prices, is higher than the cost of the station net of interest. As compared with an oil-fired conventional station, these expenses are at least four time less for the same amount of energy generated, and consist of about 70% of manpower costs. The high added value in terms of work activities, many of which at high technological levels, is a pecularity of the transformation process of nuclear energy into electric energy.

Table 13 gives the percental incidence of the expenses associated with the various fuel cycle phases, including interest, referred to the present economic conditions. It can be noted that

Table 13

Nuclear Fuel Cost
Percental Breakdown According to the Fuel Cycle Phases

	%
Natural uranium	41.0
Conversion	2.0
Enrichment	30.8
Fabrication	15.4
Transport and related expenditures	2.0
Spent fuel transport and reprocessing	15.1
Plutonium credit	−6.3
	100

'natural uranium' constitutes the most expensive item and therefore its cost variations reflect remarkably on the kWh production cost.

Another main component is 'enrichment', followed by 'fabrication' and 'reprocessing'. It should be noted that the last two components bear by the same amount on the kWh cost, notwithstanding the fact that the unit cost of reprocessing is about twice the 'fabrication' cost. This is due to the different times in which the two activities are to be performed in respect of fuel utilization in the reactor; this difference determines an interest of opposite sign.

Finally, a few interesting observations can be made on the investment in industrial plant necessary to the various phases of the fuel cycle.

It is a well-known fact that in the optimal and well-balanced system of nuclear stations and industrial facilities associated with the fuel cycle, the investment required by the industrial facilities does not exceed 10-15% of the investment for nuclear stations. At least half of the investment associated with the nuclear fuel cycle is due to the enrichment plant; the other half is more or less equally shared by natural uranium production and reprocessing.

The amount of investment for fuel fabrication, though remarkable in absolute terms, does not bear appreciably on the whole system.

5.2 Natural Uranium Procurement

Any problems relating to the procurement of raw materials, or of industrial products or services, cannot prescind from an analysis first of the relevant demand-and-offer situation in a given market, and then of other factors of various nature that may influence that market.

For natural uranium, like for almost all raw materials, reference has to be made to the world market, both because the activities of the main producers cover every geographical area, and because the present state of economic integration among countries renders obsolete any possibility of autarchy, albeit limited to such an important material as uranium. This policy, however, does not alter the fact that the countries rich in uranium sources are in a better strategic position as concerns the exploitment of the nuclear source.

Based on assumptions similar to those given in Section 2 for European countries, and on the expectations that light-water reactors will cover about 90% of the total installed capacity as

2000, the most recent OECD evaluations of the world natural uranium requirements lead to the following estimates, expressed in thousands of tons of uranium:

	Yearly Demand	Cumulative Demand
1980	41	128
1985	71	423
1990	102	873
1995	134	1,477
2000	178	2,276

The aforesaid estimates do not take into account the recycle of the plutonium and uranium recovered through reprocessing. However, it should be borne in mind that from 1985 recycling of these materials would entail an increasing saving on the cumulative requirements from 10% in 1990 to 20% in 2000. Incidentally, the requirements of the European OECD countries will decrease from 40% in 1980 to 30% in 2000 of the world requirements.

In planning to meet the world requirements, not only the ascertained or estimated reserves are to be taken into account, but also the actual production capabilities available in the period of time considered. As of January 1, 1977, the uranium world reserves, broken down according to the OECD classification, were:

- 1,650,000 tons of ascertained uranium (Reasonably Assured Resources), and
- 1,510,000 tons of estimated uranium (Estimated Additional Resources)

at an extraction cost lower than 30\$/lb U_3O_8, and

- 540,000 tons of ascertained uranium (RAR), and
- 590,000 tons of estimated uranium (EAR)

at an extraction cost ranging from 30 to 50 \$/lb U_3O_8.

It is worth noting that, as compared to the 1975 OECD evaluation, the total ascertained and estimated resources increased by about 30%, owing to the resumption of the research and prospection activities in the past years associated with the implementation of the nuclear programs in several countries. Even though at present many programs have been sized down remarkably, the research and prospection activities are still very intense (e.g. the NURE program in the USA) and will certainly lead to the discovery of new important resources.

Figure 4 shows the countries where these resources are located, together with the magnitude of the major reserves. The USA,

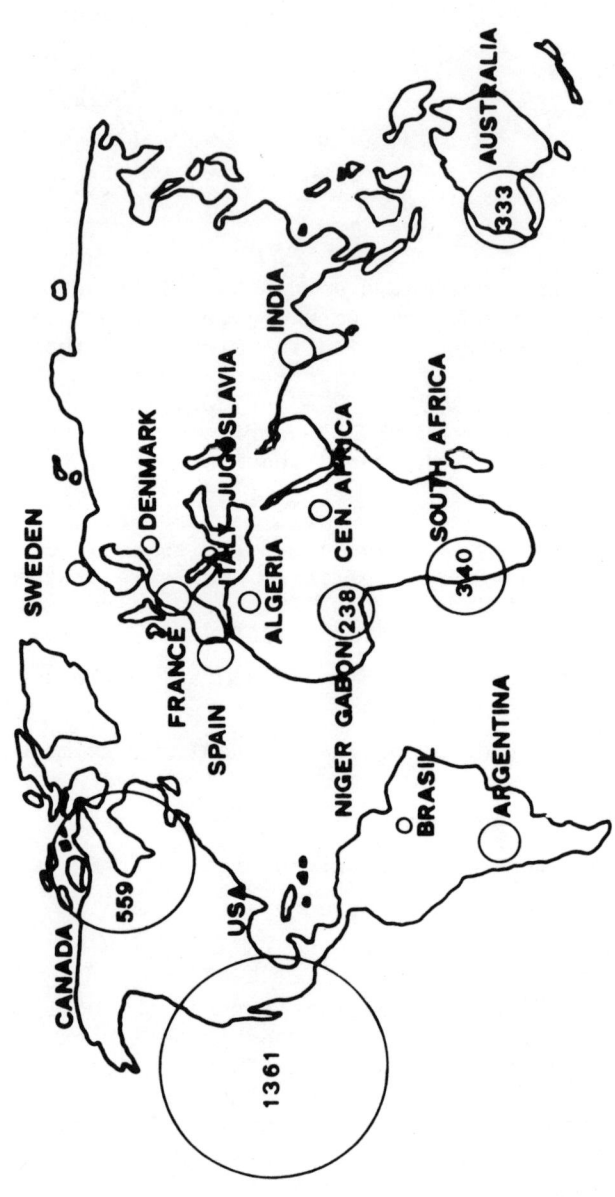

Figure 4. Ascertained and estimated uranium resources in the world (thousands of tons).

Canada, South Africa and Australia possess about 80% of the total reserves. An evaluation of the importance of these reserves should take into account the actual yearly production capability. In fact, the maximum possible production of uranium from the ascertained deposits economically exploitable at an extraction cost lower than 30 $/lb U_3O_8, is evaluated to be about 53,000 t/year in 1980, 92,000 t/year in 1985 and 110,000 t/year in 1990. Table 14 gives the breakdown among the main producer countries.

Table 14[5]

Attainable Uranium Production Capabilities

(tonnes U)

	1977	1980	1985	1990
Australia	400	500	11,000	20,000
Canada	6,100	7,950	12,500	11,250
South Africa	6,700	11,700	12,500	12,000
USA	14,700	22,600	36,000	47,000
Niger	1,600	4,100	9,000	9,000
Others	10,200	6,150	10,200	10,750
Total:	33,000	53,000	92,000	110,000

A comparison between demand and offer can therefore be made by comparing the total requirements and reserves on one hand, and the yearly demand and relating production capability on the other hand. These two comparisons are shown in Figures 5 and 6, respectively. From the data contained in these figures the following can be inferred:

- The ascertained reserves exploitable at less than 30 $/lb U_3O_8 are in theory sufficient to cover the cumulated requirements up to 1995; if the ascertained reserves exploitable at higher costs are added to the aforesaid reserves, the requirements are covered up to the year 2000 (curve A in Fig. 5).
- If the natural uranium requirements needed to cover the whole life of the reactors in operation at a given year are taken into consideration, it appears that the ascertained reserves in the first OECD cost category will cover the requirements up to 1987 and the combined reserves in both categories will cover up to 1991 (curve B in Fig. 5).

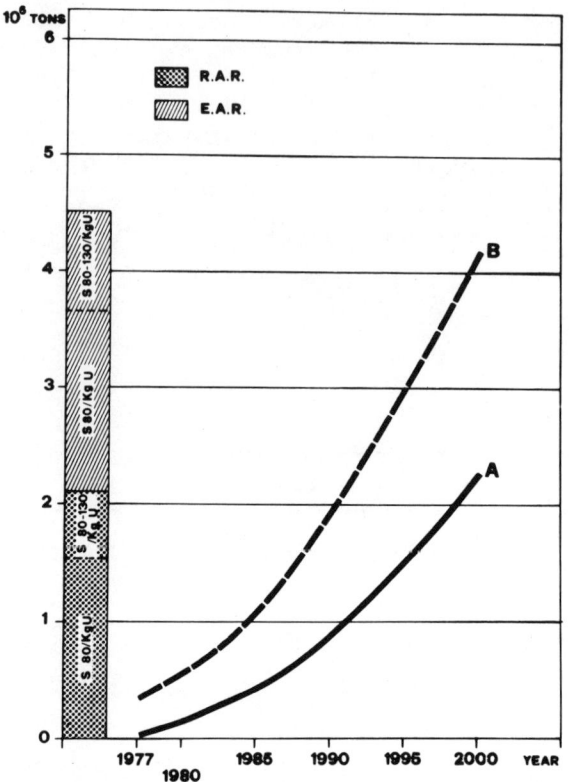

Figure 5. Cumulative uranium requirements

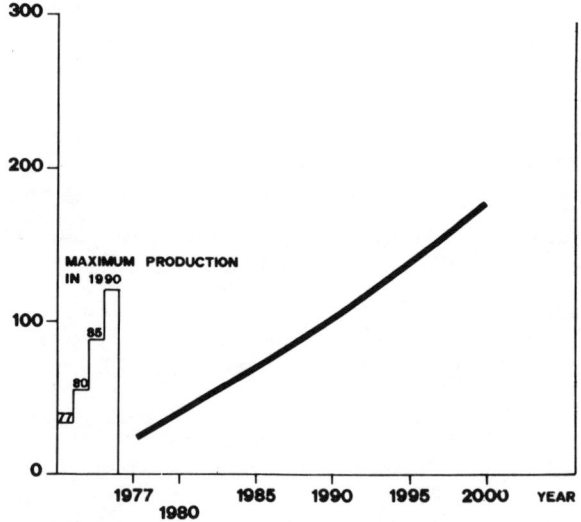

Figure 6. World annual uranium requirements (1977-2000).

THE ROLE OF NUCLEAR ENERGY

- The previously illustrated production capabilities cover the yearly requirements just beyond 1990 (Figure 6).

In view of this it is clear that every possible effort (financial and technical) is to be promptly made to assess the estimated resources, to find new resources and to develop new techniques capable of exploiting the ascertained but uneconomical deposits, that, though extensive, contain small percentages of uranium ore.

In Europe the ascertained resources are rather limited, the ore in the two cost categories totalling about 90,000 tons, to which are to be added 100,000 tons from estimated resources. Therefore, just like with fuel oil, Europe has to carry out a careful uranium procurement policy which will be subject to the uncertainties due to the export policies of the uranium producer countries, especially as regards non-proliferation.

In this context the results of the INFCE (International Nuclear Fuel Cycle Evaluation) program will acquire particular importance in that it aims at finding all the solutions that allow peaceful utilization of nuclear energy and at the same time provide the maximum security against proliferation.

5.3 Enrichment Services

Availability of the enrichment services is fundamental in the fuel cycle strategy because of the present and future predominance of light-water enriched-uranium reactors. Unlike the situation of natural uranium, starting from the coming years the European situation as regards enrichment services will be characterized by an almost total autonomy thanks to the commissioning of the EURODIF and URENCO plants that have a capacity of about 10 and 2 million SWU per year, respectively. Actually, in the coming years in Europe--and probably in the whole world--the enrichment services will exceed the actual requirements because of the delay or sizing down of the nuclear programs in a few countries. In fact, the world capacity should reach about 28 million SWU/yr by 1980, 40 million SWU/yr by 1985 and range between 60 and 90 million SWU/yr by 1990 (see Table 15), whereas the world requirements should be about 20 million SWU/yr by 1980, 35 million SWU/yr by 1985 and 55 million SWU/yr by 1990. If the build-up of enriched uranium stocks in the coming years is taken into account, the surplus of the offer over the demand will be greater than what appears from a comparison between yearly requirements and production capacity. An answer to this question, at parity of separative work, might be a decrease in the tail assays during the enrichment process with a consequent saving in natural uranium at the expense of a smaller amount of enriched uranium product as shown in Figure 7. In general, though the unit cost of enriched uranium would be higher, the lower than the optimum tail content

Table 15

World Enrichment Service Capacity
(10^3 ton SWU)

	1978	1980	1985	1990
A. Plants in operation or under construction				
– USA Dept. of Energy DOE –Gaseous Diffusion	17	21	26.6	26.6
– " " " " –Centrifugation	–	–	–	8.8
– URENCO	0.2	0.6	2	2 – 10
– EURODIF	–	6.1	10.8	10.8
1st Sub-total	17.2	27.7	39.4	48.2 – 56.2
B. Planned Plants				
– COREDIF				5.4 – 10.8
– South Africa				5
– Brazil			0.2	0.2 – 2.5
2nd Sub-total			0.2	10.6 – 18.3
C. Proposed Projects				
– Canada				0 – 9
– Australia				0.3
– Japan				0.1 – 6
3rd Sub-total				0.4 – 16.3
TOTAL	17.2	27.7	39.6	52.2 – 89.8

THE ROLE OF NUCLEAR ENERGY

would allow a reduction in the total cost (natural uranium + separative work). This solution might be the most convenient, because it allows the enrichment plant to operate at rated capacity factor and reduces the enriched-uranium iventory.

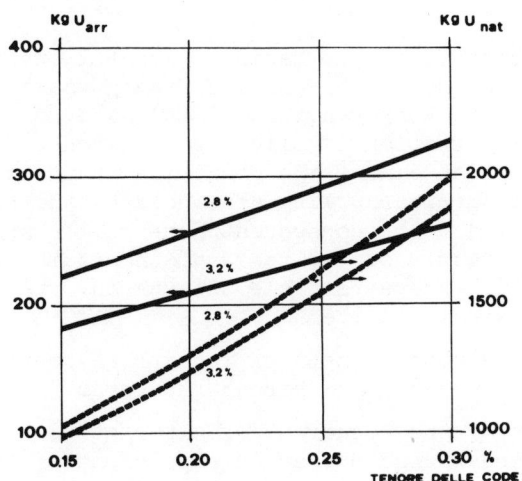

Figure 7. Natural uranium feed and enriched uranium product as a function of the tail assay and per 1 ton of SWU (for two typical enrichments).

5.4 Irradiated Fuel Reprocessing

The problem of irradiated fuel reprocessing is worth mentioning because of its implications, many of which are not strictly technical. It is a well-known fact that reprocessing of depleted fuel assemblies unloaded from nuclear stations aims at recovering reusable materials and at separating the long-lived radioactive wastes, generally called 'nuclear wastes'.

The recovery of the residual uranium, and especially of the plutonium, allows an increase in uranium exploitment thus limiting the demand. The separation of the fission products and actinides remarkably limits the amount of radioactive material (about 3% of the irradiated fuel) that necessitates the adoption of all the measures for its treatment and final disposal.

The normal reprocessing systems adopted so far are generally based on total separation of plutonium from uranium, even though subsequently the plutonium and uranium are blended again for fuel fabrication.

The risk of plutonium being used for purposes other than

utilization in nuclear power reactors was evaluated in the early days when nuclear energy started being put to peaceful uses. In fact, the utilities have always adopted physical protection measures agains nuclear materials spontaneously, also through agreements between Governments when the fissile material transport involved more than one country.

Subsequently, on the basis of general international safety considerations, the Non-Proliferation Treaty was signed and then the International Atomic Energy Agency (IAEA) safeguards were adopted. These means of international policy are, no doubt, adequate to reach the non-proliferation aim of the countries that signed the Treaty. However, the new American policy announced by the President of the USA in April 1977 and the consequent Non-Proliferation Act of March 1978, could be a result of the residual uncertainties on the actual possibility of full-scope reliable safeguards. The American policy covers the following main items:

1. Indefinite postponement of industrial reprocessing and plutonium recycle in thermal reactors.

2. Priority of the research for alternative solutions to fast-breeder reactors and postponement of the date of commercial operation of the latter.

3. Increase in the availability of nuclear fuel to the countries requiring it, in order to avoid their reprocessing the material themselves.

4. Embargo on the export of technologies and know-how relating to enrichment and reprocessing.

5. Extensive and exhaustive international discussions to allow every nation to reach its energy aims and at the same time to reduce the risk of proliferation.

This last item gave President Carter the opportunity of proposing the formation of an international group to evaluate the implications of all the fuel cycle activities on non-proliferation; this led to the foundation of the INFCE (International Nuclear Fuel Cycle Evaluation). However, the USA position is not shared by the countries that strongly depend on foreign energy sources; in fact, the EEC Parliament recently approved all the proposals on reprocessing, fast reactors and radwaste treatment submitted by its Commission.

The necessity of continuing to reprocess irradiated fuel is evident also in the report "The Windscale Inquiry" submitted by the BNFL (British Nuclear Fuels Ltd) to the British Government at the beginning of this year in order to expedite a favourable decision on the construction of a new reprocessing plant. The conclusions of this report point out that, should storage of the depleted fuel be preferred to reprocessing idefinitely, the build-up of irradiated fuel would leave important energy resources unused, and with time it would determine the serious problem of their final disposal without eliminating the proliferation risks. Therefore, it is preferable to reprocess the fuel as it is unloaded from the reactors, and to solve any problems as they arise, instead of having to cope with the same problem in a more or less near future and with the aggravating circumstances of having to get rid of an enormous amount of depleted fuel on one hand and of lacking the experience that could have been acquired in the meantime on the other.

As for the risk of proliferation posed by plutonium, there is no doubt that fast breeder reactors would contribute to reduce this risk instead of increasing it, because they burn the plutonium as it becomes available from thermal reactors. In fact, if only thermal reactors were in operation, the potential amount of plutonium available in the world would increase idefinitely.

Of the fuel reprocessing processes for fast breeder reactors, the CIVEX process seems particularly promising from the point of view of non-proliferation. In this process plutonium never exceeds a 25% concentration and, above all, it need not be completely separated from the fission products whose high radioactivity would render the product absolutely 'inaccessible'.

The European situation in the reprocessing field can be synthetized as follows:
- At Windscale, in Great Britain, the BNFL has a plant where both metal and oxide fuels can be reprocessed. The plant has two different head-ends, having a capacity of about 2000-t/year for Magnox fuel and 400-t/year for light-water reactor fuels, respectively.
- At La Hague, France, COGEMA has a 900-t/year reprocessing plant for metal fuels. In 1976 the head-end for oxide fuel was put in operation; the capacity of this head-end will be raised progressively to about 800-t/year after 1980.
- In 1974 EUROCHEMIC decided to shut down its plant in Mol, Belgium. For the time being, the cumulated radioactive wastes are being conditioned in view of the plant decommissioning or of the resumption of the activities by Belgoprocess in order to meet the requirements of the Belgian utilities.
- In Germany, on the initiative of the German nuclear power utilities, the DWK was founded for fuel reprocessing; this company has taken over all the activities previously performed in this sector and is now planning the construction of a 1500-t/year reprocessing plant in the Gorleben area. The site and part of the preliminary safety report have already been approved by the authorities. The construction permit

should be issued within 1980 and operation of the plant should begin by the end of the eighties.

6. FAST REACTORS

As is known, thermal reactors exploit about one per cent of the energy potential of uranium; in fast reactors this percentage rises to at least 60% so that the uranium available today could practically supply energy for electricity generation for at least a few centuries.

This subject is particularly interesting for European countries, because of their dependence on foreign supplies also for uranium. On the relatively optimistic assumption that Europe will have access to one third of the world reserves of uranium, it is evident that a European nuclear program based on thermal reactors alone would necessarily have a limited horizon. Hence, the need to resort to fast breeder reactors.

Since fast reactors find their justification in the so-called plutonium economy, they could become commercial in 1990 and reach a fair degree of penetration at the beginning of the next century only if the present nuclear programs based on thermal reactors and reprocessing plants are implemented in good time so as to make available sufficient quantities of plutonium to feed the fast reactors.

So far Europe has built and successfully operated numerous experimental reactors and two 250-MW reactor prototypes. A large power station--the 1200-MW Superphenix-- is being built in France by a multinational venture in which also Germany and Italy participate. Other two stations of the same rating are in an advanced stage of design in Germany and Great Britain.

Such advanced results have not yet been reached anywhere else in the Western world. Therefore, the European nuclear industry may assume a very important role in this area, provided the programs now on the board are pursued.

The areas in which it is necessary to increase the research and development effort are basically fast reactor safety and fuel cycle technologies. For obvious reasons the latter have so far been developed on a limited scale, where it is necessary to have technologies suited for application on a commercial scale.

The money effort involved in the aforesaid research and development activities, as well as in the construction of the stated demonstration plants, is very great; yet, it is warranted by the role that fast breeder reactors will have to play in the European context as the only valid alternative to meet the future energy requirements.

REFERENCES

1. World Energy Conference, Istanbul (1977).
2. OECD, "Energy Balance of OECD Countries - 1974-1976", Paris (1978).
3. K.R.Willaima, "Projected Energy Requirements upto the Year 2000" International Symposium on Uranium Supply and Demand, London (1977).
4. ENI Statistical Summary (1976), Nucleonics Week.
5. OECO-IAEA, "Uranium: Resources, Production and Demand" (1977).

BIBLIOGRAPHY

1. H.Smith, R.B.Toombs, "Development of Conventional Energy Resources", 10th World Energy Conference, Istanbul (1977).
2. A.M.Angelini, "I Problemi della Produzione di Energia Elettrica: Un Approccio Sistematico". Accademia Nazionale dei Lincei, Roma 1978.
3. OECD, "Nuclear Fuel Cycle Requirements", Paris, (1978).
4. A.M.Angelini, "La Qualita in Campo Nucleare". XXI Nuclear Congress Rome (1976).
5. C.Salvetti, "Qualificazione dell'Industria Nazionale nel Settore Nucleare", Giornate dell'Energia Nucleare, Milan (1976).
6. S.Villani, "L'Energia Nucleare nella Prospettiva Europea".
7. European Economic Community. Documents of the European Parliament Sessions, 1977-1978. Doc. 519/77.

SOME TOPICS IN REACTOR PHYSICS

Ugo Farinelli

CNEN-CSN Casaccia

Santa Maria di Galeria, Rome, Italy

1. INTRODUCTION

Reactor Physics is an applied science. Its value rests in its ability to make predictions and to solve problems that are of practical interest. Its purpose is to deal with the neutron and gamma problems that together with heat transfer, materials, thermohydraulics and other engineering problems (and in strict interconnection with them) have to be solved in order to design and to operate, safely and economically, nuclear reactors and especially those of greatest practical impact, the power reactors needed for electricity production.

All this seems very trivial, but actually it must not be obvious if so much of the work that is carried out in reactor physics seems to have little connection with the final purpose, as we shall exemplify a little further.

Actually, it sometimes seems that applied physics problems fall into two categories: those that can be solved and those that are useful, and that the intersection between these two sets is void. Therefore, it is not strange that many people prefer to use their skills for problems that can be solved, and perhaps in an elegant way, rather than fight for compromise in trying to get approximate answers by the quickest and most effective way to the mundane preoccupations of reactor designers and operators. I am of course exaggerating to carry up my point, but there always seem to be everywhere discriminations against 'dirty' physics (for instance, 'clean' experiments are called those that are remotest from the actual practical situations). Therefore there is a tendency to evolve reactor physics (as many other applied sciences) by their own internal dynamics rather than by the questions that are put from the outside. Quoting from Alice in Wonderland, the Cheshire Cat said to Alice, "If you don't know where you want to go, then it doesn't really matter which road you take".

However, the progress of reactor physics is eventually shaped by three main factors (and I think in this order of importance the changes in the questions that are asked, the evolution of the 'environment', or of the instruments that are available (I am thinking especially of the computers), and finally the internal evolution of reactor physics as a science in itself. I shall try to say a few words on each of these three points, and to interpret some features of the history of reactor physics in terms of these factors.

2. A SHORT HISTORICAL REVIEW OF REACTOR PHYSICS

2.1 The Heroic Period (1942-1955)

In the heroic times of the beginning, feasibility was the question. Can one realize at all a chain reaction in these or those general conditions? Data were scarce, and one had to make the most of the scanty evidence from cross section measurements, from integral results, from theory, from educated guesses. Practically no computer was available (one of our smallest pocket calculators has today a capability surpassing by far that of the mechanical calculators that were then available): so one had to resort as much as possible to analytical solutions. Some of the basic achievements of that period are linked to the names of the most outstanding physicists of the time: Einstein (for instance for the slowing-down theory in a hydrogeneous medium), Fermi (for a large number of basic models, like the continuous slowing-down model), Wigner (cell theory, thermalization and much else, Von Neumann, Placzek and so many others. All these contributions consisted in very simple (and genial) physical intuitions that brought to simple analytical forms that could be solved directly or with a minimum of numerical calculations; the answers were precise enough for the basic feasibility problems to be solved.

2.2 The Exploratory Period (1956-1967)

A second period is characterized by a broad investigation of a very large variety of reactor concepts. In principle, very many combinations of fuel, moderator, coolant, structural materials etc. could give rise to a viable reactor. It was an illusion of the late 50's and most of the sixties that neutronic calculations and engineering conceptual design (or at most experimental reactors with relatively small power) could give an answer in terms of 'the best' possible reactor. An amazing set of reactor concepts was explored in that period: I recall as a few scattered examples, the spectral shift (H_2O-D_2O) reactor of B&W; the D_2O moderated - Na cooled reactor in South Africa; the liquid plutonium fast reactor experiment at Los Alamos; the D_2O moderated - gas cooled reactor in Czechoslovakia; the molten salt, thorium cycle reactor at Oak Ridge; the graphite sodium reactor at Atomics International; the organic moderated and cooled reactor in the US, USSR and Italy; the steam cooled fast reactor of General Electric; the intermediate spectrum thorium reactor sponsored by the South-West utilities in the US; the boiling-heavy water reactor built in Norway in a joint European venture (and still operating); the gas-suspension or water-suspension reactor concepts explored in the Netherlands; and many, many others.

In order to study these concepts, one not only needed to assess feasibility, but also to extract a number of economic and safety parameters that could allow some comparison between the various reactor concepts. The simplified analytical methods that had been developed especially for well-thermalized reactors, were no longer adequate; but in this period electronic computers appeared and developed rapidly to a size, speed and complexity that allowed the solution of very large numerical problems (in particular, as we shall see, systems of linear algebraic equations). Also the experimental data base on which calculations could be based and in particular the nuclear data, both from differential and integral measurements, expanded enormously.

The evolution of the mainstream of reactor physics in this period, in my opinion, is characterized by the translation of problems numerical form, and by the development of methods that were better adapted for this translation (for instance the S_n methods).

Time showed that the idea that a decision could be made a priori, objectively, on which reactor concept was 'the optimum' was a loser. Very few reactor concepts have survived (PWR, BWR, CANDU, RBMK(?) and LMFBR). But there is no reason to believe that these are the 'best' concepts; this is far from being proven. These are the concepts for which enough money has been concentrated to make a real breakthrough to commercialization. For different reasons - the fallout of military research, industrial competition, marketing, application of know-how gained in other industrial areas etc - enough research, development, engineering, experimentation, demonstration and prototype reactors have been funded, and last not least there has been a dumping of prices for the first installations produced in order to promote a large scale sale, and therefore, on the long range, a recovery of the initial investment over a large number of systems. It has been shown that at least 3 billion dollars (and perhaps much more) have been spent to bring each successful system to commercialization. And the situation may get even worse in the future. This concentration of funding, and of work, is the real reason of the breakthrough of the systems that have had success. There is no reason to believe that at least some of the other prospective systems would not have had the same success, had the same amount of funding and effort been spent on them.

2.3   The Consolidation Period (1968- ?)

This brings us into the third period of nuclear energy and reactor physics. Efforts, after the late 60's, were concentrated essentially on a few reactor types: those that had started gaining commercial success (BWR, PWR, CANDU) and a very few new candidates (LMFBR, HTGR) for which this same concentration of efforts gave good probability of future success.

At this point, feasibility was of course out of the question. The main effort went into optimizing these reactor types - but not paper-type, mathematical theory optimization; rather a very practical comparison of the actual detailed options for design and for operation. A number of variations to the basic reactor concepts were studied in this period, and some were actually routinely introduced: for instance, the steady increase in unit power for the reactor (from 400 MWe to 1200 MWe and more); the increase in burnup (from .8 or 1% to about 3% for LWR's); the rod-cluster control in PWR's; the burnable poisons, especially Gd for BWR's and more recently borated pyrex for PWR's; the adjuster rods and the two-phase control systems for CANDU's.

Pu recycle in thermal reactors has been widely investigated also with full-scale loading in a reactor, although it has not become a favourite option, at least for the time being. But most of the effort has actually gone into detailed design, in order to improve economics and safety features. Just to pick up one specific example, fuel elements for a BWR reactor are made up of a certain number of pins (an 8×8 square is a common choice); some of these pins contain burnable poison to control long-term reactivity; one may be used for structural purposes (to keep in place the spacer grids); the other pins may have different enrichments in order to reduce the power peak at the outer boundary of the element, where the water gap (needed for the cruciform control rod) introduces extra moderation. How many different enrichments should one have? And where should the different types of pins be located in the fuel element? In order to answer these questions (and there is no 'absolute' answer, no optimum of optima) one has to keep in mind a number of conditions, some of them very practical. Again, I will only mention a few. The flux-peaking changes with time, as a function of the burnup, which is different for the various pins. It makes no sense to go to a lot of effort to optimize at zero-life if this makes the situation worse at the end-of-life. The flux peaking changes with height in the fuel element, due to the different density of the coolant and the presence of steam. It changes with the position of the control rods (and this is a large effect) and therefore it depends on the way in which the reactor is operated. Optimization through life and height of enrichment and control rod pattern designed to keep the average (or the maximum?) power peak as low as possible is generally not the way to obtain the maximum burnup (or the minimum average initial enrichment). Shall we optimize for power or for energy?

The cost of having different enrichments increases with the number of enrichments. Where is the trade-off? (A new solution, which if adopted is going to put the optimization problem in a new, if not easier, light is a proposal by the fuel producers not to manufacture fuel to customer's specifications but to provide a fixed set of enrichments - probably differing by 0.2% steps - to fabricate pins that the fuel designer may combine to his fashion in a fuel element). If one reprocesses the fuel, it is probably not justifiable on economic grounds to separate different fuel pins, so that the whole fuel element is going to be reprocessed in one single batch, and one loses the entropy involved in differences of isotopic compositions. Finally, the number of different enrichments may be limited by safety reasons: in order to avoid to introduce the wrong pin in a certain position of the elements, the fuel pins and the grids have matching keys and latches; but the number of keys sufficiently different to ensure that no forcing of a wrong pin is possible, is limited.

In this third period, the computers have become much larger, much cheaper, much more effective and universally available. This was an absolute condition in order to be able to perform those detailed calculations that were needed to solve the problems mentioned above.

The main reactor physics effort in this period has been that of setting up standardized calculational systems and appropriate data bases that could be called upon for design and operational needs. These methods had to be relatively quick (because of the necessity of performing a very large number of detailed calculations) but at the same time sufficiently accurate, since one was looking at relatively small effects. The general answer to these (apparently conflicting) requirements were semi-empirical methods, adjusted to experiments (both critical experiments and data from operating power reactors) or to the results of a few benchmark calculations with more fundamental methods. The success of these methods was also linked to the fact that the range of systems to be studied (within each of the 4 or 5 reactor types investigated) was rather limited. As I will mention later, all these methods were based on some sort of synthe--sis (space, energy, space-energy).

2.4   Is there a Fourth Period?

Have we entered a fourth period of nuclear energy and reactor physics: It may be too early to be sure - but there are certainly symptoms that this may be the case. It may be the period of 'implementation' - the period in which a large number of power reactors are put into operation in more and more countries. But it is certainly also the period in which in many countries opposition to nuclear power is rising and getting more powerful. Most of us who have worked on nuclear power have been brought to believe by our personal experience that this is the safest option among those actually available today: and so we have the difficult task to try and have the others

share this conviction. It is, I say, a difficult task, and not only
on psychological grounds, but also on technical grounds. If many
thousands of reactors are going to be built, we cannot ignore events
that have an even extremely small probability to happen, if they are
going to have very large negative consequences. We therefore have to
study situations that may be very far from the reference situations
that we usually meet in design and operation, and for which very
little experimental evidence exists. In general, we are required to
<u>validate</u> our predictional systems, to show to others than ourselves
that they have a high degree of reliability for all the situations
that we consider. The arguments of 'quality assurance' that are domi-
nating the nuclear power scene have not spared reactor physics. This
means that semi-empirical methods, which rightly continue to hold the
field, need more back-up from fundamental methods and from basic
nuclear data, and in some cases also from some sophisticated experiments.

Some people think that in the present phase safety considerations and particularly safeguards and non-proliferation issues
will bring some old concepts and some new ones in the focus of investigations. I don't agree with them. I think that the essential part
of the work will continue on the present commercial reactor types
and on LMFBR's, with variations and improvements but, at least for
some time, without major changes, and that our most important tasks
at the present time are implementation and validation.

3. CALCULATIONAL METHODS IN REACTOR PHYSICS

3.1 Boltzmann's Equation and Its Integration

In principle, the great majority of reactor physics problems could be solved by integrating Boltzmann's equation:

$$\frac{1}{V}\frac{\partial \Phi}{\partial t} = -\vec{\Omega}\cdot\mathrm{grad}\ \Phi - \Sigma\phi + S + \iint \Phi'\Sigma_s dV'd\Omega' \qquad (3.1)$$

One has to add equations that link the source with the
flux, or that link the coefficients with Φ in the case of feedbacks.
In principle, the only phenomena that are not taken into account by
this equation are the fluctuations in the neutron population (Eq.
(3.1) is deterministic, while the nuclear phenomena are probabilistic) and the more accurate kinetic behaviour which needs consideration of delayed neutrons.

However, this is a rather formal statement - like saying
that all problems in electricity or electronics reduce to the integration of Maxwell's equations. When one comes to practical problems,
they have such a complexity that not only Boltzmann's equation cannot be integrated analytically (which is quite obvious) but it cannot
be integrated numerically either, even with the largest and fastest
computers available. I think this is a very important fact to be born

TOPICS IN REACTOR PHYSICS

in mind, and that explains why reactor physics is a continuous search for better approximations and more effective models. In order to understand why straightforward numerical integration in a complete practical case is impossible we must look at some numbers.

A LWR (just to make an example) has about 40,000 fuel pins. In order to perform a complete calculation, one should represent each fuel pin individually (even fuel pins that are equivalent at the beginning of life will not be equivalent as burnup goes on). Each fuel pin has axial variations, and in order to study it one should subdivide it into a certain number of regions (may be 10 may be 100 according to the degree of detail desired). Each pin is not uniform, there are variations of the flux (and eventually of the composition) going through the fuel, and then the cladding and then the moderator. The detailed description of a reactor might require as much as one million space points! But the situation concerning energy is not much better; the amount of detail which is available in the description of a cross section (for instance of U-238) is such that it would require several thousand energy points to represent it adequately. The dependence on angle adds another dimension to the problem. There is no way to solve a problem with several billion variables (product of space points by energy points by angular points) even with an idealized computer of the years to come.

3.2 Space and Energy Synthesis

All the solutions that are adopted to overcome this problem have to do with synthesis. There are various types of syntheses, and we will consider some of them here. We will neglect here the angular dependence of the flux (as in diffusion theory) which will be briefly mentioned further on.

The problem would be greatly simplified if the flux could be separated into energy and space:

$$\phi(E,\bar{r}) = f(E)\cdot\phi(\bar{r}) \tag{3.2}$$

This is unfortunately impossible, except for highly idealized problems of no practical value (homogeneous bare reactor). The method that is actually adopted is a succession of synthesis steps. For instance, the first may be a synthesis in energy, relative to a space-independent system, that allows to set up an amenable energy grid; the second may be a simplified one-dimensional treatment of the individual fuel pin to evaluate the effect of resonances; then one may have a further energy condensation at the level of the fuel pin to arrive to a few energy group representation; then a space synthesis, consisting in the calculation of the fuel element (in two dimensions and few groups); then a calculation of the whole reactor, represented by the different fuel elements. If this last calculation is made in three dimensions, another synthesis is generally operated:

either by further reducing the number of energy groups (1 to 4), or by synthesizing the space dependence in the radial and in the axial directions.

The first step (which is common for fast reactors, but is only now starting to be used in thermal reactors) consists in starting from the basic collections of evaluated nuclear data files (such as ENDF/B) and solving in some way the time and space independent expression of Boltzmann's equation:

$$(D B^2 + \Sigma)\phi = S + \iint \Sigma_s \phi' dv d\Omega' \qquad (3.3)$$

The first term represents the leakage in a bare, one-dimensional homgeneous system. This can be done in many ways, which we have no time to discuss here.

The second step - evaluation of the resonance self-shielding effect - is performed considering a single pin surrounded by its moderator (the elementary cell). It is a one-dimensional calculation and it uses some simplified form of transport theory - generally the integral transport or collision probability theory. Since many types of cells are present, the calculation is repeated for each type of cell (at least, for those which are sufficiently different to justify a separate calculation). A similar type of calculation is made for thermal reactors to find the space-energy distribution of the thermal neutron flux. Various degrees of sophistication can be used in this step: for instance, the thermalization of neutrons can be described in detail (and this may be important for systems containing plutonium). The space distribution again uses some simplified form of transport theory.

These calculations suppose that the result is not influenced by what happens in the surrounding cells: this is a first example of the approximations that are necessarily introduced by the synthesis procedure. For resonance calculations, a correcting factor close to 1 (the so-called Dankof factor) is generally introduced to take account of this fact. For thermal neutron flux calculations, it may in some cases be considered necessary to calculate more than one cell at a time (4 or 9 cells) to take into account interference effects. These multi-cell methods greatly complicate the procedure, not only because it is no longer a simple one-dimensional process, but also because it is necessary to consider a large number of combinations of adjoining cells. However, this is sometimes done for instance to deal with the fuel pins surrounding a pin of burnable poison or the corner pins in a fuel element, where large perturbations exist

The result of these calculations are few-group constants (anything from 2 to 30 energy groups, according to reactor type and calculational procedure). These constants are used in the following step, which is generally the calculation of a fuel element, or at most of a group of fuel elements (for instance the 4 elements surrounding the control rod in a BWR). This calculation is generally done by diffusion theory, in two dimensions and using a small number of mesh points (e.g. 4) per cell. Again, this procedure implies that one considers that the flux distribution inside a fuel element can be calculated independently of the fuel elements surrounding it, which is an approximation with a limited validity.

Suppose that we want to calculate, as a final result, the three-dimensional flux distribution in the reactor. This will generally require a further reduction of the number of energy groups which will be performed by weighting the cross sections over the flux distribution in each element.

The final 3-D calculation can be performed by straightforward diffusion theory, or it may involve a further space-synthesis process. This space synthesis can be obtained in several ways. One is to express the axial distribution $\phi(z)$ as the sum of a limited number of orthogonal functions:

$$\phi(z) = \sum_{i=1}^{N} a_i f_i(z) \tag{3.4}$$

and finding for each region in the (x,y) plane the set of values of the coefficients a_i that best approximates the solution to the diffusion equation (this is often done by applying variational principles). The finite elements technique, borrowed from structural mechanics, is another route that has recently been explored to alleviate.

In conclusion, several steps of synthesis are needed in the calculation of a reactor. Each of these steps introduces approximations and errors of which the reactor physicist must be aware. The best calculation will always be a compromise of the various approximations. It makes very little sense to aim at obtaining a terribly high degree of detail or accuracy in one of the variables, if this will be destroyed by one of the following steps. In this sense, the calculations that make use of the most detailed energy representation for the nuclear data, but neglect the actual geometrical complications of the real system, are unbalanced answers, and so are the transport calculations that assume simplified geometries or unrealistic simplifications for the cross sections.

We shall now make some considerations on the approximations of the transport equation.

3.3   The Diffusion Equation and Its Recent Developments

The diffusion equation is the well-known, most used approximation. It can be obtained from the Boltzmann equation or it can be derived directly. In its time independent form it says that, for a given energy group i:

$$\text{PRODUCTION OF NEUTRONS} = \text{LEAKAGE} + \text{ABSORPTION} + \text{SCATT.OUT} - \text{SCATT.IN}$$

$$S^i(\bar{r}) = \text{div}\left(D^i \text{ grad } \phi^i(\bar{r})\right) + \left(\Sigma_a^i + \Sigma_r^i\right)\phi^i - \sum_{j<i} \Sigma^{j \to i} \phi^j \quad (3.5)$$

If the neutrons are produced by fission,

$$S^i(\bar{r}) = \chi^i \sum_{j=1}^{G} \nu_j \Sigma_f^j \phi^j(\bar{r}) \quad (3.6)$$

where χ^i is the probability that a fission neutron has an energy within group i, ν_j is the average number of neutrons produced in a fission originated by a neutron having an energy in group j, Σ_f^j is the macroscopic fission cross section in group j.

If one combines (3.5) and (3.6), using condensed notations

$$L \phi - A \phi + T \phi = \frac{P}{K} \phi \quad (3.7)$$

where L is the 'leakage' operator, A the absorption, T the energy transfer operator, P the production operator. Since (3.7) is an equilibrium equation, describing a critical reactor, then we have to divide P by K (the usual K-eff) to obtain an equation that admits solutions also for non critical reactors (K ≠ 1). In this case, K is the eigenvalue of the equation (but many other eigenvalues could be introduced instead - for instance the concentration of boron in the moderator necessary to make the reactor critical, or the critical position of the control rods, etc.).

Many properties of the system (3.7) have been investigated in general terms; for instance, the consideration that it is not sel adjoint make the development of perturbation methods rather interesting.

But I should like to point out some trends in diffusion theory. The deduction of the diffusion equation is made under certai conditions that restrict its applicability. For instance, the diffusion equation does not work when the absorption cross section is not much smaller than the scattering cross section; when scattering is strongly anisotropic; in the vicinity of singularities, like point sources, or surfaces separating materials with different constants; in the presence of voids etc. Practice has shown that some of these limitations can be ignored in many cases (for instance the diffusion equation is normally used to predict neutron fluxes close to boundaries, and only in few cases the results are appreciably incorrect) But the diffusion equation is so convenient to use that it has been tried to extend its applicability beyond the limitations inherent in

its deduction..The weak point of the equation lies especially in the
diffusion coefficient D. Some of the improvements have therefore concerned the expression for this coefficient. For instance, in cases
in which there is preferential leakage of neutrons in one direction
(as in a fuel channel with a high void content) one can introduce a
diffusion coefficient which is higher in the axial direction than in
the other directions. By appropriate definition of this anisotropic
diffusion coefficient, it is possible to account for effects that
would otherwise be ignored by diffusion theory. Another anisotropic
diffusion theory calculates point-wise diffusion coefficients in the
three directions that depend not only on the material at the point
being considered, but also on the other materials that can be met at
a few mean free paths of distance. By doing this, one not only removes the limitations at boundaries, but also allows empty regions
to be considered. Finally, in order to deal with shielding problems
- where the inability of diffusion to account for anisotropies in the
flux makes it inapplicable in most cases - several alternatives have
been devised, from the removal-diffusion theory (where the neutrons
that have undergone no collision or only a collision at small angle
are considered separately as a directional flux, and only after the
first collision they make up the isotropic source for a successive
diffusion calculation) to the recent British methods using ad hoc
diffusion coefficients that have been evaluated separately for a
number of typical configurations by means of Monte Carlo calculations.

3.4   Monte Carlo Methods

Monte Carlo itself is a very attractive approach to solving
the Boltzmann equation. While the time taken for a diffusion calculation increases very rapidly with the number of dimensions and the
complexity of the representation, the time taken by Monte Carlo is
only mildly dependent on the complexity. Therefore Monte Carlo methods
are especially adapted to problems that involve very complicated
geometries and multi-variable siutations. This is the reason why most
Monte Carlo codes are capable of handling three-dimensional geometries with any degree of complication. The problem becomes the practical one of how to represent these geometries in the input of the
code, and a lot of improvement has been achieved in recent times in
simplifying the task of the user (e.g. combinatorial geometries).
Moreover, a very detailed structure of the cross section representation can be used without unduly affecting computational times. A
continuous representation of the energy variable is possible and is
included in some of the widely used Monte Carlo codes. Continuous
representation, or arbitrary subdivision, of the angular variable is
also possible.

A limitation of the Monte Carlo method is that it can only
supply a limited number of integral results at a few selected points
in the system. Monte Carlo is a statistical approach; its precision

relies on the number of neutron histories, and in particular on the number of events of a given type that are observed. Its precision therefore increases slowly with the number of histories (as $\sqrt{n}$) and the time needed to reach a good accuracy may be prohibitive. Due to the fact that relatively simple calculations are repeated a very high number of times, there is a large incentive to reduce as much as possible the time required by each step of the calculation (for instance, the generation of random numbers). In some cases, straight forward application of the Monte Carlo method would lead to unacceptable computing times even for very low precisions. A typical case is the study of a shield. If one starts from the source neutrons in the core, since the attenuation is of a factor 10^{14} or more, it woul be necessary to run at least 10^{14} stories to have the faintest indication of the leakage through the shield. One can then apply a numbe of the so called 'variance reduction techniques', that by biasing the probabilities of event in order to favor the neutron histories that are more likely to be of interest, but at the same time by assigning a weight to each history that compensates for that bias allow a great reduction in computation time. The most current variance reduc tion techniques are: importance sampling, in which an altered probability distribution is used in selecting the events: 'important' events are selected more frequently; among importance sampling techniques, path stretching according to the particle's direction and source energy biasing may easily reduce the computing time by an order of magnitude. The expected value technique consists in replacing a part of the random walk by an analytical estimation of its average value. Russian roulette allows early termination of unimportant histories, while splitting improves the estimates for importan histories.

All these techniques have two problems: first of all, they require some knowledge of the importance (and calculating the importance is just as difficult as calculating the flux, only a much lowe precision is required). Secondly, it becomes more difficult to evaluate the accuracy of the results one obtains, since if the importance functions are not precisely known, the variance reduction techniques alter the normal nature of the final distribution and therefore the assessment of the probable error becomes very questionable, althoug the mean value is hopefully not changed.

3.5   Conclusions

A numerical integration of Boltzmann's equation that keep account of the angular dependence of the flux is more time consuming than using the diffusion equation (by a factor that may range betwee 2 and 50!) but modern computers, and optimization of numerical techniques, has made this acceptable in dealing with one and two-dimensional problems. A full scale 3-D transport code has been recently announced but not yet distributed. The great majority of the transport codes uses Carlson's S_n method, which consists in dividing the angle that characterizes the direction of the flux in a discrete number of angular intervals (generally with equal spacing of cos θ). This method lends itself to numerical computation better than the expansion in Legendre polynomials. A number of difficulties, of analytical or numerical nature, have been experienced (the diamond difference breakdown, the ray effect etc.) and work is under way to get rid of them in all circumstances and in a reliable manner.

There are other methods, that we have no time to discuss or even to mention, to approximate the solution to the Boltzmann's equation. I will just recall here (in addition to the collision probability methods which in some form or another are used in most calculational methods at least in one step of the synthesis procedure) the *heterogeneous methods*. These methods, originally developed in the Soviet Union by Feinberg and Galanin, are well adapted to channel-type reactors, like the heavy water reactors. They consider the fuel elements as lines immersed in the moderator that are sources of fission neutrons and sinks of thermal neutrons. These methods, although developed before the advent of the computers, are well adapted to numerical evaluation. With some improvements and complications (like dipole or multipole terms in the neutron source) they can be used very effectively in studying problems of heavy water reactors.

4. TRANSLATION INTO NUMERICAL PROBLEMS AND THEIR SOLUTION

4.1 Discretization of the Diffusion Equation by Integration

The following considerations, although made for reactor physics applications and in this context, are actually applicable to a great variety of problems in applied and in pure physics, with a minimum of changes.

The diffusion equation, even in its simplest form, time-independent, for a single energy and with assigned source:

$$-\text{div } D(\underline{r})\text{grad } \phi(\underline{r}) + \Sigma(\underline{r}) \phi(\underline{r}) = S(\underline{r}) \quad (4.1)$$

cannot be integrated analytically, unless D, Σ and S are constant or have very special space dependence, and the geometry of the system and the boundary conditions are extremely simple. One therefore has to transform this problem into a numerical problem, so that it can be handled by a computer.

This can be done in several ways, but the most common procedure is to transform this equation into a system of algebraic equations (finite difference equations). This again can be done in different ways (power series, integration, variational methods....). I will briefly outline the method by integration.

Let us consider for the moment a one-dimensional problem, so that the only space variable is x. (4.1) becomes:

$$-\frac{d}{dx}\left[D(x)\frac{d\phi}{dx}\right] + \Sigma(x)\,\phi(x) = S(x) \qquad (4.2)$$

$\phi(x)$ is the function we want to calculate.

Let us suppose we want to integrate this equation between 0 and X and that the boundary conditions are

$$\phi(0) = \phi(x) = 0 \qquad (4.3)$$

Let us subdivide the region 0-X into N intervals by means of N-1 points x_i and let us consider the two intervals about x_i (see figure)

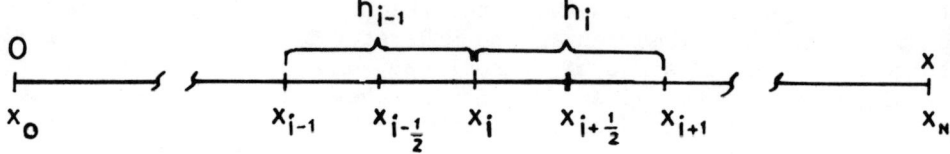

The widths of the two intervals:

$$h_{i-1} = x_i - x_{i-1}; \qquad h_i = x_{i+1} - x_i$$

will in general be different. Let us also consider the middle points of each interval, which we will call conventionally $x_{i-\frac{1}{2}}$ and $x_{i+\frac{1}{2}}$. We shall make another assumption: if there are discontinuities in D, Σ or S (which is a rather common case in practice) we will choose the points (the 'mesh points') so that all discontinuities fall on one or another mesh point.

Let us now integrate Eq.(4.2) in the interval between $x_{i-\frac{1}{2}}$ and $x_{i+\frac{1}{2}}$. One gets

$$-D_{i+\frac{1}{2}}\left(\frac{d\phi}{dx}\right)_{i+\frac{1}{2}} + D_{i-\frac{1}{2}}\left(\frac{d\phi}{dx}\right)_{i-\frac{1}{2}} + \int_{x_{i-\frac{1}{2}}}^{x_{i+\frac{1}{2}}} \Sigma(x)\,\phi(x)\,dx = \int_{x_{i-\frac{1}{2}}}^{x_{i+\frac{1}{2}}} S(x)\,dx \qquad (4.4)$$

We have indicated with an index K the value of a variable at point x_K. These values at mid-points in the intervals are always defined, because if there are discontinuities they fall at mesh points. The second integral can be performed, if necessary (in case of a discontinuity) by splitting the integral at the mesh point:

$$\int_{x_{i-\frac{1}{2}}}^{x_{i+\frac{1}{2}}} S(x)\,dx = \int_{x_{i-\frac{1}{2}}}^{x_i^-} S(x)\,dx + \int_{x_i^+}^{x_{i+\frac{1}{2}}} S(x)\,dx = \hat{S} \qquad (4.5)$$

Let us introduce the approximations:

$$\left(\frac{d\phi}{dx}\right)_{i+\frac{1}{2}} \simeq \frac{\phi_{i+1} - \phi_i}{h_i}$$

and

$$\left(\frac{d\phi}{dx}\right)_{i-\frac{1}{2}} \simeq \frac{\phi_i - \phi_{i-1}}{h_{i-i}} \tag{4.6}$$

These approximations are always acceptable provided we have chosen the intervals small enough since the flux is continuous for physical reasons and it must even be slowly variable for diffusion theory to be applicable; for instance it is sufficient (but not necessary) to take all intervals with a width of the order of D. As far as diffusion theory is applicable (and therefore $\Sigma_a < 1/D$) ϕ will not change very much over a distance D, and it will be generally adequate to assume a linear variation over this distance.

This same consideration of a slow variation of ϕ allows us to perform the other integration in (4.4) without knowing $\phi(x)$ in detail:

$$\int_{x_{i-\frac{1}{2}}}^{x_{i+\frac{1}{2}}} \Sigma(x) \phi(x) \, dx = \hat{\Sigma} \, \phi_i \tag{4.7}$$

Equation (4.4) thus becomes:

$$-\frac{D_{i+\frac{1}{2}}}{h_{i-1}} \phi_{i-1} + \left(\frac{D_{i-\frac{1}{2}}}{h_{i-1}} + \frac{D_{i+\frac{1}{2}}}{h} + \hat{\Sigma}_i\right)\phi_i - \frac{D_{i+\frac{1}{2}}}{h} \phi_{i+1} = \hat{S}_i \tag{4.8}$$

If we write such an equation for all the couples of adjoining intervals, we end up with N-1 unknowns (the values ϕ_1, ϕ_2, $\cdots$, ϕ_{N-1} of the flux).

This can of course be written in matrix form:

$$A \phi = S \tag{4.9}$$

where ϕ and S are line vectors and

$$A = \begin{pmatrix} \frac{D_{\frac{1}{2}}}{h_0} + \frac{D_{1+\frac{1}{2}}}{h_1} + \hat{\Sigma}_1 & -\frac{D_{1+\frac{1}{2}}}{h_1} & 0 & 0 & \cdots \\ -\frac{D_{1+\frac{1}{2}}}{h_1} & \frac{D_{1+\frac{1}{2}}}{h_1} + \frac{D_{2+\frac{1}{2}}}{h_2} + \hat{\Sigma}_2 & -\frac{D_{2+\frac{1}{2}}}{h_2} & 0 & \cdots \\ 0 & -\frac{D_{2+\frac{1}{2}}}{h_2} & \frac{D_{2+\frac{1}{2}}}{h_2} + \frac{D_{3+\frac{1}{2}}}{h_3} + \hat{\Sigma}_3 & -\frac{D_{3+\frac{1}{2}}}{h_3} \\ \vdots & \vdots & \vdots & \vdots \end{pmatrix} \tag{4.10}$$

This matrix has non-zero terms only in correspondence to the principal diagonal and the two diagonals immediately adjoining it.

Schematically:

$$A \equiv \begin{pmatrix} & & 0 \\ & & \\ 0 & & \end{pmatrix}$$

4.2 Extension to Two and Three Dimensions

When, instead of just one dimension, we consider two dimensions, for instance x and y, in addition to the derivatives made with respect to x we shall also consider those with respect to y. When we discretize the equation, in a similar way to what we have done previously, by choosing an appropriate number of mesh points in the plane, the value of the flux in each point will be linked not only to the values of the flux in the mesh points immediately to the left and to the right, but also immediately above and below the point we consider.

If we introduced two indices to identify each mesh point, A would have 4 dimensions and things would be difficult to handle. We can continue to use vectors to represent ϕ and S if we adopt the following convention. We use just one index, and we number orderly the mesh points in the plain following one after the other the lines parallel to the x axis. This implies that the mesh points are chosen along such lines (although these lines may have variable distances from each other) and that the points are also aligned along the y-direction.

1	2	3	4	5	6	7
8	9	10	11	12	13	14
15	16	17	18	19	20	21
22	23	24	25	26	27	28
29	30	31	32	33	34	35
36	37	38	39	40	41	42

The derivative with respect to x in x_i will still contain the terms in (i-1) and (i+1); the derivative with respect to y, on the other hand, will contain the terms immediately above and below point x_i, that is (i-K) and (i+K) where K is the number of vertical lines. The matrix A will be larger (at least, if the total number of mesh points is greater, which will be the normal case) but only slightly more complicated, in that it will contain in addition to the three non-zero diagonals of the one-dimensional case, two more non-zero diagonals, K lines away from the main diagonal (i.e. these representing the (i-K) and (i+K) terms).

In a similar way, it is possible to extend the same method to three dimensions.

$$A \equiv \begin{pmatrix} & & 0 \\ & 0 & \\ & 0 & \\ 0 & & \end{pmatrix}$$

Two new non-zero diagonals will appear in A.

4.3   The Energy Dependence

We can further complicate the problem by introducing the energy dependence. Let us subdivide the energy range of interest in G intervals (the energy groups) and suppose that we know the average values of the cross sections in each group. We shall then have G diffusion equations:

$$-\mathrm{div}(D^i \,\mathrm{grad}\, \phi^i) + (\Sigma_a^i + \Sigma_r^i)\phi^i - \sum_{j<i} \Sigma^{j \to i} \phi^j = S^i \qquad (4.11)$$

$$i = 1, 2, \cdots, G$$

where the upper index refers to the energy group, Σ_a^i is the absorption cross section, Σ_r^i is the removal cross section, representing the neutrons slowed down out of group i by elastic or inelastic scattering, $\Sigma^{j \to i}$ the transfer cross section by scattering from group j to group i, and the limitation j<i in the summation indicates that we do not consider upscattering (in usual notation, group number 1 is at the higher end of the energy scale, group G at the lower end). We still suppose that S^i is assigned.

It is not possible to arrive at discretization by writing a system $A^i \phi^i = S^i$ for each group because the exchange terms connecting the various groups appear in the summation of (4.11). But it is possible to write one system only $A\phi = S$ that includes all energy groups. To do this, we define a flux vector that includes orderly all the values of the fluxes of group 1 in the various mesh points, then all those of group 2 and so on up to group G:

$$\phi \equiv (\phi_1^1, \phi_2^1, \phi_3^1, \cdots \phi_N^1, \phi_1^2, \phi_2^2, \cdots \phi_N^G) \qquad (4.12)$$

The same procedure will be followed for the source S. If there are N mesh points and G energy groups, the dimension of these vectors will be N × G.

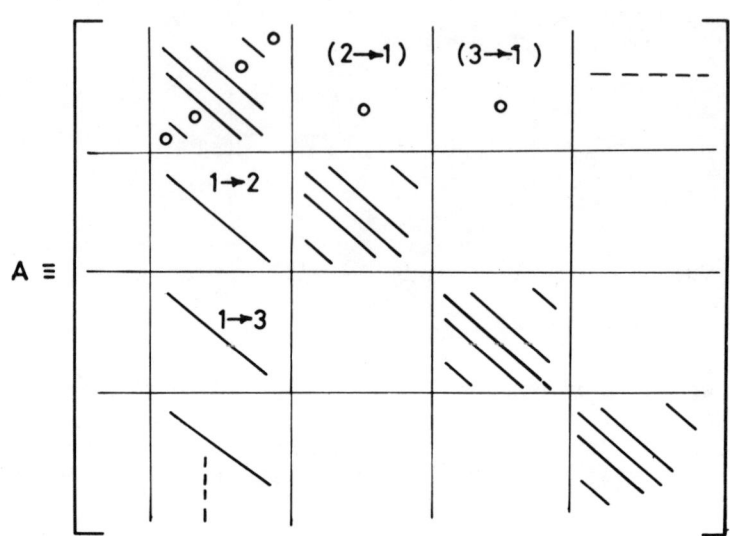

Correspondingly, the matrix A will have $(N \times G)^2$ elements, a limited number of which will be non-zero as schematically indicated in the figure. We can think of it as formed of submatrices; those along the main diagonal represent the diffusion of neutrons within each energy group. The other submatrices will have only non-zero elements if they are below the main diagonal, and then only on their own main diagonal: such elements represent the transfer of neutrons from one group into another in the same space point.

4.4   The Iterative Solution of the Numerical Problem

Having transformed the differential diffusion equation (or equations) into an algebraic system is only the beginning. To say that the solution of $A \phi = S$ is

$$\phi = A^{-1} S \qquad (4.13)$$

is a purely formal statement. In a typical case that we usually have to deal with, there may be about 10,000 mesh points (for instance 100 × 100 in 2 dimensions, or 30 × 30 × 10 in 3 dimensions) and 5 or 10 energy groups. ϕ and S are then a 100,000 components vector and A has 10^{10} elements. Even if only about a million of these components are likely to be different from zero, this no longer holds for A^{-1}.

It is necessary to find a 'direct' method, which does not go through the inversion procedure: the 'conventional' inversion by the Kramer method involves something like N! operations for a matrix of order N. This is impossible even for relatively small matrices; if N = 20, N! = 2.4×10^{18}. Even with a very fast computer, this would require several thousand years. For N = 50, the time required would be much longer than the age of the universe. Moreover, these methods would bring to cumulation of errors, and round-offs would make any number just random in a very short time. So another requirement for our method is that errors do not cumulate.

The solution to this problem is in the iterative methods. Their principle is the following.

Let us split A into two matrices:

$$A = M - N \qquad (4.14)$$

chosen such that M^{-1} is easily known. Our equation becomes:

$$M\phi = N\phi + S$$

or

$$\phi = M^{-1}(N\phi + S) \qquad (4.15)$$

The iterative proce starts with a 'guess' flux vector, $\phi^{(0)}$; this is put in the second mumber and one obtains a 'first approximation flux' $\phi^{(1)}$:

$$\phi^{(1)} = M^{-1}\left(N\phi^{(0)} + S\right) \qquad (4.16)$$

If our 'guess' had been correct (which is of course extremely unlikely) then $\phi^{(1)}$ would be equal to $\phi^{(0)}$ and it would be the solution of the problem. Since this will not be the case, we introduce $\phi^{(1)}$ - at the right-hand side of the equation (4.15) to obtain $\phi^{(2)}$, the 'second approximation flux':

$$\phi^{(2)} = M^{-1}\left(N\phi^{(1)} + S\right)$$

The procedure is then repeated; it can be shown that under very general conditions (that are always satisfied in our problems) one reaches convergence, that is to say that if one continues to iterate, the flux at the first member becomes more and more similar to that at the second member and therefore to the solution. More exactly, one will always find a number I_{min} of iterations such that for $I \geq I_{min}$

$$\frac{|\phi^{(I)} - \phi^{(I-1)}|}{|\phi^{(I)}|} < |\varepsilon| \qquad (4.17)$$

ε being an assigned quantity, as small as wished. Eq.(4.17) is called the convergence criterion, and it is tested after each iteration; when it is satisfied, the iterative process is terminated.

How do we split A into M and -N?

Let us consider A as formed by its main diagonal d, a part below it, e and a part f above it; and let us construct three matrices that are equal to A for d, e and f respectively and zero elsewhere:

$$A = \begin{pmatrix} & & f \\ & d & \\ e & & \end{pmatrix} \quad D = \begin{pmatrix} & & 0 \\ & d & \\ 0 & & \end{pmatrix} \quad E = \begin{pmatrix} & & 0 \\ & 0 & \\ e & & \end{pmatrix} \quad F = \begin{pmatrix} & & f \\ & 0 & \\ 0 & & \end{pmatrix} \quad (4.18)$$

We shall consider three splittings of A.

The Jacobi method consists in putting

$$M = D; \quad -N = E + F \tag{4.19}$$

The Gauss-Seidel method puts

$$M = D + E; \quad -N = F \tag{4.20}$$

and the more general relaxation methods assume:

$$M(\lambda) = \frac{D}{\lambda} + E \tag{4.21}$$

where λ is a positive parameter.

Let us consider them in practice.
Jacobi assumes for the first component of the vector (suffix 1):

$$[M \phi^{(0)}]_1 = a_{11} \phi_1^{(0)} \tag{4.22}$$

$$\phi_1^{(1)} = \frac{1}{a_{11}} \left\{ S_1 - (a_{12} \phi_2^{(0)} + a_{13} \phi_3^{(0)} \cdots + a_{1N} \phi_N^{(0)}) \right\} \tag{4.23}$$

and for the second component

$$\phi_2^{(1)} = \frac{1}{a_{22}} \left\{ S_2 - (a_{21} \phi_1^{(0)} + a_{23} \phi_3^{(0)} \cdots + a_{2N} \phi_N^{(0)}) \right\} \tag{4.24}$$

and so forth.

In general at the K-th interation:

$$\phi_\ell^{(K)} = \frac{1}{a_{\ell\ell}} \left\{ S_\ell - \sum_{m=1}^{\ell-1} a_{\ell m} \phi_m^{(K-1)} - \sum_{m=\ell+1}^{N} a_{\ell m} \phi_m^{(K-1)} \right\} \tag{4.25}$$

TOPICS IN REACTOR PHYSICS

and the convergence criterion will depend on the sign of

$$\Delta = \frac{\sqrt{\sum_{i=1}^{N} (\phi_i^{(K)} - \phi_i^{(K-1)})^2}}{\sqrt{\sum_{i=1}^{N} \phi_i^2}} - \varepsilon \qquad (4.26)$$

The Gauss-Seidel approach is actually very similar to Jacobi's; in evaluating the components of the vector in approximation K, instead of using all the components of the flux in approximation K-1, it uses those components in approximation K that it has already calculated (i.e. the components of index smaller than the current one):

$$\phi_\ell^{(K)} = \frac{1}{a_{\ell\ell}} \left\{ S_\ell - \sum_{m=1}^{\ell-1} a_{\ell m} \phi_m^{(K)} - \sum_{m=\ell+1}^{N} a_{\ell m} \phi_m^{(K-1)} \right\} \qquad (4.27)$$

The relaxation method is just an interpolation or an extrapolation of these two methods:

$$\phi_\ell^{(K)} = \frac{1}{a_{\ell\ell}} \left\{ S_\ell - \sum_{m=1}^{\ell-1} a_{\ell m} (\phi_m^{(K-1)} + \lambda \, [\phi_m^{(K)} - \phi_m^{(K-1)}]) + \sum_{m=\ell+1}^{N} a_{\ell m} \phi_m^{(K-1)} \right\}$$

$$(4.28)$$

The choice of the iteration method will influence the speed with which convergence is reached, but not the final result. In order to use relaxation methods to their best, one should be able to predict the behaviour of the flux shape from one iteration to the next. Some sophisticated methods, and some empirical recipes, are available on how to choose (and change) λ in order to speed up convergence. For instance, one sometimes assumes $\lambda = 1$ for the first iterations and then one chooses an appropriate value of λ according to the behaviour of the flux in the first iterations. In some codes, λ is changed at each iteration.

How many operations are needed at each iteration? For a 2-dimensional, one-energy group case with 10,000 mesh points, we have 5 non-zero elements for each matrix line.

For each calculation like (4.28) we have about 10 operations. This for each mesh point; so the total number of operations is about 100,000. This has to be repeated for each iteration. The number of iterations needed may be in a typical case between 50 and 100. So the total number of operations is of the order of 10^7, or about 10 seconds for an average computer.

When we have many energy groups, it is not necessary to

calculate them all at the same time. For instance, in the case of 2 groups:

$$A = \begin{pmatrix} A_{11} & 0 \\ A_{12} & A_{22} \end{pmatrix} \quad (4.29)$$

and the system can be written as

$$A_{11} \phi^1 = S^1$$
$$A_{22} \phi^2 = S^2 - A_{12} \phi^1 \quad (4.30)$$

and the two equations can be solved successively.

4.4   The Eigenvalue Problem

In a multiplying medium, as we have seen, the source is:

$$S^i = \chi^i \sum_{j=1}^{G} \nu^j \Sigma_f^j \phi^j \quad (4.31)$$

which we will represent formally by

$$S = \sigma \phi \quad (4.32)$$

where σ is the source, or production, operator in matrix form.

The system to be solved will be:

$$\left(A - \frac{\sigma}{K}\right) \phi = 0 \quad (4.33)$$

where $K = K_{eff}$ is the eigenvalue of the problem.

To solve this system one has to call upon a somewhat more complex iterative procedure.

One assumes a guess flux $^{(o)}\phi$ and a guess eigenvalue $^{(o)}K$ (for instance $^{(o)}K = 1$). One thus calculates a guess value for the source

$$^{(o)}S = \frac{\sigma}{^{(o)}K} {}^{(o)}\phi \quad (4.34)$$

One then solves this equation by the iterative procedure seen before; that is one starts by putting

$$\phi^{(o)} = {}^{(o)}\phi$$
$$\phi^{(1)} = M^{-1} (N {}^{(o)}\phi + {}^{(o)}S) \quad (4.35)$$

and continues to iterate until (after a sufficiently large number I of iterations) one calculates the solution $\phi^{(I)}$. We shall call these iterations 'internal iterations'. The result

$$\phi^{(I)} = {}^{(1)}\phi \tag{4.36}$$

can be considered as the value of the first 'external iteration'. One also needs a first iteration value for K. The common choice is

$$^{(1)}K = {}^{(0)}K \frac{|\sigma\ {}^{(1)}\phi|}{|\sigma\ {}^{(0)}\phi|} \tag{4.37}$$

that tends to reduce the difference between $^{(1)}S$ and $^{(0)}S$.

With the first iteration source $^{(1)}S$ one solves the equation

$$A\ {}^{(2)}\phi = {}^{(1)}S \tag{4.38}$$

by an appropriate number of internal iterations, and assuming of course $\phi^{(0)} = {}^{(1)}\phi$ as a guess flux. This procedure is iterated until, after a sufficiently large number, M, of cycles of internal and external iterations one finds values for the flux and the eigenvalue such that:

$$\left|\frac{{}^{(K)}S - {}^{(K-1)}S}{{}^{(K)}S}\right| = \frac{\left|{}^{(K)}K^{-1}\sigma\ {}^{(K)}\phi - {}^{(K-1)}K^{-1}\sigma\ {}^{(K-1)}\phi\right|}{\left|{}^{(K)}K^{-1}\sigma\ {}^{(K)}\phi\right|} < \varepsilon_e \tag{4.39}$$

where ε_e is the convergence criterion one has chosen for the external iterations (convergence on the source is the most common but not the only possible criterion).

This procedure, with external and internal iterations, obviously stretches the times as compared with a non-multiplying medium. However, the total time is not the time one had before multiplied by the number of external iterations; it is smaller, because it is not necessary to calculate with good accuracy the flux in the first iterations, since this flux is only used to obtain a new approximation for the source. The time can so be reduced; for instance, in the case of 10,000 mesh points, 30 energy groups and assuming that one needs about 30 external iterations, the computing time can be restricted to about 15 or 20 minutes.

A limiting case of this method is the alternate iteration method, by which only one internal iteration is made for each external iteration (in other words, the value of the source is corrected at each internal iteration).

Schematically, the procedure is the following:

$$(o)_S = \frac{\sigma \, (o)_\phi}{(o)_K}$$

$$\phi^{(1)} = M^{-1} \left((o)_S + N \, (o)_\phi \right)$$

$$(1)_K = (o)_K \frac{|\sigma \, \phi^{(1)}|}{|\sigma \, (o)_\phi|} \qquad (4.40)$$

$$(1)_S = \frac{\sigma \, \phi^{(1)}}{(1)_K}$$

$$\phi^{(2)} = M^{-1} \left((1)_S + N \, \phi^{(1)} \right)$$

$$(2)_K = (1)_K \frac{|\sigma \, \phi^{(2)}|}{|\sigma \, \phi^{(1)}|}$$

and so on. The convergence criterion may still be (4.39). This method allows in many cases important time savings, but in very complex cases there may be difficulties in reaching convergence.

5. MODULAR CODING SYSTEMS

5.1 Introduction

Modular Codes and Modular Coding Systems are not a new subject, but they are certainly an open subject, which is presently being discussed in many places. There are, as we shall see, many incentives to setting up computational capabilities in the nuclear energy field (and in many other fields) that are made up of a certain number of 'modules', or relatively small codes, that can be connected with each other in many different ways, and that allow a very flexible use of the computer.

I shall try to show that there is no unique, 'best' solution to the problem of modular coding and that the choice among the very different approaches that are available should be made on the basis of the particular conditions met in a real situation: the type of incentives, the computer environment available, the problems to be dealt with etc.

5.2 Definitions

I shall start by giving a few formal definitions. These definitions were agreed at a specialists' meeting on modular systems which was convened a few years ago at Ispra under the auspices of the NAE Committee on Reactor Physics and the NEA Computer Programme Library.

Module

'A module is a calculational algorithm transforming input data, having a physical meaning, into output data, also having a physical meaning, according to a simply defined physical process".

INPUT ⟶ | MODULE | ⟶ OUTPUT

This means that we do not consider as a module, in our language, a manipulation of data that will not change their physical meaning, like for instance, re-ordering the data or changing their format.

Modular System

"A modular system is a system that allows the execution of a sequence of modules with various PATHS to be selected either at loading or/and at run time, and that provides the flow of information between modules".

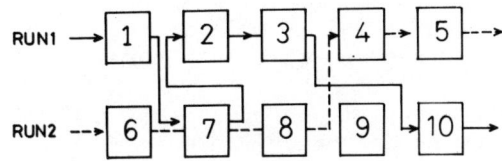

In order to qualify for a modular system, the sequence of modules to be executed must not be fixed; it may be specified at load time, or it may be chosen by the system itself, according to instructions, as a function of the results obtained at each step of the calculation.

Interface

"An interface is the collection of data which is required by the operation of a module and/or produced by the operation of a module and which is available for use by other modules".

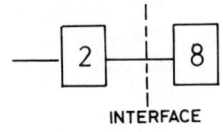

For instance you will have an interface between module 2 an module 8, which is the collection of data produced by module 2 and which are used in module 8 in the following calculation.

Data Pool

"The Data Pool is the realisation of one or more interfaces between modules relevant to a particular problem".

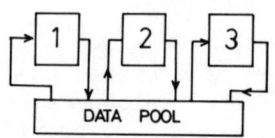

5.3 Purposes of a Modular System

There are many possible motivations for setting up a Modular Code System, which in general fall into three categories:

1st: AUTOMATE the sequential performance of a number of codes without time-consuming and error producing manual operations:

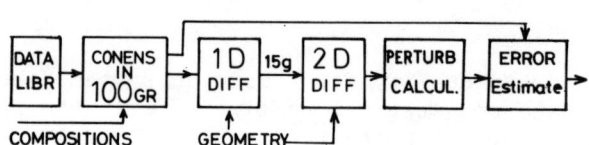

2nd: BREAK-DOWN large programs into smaller units that can be accommodated into a smaller computer

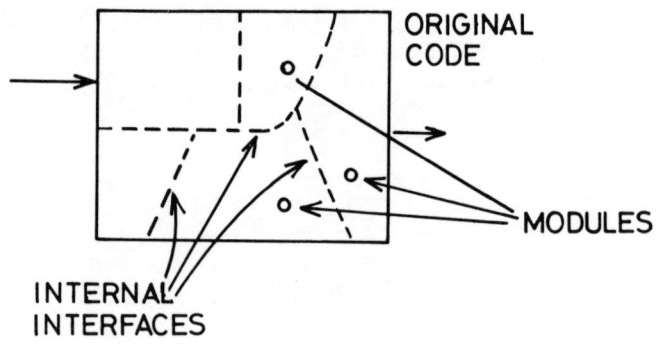

3rd: REDUCE programming efforts and duplications by making different codes compatible, and reducing the size (and the physical problem involved) in each code; make the codes easier to check and to change; reduce input-output efforts.

Historically, the first motivation came first and remained for a long time the most important one. What happened was that one had to run routinely several codes one after the other, using the output from one code as the input for the next. This generally required manipulation of cards and tapes, re-arrangement of data, changes in formats etc. All these manual operations took time and were likely to introduce errors. So it was natural to try and automate this sequence of operations. As long as the sequence remained constant, we did not have a 'modular system' but a 'linked-codes system' or a 'chain of codes'; however, once the interface problems were solved for a number of codes, the next step to a flexible path through the codes was a natural and not too difficult one.

The second motivation for a modular system is to break down large programs so that you can accommodate them on a small computer, but it is not segmentation. I would like here to remark the difference between Modular Coding and Segmentation. Segmenatation is a common practice in programming. It does not require a segment to be a module in the sense that it does not necessarily have a physical meaning. The operation may be a purely mathematical one, to switch in and out of the fast memory a set of data, like when one considers one group after the other in a diffusion calculation. Even if segmented, this is just one module, because it performs just one physical transformation. If the possibility exists, it is most proficient to reduce the size of a program by modularization than by segmentation, because each module can be treated and used independently, and this allows a much greater flexibility.

The third motivation is to reduce the programming efforts. One reason is that modular coding makes codes easier to check and to debug, because you have smaller codes and at the same time codes that have physical meaning, so that you can check each code by itself not only from the mathematical point of view but also as to the physical content. Also when you want to change a part of a calculational scheme you just have to change one module and not a very large code.

For instance, in the U.S. this third motivation was the most important in going to modular coding. Another associated reason in the U.S. concerns codes which are not released by the industries because they contain some proprietary information. Now, the proprietary information is likely to be contained only in a limited part of the code - for instance what thermo-hydraulics model was used for the slip correlation in a BWR. Instead of classifying one large code, things are made much simpler if we have a system in which just one module is classified or commercially protected. It is sufficient to take this module out or replace it with a dummy module or with a generally available option, and all the rest can be made available and people can put in their own modules for that part which is missing. Finally the modular coding reduces very much the Input/Output routines, how to print, how to read data. You just have to face the problem once when you are designing the modular code system, and not every time you are writing a module, because you will have a special routine for input and output of the whole system.

So these are three rather different and somewhat conflicting motivations because if you want to establish for instance what is the optimum size for a module, this will vary depending on the relative importance you attribute to the three motivations.

5.4   The Main Options for Modular Systems

Correspondingly, but not in a one-to-one relation with motivations, there are three main different solutions to the problem of modular coding. In fact, what in my opinion draws the basic distinction among the various solutions for modular coding is the way in which the problem of interfaces is dealt with.

The first solution is what we call *Lubricating Routines*. These routines are small codes that perform data handling operations that make the output of one code compatible with the input requirements for another code. In general, the output of a code has a different format, a different organisation of the information than is required by the input of the other code. The handling of the data which is needed, is particular to the couple of modules we are considering.

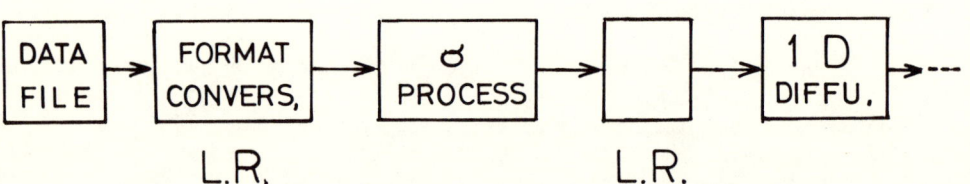

For instance you pass from the Data File to the cross section processing module by means of a format conversion which is a lubricating routine. From the output of the cross section processing you enter the 1-D diffusion code by changing, re-arranging in general the cross sections and by introducing the geometry of the system. This is another lubricating routine.

This type of system was adopted, for instance, in one of the first modular systems used in Nuclear Physics and Nuclear Calculation which is the CARONTE System developed at Ispra.

The second alternative is to define *Rigid Interfaces*. All possible input and output data are defined physically, dimensionally and as format once and for all and stored and retrieved at a fixed location.

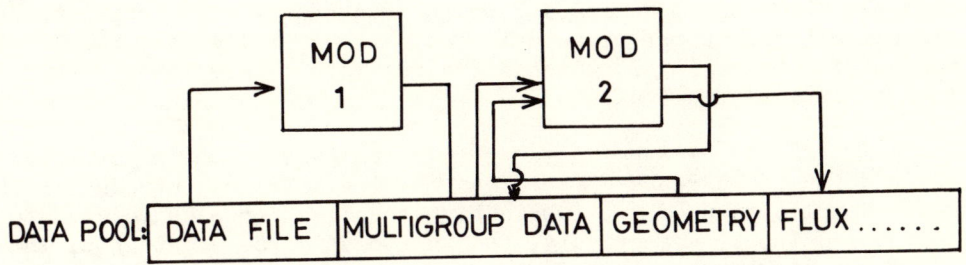

The cross sections will always be written in a certain sequence, in a certain way according to fixed rules; they will always be found in a fixed part of the data vector. Actually, in order to define the rigid interfaces, you will define essentially the Data Pool which is a vector and you know then in what point of the vector exactly a certain piece of data will be found. All modules will have to conform to this definition. They will read their inputs and write their outputs according to these rigid specifications.

This solution is being used in number of codes and modular system. For instance in the U.S. there has been a constant effort to define rigid interfaces for fast reactor calculations.

The third system is to define *Flexible Interfaces,* which means that one still has a data pool, a data vector, but the content of this vector is not fixed, it will change from time to time, from problem to problem according to what is being processed. The data pool (or the interface) itself contains information on the sequence, format, type of data it contains.

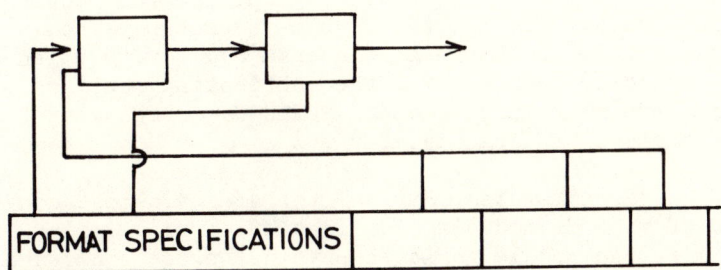

Each of these three approaches has advantages and disadvantages.

The approach of the 'Lubricating Routines' is the simplest if one wants to use already existing stand-alone codes without changing anything. It has two main disadvantages:

(1) the lubricating routine is characteristic of both the code from which the output comes and the code for which the input is prepared; if one increases the number of codes in the modular system as N the number of lubricating routines increases nearly as (N!). (Actually this is not correct, because you cannot combine all the modules, there are some modules which have no physical reason to be interfaced, the number of actual combinations is limited; but the number of lubricating routines increases rather more rapidly than N in any case, say as N^2).

(2) there is no 'data pool' or no general way of storing information to be re-used (this can be alleviated by introducing the extra complication of two lubricating routines for each module, for output and input respectively, feeding to and from a data pool in a standardized form: but this is actually a more complicated way of realizing a system with rigid interfaces, with all the difficulties and few of the advantages.

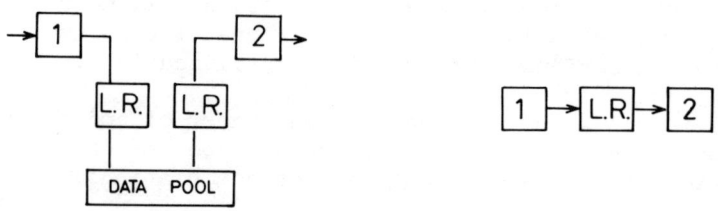

The system of lubricating routines (as evidenced by the experience with CARONTE) is well adapted if one is dealing with codes made by others, can accommodate them on his computer (does not have to break them down) does not require too much flexibility, is ready to spend a consistent amount of time in writing lubricating routines each time a new module is added; the result is probably machine dependent due to the particular nature of the data dealt with which suggest the use of machine language for the lubricating routines (it is extremely cumbersome to use standard FORTRAN for changing formats, rearranging strings of data etc.).

The second system, Rigid Interfaces, is most convenient from the point of view of reducing duplication of programming effort. One can have just one module for each physical processing, and this is changed only when a substantially more efficient algorithm is developed for the same function. Rigid Interfaces have been defined in the U.S. (they are called Standard Interfaces there), they are adopted for instance to a large extent in the ARC modular system at Argonne, and rigid interfaces are also present in smaller modular codes like the WIMS system at Winfrith.

However, standardizing interfaces is a much more difficult job than it appears at first sight.

One should never underestimate the difficulty in standardisation if this has to do with more than one laboratory, and if you have to reduce the programming effort, it most certainly has to do with many laboratories. You have to take a lot of decisions that are more or less indifferent decisions; it does not really matter if you decide one way or the other. Now, this kind of decision is often the most difficult to take, because people argue for months about things that have a very small edge one above the other. If relative advantages and disadvantages are small, then one has to discuss it all the more. That is the main point. In the U.S. the standardisation of the interfaces for fast reactor calculations took years and the most difficult decisions were really when there was no great significance in the choice to be made. Let me give one example here. Should we write the flux at the mesh points or at the middle points between mesh points? There are diffusion codes that use one option and diffusion codes that use the other. There are small points in favour of either one. Many other examples could be made, for instance on which cross section definition to use. You can always choose among several sets of data that give you exactly the same information, but you have to pick up just one and that is terribly difficult. This concerns the physical content, but of course also the format and the sequence and so on. One could for instance arrange the cross sections by materials, or by cross section type, or by energy group.

Then there are many other difficulties, which are more fundamental. There is one extreme example which I can mention; it concerns the DOMINO code. The DOMINO code is in a way a lubricating routine, that connects the two-dimensional SN transport code DOT, to the three-dimensional MORSE Monte Carlo code. In some shielding problems, you cannot calculate the whole reactor by Monte Carlo. You make first a two-dimensional transport calculation and obtain the flux in, say, a cylindrical geometry. You then have to use this flux as an input in a 3-dimensional calculation, for instance to study the neutron streaming along a bent coolant pipe going through the shield and out of the reactor. Well, this is difficult. You have to define how to change the mesh; from R-Z to X-Y in a different direction and so on. It involves some physical consideration as well. So this is something that is much more than just changing the format. There are many more examples, of a lower category, but they involve choices which are not purely formal; they may involve sampling, or interpolation; there is some physics, some approximations that you have to take into account. So the problem of defining rigid interfaces, which is probably the most attractive if you start from scratch and if your motivation is that of reducing programming effort, is not the easiest. It may be very difficult and it is not flexible. If you have forgotten something, you may have to change the whole of the interface specifications.

The flexible interface approach looks more attractive, but practically involves many of the difficulties of the rigid interfaces. This is so because the interpretation of the data pool which is contained in the data pool itself cannot be arbitrary, there must be some key on how to read these instructions that must in itself be pre-determined. One real point of advantage is that the size and composition of the data vector is to a large extent arbitrary, and therefore it is relatively easy to add new quantities in the data pool. On the other hand, the programming of the modules is somewhat more cumbersome; there is a necessity of redefining the indices of the variables according to the content of the data pool.

There are examples of flexible interfaces, e.g. a reactor shielding code system in the U.K. is based on the concept of flexible interfaces.

Rigid and flexible interfaces are more convenient than lubricating routines for developing new codes or for breaking down an existing code into smaller parts.

5.5   Modular Systems and the Computer Environment

The sequence of modules, the pathways to be called to the fast memory and the handling of the interfaces is carried out by what is called the *Driver Program*, at least a part of which must be resident in the fast memory of the computer in order to switch from one module to another.

There are several possibilities for the Driver program. It can be written in JCL, the Job Control Language. Or else it can be made a part of the supervisor. It is also possible to write the Driver in FORTRAN Language, although it may not be easy because of the limited capabilities of FORTRAN in dealing with strings of data. Or else one can use, or develop, special languages for writing this Driver. There are special languages like FLOWTRAN, DATATRAN etc. that are especially adapted for dealing with this sort of problem. The choice depends very much on whether you want the modular system itself to be machine compatible, to be portable. If you want to write a modular code system which you can transfer from one computer to another then the obvious choice is to write the Driver in FORTRAN. I can say that historically the problem of portability of the modular system was considered very important at the beginning. It is much less important today. Why? First of all the main cause of non-portability is in the modules and not in the modular system. There is no reason to make the driver program portable if the modules are not portable from one computer to another. It is very difficult to make modules completely portable. One should use Standard FORTRAN, which is a minimum FORTRAN language common to all computers, that does not use any of the convenient extra features of the various FORTRAN 'dialects' which are implemented on the different machines. This is already a limitation which not everybody is happy to conform to. Other problems may come from the different word length in the various machines, and therefore from the different needs for double precision. Also the size of the memory which is available may be different from one case to another, and therefore require a different degree of segmentation. This problem can be overcome by adopting a maximum size for the modules (in one instance, 200 Kbytes) which can be accommodated in practically any computer on which the modular system will be implemented, and to proceed to a segmentation of the codes that respects this limitation. There may also be special problems for computers using virtual memory.

All this explains why, where portability is the main objective, the emphasis is on the modules rather than on the driver; once the modules are fully compatible with various computers, it is not difficult to write a Driver Program, specialized for each machine (and for the way in which this machine is used), most probably in JCL, having the appropriate interface with the supervisor.

Other problems that have been met with modular systems concern for instance the possibility of interactive control at run time. It may be very convenient to have some selected information visually displayed during execution, and to use it to choose the path to be followed in the computation. The JOSHUA modular system at Savannah River, for instance, allows for such a possibility. Of course, this kind of option is likely to be very much machine-dependent.

Other choices that have to be made are whether the computer will be entirely devoted to the modular system or be shared by other computing tasks. Furthermore, if the machine works in a time-sharing mode, be it a dedicated or a shared machine, one could run concurrently more than one module relative to a same problem (when a 'parallel' execution is possible) or else share the execution only of modules relative to different problems. Finally, one could interface the Operating System (OS) with the modular system, or directly use the OS to drive the MS, or write a new OS. An OS allowing for dynamic allocation would be the best, but is certainly very difficult to develop.

5.6 Conclusions

What should one do about modular systems in a developing country? There is no doubt that motivations and solutions are likely to be different from those in, say, the US. In the US the main effort is standardization, in order to reduce programming efforts. The availability of many very large computers makes computing efficiency relatively less important than portability. Standard FORTRAN and rigid interfaces may be the best solution there.

Rigid interfaces may again be the best solution for a limited MS, that only deals with a well-defined and relatively narrow class of problems. On the other hand, if one wants to set up a MS in which one includes modules for neutronics, thermal hydraulics, heat transfer, structural mechanics, safety and so on, then the problem of defining rigid interfaces becomes unrealistically complex.

In a small country which has limited hardware capabilities I think that the main emphasis should be on making the code compatible with the machine, breaking down codes to a size which can be accommodated and run effectively in the machine. I do not think one has to worry about compatibility too much. It would certainly be a problem if you start a network that involves many other small countries where there is a variety of computer types. So it may not be a negligible preoccupation for small countries, but I think that this point should be given second priority with respect to other types of problems. It is probably advisable to use FORTRAN at the highest level allowed by the computer in order to exploit all the possibilities given by the machine, and to tailor the modules to the size of the computer.

It is very important to make full use of the fast memory of the machine and at the same time it is certainly very useful to avoid too much run-in and run-out from the core of the computer, if it is possible. It may well turn up that the optimum size for modules is larger in a developing country than in an advanced country, where one uses large computers in the multi-programming mode, with a sophisticated allocation of memory, so that smaller modules may get higher priority in the execution and at the same time little CPU time is lost in running programs in and out of the fast memory because at the same time other programs will be executed. In a smaller computer, with fewer competing programs in parallel, it may be more important to reduce the total turn-around time and to reduce data transfers.

The most convenient approach to MS in a developing country is, in my opinion, a rather flexible one; it is not necessary to commit onself uniquely to one of the three types of solution that I have mentioned; a combination of all three may be the best, and even complete automation may not be necessary. I do not see any reason other than philosophical against mixing these systems. For instance, it may be convenient to define rigid interfaces for the most commonly used codes, so that passing from one to another is made easy. Other codes that are used more seldom could use lubricating routines or even a flexible interface.

It is also good to spend some time in thinking of a good Overlay or Modular Breakdown Procedure for breaking large computer codes down into modules which are compatible with your computer size and doing it by preserving the physical content of the module. This would mean trying to divide the code into parts each of which has a physical meaning, and is therefore a module in the sense that I have defined before. It is also advisable to discuss choices on MS's (for instance interfaces) with all the groups of people that are involved, for different disciplines (like reactor physics, thermal hydraulics, mechanics, safety etc.) and, if such is the case, from different institutions or countries; this can best be done by setting up an ad-hoc group, even if working only part-time on this task, that is responsible for preparing the detailed specifications for the modular system or systems, and that co-ordinates the various choices.

6. REACTOR PHYSICS REQUIREMENTS IN A DEVELOPING COUNTRY

6.1 Introduction

What are the requirements of Reactor Physics in a developing country? (for some aspects the same considerations will apply to a small developed country). We shall assume that the countries we are talking about have started developing or are going to develop a Nuclear Power programme; of course, there are reactor physics problems connected with research reactors as well but it is a slightly different subject, and I shall not deal with them here.

First of all, it is very unlikely that a developing country or any other country, at least for the time being, should find it convenient to develop a new type of reactor. I have mentioned that any new reactor type can be brought to commercialization only by the concentration of tremendous amounts of effort and money. I have mentioned the figure of 3 billion dollars from one reactor prototype to commercialization. It is probably much worse by now. So that there hardly seems to be any justification for developing a new reactor type: the motivation cannot be the fact that a new reactor type looks more attractive on paper. This is not a valid motivation. Any reactor type may look attractive on paper and it may also have the potential of being developed to an interesting operating stage, but there is little correlation between paper studies or laboratory tests and the commercial success of a reactor type. In any case, the effort which is needed to bring a new reactor to commercialization, I think, is hardly justified by the benefit one can hope to gain.

The reactors that have already passed the threshold of commercialization are the Light Water Reactors and the CANDU Reactors. They seem to be adequate for any situation on the short and medium range. These reactors can be operated on a number of different fuel cycles.

For instance, the security of fuel procurement may push towards the utilization of a natural uranium cycle in a CANDU reactor; a thorium cycle could in principle be used both on a CANDU or on a LWR. Slightly enriched uranium cycles can be used on either CANDU's or LWR's, both in a once-through option or with reprocessing. Plutonium re-utilization can be made on both reactor types.

On the longer range the Sodium Cooled Fast Breeders (LMFBR) are, in my opinion, the only alternative that justifies research on a large scale at the present time. In the years to come, other reactor concepts may qualify for a large-scale investigation, like for instance the Gas Cooled Fast Reactor, but I think one should demonstrate first the viability of the LMFBR concept before one concentrates a lot of research effort in gas cooled breeders. The High Temperature Gas-cooled Reactor (HTGR) has come to be considered, and I think rightly so, as something which comes after the LMFBR, because the main advantages it may have are those of being able to generate higher temperatures, that could be used in some industrial processes This is probably again a second phase after full scale demonstration of Fast Breeders for electricity production. Finally, it would not seem advisable to deploy in a country LMFBR's until it has installed proven thermal reactors on a large scale.

Therefore, even if a developing country, or a small country or an association of such countries should decide to design their own reactor, I see no reason why this reactor should not belong to one of the established types.

6.2 The License Regime

Even within the choice of a proven reactor type, there is a whole range of possibilities, from buying a turn-key plant to developing such a reactor in a completely independent way. I think it would be very useful for anyone who starts a nuclear power program to study in detail the license agreements, and the way in which they work. Of course there are many differences between one reactor vendor and another (and these differences should play a significant role in the choice of a reactor type), but there are also common features, and there is always a broad margin within each license agreement for an independent action by the buyer or the licensee. For instance, Italy has a license to build BWR's from General Electric, another for PWR's from Westinghouse and one for CANDU reactors from AECL. We have studied these agreements and the Italian policy is to work gradually towards an independent capability of designing and building these reactors, and there is very much that can be done towards this goal within the license agreements (which in any case are periodically re-negotiated).

The cost of the license itself amounts to a small fraction (a few percent) of the cost of the reactor, so that economic incentive or even balance-of-payment considerations would not be a primary factor in moving out of a license regime. But the situation is actually much more complex than that and the final price that is payed to the vendor is much higher than just the license fee.

The way in which this mechanism works depends on a complicated interplay between at least four parties: the vendor, the national

nal industry, the utility and the safety licensing authorities. One
example is provided by the continuous modifications undergone by the
plant during the design and construction period. One of the reasons
is that the buyer is very often not happy with a design that is not
up-to-date due to the long construction times. Modifications to the
designing are developed by the vendor in the meantime, and although
the buyer does not have to conform to them, most of the times he will;
and this costs money, to change the design, to adapt the construction
plans to these changes, possibly to re-examine some of the safety
analysis. Most of the license agreements foresee a 'consultation'
mechanism: the licensee can ask the vendor to perform a calculation
or an analysis; this is charged as a function of the man-hour involved (of course at a high price). The buyer does not have to consult
the vendor, but the circumstances are such that he invariably does:
for modifications in the design, for extra questions put by the safety
authorities, for changes required by the manufacturing industries
etc. At the end, the cost of consultations amounts to a substantial
fraction of the cost of the plant, and is generally much higher than
the cost of the license itself.

Even consultations can be of different type: one can ask the
vendor to perform a certain calculation; or he can make his own calculations and then send them to the vendor to be revised and commented; or he can propose a 'joint effort' to the vendor and send his
personnel to perform the calculations under the vendor's supervision.
Each of these steps of course involves a different degree of independent development, and amount of self-confidence of the licensee; and
the relative cost of consultations decreases proportionally. One soon
finds out, for instance, that it already makes a lot of difference
if one is able to make very specific questions instead of general
ones. So you see that by learning how to perform design and calculational tasks one can get rid of most of the weight of the license.

There are other considerations, of course. One is the so-
called Quality Assurance (QA) of the design. Even if a licensee is
able to perform a certain task, it often happens that the safety
authority prefers to have it backed up by somebody who has shown on
a large scale to be able to do it. So the theoretical ability of
performing the design is not always enough: it is also needed to
show the capability of doing so with a certain degree of confidence,
with a certain quality. So in parallel with the development of design
capabilities, one should always devote much attention to the qualification of such capabilities.

Another problem which is met in license agreements is the
fact that the system analysis is performed by the vendor in such a
way as to show the need for the components that he has developed, or
those that he usually adopts. Therefore, a license agreement for the
reactor system should always be discussed in parallel with an agreement for the components, and the route to an independent system

capability must be paralleled by an effort towards independent component manufacturing, the two being strongly interrelated. And components are still not enough: the know-how on materials is just as important.

Other problems derive from the limitation of guarantees when components that do not derive from the vendor are used. This applies even to the fuel in the case of a turn-key plant: in case of malfunctioning of the plant it is up to the utility (and generally not easy) to demonstrate that it is not due to the fuel that has been used.

So I think in conclusion that even if a country wants to arrive at an independent reactor system, the license system is a very good starting point, and that the present license agreements (if properly negotiated) allow for a gradual development of an independent capability. For instance, in Italy a number of large components (like the pressure vessel) are already out of the license agreement, and the system analysis capability is already quite advanced.

6.3   Reactor Physics Tasks

Much before one even thinks of designing his own power reactors by himself, there are many tasks a reactor physicist is called to.

If a country buys a reactor, the first thing it should know is what it is buying, whether it conforms to its necessities, in other words to properly prepare the ask for bids. A minimum follow-up of the specifications, of the main options in the design, of the prediction of the performance of the reactor is certainly needed.

One of the worst things one can do is to ask a vendor 'what do I need?'. Unfortunately, in the reactor business this is done very often. So the first thing a country buying a reactor should do is to develop its own specifications according to its own needs.

The second point is Safety Analysis. I am convinced that each country installing a reactor should make a safety analysis which is at least partially based on its own capabilities. This is not only because the geographical situation is different: for instance the frequency spectrum of the reference earthquake which is used in designing a reactor is different in Italy and in California; the possibility of discharging certain radioisotopes in the environment can be very different from one place to another. But it is not only the geological or geographical environment in the strictest sense that matters: it is also, for instance, the industrial environment, the relative capability or specialization of a construction industry, its possibility of meeting certain specifications; it is also the 'cultural' environment that influences the type of maintenance and of operation of a plant. For instance, I have often noticed that in less developed countries the operation of a reactor is generally done by people with higher qualifications than in a highly developed country. It may be a question of relative cost of manpower, or of confidence, or tradition: but the results are quite different. The training of an operator in a developed country tends to be more rigid; the reactor operator should follow certain rules very closely, both in case of normal operation and accidental conditions; he does not have to understand what happens and in a way he becomes part of the equipment. In developing countries it is more common to require from the operator to understand what is happening; this may lead in certain cases to the operator not doing exactly what he is supposed to do by the rules, but in other cases it makes the operator to understand more readily what is going on and act perhaps in a more responsible way, so that I think it may be a better solution. But in any case this is part of the environment, and the difference in environment, for better or worse, has to be taken into account when you make a safety analysis (especially if you think that the majority of accidents so far experienced in reactors were due to errors of operation rather than to system malfunctioning). Finally, one should not rely entirely on other people in such a thing as assessing his own safety.

Operation presents a number of problems for the reactor physicist quite apart from accidents. Large accidents (such as a LOCA) are fortunately highly improbable, while small problems are met very frequently during reactor operation. For instance, if some instruments deviates from the predicted value, one should be able to understand what is going on, whether the instrument is not working properly, or the plant is not working properly, or the model used to make the prediction was not adequate. It is important to develop a capability of making rapid calculations in order to assess these situations.

There are many possibilities of changes in the operational schedule that call for a re-assessment of the situation in order to make the appropriate decisions; for instance, one may have an unscheduled shutdown, one could use this shutdown for some operations on the fuel. Another flexibility which is generally kept by the reactor operator as an open option is the cycle stretch-out (operation at lower power after the exhaustion of the reactivity margin at nominal power), and this too requires reactor physics evaluations.

Another example is when one asks for a bid for fuel elements to reload the reactor. If one receives two bids that, for the same physical life, offer two different enrichments, the lower enrichment fuel costs less. Is one going to trust the calculations behind the lower bid, and risk an early shut down of the reactor (the guarantees are based on the cost of the fuel and not on the much higher cost of reactor inactivity)? Or is one going to accept the higher cost of the more enriched solution? I think that in a case like this the operator should be able to make his own independent assessment, or at least to evaluate the calculations made by others.

There are many other miscellaneous problems involving reactor physics for the operation of a reactor, or for the rest of the fuel cycle that we do not have time to review here.

6.4   Code Development

We have covered much of this subject when discussing Modular Code Systems. I think it is worthwhile to use existing codes as much as possible: there will always be enough software development needed without duplicating what has already been done. The most successful codes now in use incorporate so much development and improvement in the numerical techniques and in programming, that even with a better analytical approach one generally obtains a lower efficiency. By this I am not trying to discourage local talent, and good ideas should be pursued, but one should not have the illusion that it is relatively easy to arrive at an optimised, efficient code. I think that before one engages in any large scale efforts, one should try to make some cost-benefit analysis; this is very important, but seldom done, because the answers are cruel most of the time.

6.5   Experiments

How much experimental activity connected with reactor physics should be done in a country that starts a nuclear power program?

Let me say first of all that I think that an experimental reactor (be it a research reactor or just a critical facility) is certainly justified. Even it it does not generate useful results, it gives one a feeling that is difficult to acquire otherwise. Although there is always a tremendous difference, even from the purely neutronic point of view, between a critical facility and a power reactor, and this difference should never be underestimated, I think that having a small reactor available, and being able to operate it and to make some changes in it is helpful for a future power reactor designer.

It is more difficult to answer the question whether new and valuable information in reactor physics can be extracted from low power experiments, especially as proven reactor types are concerned. I believe there is still a margin for some useful experimental information that could be derived from experiments carried out in a developing country, but it is certainly not easy to find the right experiment. Most, though not all, of the simplest experiments in thermal reactor physics have been done, and the results are available: rather than re-doing them, I think it is profitable to spend the time in collecting the results and in re-analyzing them. Some new interesting experiments however are possible; for instance, in the last years it has become clear that the situation with shielding calculations, even for LWR's, was not as good as one thought, and that a few good experiments on simple streaming geometries could help in calibrating the experimental procedures. 'Benchmark' experiments in this and in other fields may be useful in testing codes and nuclear data.

6.6 Calculational Schemes and Nuclear Data

Although there is a growing tendency to perform thermal reactor calculations starting from basic data and very sophisticated theoretical methods, the results thus obtained are still far less accurate than those of semiempirical methods adjusted on experiments. The two approaches are actually more complementary than alternative: the basic approach is best suited to test the nuclear data against simple and precise integral experiments, and then to generate theoretical benchmarks; the simpler models are then calibrated against more complicated experiments and theoretical benchmarks, and finally applied to actual reactor calculation, both for design and for operation purposes.

Also in view of the much larger effort required by the basic methods, I believe it is more justified for a developing country to concentrate on the simpler models, which are of more direct applicability to practical problems. Experimental results, also on operating power reactors, are available, and one can develop methods, or adapt them to his needs, by correlating his predictions with these experiments. There are many ways of doing this. One way, which I think is the worst, is to apply correction or bias factor to the final result; I remember a method recommended by a vendor some time ago, where neutronic calculation were performed in a rather basic way in order to determine the power distribution in the core, and then the final result was multiplied by a space dependent factor, that went from 0.8 to 1.2, obtained by the comparison of calculations and measurements in similar reactors! I think that by doing this, instead of preserving whatever physical meaning was present in the original calculations, one simply destroys it I believe it is much better to try and identify in a physical way the origin of the discrepancies and try to correct them either in the calculational model or in the nuclear data used. There is a lot of improvement that can be done just by changing the nuclear data within their stated uncertainty.

The main features of the simpler calculational models should be simplicity and effectiveness (never introduce complications unless they are necessary to reach a certain accuracy in a given set of problems) but at the same time preservation of the physical meaning.

For instance, for LWR calculations in Italy we generate the cross sections by a code (RIBOT) which uses simple interpolation formulas. We use 2 to 5 groups, depending on whether you have pluto-

nium in large quantities or not, and the cross sections are just
written as algebraic expressions. The input quantities are ratios
of U-235, U-238, Pu-239, 1/v absorber and Hydrogen contents. By introducing these five parameters one is able to reproduce the cross sections in five groups to a very good accuracy. This takes a fraction
of a second. You can generate all the cross sections for a large
number of pins in a few seconds and even include burnup. These group
cross sections have been adjusted to experiments or results of sophi-
ticated methods. Then we have a number of recipes to go to the cell
homogeneization. And these also have been checked by experiments and
by comparison with theoretical benchmarks. For instance, we are carrying out a certain number of theoretical benchmarks by Monte Carlo in
fuel elements. The Monte Carlo calculation may take two hours of
computing time, but it is done just once, and one gets a very accurate power distribution. Even for a few cases (for instance for void
contents higher than those obtained in normal operation of a BWR)
this is acceptable, and a number of good theoretical benchmarks are
thus made available. It is important to supplement these with actual
data from operating power reactors, that supply the ultimate check
of the methods, at least as normal operation conditions are concerne

NUCLEAR NON-PROLIFERATION

Raymond L. Murray

Department of Nuclear Engineering

N.C. State University, N.C., U.S.A.

1. POLITICAL ASPECTS OF NUCLEAR NON-PROLIFERATION

It is very difficult to do justice to a subject as broad and complex as this in a period of an hour or two. I shall therefore limit myself to the most important aspects of the subject. My remarks in this seminar will consist of two parts, first, the political aspects of non-proliferation and second, some of the academical aspects of the subject. Let me start by giving some definitions, discussions of nuclear weapons and the motivation for using them, type of protection against such action, and U.S. policy and study program and the international situation. I will say a little also about reprocessing of nuclear fuel and its relation to the fuel supply problems and as well comment on the breeder reactor.

As most of you know, proliferation refers to the spread of nuclear weapons, either among countries of the world or among individuals or groups who have illegal acts in mind, such as terrorist activities. One can identify a variety of the latter types ranging from people who are mad or well-organized sub-national groups. There are, of course, many reasons countries might wish to achieve nuclear weapons capabilities. Prominent among these are prestige and protection.

- their perception of the utility of nuclear weapons to their security needs or prestige;
- the confidence they have in existing security arrangements including alliances;
- their views on nuclear explosives for 'peaceful' purposes; and
- the degree of resentment they feel toward the fact

that the nuclear weapon states continue to add to their nuclear arsenals.

Information on nuclear weapons remains classified, at least as to the mechanism that is involved. The open literature states that a nuclear device may consist of two parts that are suddenly brought together to form a supercritical assembly presumably by the use of conventional explosive forces. The other possibility is the implosion technique in which a sphere of fissile material is greatly compressed by explosive material into a highly supercritical assembly. It is noted that the critical mass of U-235 without reflector is about 50 kg. A reflected sphere has a mass of about 15 kg, while the critical mass of plutonium is somewhat smaller, about 10 kg. We find also that the content of the isotope plutonium-240 is crucial to the successful application of a weapon. The neutrons that come from spontaneous fission of Pu-240 tend to premature and ineffective detonation of the device.

In order to minimise the spread of nuclear weapons, an international agreement has been signed by a large number of countries. This non-proliferation treaty states that the country which signs will not seek to obtain nuclear weapons by any means and will allow the International Atomic Energy Agency to inspect its facilities to assure that no improper activity is taking place. There are a few countries, for example France, that have not signed the treaty. It is not clear how effective such an agreement is because there are no penalties for violating the treaty.

The types of safeguards that are used to minimise the chances of proliferation are of two types. First is the accounting for material on a frequent basis to be able to detect any diversion shortly after its occurrence. The second is physical security which involves conventional barriers and guards, but also a variety of detection systems, particularly nuclear in character. A great variety of devices and systems have been considered to achieve protection against attack or against clandestine action. Examples are familiar things such as well-lighted areas, strong fences and TV cameras. Others are beams of electromagnetic waves the interception of which causes an alarm, flashing lights or extremely loud sound that cause disorientation, and gases that temporarily disable an intruder. For transportation of nuclear materials in addition to the conventional armored cars there are vehicles that automatically become inoperative upon attack.

Personnel identituation can be checked by various techniques from visual comparison of badge pictures with the employee's face to personalized entrance cards to fingerprint or even voiceprint checks. The realiability of personnel can be enhanced by psychological tests.

President Carter in early 1977 as a part of his energy

policy emphasized the need for controlling proliferation and took a
number of actions that were regarded as beneficial for such purposes.
One was to suspend reprocessing of nuclear fuel in the U.S. on the
grounds that it furthered the availability of plutonium which can be
used as a weapon and to actively discourage such work abroad. The
second was to recommend termination of the demonstration fast breeder
construction program in the U.S. The keystone of U.S. policy is that
all efforts should be undertaken to delay the acquisition of nuclear
weapons by all countries who as yet do not have them. It is realized
that if a country were determined to have the weapon it could achieve
it. A third point of the U.S. policy is that a large-scale study of
alternative nuclear systems should be initiated to find ways that
nuclear power can be made available with less chance of diversion
of nuclear materials.

The U.S. has set in progress a program by the initials
NASAP standing for Nuclear Alternative System Assessment Program. In
that program there is a great deal of attention being given to the
development of reactors and fuel cycles employing the element thorium
from which the fissile fuel U-233 can be bred. In addition, the program examines the benefit to be achieved from energy centers, where
all of the processes in the production of nuclear energy are carried
out, or fuel cycle centers in which enrichment, reprocessing, and
fuel fabrication are done. The U.S. also proposed the program labelled INFCE which stands for International Fuel Cycle Evaluation. This
international cooperation activity includes studies by groups of
nations or a number of important subject areas, listed below:

- Nuclear fuel and heavy water availability
- Enrichment availability
- Assurance of long term fuel cycle
- Reprocessing, plutonium handling, and recycle
- Fast breeders
- Spent fuel management
- Waste management and disposal
- Advanced fuel cycle and reactor concepts.

These studies are expected to lead to ways in which nuclear power
can be made available with a minimum chance of nuclear proliferation.
The U.S. Administration says that reprocessing and the breeder programs are being postponed but not necessarily abandoned.

As you undoubtedly know there is considerable disagreement
in the U.S. with President Carter's policy as it refers to reprocessing and the breeder, particularly on the part of the industrial
community. The logic that they provide is that one can get additional
energy by recycling the plutonium as fuel and by recycling the uranium through the isotope separation process in order to extract

additional U-235. They noted that the cost of uranium continues to increase as the more abundant ore becomes exhausted and that it is desirable to continue using the converter reactor for as long as possible in the interest of helping to solve the energy problem. With regard to the breeder, they feel that it is desirable to test its commercial benefit and put it into operation in order to better utilize natural resources. In addition, a great deal of concern surrounds the accumulation of spent fuel coming from reactors. Storage facilities provided by the power companies are wholly inadequate to meet requirements over a long period. The U.S. Government has announced its intent to build appropriate facilities to receive the spent fuel. There is of course a question as to how soon those can be made available. The general subject of waste treatment and disposal is somewhat clouded by the policy of storing fuel in spent form since most of the previous research and development assumed that reprocessing would be done. The problem of waste disposal has been a aubject of considerable concern to the average person in the U.S.

On the international scene there has been considerable reaction to the policy of the U.S., especially among countries that are committed to nuclear energy in its ultimate form namely involving enrichment, reprocessing, refabrication and breeding. A number of countries do not have adequate supplies of other fuels and believe that they must rely on nuclear energy. Failure to reprocess or use breeders is regarded by many countries as being in opposition to the development of nuclear power. There is a certain amount of cooperation underway with the INFCE program but it is not yet clear how many countries in the world will adopt the conclusions that come from those studies. Finally, there is a new point of view that has been advanced by some thoughtful persons:

> 'All the considerations we have discussed here show up as destabilizing routes in our second proliferation scenario. Not only does country Y find logical incentives to install domestic nuclear-fuel facilities, but also it perceives a world more fragmented and less secure. Feeling less secure itself, it naturally imagines others feeling the same way and hence it must increase its own security unilaterally. Escalation of uncertainty leads to escalation of international instability; a program originally intended by the U.S. to decrease tthe dangers of nuclear proliferation inadvertently has the opposite effect. Meanwhile the U.S. isolates itself from the mainstream of world nuclear policy, and its ability to favorably affect that policy diminishes.'

In this view then the U.S. action seems to be counterproductive. The authors of the article also point out seven ways in which a country can achieve nuclear capability. These are listed below in abbreviated form:

Sources of weapons material*

1. Research reactor + small reprocessing plant
2. PU production reactor + small reprocessing plant
3. Spent fuel from commercial fuel cycle + small reprocessing plant
4. U enrichment by centrifuge
5. U enrichment by laser
6. U enrichment by nozzle
7. HEU or PU by black market

At this time considering all of the factors that are involved, it is not clear what the long term future holds in the area of nuclear proliferation. The only recommendation that I should like to make is that every effort should be made to improve communication and understanding among nations during this crucial period of uncertainty.

2. ACADEMIC ASPECTS OF NUCLEAR NON-PROLIFERATION

I should like now to look at some of the more technical aspects of the subject such as various fuel cycles and the radiation problems involved in the cycle using thorium. As we know, the fuel cycle of a nuclear reactor consists of several steps starting with the exploration for ore of a fuel such as uranium and ending with the final disposal of radioactive fission products as waste. The chart of Figure 1 lists steps in the 'front end', 'middle' and 'back end' of the fuel cycle.

'Front End'
    Exploration
    Mining and Milling
    Chemical Conversion
    Enrichment
    Fabrication of Fuel Assemblies

'Middle'
    Fuel Installation, Rearrangement, and Removal
    Energy Production in the Nuclear Reactor

'Back End'
    Storage for Radioactive Cooling
    Transportation of Spent Fuel
    Reprocessing and Recycling
    Waste Storage and Disposal

Figure 1. The Nuclear Fuel Cycle.

―――――――――――――
*David J. Rose and Richard K. Lester, 'Nuclear Power, Nuclear Weapons and International Stability', Scientific American, April, 1978.

The conventional concept of the uranium fuel cycle has been to recycle uranium from spent fuel through the isotope separator, to store the plutonium for future use in the breeder reactor, and to get rid of the wastes. Figure 2 shows this cycle schematically. The consequence of President Carter's ban on reprocessing led to the cycle known as 'once through'. Depending on whether the spent fuel is discarded or merely stored temporarily the terms 'throwaway' and 'stowaway' are used. Figure 3 shows this simple cycle. Prior to this US action, plans had been laid to utilize the plutonium produced in a converter reactor as a supplementary fuel. The spent fuel would be reprocessed to separate U, Pu, and fission products from each other. The Pu would be recycled to a fabrication plant to be blended with uranium as a mixed oxide (MOX). The U would be re-enriched, and the radioactive wastes converted to a solid fused with glass, and disposed off deep underground. This so-called plutonium recycle mode is shown in Figure 4.

Several alternatives have been proposed. One is coprocessing of U and Pu, in which the plutonium is mixed with and thus diluted by a large amount of uranium, about ten per cent of the total U, the rest being recycled to isotope separation. Since the Pu is diluted by a large bulk it is less easy to divert for illegal purposes. It is not accessible for use in a weapon without chemical separation of the two elements. Following the coprocessing would be the co-fabrication into mixed oxide.

In fuel fabrication there are some new methods that show promise of eventually replacing the pellet preparation method now used in industry. One is called 'Sphere-Pac' which combines the 'sol gel' process and vibratory compaction. Note that the sol-gel method as applied to fuel involves the mixture of uranium nitrate with a solution of material called 'hexa'* and urea an ammonia compound**, and the mixture is dispersed in air as droplets which are caught in a hot organic liquid where they solidify within seconds. The solid is washed with ether and ammonia, and dried in air. It consists of $UO_3 \cdot 1/3 NH_3 \cdot 5/3 H_2O$, and is reduced and sintered to UO_2 at 1200°C. The particles are almost perfect spheres. Three sizes are used to get very compact dense fuel: 800-1000 micrometers, 80-100 micrometers, < 45 micrometers. An electrical shaker provides the vibratory compaction, with 5 kg payload, acceleration of 125 g and frequency range 5 to 5000 cycles per second. Densities achieved are 87-89% of theoretical.

Another alternative is 'spiked' fuel, in which plutonium (or highly enriched uranium) is mixed with a radioactive material that emits a sufficient number of gamma rays to serve as a strong deterrent to theft. Isotopes that could be used are cesium-137 or

---

* Hexamethylenetetramine
** NH_2CONH_2

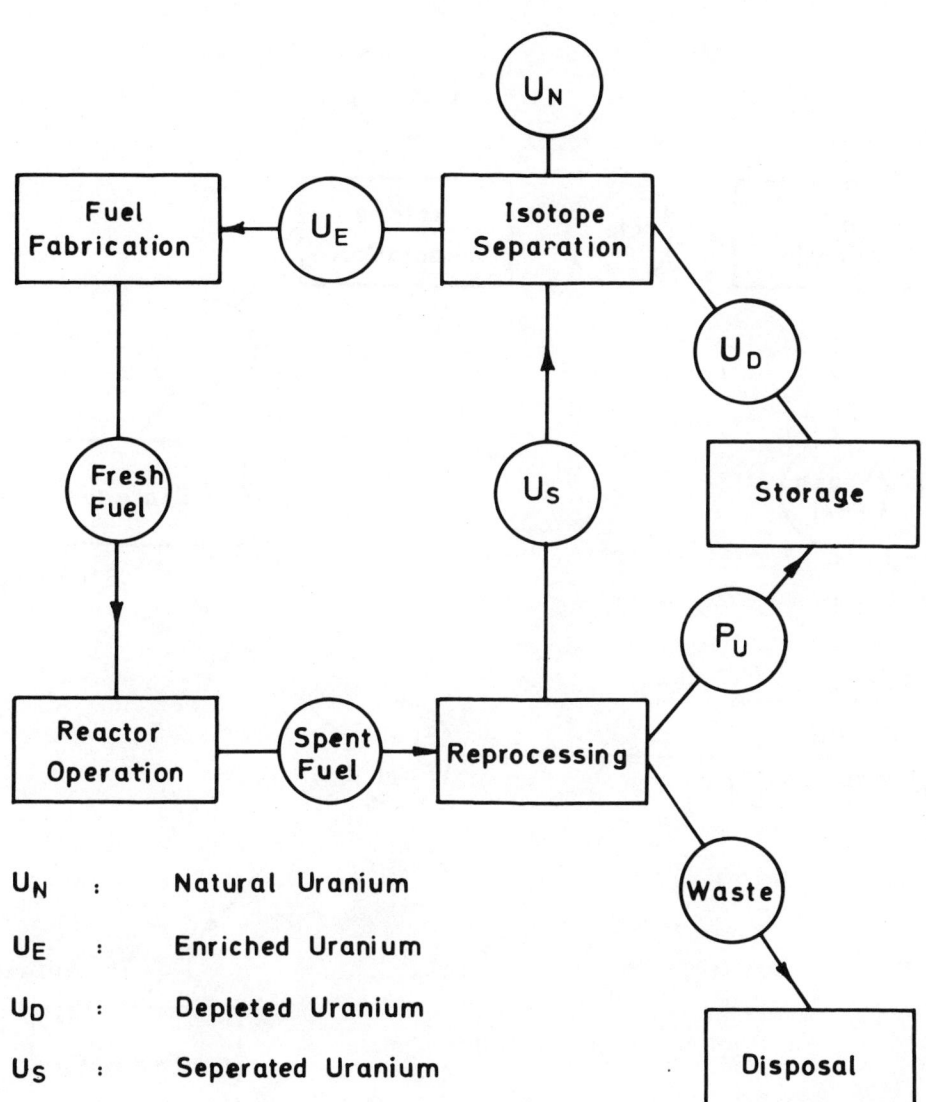

Figure 2. Conventional Fuel Cycle.

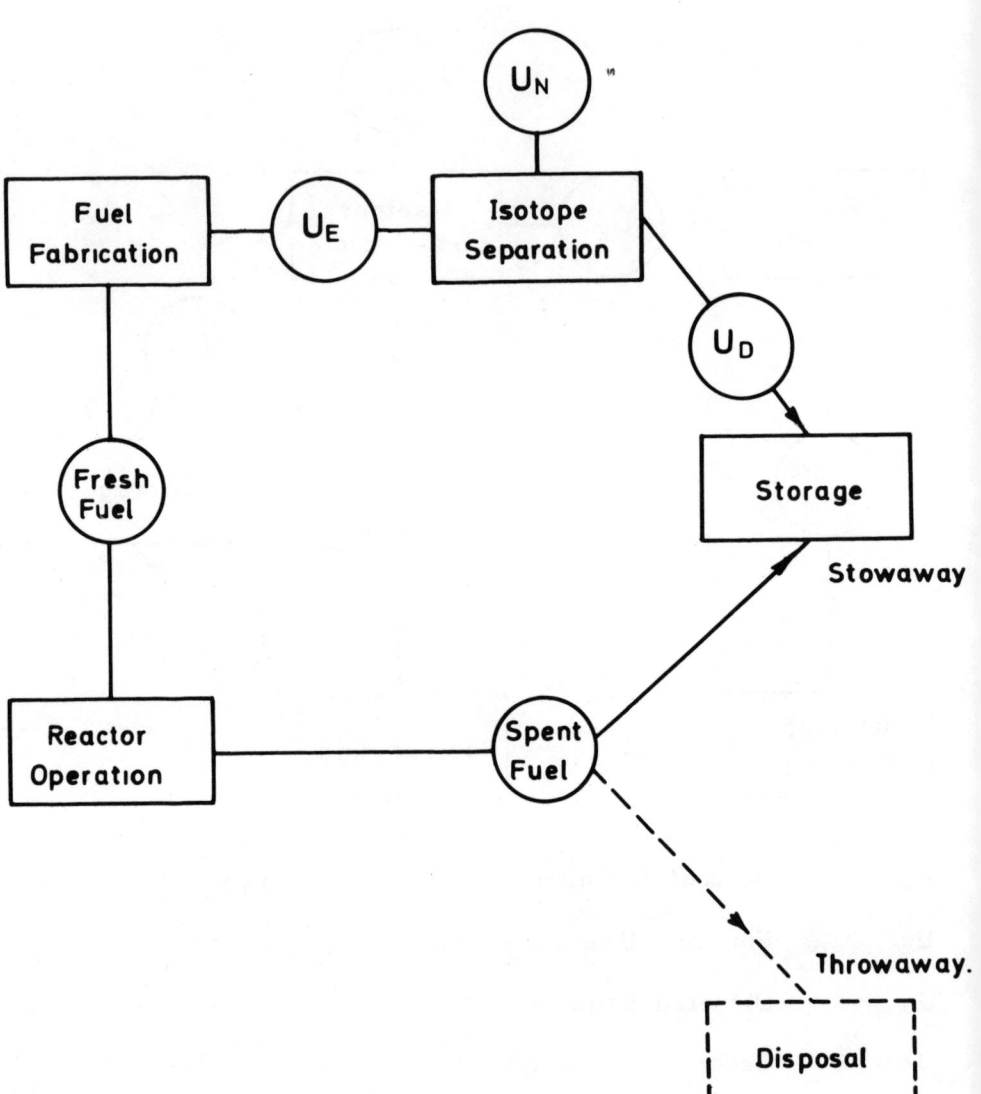

Figure 3. Once-through Fuel Cycle.

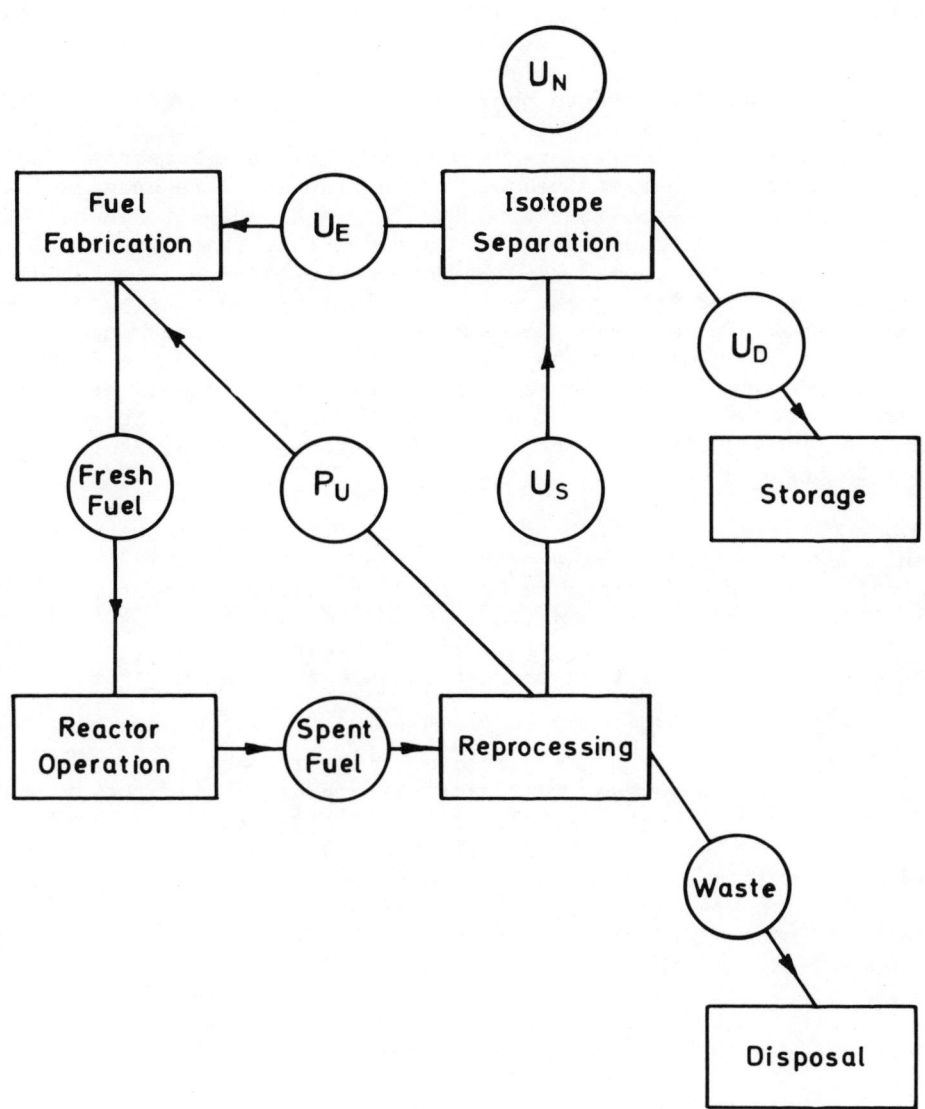

Figure 4. Plutonium Recycle.

cobalt-60. Alternatively, fresh fission products could be left with the recycled plutonium. Figure 5 shows this particular cycle. A very recent concept called CIVEX involves recycling to the fabrication process a mixture of U, Pu, and fission products (see Figure 6), thus achieving the benefits of both dilution and spiking throughout the cycle.

Let me now turn to the subject of the thorium cycle. This element as you know is fertile. Absorption of a neutron in Th-232 followed by electron emissions results in the fissile isotope U-233. Figure 7 shows the nuclear reactions. The latter is regarded as a candidate for a non-proliferating fuel cycle because it can be diluted with natural U or depleted U to yield a fuel that is very similar to the slightly enriched U used in modern converter reactors. Such fuel is identified as 'denatured' in analogy to the substance denatured alcohol. It has been proposed that U-233 could be generated in reactors within a tightly controlled energy center and the denatured fuel shipped out for use in reactors elsewhere. It should be noted however that some plutonium would still be produced in the reactor and thus if U-233 is to be recovered from spent fuel something still needs to be done about Pu.

Turning to the alternative cycle involving thorium-232 and uranium-233, we find a rather different situation. Th-232, the only naturally occurring isotope of thorium, is weakly radioactive with half life of 1.41×10^{10} years. Also U-233 has a value of t_H much longer than that of Pu-239, viz., 162,000 years. However, some by-products of reactor neutron irradiation are troublesome. First of all, the intermediate isotope protactinium-233, $^{233}_{91}$Pa, has a 27.4 day half life, and too quick a reprocessing of spent fuel would put U-233 in the waste stream. Then some n, 2n reactions in Th-232 and U-233 yield U-232 which has a 72-year half life and decays into Th-228. The latter has a 1.9 year half life. Far down the chain are some hard gamma emitters. Thus, any separated U-233 will be accompanied by U-232 which yields these isotopes making refabrication of the fissile element difficult. Any separated Th-232 will be accompanied by some Th-228 and the same problem exists for recycling thorium. The isotope U-233 bred from Th-232 is thus accompanied by some high

$$^{232}_{90}\text{Th} + ^{1}_{0}\text{n} \longrightarrow ^{233}_{90}\text{Th} + \gamma$$

$$^{233}_{90}\text{Th} \xrightarrow[22\ m]{} ^{233}_{91}\text{Pa} + ^{0}_{-1}\text{e}$$

$$^{233}_{91}\text{Pa} \xrightarrow[27\ d]{} ^{233}_{92}\text{U} + ^{0}_{-1}\text{e}$$

Figure 7. Production of U-233.

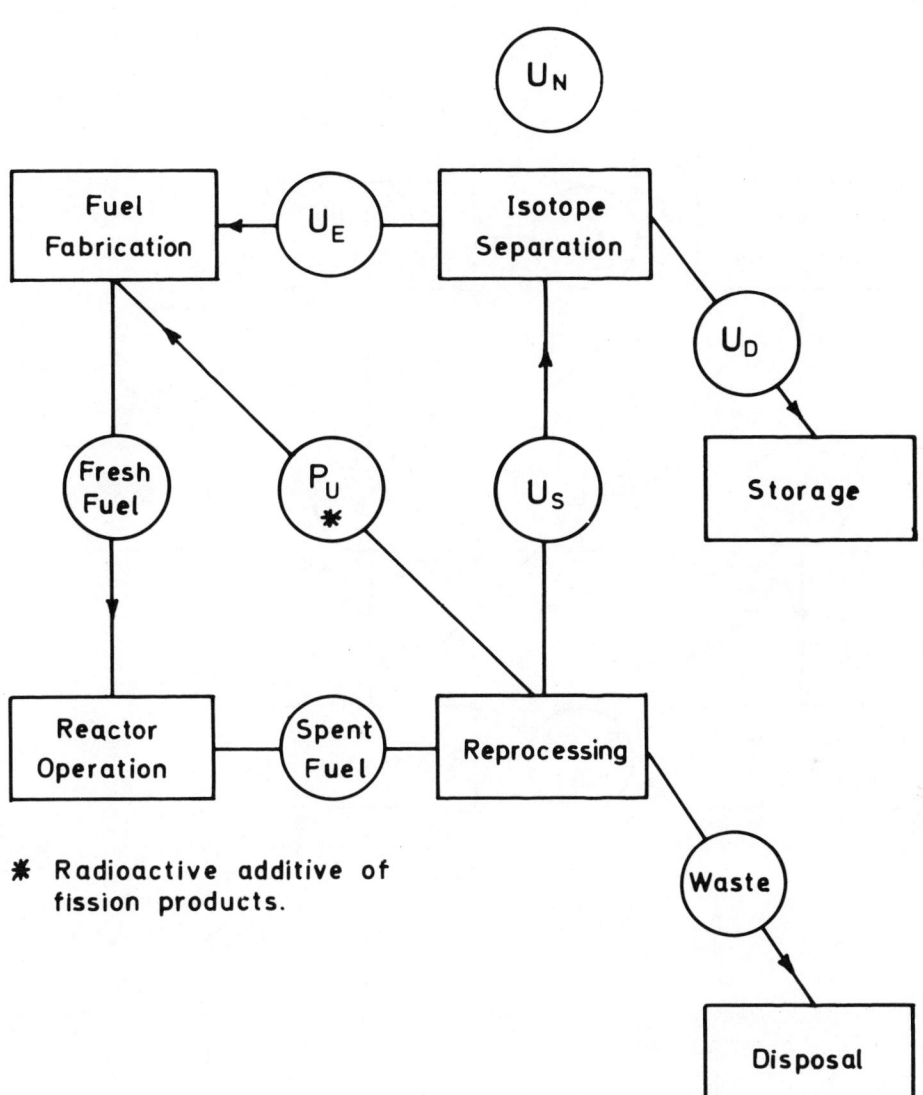

Figure 5. Spiked Plutonium Recycle.

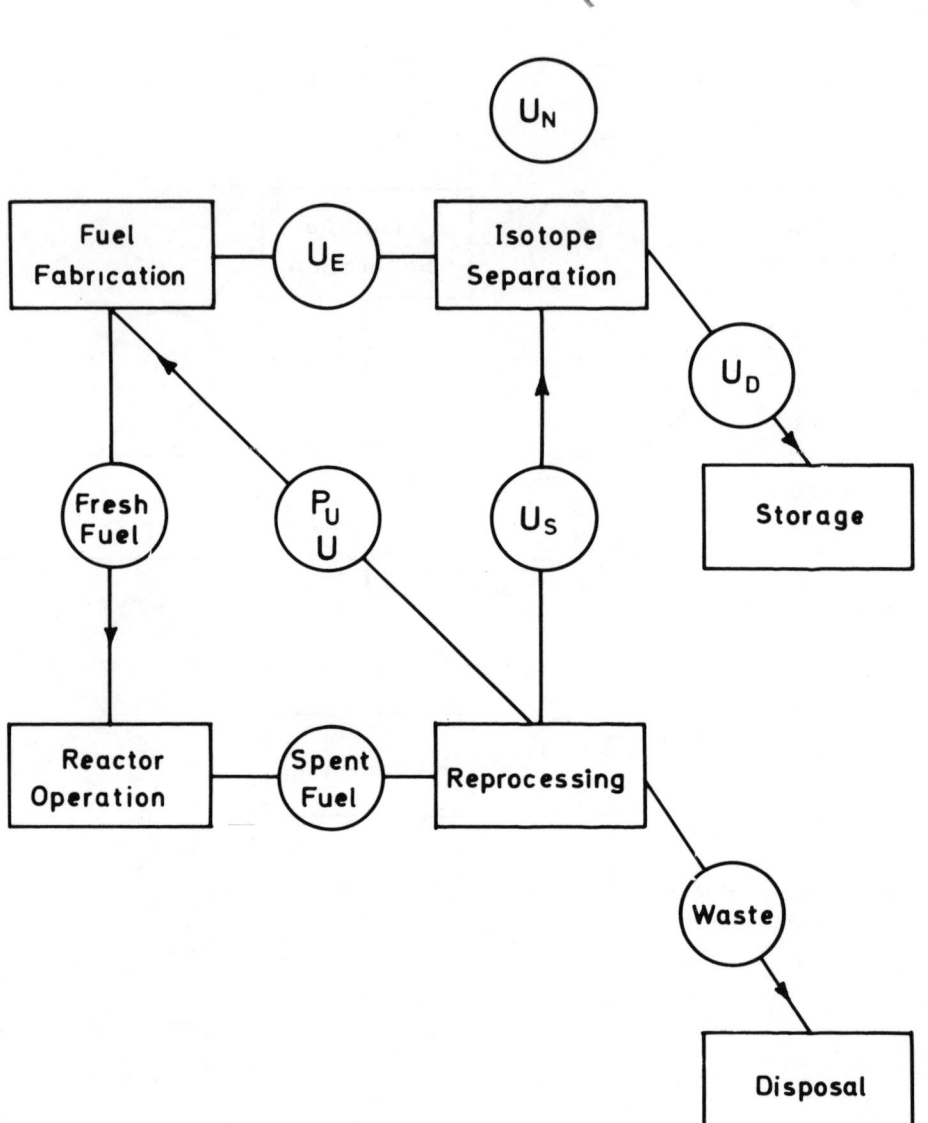

Figure 6. Coprocessed Uranium and Plutonium Recycle.

energy gamma rays. These can be regarded as a benefit or a detriment depending on the point of view. The presence of penetrating gamma rays will deter any one seeking to divert the fissile material. At the same time, the reprocessing, refabrication, and fresh fuel handling require greater shielding and there is more chance for personnel

$$^{232}_{90}\text{Th} + ^{1}_{0}n \longrightarrow ^{231}_{90}\text{Th} + 2\,^{1}_{0}n \quad \text{Threshold 6.3 MeV}$$

$$^{230}_{90}\text{Th} + ^{1}_{0}n \longrightarrow ^{231}_{90}\text{Th} + \gamma \quad \text{5 to 60 ppm 23 barns}$$

- -

$$^{231}_{90}\text{Th} \xrightarrow{25.6h} ^{231}_{91}\text{Pa} + ^{0}_{-1}e$$

$$^{231}_{91}\text{Pa} + ^{1}_{0}n \longrightarrow ^{232}_{91}\text{Pa} + \gamma \quad \text{212 barns}$$

$$^{233}_{91}\text{Pa} + ^{1}_{0}n \longrightarrow ^{232}_{91}\text{Pa} + 2\,^{1}_{0}n$$

$$^{232}_{91}\text{Pa} \xrightarrow{1.3d} ^{232}_{92}\text{U} + ^{0}_{-1}e$$

$$^{233}_{92}\text{U} + ^{1}_{0}n \longrightarrow ^{232}_{92}\text{U} + 2\,^{1}_{0}n$$

Figure 8. Reactions leading to U-232.

Figure 9 shows the decay chain of isotopes leading from $^{232}_{92}\text{U}$ to $^{208}_{82}\text{Pb}$.

$$^{232}_{92}\text{U} \xrightarrow[72\,y]{\alpha} ^{228}_{90}\text{Th} \xrightarrow[1.91\,y]{\alpha} ^{224}_{88}\text{Ra}$$

$$\xrightarrow[3.64\,d]{\alpha} ^{220}_{86}\text{Rn} \xrightarrow[55.3\,sec.]{\alpha} ^{216}_{84}\text{Po}$$

$$\xrightarrow[0.145\,sec.]{\alpha} ^{212}_{82}\text{Pb} \xrightarrow[10.64\,h]{\beta} ^{212}_{83}\text{Bi}$$

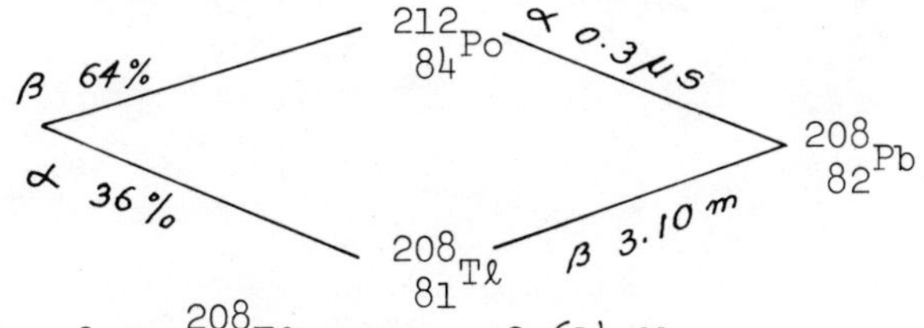

$^{212}_{84}\text{Po}$, α 0.3 μs

β 64%

α 36%

$^{208}_{82}\text{Pb}$

$^{208}_{81}\text{Tl}$, β 3.10 m

γ-rays from $^{208}_{81}\text{Tl}$ ..... 2.614 Mev.

Figure 9. U-232 Decay chain.

exposure to radiation. The basis of the radiation problem is of scientific and engineering interest. We display the nuclear reactions that ultimately yield the 2.6 MeV gamma emitter Thallium-208. Figure 8 shows the equations that produce $^{231}_{90}\text{Th}$, $^{231}_{91}\text{Pa}$, $^{232}_{91}\text{Pa}$, and $^{232}_{92}\text{U}$.

I think it is very desirable to study the thorium cycle in order to open up a new nuclear resource for use.

UNIFIED NEUTRON TRANSPORT THEORY

Raymond L. Murray

Department of Nuclear Engineering

N.C. State University, N.C., U.S.A.

I remember very well my first encounter with neutron transport theory. It was in 1948 or 1949 at Oak Ridge, where I was involved in critical mass calculations for a plant that performed chemical processing of uranium solutions. We had been using semi-empirical formulas provided for us by our consultant Richard Feynman, whose later accomplishments you know, and by Eugene Greuling author of a classic paper on the theory of the water boiler reactor. I was fortunate to be able to attend a series of lectures given by Dr.A.M.Weinberg at a reactor course attended by the staff of Admiral Hyman Rickover. President Carter may have been one of my classmates. Prior to these presentations by Dr.Weinberg, our calculational models were based on simple diffusion theory, coupled with some elementary neutron slowing concepts. I was amazed to see the complexity of the development of the spherical harmonics solution from the general transport equation involving spatial, energy, and angle coordinates. As many of you know, Dr.Weinberg and his co-author Dr.Eugene Wigner extended the material in these lectures to form their famous text book. However, the difficulties I and other students had in appreciating the theory led me to believe that for pedagogical purposes it was preferable to establish a strong foundation in reactor physics based on the neutron balances in diffusion theory and later to introduce transport theory as a generalization. In my own text 'Nuclear Reactor Physics' which came out in 1957, I used this approach. Dr. Allan Henry, who now has his own fine text in reactor analysis agrees. You see, what we are comparing are the two choices usually available in science and engineering either (a) to go from the general to the specific limited approximation or special case or (b) to supply a succession of corrective improvements or extensions to a simple first order theory. I suppose that all education and research is actually a combination of these techniques.

Let us look back to the reactor theory, analysis, and design of the World War II period. Much of the work that led to operating reactors was based on the familiar four-factor formula $K_\infty = \varepsilon p f \eta$ (or its generalization to six factors including leakage effects As you know, the product of factors parallels the sequence of events from birth to death of a neutron and the factors in turn are averages over histories of many neutrons. This formulation was invoked simply because the complete solution of the Boltzmann transport equation was too difficult to achieve, understand, and use. Indeed, even after thirty years have elapsed and many fine papers and books have appeared in the literature, the problem of connecting theory and actual reactor still persists. This inherent complexity, associated with the multiple variables of phase space, led to the development of a large variety of special models. First was the one-speed approximation to transport theory, which can only be rigorous in the case of a constant cross-section, a rare situation indeed. Then we had age theory in which the angular dependence was eliminated and several approximations made to yield a position-energy dependent equation to be solved for the neutron flux in the epithermal region. At still another level, emphasis was placed on the spatial aspects of neutron attenuation, with collision probabilities and escape probabilities derived from simple geometrical considerations. The interaction of neutrons with moving nuclei was treated so differently in the two regimes lebelled 'thermal' and 'resonance' that the basic equivalence was obscured. Indeed the formalisms of thermalization theory and the Doppler effect are classically quite different.

Over the years my students and I have been attempting to achieve increased understanding of the transport equation and of its solutions for various media. Progress has been slow because of our own limitations and the difficulty of the subject. At the same time, we experienced the effect of the high speed computer on neutron calculations and on analysis in general. The need for detailed mathematical manipulation based on and yielding physical insight has tended to disappear. Those of us older people who were educated in the era when analysis was all important would like to see a revival of respect for and need of analysis. This is very interesting subject for thought and discussion.

I should like now to mention a few of the efforts we have made to help unify transport theory as it applies to neutrons. At one end we have the general integro-differential equation with many variables applied to a system containing many materials and many regions. At the other end we have the time and space dependent neutron and isotope distributions. The gap between these is very large.

The first topic is a conceptual view of the levels of transport theory. Consider first the equation governing the steady state flux, where r and Ω are vector quantities, $\phi(r,v,\Omega)$.

UNIFIED NEUTRON TRANSPORT THEORY

Integration of the equation over each of the variables position, speed, and direction, yields three equations that refer to fluxes with lower resolution $\phi(\mathbf{r},\Omega)$, $\phi(r,v)$, and $\phi(v,\Omega)$. These in turn may be integrated over one more variable, to obtain $\phi(r)$, $\phi(v)$ and $\phi(\Omega)$. Finally, integrating over the remaining variables leads to a total system neutron flux ϕ_T. This sequence can be displayed schematically by the 'box' diagram in Figure 1.

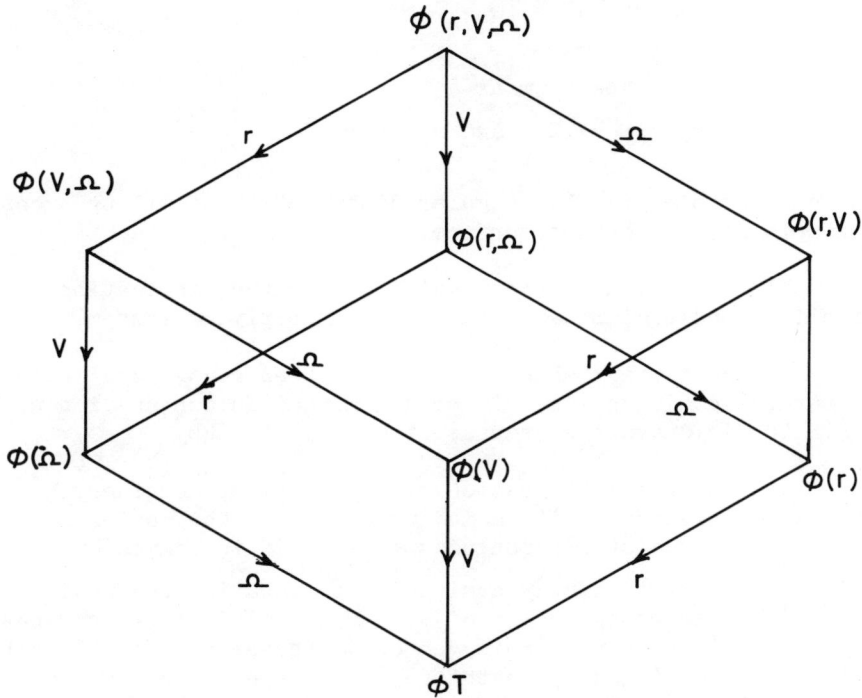

Figure 1. Integration paths.

In principle, the solutions of the equations at various levels should ultimately lead to the same ϕ_T but any approximations made in the formulas or solution methods will lead to disparities. There could be as many as six different results depending on the order of the treatment.

We can correlate the diagram with some standard models. The path
$$r\,v\,\Omega \xrightarrow{v} r\,\Omega$$
is of course one speed transport. Then
$$r\,v\,\Omega \xrightarrow{\Omega} r\,v$$

is the generalization that embraces age theory. Then

$$r v \Omega \longrightarrow v \Omega$$

is rarely encountered. The operations

$$r \Omega \xrightarrow{\Omega} r \quad \text{and} \quad r v \xrightarrow{v} r$$

lead to the spatial distribution within say a neutron group. The processes

$$v \Omega \xrightarrow{\Omega} v \quad \text{and} \quad r v \xrightarrow{r} v$$

lead to the neutron spectrum at a point, while

$$v \Omega \xrightarrow{v} \Omega \quad \text{and} \quad r \Omega \xrightarrow{r} \Omega$$

would in principle give the angular distribution of all neutrons, for example at an emergent surface.

The equation in the total flux has the implication of a criticality condition in a steady state multiplying system.

Some of these alternatives have been recognized in the literature, e.g. in the use of the ABH method in conjunction with SOFOCATE to yield average cross sections of LEOPARD.

As another illustration of unification, consider the reaction rate of neutrons with a medium for which the moving nuclei are in maxwellian distribution of speeds v. It is convenient as we show in our paper of a few years ago[1], to visualize the neutrons as targets bombarded by n_o nuclei of mass M in an angular distribution of nuclei $n_o\, m(v,\vec{\Omega})$ with a unit vector $\vec{\Omega}$ representing the direction of motion of nuclei and v_r is the relative speed. The function m is merely $1/4\pi$ times the speed distribution. For a neutron number density $N(v)$ where v is the neutron speed and a cross section $\sigma(v_r)$. The total reaction rate is

$$R(v) = n_o\, N(v) \int_V m(V,\vec{\Omega}) d\,V \int_\Omega v_r\, \sigma(v_r) d\Omega$$

The procedure for carrying out the integration is reviewed. Noting $d\Omega = \sin\theta\, d\theta\, d\phi$, symmetry permits integration of ϕ. Integration over v_r instead of $\mu = \cos\theta$ is performed. Then the order of integration is reversed, from v_r, V to V, v_r. Finally, reduced variables are introduced

$$y = \beta v_r \, , \quad x = \beta v$$

where

$$\beta^2 = \frac{M}{2KT}$$

UNIFIED NEUTRON TRANSPORT THEORY

leading to a general relation applicable to any form of cross section,

$$R(v) = \frac{n_o N(v) v}{\sqrt{\pi}\, x^2} \int_0^\infty dy\, y^2\, \sigma(y) \{\exp[-|x-y|^2] - \exp[-(x+y)^2]\}$$

Three special cases are of interest:

(a) Nuclei with absorption cross section proportional to $1/v$. Insert

$$\sigma = \sigma_a(x)\, x/y$$

and integrate to obtain the familiar

$$R(v) = n_o\, v\, N(v)\, \sigma_a(v).$$

(b) Nuclei with constant scattering cross section σ_s. Integration gives

$$R(v) = n_o\, N(v)\, \sigma_s\, V(x)/\beta$$

where the classical scattering function is

$$V(x) = (x + 1/2x)\, \mathrm{erf}\, x + 1/\sqrt{\pi}\, \exp\cdot(-x^2).$$

(c) Nuclei with resonance capture obeying the Breit-Wigner formula,

$$\sigma_a(E) = \sigma_{ao}\, \sqrt{E_o/E}\, \left[1 + \frac{2}{\Gamma}(E - E_o)^2\right]^{-1}$$

where E is the neutron energy in the laboratory system, E_o is the energy at which the peak σ_{ao} appears and Γ is the width of the resonance. The reaction rate may be written to reveal a generalized Doppler broadening function

$$\overline{\sigma_a(E)} = \sigma_{ao}\, \sqrt{E_o/E}\, \psi_G$$

where

$$\psi_G = \xi/2\sqrt{\pi} \int_{-\infty}^{\infty} \frac{dW}{1+W^2}\, \exp(-|x-y|^2)$$

with

$$\xi = \Gamma/\Delta$$

and

$$\Delta = \sqrt{4m\, KTE/M}$$

The simplification comes from the use of the condition

$$\sigma(-y) = -\sigma(y)$$

As a third illustration of unification, let me discuss the treatment of the thermalization of neutrons in two quite dissimilar media - atomic hydrogen gas and a heavy monatomic gas. The classic paper of Wigner and Wilkins[2] serves as the starting point for any study of the interaction of neutrons with particles in a maxwellian distribution. To first order in the ratio of 1/v absorption cross section of neutrons at a reference energy kT and the epithermal cross section (assumed constant), the flux distribution in reduced energy $\varepsilon = E/kT$ can be written generally as

$$\phi(\varepsilon) \simeq \frac{2}{\sqrt{\pi}\,\Gamma} \varepsilon\, e^{-\varepsilon} + \frac{K(\varepsilon)}{\varepsilon}$$

where

$$\Gamma = \frac{\Sigma_a(kt)}{\xi\, \Sigma_s}$$

where ξ is the logarithmic energy loss per collision, with value 1 for hydrogen and approximately $2/A$ for a very heavy gas, where A is the ratio of nucleus and neutron masses. The maxwellian form $\varepsilon e^{-\varepsilon}$ is joined to the epithermal 1/E spectrum by a function $K(\varepsilon)$ which is zero at $\varepsilon = 0$, and unity for $\varepsilon = \infty$. Explicit series expansions for $K(\varepsilon)$ are available for both cases for small and large arguments. The trends are expressible as follows:

ε	A	1
small	$-\varepsilon^2 + \frac{5}{3}\varepsilon^3 - \cdots$	$\frac{2}{\sqrt{\pi}}\left(\frac{4}{3}\varepsilon^{5/2} - \frac{52}{45}\varepsilon^{1/2} + \cdots\right)$
large	$1 + \frac{3}{2\varepsilon} + \frac{17}{4\varepsilon^2} + \cdots$	$1 + \frac{1}{2\varepsilon} + \frac{6}{\varepsilon^2} + \cdots$

Figure 3 shows the calculated spectrum for a heavy gas of carbon, A = 12.

I should now like to end with some discussion on the philosophy of reactor analysis.

I believe that one of the roles of reactor physics is to help interpret processes, and incidents in the practical production of electrical energy. In particular, it is important that reactor physics be a useful tool for a plant engineer or technician providing both qualitative and quantitative understandings. Neutron transport theory, as the ultimate basis of reactor physics should not be restricted to the elegant mathematical solutions for idealized geometries but should contribute to real situations. Let me illustrate. The classical concept of criticality of a reactor is that there is no change with time in the neutron population. The fact is that no reactor operates steadily under such a condition. Since reactor power depends mutually on neutron flux and fuel content, and since all reactors use fuel, the flux must increase with time. Indeed, the problem is even more complicated - the rate at which the fluxes change vary with position in the reactor. I think the mathematical analysis that has been done in the area of transport theory has dealt with the critical reactor, a non-existent entity.

It is a fact of reactor physics that even simple realities cause great mathematical difficulty. We can illustrate by formulating the problem of a reactor with fuel burnup, assume a two group neutron diffusion model but with strong thermal absorption and hence negligible leakage. The governing equations are thus

$$D_1 \nabla^2 \phi_1 - \phi_1 \Sigma_1 + \phi_2 \Sigma_{f2} \nu\varepsilon = \frac{1}{v}\frac{\partial \phi}{\partial t}$$

$$\phi_2 \Sigma_2 \cong \phi_1 \Sigma_1 p$$

Let us assume that the only isotope burned is U-235, according to

$$\frac{dN_u}{dt} = -\phi_2 N_u \sigma_{au}$$

also we assume the core to be unreflected such that the fluxes are zero near the boundary. The final constraint is that the power should be constant,

$$\int \phi_2 N_u \sigma_{fu} w \, dv = P$$

For a given reactor size and initial fuel distribution we wish to find the proper amount of uniform and adjustable control absorber as a function of time.

I submit that this neutron physics problem although greatly idealized in its own way, is a much more relevant problem than the solution of a critical two region slab reactor by transport theory with anisotropic scattering.

I would hope that all of us who are interested in reactor physics and transport theory could make a greater contribution to the practical understanding of reactor behavior.

REFERENCES

1. Raymond L. Murray, Nucl. Sci. and Eng., 26, 362 (1966).

2. E.P. Wigner and J.E. Wilkins Jr, Report AECO-2275 (1944).

PART II

PHYSICS AND TECHNOLOGY

AMORPHOUS SEMICONDUCTORS

William Paul

Division of Applied Sciences, Harvard University

Cambridge, Massachusetts 02138, U.S.A.

INTRODUCTION

In this brief series of lectures I should like to give you an account of the principal structural and electronic properties of amorphous semiconductors.

The first section will deal with the main methods of preparation of the disordered material and the characterization of its structure. Disorder in general, and the main types of topological disorder in particular, will be discussed and related to the basic molecular unit in the structure. It will be shown to what extent a reliable statistical description of the material can be established through an iterated comparison of the experimental diffraction results and the predictions of ball-and-stick structural modelling. Finally, I shall discuss, briefly, the identified or postulated structural defects which dominate many of the electronic properties.

The second section will deal with the electronic band structure we might expect to find based on the structural model. The intrinstic band structure of the fully-coordinated, defect-free network must be carefully distinguished from the states produced by a variety of natural defects. The consequences of the electronic density of states in the vicinity of the energy gap between valence and conduction bands for the optical, transport and recombination processes will be outlined. The possible importance of the electron-phonon interaction in determining many of these properties will be emphasized.

In section 3 I shall try to describe the current situation in research on amorphous silicon and amorphous silicon-hydrogen alloys, which are very much investigated since the demonstration that they could be easily doped n and p type and made into surprisingly good photo-detectors.

Finally, in the Section 4, I shall discuss another large and important class of amorphous semiconductors, the chalcogenide glasses. Here there is presently a very exciting model for the fundamental physical behavior and also a variety of interesting switching, memory and imaging effects that have - and some think will continue to have - a very significant impact as useable cheap devices

1.1 Preparation

In this Section I shall arbitrarily exclude the large class of disordered semiconductors represented by the insulating oxides (such as SiO_2), polymeric substances and semiconducting liquids. We may divide the rest into two large groups[2]: those based on tetrahedrally-coordinated materials of the Si family (for example, amorphous C, Si, Ge; 3-5 compounds; and alloys of <u>all</u> chemical compositions), and those containing one or other of the two-fold coordinated chalcogens (for example, amorphous S, Se, Te, As_2X_3, Ge_yX_{1-y} where $X \equiv S$, Se or Te; ternary and quaternary chalcogenides such as Si-Te-As-Ge alloys)*. Roughly speaking, materials of the first group cannot be prepared by cooling from a melt, whereas materials of the second group quite often can be. The method of preparation, it should, however, be emphasized, is not the main element determining the difference in properties between these two groups of substances.

When a liquid is cooled, very often a crystalline solid is the result. There are, however, at least two cases where the system makes a transition from the liquid phase to a phase which is structurally rigid, but possesses no long-range order. The first is a gel the second a glass. In the former, liquid solutions of long chain polymer molecules can be made to transform suddenly into a phase exhibiting rubber elasticity either by gradually increasing the number of chemical cross links or by raising the density and relying on the natural entanglements of the chains. At a lower temperature such systems undergo a further transition into a glassy phase of the more usual kind.

The second case is the one appropriate to the present discussion. When certain materials - SiO_2, Pd_4Si, glycerin or As_3Se_3 - are cooled from the melt, a transition into a rigid stable phase with no long-range structural order can take place. The dependence of the volume on temperature, shown in Figure 1, illustrates what takes place. Instead of a first-order transition from liquid to

---

*A third group comprises elements such as As which, as might be expected, are intermediary between the two groups mentioned.

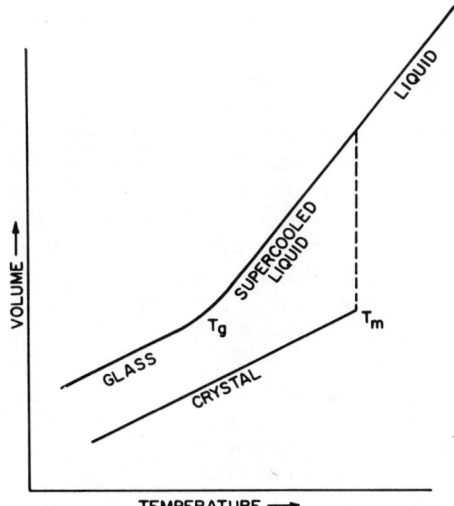

Figure 1. Dependence of volume on temperature as either a crystal or a glass is cooled from a melt.

crystal at the melting temperature T_m, the material supercools to well below T_m. Near the so-called glass transition temperature T_g (whose value depends on the sample history and the rate of cooling) the viscosity, volume and other parameters change rapidly, although not discontinuously, according to an empirical law (Fulcher's Law)

$$\zeta^{-1} = \zeta_o^{-1} \exp[-A/(T-T_o)]$$

where T_o is about 20% lower than T_g.

There appears to be no fully accepted theory of the glass transition and no explicit characterization of the glassy state. Thus, although the glass is relaxing toward a more stable state, it is not clear what that state is. The two possibilities are (1) the crystalline state. On this model supercooling slows the transition to a stop and the frozen-in disorder of the liquid state is the glass disorder. (2) An ideal glassy state in internal metastable equilibrium, possessing zero entropy[3].

For the purposes of the present discussion, we note the following. There are examples of all kinds of bonding in glasses: covalent (As_2S_3); ionic ($ZnCl_2$); metallic (Pd_4Si); hydrogen ($KHSO_4$); van der Waals (Se). A wide range of electrical conductivity is found; thus, SiO_2 is a dielectric, Pd_4Si a metal, and As_2Se_3 a semiconductor. Only the semiconductors such as As_2Se_3 concern us here. These are

often covalent, perhaps because their directional bonds stabilize
the disordered network and prevent the bond-breaking which would be
a necessary antecedent to crystal formation. It is generally correct
to take the view that their structure and properties are very little
different whether they are prepared by cooling from the melt or by
one of the rapid cooling methods I now describe. Therefore, we shall
dwell no more on the kinetic and thermodynamic questions raised by
the nature of the glass transition. There are other methods of producing amorphous semiconductors, which are structurally indistinguishable from glasses of the same chemistry prepared from the melt,
when that can be done. These involve a bypass of the process of crystallization through rapid solidification onto a cold substrate.
Usually these materials are in the form of thin films several hundred
Å to several microns in thickness. The methods are (1) evaporation
of the desired material in vacuo, using joule, laser, or electron-
beam heating, onto metal or dielectric substrates maintained at a
controlled temperature below that inducing epitaxy. (2) d.c. or r.f.
sputtering by bombardment of a target of the material by A ions onto
similarly temperature-controlled substrates. Incorporation of gases
other than A, such as H_2, PH_3, B_2H_6, N_2, can be used to dope the
material or to form alloys. (3) electrolytic deposition by passing
a current through a solution containing ions of the desired material.
(4) bombardment of a crystal of the material to produce disorder.
(5) decomposition of a vapor by glow discharge. As in (2) doping may
be achieved by adding other gases to the discharge.

There are many important parameters in each preparation
method which determine the eventual details of the structure, including the defect structure, and the impurity concentration. For example, the important parameters in method (2) are the type, preparation
and temperature of the substrate, the rate of deposition, the pressure
of the A sputtering gas, the geometry of the sputtering apparatus,
and the power fed into the gas plasma.

It cannot be overemphasized that the investigation of amorphous semiconductors has probably now reached the stage where attention must be paid to all of the deposition parameters, and especially
also to the chemical contaminants, if useful comparison of results
from different laboratories is to be done.

Instead of discussing how the detailed parameters of the
preparation method might be expected to influence the structure -
obviously an important topic if improved materials are to be made
- we shall rather discuss how the structural arrangement of the atoms
is determined, never mind how they got there.

1.2   The Nature of Disorder

In the perfect crystal, a basic molecular unit is repeated
on a periodic lattice in three dimensions. In the topologically
disordered solid there may also be a basic molecular unit (it could

AMORPHOUS SEMICONDUCTORS

be a single atom), but it is not possible to label the positions in space of any element of the unit in isomorphism with any infinite periodic array.

Topological disorder is therefore qualitatively different from so-called Quantitative Disorder, although the latter always occurs when topological disorder is present. A crystal is quantitatively disordered when random fluctuations in potential are superimposed on the perfectly periodic potential. This may be the result of lattice vibrations (vibrational disorder), or the random replacement of one chemical element by another (chemical disorder) or a random variation of the net electron spin on an element (spin disorder). Throughout, the periodic arrangement of the basic molecular unit will continue to be identifiable; the basic periodic lattice is not abandoned.

The basic molecular unit depends on the chemical element(s) involved and also, to some extent, on the preparatory method. Thus, for example, there may be a tendency to form a molecule in a depositing vapor and the disorder involves the nonperiodic arrangement of the molecules. Under different conditions of preparation, that molecule may not be permitted to form, either in the vapor or on the growth surface, so that chemical disorder is important, and perhaps important enough to dictate the local topology. There is a wide range of structural features for any single system, sufficient to ensure that an exact description of the atomic coordinates of the structure is not possible, and that we must rely on the statistics of the short-range order.

1.3 Experimental Determination of Structure

I shall now summarize the principal means at our disposal to determine the structural arrangement of the atoms in a homogeneous phase. The determination of the average chemical constitution would take us too far afield, and in any case, that problem is no different from the problem for any small volume of material, ordered or otherwise. The methods used to determine the structure are:

(1) Analysis of the angular diffraction pattern of electrons, X-rays or neutrons. Inversion of the angular distribution by Fourier analysis gives the Radial Distribution Function, i.e., the density of atoms as a function of distance from any one atom chosen as origin[4]. The R.D.F. of a crystal at 0 K would be a series of δ-functions (ignoring zero-point broadening). For an amorphous material, one typically finds three or four severely broadened peaks, and no distinguishable structure at larger distances.

(2) Ball-and-stick model building[5]. Starting from a basic molecular unit, which is either deduced from knowledge of the unit in the crystal or vapor, or indeed from a preliminary analysis of

the R.D.F., and following a very few simple rules, structural models are built. Measurements are then made of the R.D.F. of the model, and this is compared against the diffraction results of (1). The method is useful in eliminating unsuitable models, but is usually unable to establish in definitive fashion any unique model.

(3) Extended X-ray absorption fine structure (EXAFS). There exists fine structure on the absorption edge associated with certain X-ray region transitions which can be interpreted to give the arrangement of nearest neighbors of a particular atom of the amorphous material. This method has been applied, for example[6], to prove that As inserted into amorphous Si is predominantly three-fold coordinated.

(4) Small angle X-ray or electron scattering. Radiation scattered at angles smaller than the first diffraction peak is attributed to regions in the material where the electron density averaged over more than an atomic volume is radically different (usually very much smaller) than the average. Thus, voids ranging in diameter from 2-3 interatomic diameters up to tens of Å may be identified. The upper size limit is determined by the experimental necessity of distinguishing the radiation scattered at very small angle from that in the direct incident beam[7].

(5) Raman scattering. This technique is useful in the identification of small molecular units. For example, it can be used to identify in the amorphous binary material bonds of the A-A type in a binary compound AB possessing only A-B bonds in its crystal[8].

(6) Localized vibrational optical absorption in the infrared. Where there are light impurities in low concentration in the semiconductor (for example, H in $Si_{1-x}H_x$ alloys) the energy and distribution of these localized modes may be utilized to infer the local configuration of the impurity[9].

(7) Electron spin resonance[10] and nuclear spin resonance Just as in crystals, these may be used to give information about the local surroundings.

(8) Optical and electron microscopy. These methods may be used to characterize the larger defects in the structure.

(9) The mass density, and especially its percentage deficit from the value for the crystal, or that calculated from the structures of items (1) and (2).

AMORPHOUS SEMICONDUCTORS

1.4 Diffraction

I shall concentrate this discussion on how the first two of these techniques may be combined to give information about the structure of amorphous germanium.

My remarks on diffraction will be confined to monatomic solids[4], and they will omit any experimental details. The intensity of radiation I scattered by a point electron out of an incident beam of intensity I_o may be written

$$I = I_o \frac{e^4}{m^2 c^4 R^2} \left(\frac{1 + \cos^2 2\theta}{2} \right)$$

where 2θ is the angle between incident and diffracted beams and R is the distance from the point electron. The above units are in the c.g.s. system. Scattered intensities from material in units of

$$I_o \frac{e^4}{m^2 c^4 R^2} \left(\frac{1 + \cos^2 2\theta}{2} \right)$$

are called electronic units. The intensity in electronic units is the scattered intensity from a sample divided by the classical intensity from a point electron.

In electronic units the scattering by an atom is given by

$$I_{e:u.} = ff^*$$

with $f = \sum_n f_n$ and $f_n = \int \exp\left[\frac{2\pi i}{\lambda}(\underline{s} - \underline{s}_o)\cdot \underline{r}\right] \rho_n(\underline{r})\, dV$

where f_n is the scattering factor per electron, $\rho_n(\underline{r})$ is an electronic charge density in the atom for the nth electron, $\underline{s}$ and $\underline{s}_o$ are the scattered and incident unit vectors, and f is the atomic scattering factor. For an assumed spherical charge density.

$$f_n = \int_o^\infty 4\pi r^2 \rho_n(r) \frac{\sin k r}{k r}\, dr$$

where $k = 4\pi \sin \theta/\lambda$.

For any arrangement of atoms a similar development gives

$$I_{e.u.} = \sum_m \sum_n f_m f_n \exp\left[\frac{2\pi i}{\lambda}(\underline{s} - \underline{s}_o)\cdot \underline{r}_{mn}\right]$$

where $\underline{r}_{mn} = \underline{r}_m - \underline{r}_n$ and f_m, f_n are the atomic scattering factors. Let us, for simplicity, suppose all the atoms are identical. Then,

$$I_{e.u.} = \sum_m f^2 + \sum_m f^2 \int \rho_m(r_{mn}) \exp\left[\frac{2\pi i}{\lambda}(\underline{s}-\underline{s}_o)\cdot\underline{r}_{mn}\right] dv_{mn}$$

where $\rho_m(r_{mn})dv_{mn}$ is the number of atom centers in the volume element dv at the position r_{mn} relative to atom m.

Warren[4] has shown that under certain simplifying assumptions, this can be rewritten

$$I_{e.u.} \approx N\left[f^2 + f^2 \int_0^\infty 4\pi r^2 \left(\rho(r) - \rho_a\right) \frac{\sin k r}{k r} dr\right]$$

where $\rho(r) = \langle\rho(r_{mn})\rangle$; averaging over all $\rho_m(\underline{r})$ a displacement $\underline{r}$ from an atom and $\rho(r) \to \rho_a$ the average atom density, where r is larger than several atomic distances.

Setting

$$\frac{\frac{I_{e.u.}}{N} - f^2}{f^2} = i(k)$$

we obtain

$$k\, i(k) = 4\pi \int_0^\infty r\left(\rho(r) - \rho_a\right) \sin kr\, dr$$

from which we get

$$r\left(\rho(r) - \rho_a\right) = \frac{1}{2\pi^2} \int_0^\infty k\, i(k) \sin kr\, dk$$

or

$$4\pi r^2 \rho(r) = 4\pi r^2 \rho_a + \frac{2r}{\pi} \int_0^\infty k\, i(k) \sin kr\, dk$$

The L.H.S. is the average number of atom centers between r and r + dr, for any atom. The first term on the R.H.S. is the average atom density, obtainable from the mass density. The second term is obtained from the experimental scattering curve. I shall not discuss here experimental corrections for Compton scattering, for multiple scattering, or the process of normalization in electron units, or the correction for the so-called termination errors, because the experimental results give i(k) only over a limited range of k[11].

The quantity $J(r) = 4\pi r^2 \rho(r)$ is called the Radial Distribution Function (R.D.F.), an example of which is shown in Figure 2.

AMORPHOUS SEMICONDUCTORS

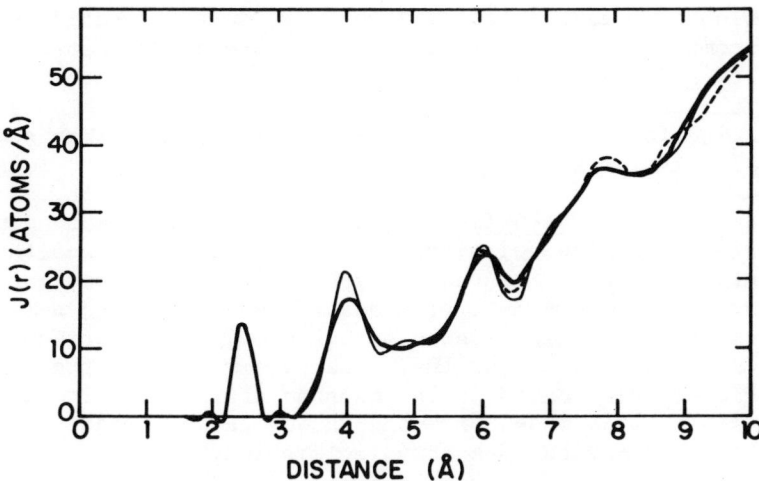

Figure 2. Experimental radial distribution function (the heavy line) of a-Ge, taken from Temkin et al., reference 11. The maximum of the first peak gives the interatomic distance, and the area under the first peak the first coordination number. The position of the second peak verifies the tetrahedral coordination. Roughly speaking, the spread of the second peak gives information on the bond angle deviation, but corrections are necessary to take account of third neighbor contributions to this peak.

Also shown are computed RDF's from the energy-relaxed 519-atom model of Polk and Boudreaux[15] and the 201-atom model of Steinhardt, Alben and Weaire[16]. The heavy line is the experimental result, the thin continuous line is for the 201-atom model and the dashed line is for the 519-atom model. The calculated curves are broadened to reduce statistical fluctuations.

1.5   Structural Modelling

Next we turn to the problem of ball-and-stick modelling of the structure[5]. The first thing to be done here is to decide on the basic unit of the structure. A very simple example is a-Ge, where the basic unit may confidently be taken to be an atom with four neighbors at roughly the corners of a tetrahedron. The basic unit, chosen in recognition of the short-range order of the nearest neighbors, may also be selected with relative ease for two-fold coordinated systems (e.g., SiO_2, Se, Te) and three-fold coordinated systems (e.g., As, Sb, Bi), but it is somewhat more difficult where fixed coordination is possible (e.g., three- or four-fold coordinated C, or a 4-6 compound such as GeTe, which has the crystal structure of crystalline As).

The following possibilities for the atomic arrangement occur:

(1) <u>Microcrystallite models</u>. The material may be composed of one or several kinds of microcrystal, in each of which the exact bond lengths and angles of the basic unit are maintained. Between the microcrystals there must be a disordered connecting network. When the microcrystallite size extends to only three or four unit cell dimensions, it is clear that a large fraction of the atoms are on the cell boundaries and in the microcrystal connecting tissue. Differentiating this model from the random network models described below becomes something of a semantic problem.

(2) <u>Amorphous cluster models</u>. The amorphous cluster is defined to be a small group of atoms (of order 100) arranged so that each atom has exactly the same short-range order, but in such a way that it does not lead to a periodic arrangement capable of filling all space. It is supposed that a very small cluster of this type may have a lower free energy than a cluster which is consistent with a crystallographic configuration. The same difficulties of fitting the clusters together with disordered connecting tissue, so as to form a macroscopic sample, occur as for the microcrystals. An example of a cluster unit is the pentagonal dodecahedron formed from tetrahedrally connected basic units.

(3) <u>Random network models</u>. The local order in the basic unit is distorted by fluctuations of finite but limited degree in the bond lengths and angles, so that one obtains a continuous random network (CRN) filling all space, without any unfulfilled bonds.

Models may be built following any one of the above general prescriptions, and the radial distribution function calculated for comparison with that obtained from diffraction experiments.

Unfortunately, in many cases, several atomic models may give an R.D.F. which is an acceptable fit to the experimentally-determined one. These distributions are still only averages over the network. More detailed comparisons are then necessary. These could be the distributions of dihedral angles or the ring statistics. The mass density of the network must correspond to the experimental mass density of material that is known to be close to full coordination (although this may be difficult to tell). There are other indirect comparisons possible in principle. Thus, the heat capacity of the network, calculated on the basis of approximate theories giving the extra elastic energy in a distorted structure in terms of the deviations of bond lengths and bond angles from their lowest energy (crys talline) values, may be compared with an experimental value derived from measurements of the heat capacity as a function of temperature. The model should permit a reasonable explanation of the effects of annealing, the kinetics of annealing and of recrystallization. Finally, the phonon energy and electron energy densities of states calculable from the coordinates of the model and a plausible Hamiltonian must correspond to those measured experimentally.

A detailed consideration of the work done along these lines would not be productive in the present context, for it seems to be the case that, while some models may be eliminated by comparison of their predictions with experiment, a unique (even statistically) model cannot be established. I shall try to illustrate this conclusion by a brief description of the result of modelling studies on amorphous germanium.

1.6   Comparison of Diffraction Results and Modelling Studies of a-Ge

Early modelling of a-Ge[5] had used a mixture of pentagonal dodecahedron and diamond-like units, which restricted the structure unduly by permitting only eclipsed or staggered configurations of the bonds between neighbor tetrahedra. Polk[12] then constructed a ball-and-stick model with a basic unit of tetrahedral symmetry to construct a continuous random network model with the following characteristics: (1) a core of five- and six-member rings which ensured the generation of a noncrystalline structure; (2) no unsatisfied bonds in the interior of the model; (3) bond length variations held to within 1% and bond angle variations held to ± 10°; (4) a choice of dihedral angle for a unit being added to the structure such that it connected with a minimum of strain. His final CRN consisted of 440 'atoms' and appeared capable of indefinite extension. Its RDF matched quite well to that found from diffraction experiments.

All later work on modelling of a-Ge using random network has been devoted to a refinement of this technique[5]. For example, the 'atomic' positions can be accurately measured by an accurately positioned laser beam. They can then be adjusted slightly by computer so as to minimize the average fluctuation in bond lengths. Alternatively, the atom positions, as originally determined from the model, may be adjusted by computer so as to minimize an expression for the elastic energy of the solid given in terms of the bond lengths and bond angles and assumed bond-stretching and bond-bending force constants.

A quantitatively different alternative to the Polk prescription was devised by Connell and Temkin[13], who built tetrahedrally coordinated units with no odd-membered rings. The coordinates of these fully-coordinated even-membered ring models could then be adjusted by computer so that the variance of the bond lengths was equal to that of the static distortion measured experimentally.

There have also been CRN models generated entirely on a computer[14], and microcrystallite models using mixtures of crystallities of different symmetries.

From the wealth of data now available, I should like to draw two conclusions. The first is that models using large microcrystals may be plausibly discounted: such models must give a peak in the RDF at the spacing for the third nearest neighbor in the crystal and such a peak is absent from the experimental diffraction data. This then illustrates how our procedure can dismiss some models.

The second conclusion is that it is extremely hard to choose between the different CRN models, even those with radically different ring statistics, such as the Polk and Connell-Temkin models. Figure 2 and 3 show comparisons of the RDF of several models

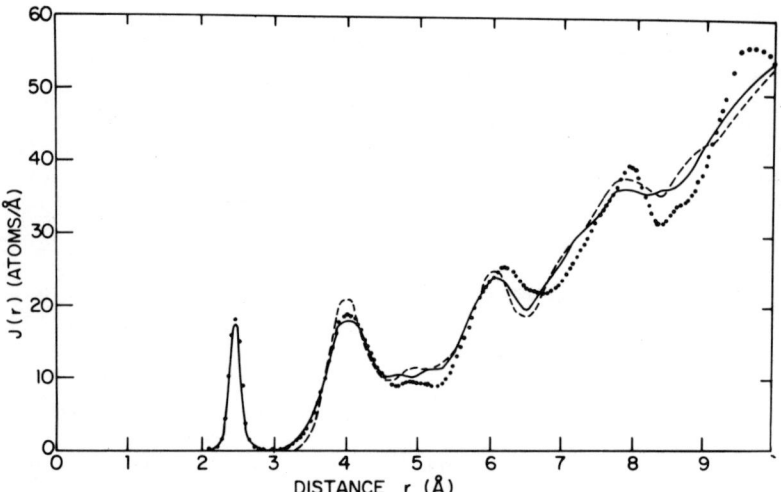

Figure 3. Comparison of the RDF's derived from the even-member ring model of Connell and Temkin and the 519-atom model of Polk and Boudreaux[15] with experimental data for a-Ge by Temkin et al[11]. The heavy line represents the experiment, the dotted line the calculations of Connell and Temkin and the dashed line the calculations of Polk and Boudreaux.

with experiment. There is, indeed, little to choose between them. The basic reason is that the initial choice of the tetrahedral unit and the restrictions placed on it in modelling, goes a long way to determining the RDF. Thus distinctions between different models must be done on the basis of a very detailed comparison of the fine structure of the models. However, it should be clear that unique choices are very difficult. My conclusion is that, at least in the case of amorphous tetrahedrally-coordinated semiconductors such as Ge and Si, structural modelling has been taken as far as it can usefully go.

So far I have concentrated on the structure of the com-completely-coordinated network with no dangling bonds. However, all amorphous tetrahedrally bonded semiconductors studied so far have voids whose density, shape and volume depend on the preparation conditions[7]. The existence of voids may be deduced from the fact that the material is less dense than the interatomic separation and structure would suggest, from the observation of small angle scattering of X-rays or electrons, and from a departure of the first coordination number (the area under the first peak of the RDF) from the value 4. The small angle scattering in a-Ge is illustrated in Figure 4. Samples of different void volume are found to have the same RDF,

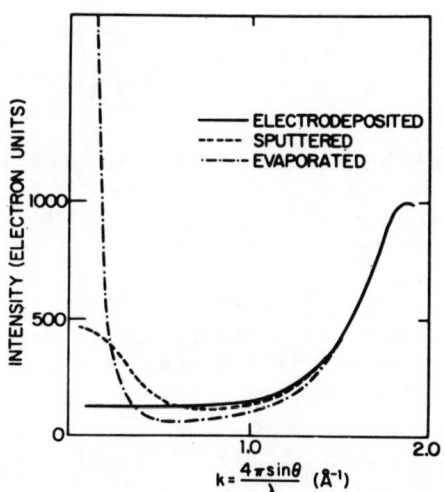

Figure 4. Illustration of small angle X-ray scattering in a-Ge. After N.J.Shevchik and W.Paul, J. Non-Cryst. Solids, 8-10, 381 (1972).

which suggests that the defective network can be approximated by a fully coordinated network of unaltered topology interspersed with voids which mildly alter the atomic arrangement on their surfaces. All of this suggests that the direct structural probe represented by the RDF does not tell us about variations in the local topology produced by defects in the random network such as dangling bonds, voids, and variations in the coordination number. These, nevertheless, may be crucial in determining the electronic properties.

Information about the structure of the defects must therefore be found, if it is to be found at all, from an analysis of the results of the more indirect types of measurement I mentioned earlier: infrared absorption, Raman, EXAFS, n.m.r., and e.s.r. It would however, take us too far afield to give a systematic account of these techniques here. I shall, however, speculate on the possible structural defects in a-Si and in the chalcogenides later on, and the electronic energy levels these produce.

2.1 Electronic Band Structure and Properties

Consideration of the electronic band structure and properties divides naturally into three parts: discussion of the intrinsic density of states in energy of the completely-coordinated, defect-free material, and of the nature of the corresponding wave-functions; discussion of the density of states in the energy gap between valence and conduction bands, produced by defects and impurities; and deduction of the transport, optical excitation and recombination properties based on the D.O.S., and their comparison with experiment. This is a tall order for such a brief review as this, and so only the principal details can be given.

2.2 Intrinsic Density of States and Corresponding Wave-Functions

The classification of electronic states by k-vector is without meaning in an amorphous semiconductor, since the coherence length of the wave-function is of the order of the interatomic distance, but it is still appropriate to examine their density and their extended or localized character as a function of their energy. There are various complementary theoretical approaches to such a description, none of them sufficiently exact to provide a band structure that can be decisively tested against experiment, but each shedding some light on the problem.

(1) <u>Tight-binding models which concentrate on the effects of topological disorder, but neglect any quantitative disorder connected with fluctuations in bond lengths and bond angles</u>.

In this approach, pioneered by Weaire and Thorpe[17] for a-Ge, it is supposed that, while long-range order is responsible for the fine structure of the density of states (including the band edge structure in crystals) the short-range order dictates the gross band structure and even the gross classification into metals and insulators. It is also supposed that the quantitative disorder in amorphous semiconductors does not decide the gross details of the valence and conduction band structure, although it will be important for consideration of the band edge states. This approach suggests right away that the valence band of four-fold coordinated semiconductors will be composed of states constructed out of sp^3 bonding orbitals. The valence band of three-fold coordinated As will be constructed out of p-bonding orbitals, since the s-p hybridization is known to be weak. By contrast, the valence band of Se and all other chalcogenides is a relatively narrow band of <u>nonbonding</u> p-states, a fact that is responsible, at least in part, for the quite different properties of this class of materials. The conduction bands of Ge are based on sp^3 antibonding orbitals, while those of As and Se are based on p antibonding orbitals. I shall discuss only the tetrahedrally coordinated material explicitly.

By adopting a very simple Hamiltonian for a four-fold coordinated system Weaire and Thorpe were able to derive a simple valence band density of states whose salient features could be tested against experiment. Their Hamiltonian was written

$$H = V_1 \sum_{i, j \neq j'} |\phi_{ij}\rangle\langle\phi_{ij'}| + V_2 \sum_{i \neq i', j} |\phi_{ij}\rangle\langle\phi_{i'j}|$$

in terms of sp^3 hybrid orbitals $|\phi_{ij}\rangle$ directed along the four tetrahedral directions. The sums are over sites i and bonds j. Since V_1 and V_2 do not vary from site to site, quantitative disorder is ignored. Thus the first term describes the matrix element between orbitals jj´ at any site i, while the second describes the matrix element at a specified bond j between orbitals on i and i´.

The wave function of the solid is written as a linear combination of all of the orbitals in an LCAO approach

$$|\psi\rangle = \sum_{ij} a_{ij} |\phi_{ij}\rangle$$

This Hamiltonian may be used to describe a periodic lattice or a CRN.

It is straightforward to show that clean energy gaps appear between the valence and conduction bands, independent of the topology of the tetrahedral network, provided a certain condition on $|V_2/V_1|$ is satisfied. If $|V_2| > 2|V_1|$, the conduction band is separated from the valence band by $2|V_2| - 4|V_1|$ and the top of each band is a p-like δ-function containing two states per atom independent of the site topology. The precise shape of the rest of the bands does depend on the topology, which has led to numerous attempts to link such structural features as the fraction of odd-membered rings with the experimentally determined density of states.

The existence of a clean energy gap may come as a surprise if one intuitively expects disorder to smear out the energy bands, but remember that quantitative disorder is explicitly excluded from the Hamiltonian! Nevertheless, the band gap has nothing to do with long-range periodicity and Bragg reflection of electrons. It comes purely from the tetrahedral coordination and a condition that $|V_2| > 2|V_1|$.

Weaire and Thorpe applied their Hamiltonian to a recalculation of the band structure of f.c.c. Ge. It has also been used to calculate the band structure for lower symmetry polymorphs of Si and Ge, and, of course, it may be applied to a disordered atomic array whose coordinates are known - such as the Polk[12] or Connell-Temkin models. Indeed, using the Polk-Boudreaux[15] version of the CRN as a basis, and elaborating the tight binding approach to use overlap integrals that depend on the angular distortions in the network, Bullett and Kelly[18] have obtained a quite impressive agreement with experiment.

Figure 5 illustrates the experimental band structures for crystalline and amorphous Ge. The sharp features associated with van Hove singularities in the density of states for the crystal are wholly absent in the amorphous phase. The gross features are, however,

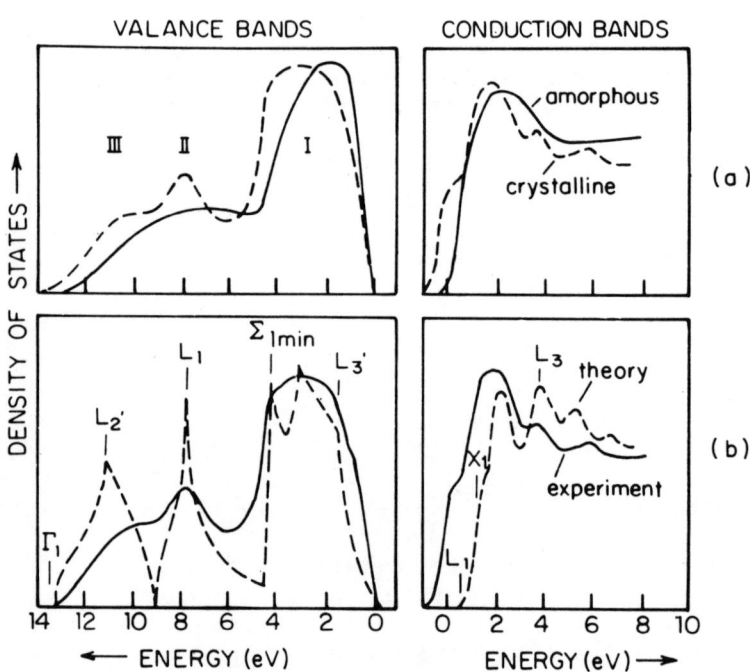

Figure 5. (a) Experimental electronic density of states of amorphous and crystalline Ge; (b) Experimental and theoretical electronic density of states of crystalline Ge. After D.E.Eastman, J.L.Freeouf and M.Erbudak, A.I.P. Conference Proceedings, Vol. 20, 95 (1974).

similar. The p-like δ-function found in the Weaire and Thorpe calculation appears as a peak in both phases, but skewed to higher energy for the amorphous phase. This is an effect which can be linked to the presence of eclipsed bonding configurations. The valley between the two lower bands (II and III) for the crystalline phase is shallower for the amorphous phase, and this effect is predicted on the Weaire-Thorpe model for CRN with many odd-membered rings. Thus we see that the gross features of the valence band structure can indeed be correlated with experiment using this Hamiltonian. Notice, though, that the conduction band cannot be treated at all, and the states near the valence band edge are treated only very qualitatively.

Analagous tight binding studies[19] of other amorphous materials such as, for example, As, Se, GaAs and As_2Se_3 have been carried out. We do not have space to discuss these, except to remark that they have similar, limited, success in correlating the gross features of the experimental and theoretical densities of states.

(2) <u>Tight-binding or empirical pseudopotential calculations, using a sophisticated Hamiltonian, applied to large unit cell, low symmetry crystalline polymorphs, which simulate the amorphous material because the low symmetry approximates that of the disordered material.</u>

This approach[20] has been applied principally to the tetrahedrally-coordinated materials. It is based on the existence in nature of several low symmetry arrangements of atoms with tetrahedral bonding, in addition to diamond and wurtzite. These low arrangements typically have two values of nearest neighbor bond length and two values of bond angle. Thus they provide periodic structures with low symmetry, large unit cells. As the symmetry is decreased, the resemblance of the DOS to that of the amorphous semiconductor improves, as expected.

(3) <u>Other approaches</u>[21], which I shall only mention in passing, involve: (a) one-dimensional random models for which the density of states in energy can be exactly calculated, and the nature of the wave-functions explored, in order to give clues about real systems; (b) three-dimensional tight binding models which neglect any topological disorder, but assume various kinds of distribution of single site energies randomly distributed over a regular array of sites. These models can be solved for the density of states vs. energy and the wave-function localization as a function of energy; (c) models with a small cluster (say, 64) of atoms in a noncrystalline array which nevertheless preserves the short-range order and has only very small bond angle fluctuations, the whole cluster being repeated periodically; (d) models which use multiple scattering theory for small clusters, each having the appropriate local order, with the clusters distributed randomly in orientation.

These approaches, while never complete in themselves, confirm the general pattern of the results described so far, to which we should add the following. States near the crystalline band edges, which are shown diagrammatically in Figure 6, are smeared in energy by the disorder and assume an increasingly localized character as their energy penetrates deeper into the former crystalline band gap.

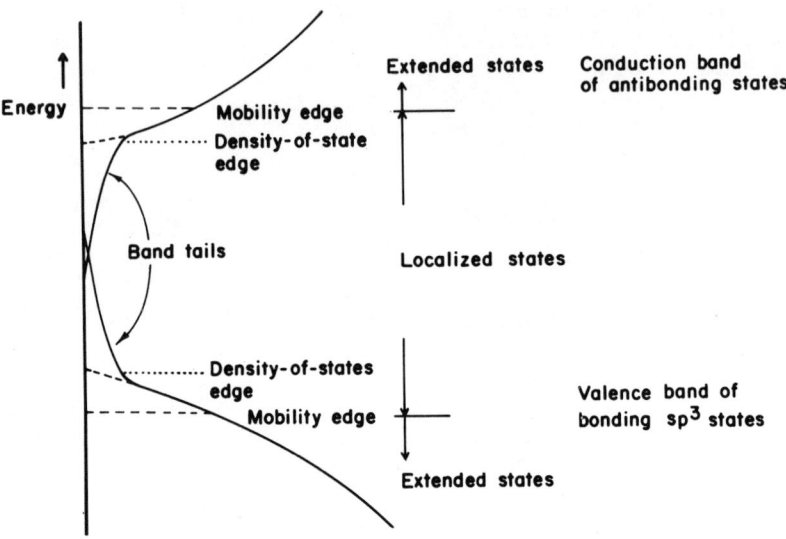

Figure 6. Model for the intrinsic density of states versus energy for a completely-coordinated tetrahedrally-bonded network, such as a-Si.

It behooves us next to say more about this very important feature of localization, first introduced by Anderson[22] in 1958.

It is generally supposed that the quantitative disorder in the potential leads to strong electron scattering and a coherence length for the electronic wavefunction of the order of the interatomic spacing. Thus the uncertainty in the wavevector is of order of the wavevector itself. This has important effects for the estimation of transport and optical properties.

It is not difficult to see qualitatively how random fluctuations in potential (quantitative disorder) can lead to tails in the band densities of states. Fluctuations in the atomic configurations are supposed to result in fluctuations in the local potential, and large fluctuations lead to large perturbations in the electronic energies. Qualitatively, it would appear that s-states would be less affected than p- or d-states by variations in the orientation of nearest neighbor atoms that leave interatomic distances the same. Thus we might expect qualitative differences in the degree of band tailing of valence and conduction bands of quite different atomic symmetries.

If fluctuations exist, it is appropriate to consider a very important paper, published by Anderson[22] in 1958, on the localization by the quantitative disorder of wavefunctions for states in a band with a continuous distribution of energies. Anderson's quantitative theory considered a three-dimensional periodic array of sites with a site energy varying randomly but with uniform probability between $-W/2$ and $+W/2$. The matrix element of his tight-binding Hamiltonian between any pair of nearest neighbor sites was $-V$, and between any more distant neighbors, zero. For large values of W/V the electronic states are localized. Anderson then considered to what extent W/V could be reduced before the probability that an electron stayed at its original site was reduced to zero at an infinite time. He found that the localized states are stable, provided

$$\frac{W}{2V} \gtrsim 2(Z-1) \ln\left(\frac{W}{2V}\right)$$

where Z is the coordination number.

Later, Mott generalized Anderson's argument by introducing the idea that the states in the tail of a band might be localized even when the states in the band center were not. This is not difficult to understand qualitatively. An electron with an energy in a band tail can only transfer to states of about the same energy. If $N(E)$ is small, such states are correspondingly further away; and so a fluctuating potential that is insufficient to localize electrons of energy E when $N(E)$ is large, may well be sufficient if the states, as in the tail, are well separated.

Thus, there will be a qualitative difference between the extended states in the center of the band and the localized states in the tail. The two kinds of states are often said to be separated by a Mobility Edge. The sharpness (in energy) of this Mobility Edge has been a subject of much debate in the literature, which we shall not pursue here. The issue is still unresolved. Mott's view is quite specific: any randomly distributed potential, random either in position or well depth, will have a range, even though small, of localized states where the state density is lowest. In this range of states the mobility will be strictly zero at $T = 0$. The obvious implication of all this is that amorphous semiconductors may have sharp mobility edges - and thus 'mobility gaps' between valence and conduction bands - even when the density of states is continuous in energy throughout the whole energy range.

The issue of localization is a very interesting one, particularly because it is an unfamiliar notion for states in the broad bands of conduction and valence states in crystals. Several comments are in order. It is not difficult to understand the localization of all of the states in a narrow band, such as that derived from core electron functions. Other examples of situations giving localized stages in crystals are the 'potential well' for the fifth (chemically unpaired) electron of a P atom in a Ge crystal at T = 0, the negative ion vacancy in an alkali halide crystal, which can trap one electron (an F-center), and the potential produced by an impurity like Au or Cu in Ge.

The situations mentioned above are recognisable local defects, and obviously the potential fluctuations in amorphous semiconductors produced by topological and quantitative disorder are not nearly as simple. How are we to recognize or define a localized state? We might assert that a state is localized if its eigenfunction decays exponentially outside of a certain region of space. This however, is hardly a good operational definition unless we can obtain $|\psi(r)|^2$ experimentally (we usually cannot). Moreover, we cannot calculate $\psi(r)$, since we never have a good definition of the potential $V(r)$. Thus, the exponential decay must be justified a posteriori.

There are several other possible definitions, but perhaps enough has been said to illustrate why Anderson's off choice - a condition for localization that the particle not diffuse away from its starting position as $t \to \infty$ - is a reasonable one, even although it cannot be related directly to an experimentally observable quantity.

Some negative statements about localized states are perhaps helpful to their understanding. First, is a Bloch state for a crystal automatically an extended state? In particular, is a deep-lying core state extended? The fact that the expectation value of $|\psi(r)|^2$ is everywhere finite suggests that the state is extended. What if we start an electron in a core state and inquire about the probability of its still being there after an infinite time? The answer is that it will diffuse away, which violates our intuition about tightly bound atomic electrons. The reason, of course, is that we cannot apply a one-electron theory which ignores electronic corelations. The electron-electron interactions in this case lead to electron localization.

Is an extended state synonymous with a Bloch state? Clearly not, as we can have an extended state with a finite amplitude on all sites of the material but fluctuating amplitude and no coherence of phase.

Is an impurity or defect state automatically a localized state? The answer is no, since small disturbances in potential produced by the defect may fail to split states off into an energy region of no states.

Finally, can extended and localized states exist at the same energy? The answer is no, but this is not intuitively obvious. It is expected that the coupling of an exponentially localized state to extended states of the same energy will alter the exponentially localized state so as to destroy the localization.

Common to all of the theories on localized states is the postulate that potential fluctuations do exist. If the fluctuations do exist, then the fact of localization is not in dispute. However, in real amorphous materials it is hard to find direct experimental evidence that band tailing or potential fluctuations or mobility edges do exist. In a real material strong electron-phonon interactions may act to modify or eliminate most of the strong fluctuations leading to extensive band tails and localized states[23]. Let us go back to the possibility of calculating band edge states, suppress our physical intuition that defects might strongly influence them, disregard the approximations made in the actual calculations to make them tractable, and then we must recognize the residual objection that the specification of the topology is usually only statistical. If local atomic correlations occur, they are not described. In the Phillips[23]-Anderson[24] model there exist local electronic and atomic shifts. The guiding principle here is that, in the formation of the amorphous solid, there is a tendency for local free energy minimization, whose range in real space will depend on the conditions of preparation, such as substrate temperature and rate of deposition. The atoms and electrons adjust in position in just such a way as to reduce the energy of the system, which is equivalent to pulling the states in the high energy tail of the valence band back into the band proper. With the potential fluctuations thus reduced, the tails of low energy at the bottom of the conduction band will also be reduced. As a result, the range of localized states at either band edge may be so small that they have a negligible effect at normal temperatures. Recently, Anderson[24] has given these qualitative ideas a quantitative underpinning with a theory of an attractive interaction between two electrons in a localized state in the presence of a strong electron-lattice interaction. Essentially, the coulombic repulsion between the two electrons that can populate an orbital (the correlation energy) is more than offset by a reduction in energy of the rest of the local system resulting from lattice deformation. This subject will be taken up again in the section on chalcogenide glasses.

2.3   States in the Gap Produced by Defects and Impurities

(1) Tetrahedrally-coordinated materials

There is ample direct evidence from the study of structure and indirect evidence from the measurement of properties to suggest that most amorphous semiconductors have a large density of localized states in the energy gap between the conduction and valence bands. In the tetrahedrally-coordinated materials, the presence of voids is well-established from structural studies. It is generally supposed that the large spin density ($\sim 10^{20}$/cc) observed in evaporated and sputtered a-Si and a-Ge is caused by electrons in void surface dangling bonds. This density is, however, more than an order of magnitude smaller than the number of atoms on the void surfaces, as calculated from the mass density deficit from the crystal and the amount of small angle X-ray scattering. We therefore are encouraged to examine what sort of electronic states can emerge from the ways in which void surface atoms can reconstruct their positions into low energy configurations. Just as for the surfaces on crystalline material, there will be weak reconstructed doubly-occupied bonds and singly-occupied dangling bonds. In Table 1 and Figure 7 we list and illustrate the possible reconstructions[25]. It should be emphasized that, although we have motivated consideration of gap states by giving evidence on <u>void</u> structural defects, the configurations to be described could conceivably occur in the formation of the material.

Table 1

<u>States in the Gap of Amorphous Groups 4 Semiconductors</u>

<u>States Caused by Departures from the Expected Coordination</u>

(a) <u>Alternative coordination and unreconstructed dangling bonds</u>

    3-coordinated group 4 atom - $(sp^3)^3$ + nonbonding (sp^3)

    filled donor levels below midgap and empty acceptor

    levels above midgap

(b) <u>Alternative coordination and bipolaron formation</u>

    (i) 3-coordinated group 4 atom - $(sp^2)^3$ + filled donor

        p-band near midgap and empty acceptor band

    (ii) 3-coordinated group 4 atom pair - one $[(sp^2)^3$ +

        empty nonbonding p-state near midgap$]^+$, plus one

        $[(s)^2 (p)^3]^-$ configuration with no gap states

(c) <u>Inequivalent coordination and weak bonds</u>

    4-coordinated group 4 atom with one weak approx. (sp^3) bond

AMORPHOUS SEMICONDUCTORS

States in the Gap of a-Si Produced by Defects

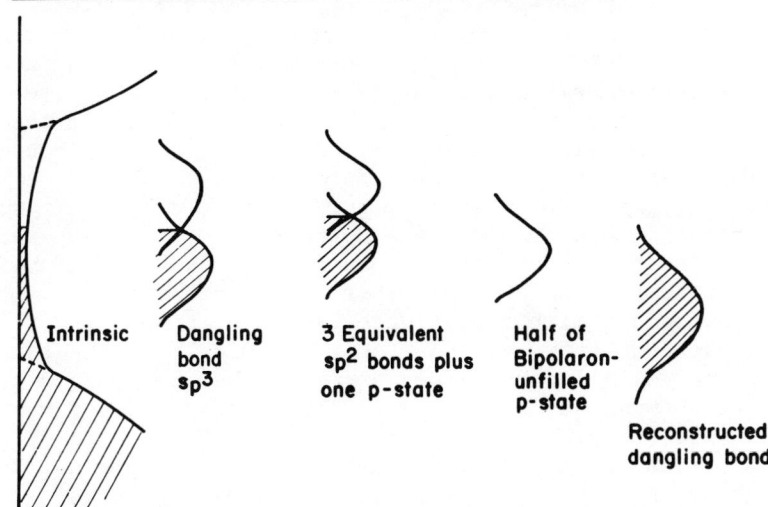

Figure 7. Model for the density of states in energy produced by different natural defects in tetrahedrally-coordinated semiconductors.

(a) <u>Alternative coordination and unreconstructed dangling bonds</u>: Imagine the lowered coordination caused by a classic 'dangling bond': an unpaired electron associated with an atom that has only three tetrahedral (sp^3) bonds. In the unlikely event that there is no atomic rearrangement whatever, we may suppose that this dangling bond introduces a narrow band of donor levels in the lower half of the pseudogap, and a corresponding narrow band of acceptors in the upper half, the two bands being separated in energy by the correlation energy of the two electrons in the sp^3 orbital. Although the single dangling bond cannot occur in a crystal, it is at least conceivable in the disordered network and can be formed without additional strain to the network.

(b) <u>Alternative coordination and associated bipolaron formation</u>: The prototype of this defect seems to occur in crystals either as a result of the existence of interstitials or in the reconstruction on certain crystal surfaces. The outcome in both cases is a 3-coordinated group 4 atom which may, however, exist in two different states of hybridization. The model for the self-interstitial is of a pair of group 4 atoms placed along the (100) direction at equal distances from a normal lattice site, and replacing the single atom at that site; each atom makes bonds to two of the nearest-neighbor atoms of the chosen lattice as well as to its partner in the inserted pair. The 3 bonds have sp^2 hybridization, which leaves one electron from each of the pair in a nonbonding p orbital. The expected electronic energy levels give an sp^2 band unlikely to be differentiated from the normal sp³ valence band, a filled one-electron (donor) p-band near mid-gap, and an empty (acceptor) band.

The model for the 2×1 reconstructed (111) surface suggests a different possibility, which we consider to be quite likely both in the reconstructions on internal void surfaces in the amorphous material, and in the disordered network itself without voids. Energy minimization on the crystal surface occurs when alternate atoms shift slightly up or down with respect to the planar surface, and an electron transfers from the down-shifted to the up-shifted atom. The down-shifted positively-charged atom adopts three-fold coordination with bonding orbitals approaching sp² symmetry and having energies coincident with the valence band; the nonbonding orbital is almost p-like and empty, and lies near the middle of the band gap. The up-shifted negatively-charged atom also has three-fold coordination but with pyramidal bonding orbitals that approach p symmetry and again have energies coincident with the valence band. The two nonbonding electrons reside in an orbital that is more s-like (than sp³) and have energies that also lie in the valence band.

(c) <u>Inequivalent coordination and weak bonds</u>: The final possibility we consider is that local free energy reduction may be achieved by atomic movements leading to electron pairing in weak or long bonds. The crystalline prototype of this is the slight reconstruction at a divacancy. No precise orbital hybridization description is possible, but it seems likely that these weak bonds will produce a filled band of states overlapping the top of the valence band and stretching into the lower half of the gap, and a corresponding empty band of antibonding states in the upper half of the gap. These states are doubly occupied, with no spin, in the ground state.

These arguments suggest a complicated distribution of filled, half-filled and empty states within the pseudogap, all stemming from natural defects. To these must be added localized states produced by impurities. In order to keep the subject within bounds, I shall consider only the effect of doping the material atoms from group 5 and group 3, the classic 'hydrogenic' donors and acceptors. See Table 2. It is then instructive to realize that, because the coordination of the impurity atom may be either 4 or 3 in the disordered network, a different set of states results than for the crystalline case.

(d) <u>Group 5 or group 3 atom in four-fold coordination</u>: In crystalline material, this produces a hydrogenic donor or acceptor with ground state binding energy

$$\varepsilon_b = \varepsilon_H (m^*/m_o)(1/K^2)$$

and orbital radius

$$r_b = r_H K(m^*/m_o)$$

Table 2

States in the Gap of Amorphous Group 4 Semiconductors

States Caused by Chemical Impurities

(a) 4-coordinated group 5 atom - $(sp^3)^4$ + antibonding (sp^3)

(classic donor,

$E_n \gtrsim E_H \frac{1}{n^2 K^2}$)

(b) 4-coordinated group 3 atom - $(sp^3)^4 \cdot$ + bonding (sp^3) hole

(classic acceptor)

(c) 3-coordinated group 5 atom - $(s)^2(p)^3$

(bonding band near top of

valence band)

(d) 3-coordinated group 3 atom - $(sp^2)^3$

(empty p-like acceptor

near midgap)

where ε_H, r_H are the binding energy and orbital radius of the ground state electron of hydrogen, m* is the effective mass at the appropriate band edge, m_o the free electron mass, and K an appropriate effective dielectric constant determined by the ability of the host atoms to screen the Coulomb attraction between the charged impurity and the nearly-free carrier. In crystalline Ge, for example, where the lowest energy conduction band minimum in the (111) direction has principal effective mass tensor components $m_\ell^* \approx 1.6 \, m_o$ and $m_t^* \approx 0.08 \, m_o$, even the ground state orbit is so large that K can be taken to be the high frequency dielectric constant. The calculated donor $\varepsilon_b \approx 0.01$ eV, which is in rather good agreement with observed values, and the corresponding $r_b \approx 45$ A. The calculated values for Si, under the same assumptions, are $\varepsilon_b \approx 0.03$ eV and $r_b \approx 20$ A, but the observed ε_b's for donor impurities are about 0.05 eV. Evidently, the larger effective mass and smaller dielectric constant for Si require a more exact treatment of the electronic shielding by valence electrons for the impurity electron ground state.

It is reasonable to suppose that a similar impurity state will exist in a-Ge and a-Si, but ε_b and r_b may be quantitatively different.

(e) <u>Group 5 or group 3 atom in three-fold coordination</u>: The above description is qualitatively similar to the doping of the four-fold coordinated crystal. There is, however, a strong likelihood that the group 4 host will, on occasion, adopt a three-fold coordination. This is even more likely if we incorporate group 5 or group 3 elements, which often prefer this coordination in their molecular and elemental crystalline forms. Thus, a Ga atom may adopt a planar three-fold coordination using sp^2 hybridized orbitals. The remaining p-state is thus empty and probably lies near the center of the gap. The filled states are in the valence band. Thus Ga acts as a deep acceptor.

As atoms may also be incorporated in three-fold coordination using p orbitals in a pyramidal configuration, with the two remaining electrons in a lone pair nonbonding s-state. The s and p filled states all lie in the valence band and the empty, antibonding states all lie in the conduction band. In this situation, As does not dope.

There is, therefore, an asymmetry in the discussion of Ga and As atoms which did not exist in the case of the crystal.

(2) <u>Group 3-5 materials</u>

A similar approach can be taken to prediction of the states in the gap produced by natural defects and common impurities in disordered 3-5 semiconductors such as a-GaAs[26]. We shall not, for reasons of space, pursue that here. However, two brief comments on new features seem worthwhile. First, both of the constituent elements of the 3-5 may occasionally adopt three-fold coordination. Second, like-atom or 'wrong' bonds may occur. If this is a frequent occurrence, so that the like-atom bond pairs are not isolated, then they may establish a band of states whose energy might be such as to dominate the D.O.S. at either the valence band or the conduction band edge. If the like-atom pairs are isolated, then they may probably be considered as antisite defects. Then, As like-bonds act as deep donors and Ga like-bonds as deep acceptors.

(3) <u>Chalcogenide materials</u>

The chalcogenides deserve special consideration, because the combination of a low average coordination number and the flexibility in bonding afforded by the existence of nonbonded p-electrons can lead to almost complete elimination of single spin states from the band gap. This has a dramatic effect on the optical and transport properties.

The direct structural evidence for defects in chalcogenides is, to my knowledge, nonexistent. Defects are postulated to exist that possess certain properties that fit the experimental observations: no net spin, no hopping transport, and no optical absorption at low infrared energies[27].

One approach to a formal catalog of defects is to suppose that atoms in chalcogenides may adopt an <u>alternative coordination</u>, an <u>inequivalent coordination</u> or a <u>transferred coordination</u>. The first two of these were discussed under the heading of Tetrahedrally Coordinated Materials. The third may occur when an atom takes up another of the dominant coordinations in the network instead of its expected one. For example, As in As_2Se_3 may adopt a two-fold coordination instead of three-fold because Se has two-fold coordination. Such a circumstance will be favored by the way in which the network is formed, but it is necessary to consider the cost in energy of the transferred coordination.

In the disordered element, for example, Se, minimization of the local free energy may lead to <u>alternative coordination</u> numbers of 1 and 3 as well as the preferred 2. Also, the bonds involving any one atom may have significantly <u>inequivalent</u> lengths and strengths. The corresponding electronic configurations in the tight binding limit are (1) two-coordination: two s-electrons of low energy, two p-electrons bonding to nearest neighbor atoms, two lone-pair nonbonding p-electrons; (2) three-coordination: two s-electrons of low energy, three p-electrons bonding to nearest neighbor atoms, one p-electron in an antibonding state of high energy; (3) one-coordination: two s-electrons of low energy, one p-electron bonding to a neighbor atom, three nonbonding p-electrons.

Transferred coordination is very likely in amorphous chalcogenide compounds. Three-coordination gives one electron in an antibonding state, i.e., the chalcogen is likely to act as a donor, and one-coordination provides an empty state in one p-orbital, i.e., the chalcogen is likely to act as an acceptor. If the three-fold and one-fold configurations are both present, electron transfer will occur, leading to x^+ and x^- centers each of zero net spin. Such centers constitute one way of explaining the salient observations in chalcogenide glasses (there exist exceptions) of no spin resonance, no paramagnetic susceptibility and no hopping conductivity, but substantial optical absorption in an exponential (Urbach) tail and photoluminescence displaying strong Stokes shifts. Charged centers of no net spin will also result from compensating chemical donors and acceptors if only one of the 'odd' combinations is present. In neither case is it clear how this would universally give near exact compensation. There have been at least two detailed models published which achieve this. Kastner, Adler and Fritzsche[28] have estimated the relative energies of several defect states involving chalcogens and deduce that the (x^+, x^-) combination discussed above provides the lowest energy defect. An additional new element in the KAF analysis is the suggestion that the x^+ and x^- occur in so-called intimate pairs, i.e., next to each other in space. This last is required to explain the detailed behavior of spectra, and it also rationalizes the <u>kinetics</u> necessary to assure equal densities of the odd chalcogen configurations.

Street and Mott[27], on the other hand, postulate that the reaction $2D^\circ \rightarrow D^+ + D^-$ is exothermic, where the superscripts denote the net charge on an (initially) dangling bond. The correlation energy increase on formation of the D^- is supposed to be smaller than the elastic distortion energy decrease on formation of D^+ and D^-. No calculation of this decrease is done; the model is justified by its success in explaining experimental results, principally on spin resonance and photoluminescence. It is evident that new, complex relatively weak bonds are being formed which make the designation 'dangling bond' inappropriate, however useful it may have been as a starting point for the model. In fact, the KAF and SM models are identical if one supposes that the necessary SM reconstruction gives the one- and three-fold coordination of the KAF model.

It seems to me that these models, which at the moment of writing dominate analysis of the properties of chalcogenides, are unlikely to be unique. Other aspects, such as the effects of donors and acceptors, like-atom bonds and bonds involving the lone pair chalcogen p-electrons, may be added and intermingled. Local free energy minimization may demand one of several alternative coordinations and one of several different distortions, even although there is a strong argument for the dominant defect being the lowest energy (x^+, x^-), or, in their notation, the (C_3^+, C_1^-) pair KAF identified.
Experiment and theory suggest a general rule that electron pair formation is as universal as the usual or average coordination and the kinetics of sample preparation will allow. This is the general content of the theory by Anderson. We shall take up the subject of electron pairs and defects in the chalcogenides again in the last section.

2.4   Transport, Optical Excitation and Recombination Properties

In a systematic review with no restrictions on length, one would now go on to deduce the various properties from the model of the electronic density of states and their extended and localized wavefunctions. These could then be compared with experimental results. Instead, I propose to concentrate on two case-studies, one on a-Si and the other the amorphous chalcogenides. This will permit me to give an exposé of current research in these fields, with all the unsolved problems, while at the same time I can relate the data to the models presented so far.

3. AMORPHOUS SILICON AND AMORPHOUS SILICON-HYDROGEN ALLOYS

Most of the current research on amorphous silicon is devoted to the study of material produced by r.f. sputtering in argon, to which H has been deliberately added, or by the decomposition of silane, SiH_4, which almost certainly leads to the incorporation of hydrogen. Let me explain how Si-H alloys arrived at their present position of prominence.

The structural studies on evaporated and sputtered a-Si established the presence of a large number of defects, and I have already discussed how these may be expected to give a spectrum of filled, half-filled and empty electronic levels spanning the energy gap. Several methods have been tried to obtain the metastable topologically disordered but perfectly coordinated network: evaporation in ultrahigh vacuum, deposition at the highest substrate temperature not leading to crystallization, annealing after preparation. The most dramatic effects result from the deliberate or fortuitous addition of hydrogen in the preparation of the a-Si[29] or a-Ge. The hydrogen compensates the dangling bonds, and perhaps more importantly, obviates the necessity for the reconstruction of weak bonds. It therefore shifts states from the pseudogap deep into the valence-band. The effect on the properties is dramatic: (1) the absorption edge shifts by about 0.5 eV to higher energies, as the removal of gap states eliminates low energy optical transitions; (2) the spin density produced by single electron states is reduced by three or four orders of magnitude; (3) the conductivity attributable to hopping transport, which is proportional to some power of the gap state density, similarly is reduced by many orders of magnitude. The transport more often proceeds after excitation of the carriers to some relatively high mobility extended state. (4) Efficient photoconductivity and photoluminescence are produced, which must involve a considerable increase in the excess carrier life-time. (5) The Fermi level may then be shifted through the gap by addition of n or p type dopants such as P or B, and in consequence, p-n junctions and Schottky barriers with good rectification characteristics may be made.

This brief summary might suggest that we have found what we sought: a way to simulate the intrinsic band structure near the band edges in the ideally-coordinated material. At first thought, it would appear that the maximum reduction of gap states would occur for a substrate T_s as high as possible without crystallization and an H-incorporation that eliminates all dangling bonds measured by spin. This is a gross over-simplification, It turns out that a-Si with included H presents a frustrating but challenging jungle of complicated properties on closer inspection. I now propose to describe a series of systematic experiments done in my own laboratory at Harvard on amorphous Si-H alloys prepared by r.f. sputtering[30]. Similar difficulties, however, are faced by workers preparing Si from SiH_4.

Although it may seem a little out of place to include a review of current problems in a discussion such as this one, I hope it will serve the useful purpose of conveying the state of the art and also providing a vehicle for the discussion of the principal properties.

We started out to incorporate just the right amount of hydrogen to compensate the defects. There are many parameters of the sputtering method: substrate temperature, deposition rate, argon pressure, hydrogen pressure, r.f. power, substrate bias, and they all are important. However, we first made the matrix of samples illustrated in Figure 8 by varying the substrate temperature T_S or the hydrogen partial pressure P_H while keeping all other parameters fixed.

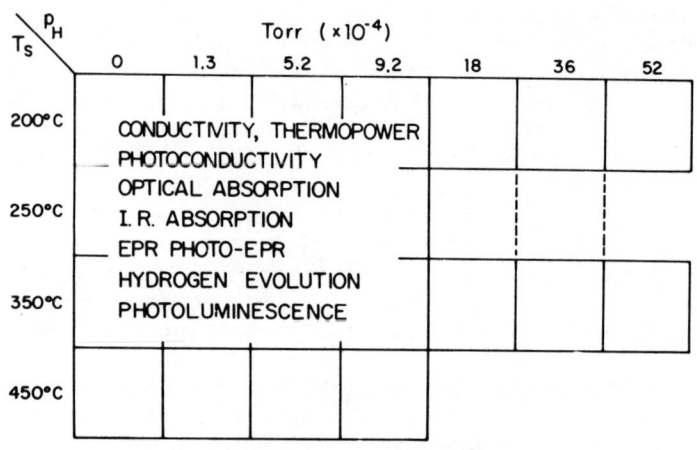

Figure 8. Matrix of samples of a-Si produced by r.f. sputtering at various substrate temperatures T_S and partial hydrogen pressures p_H. The experiments performed on each batch of samples are also listed.

First, let me treat the effect of varying T_S at a finite fixed P_H - a column of the matrix. The result is easier to state than to explain: when the temperature is raised above 200-250°C, fewer H atoms are incorporated in the sample. There are several ways of estimating the H content, and the two we use most are evolution of the H when the sample is heated, and analysis of the intensity of the infrared vibrational absorption. In the former method, a known mass of the sample is heated in a vacuum system of known volume, and the rise in gas pressure is measured as the gas is evolved from the sample. The maximum pressure reached after all the gas is evolved (easily judged if one proceeds to crystallization) is used to estimate the original atomic percentage of H in the sample, while the profile in temperature of the pressure of the evolved gas gives - one hopes - clues about the bonding of the H in the semiconductor. I shall return to the second method presently.

The effect of incorporating less H at higher T_S shows up in all of the measured properties. Figure 9 shows the effect on the absorption edge. At fixed p_H, the edge first shifts to higher photon energies with T_S and then reverses. Now, the absorption edge of an amorphous tetrahedrally-coordinated semiconductor typically is composed of three parts. (1) For absorption coefficients $\alpha < 10^2 \text{cm}^{-1}$ or so[31], the electronic transitions probably involve mostly defect states. Often this part of the edge has an exponential, Urbach-law shape which has been associated with the presence of electric fields in the material. (2) For absorption coefficients $10^2 < \alpha < 10^4 \text{ cm}^{-1}$ the electronic transitions involve both the states in the gap and those at one or other of the band edges. (3) For $\alpha > 10^4 \text{ cm}^{-1}$, band to band transitions dominate. It is traditional in this regime to write

$$\alpha \hbar \omega = A(\hbar \omega - E_o)^2$$

and to plot $(\alpha \hbar \omega)^{\frac{1}{2}}$ vs $h\nu$ to obtain an optical energy gap E_o, but there is no real justification for this arbitrary procedure. The expression for α corresponds to transitions between two bands of states, in each of which the state density increases parabolically

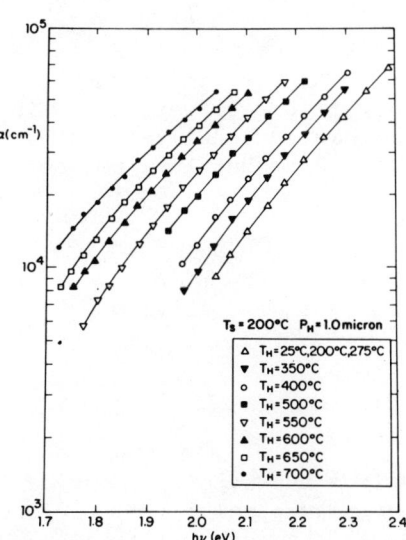

Figure 9. Effect of varying the substrate temperature T_S at fixed hydrogen pressure p_H on the absorption edge of a-Si-H alloys. After E.C. Freeman & W. Paul to be published.

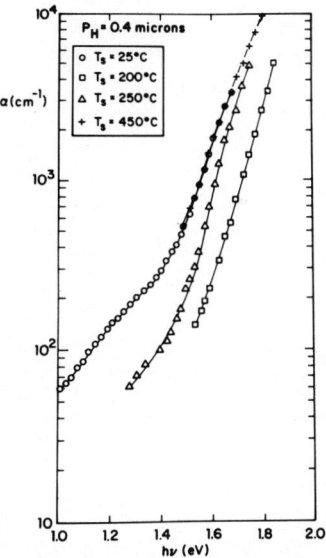

Figure 10. Effect of annealing an a-Si-H alloy to successively higher temperatures. Hydrogen evolves and the absorption edge shifts to lower energies. After E.C. Freeman & W. Paul, to be published.

away from the band edge, when the matrix element for the transition is independent of energy. We do not, however, have any knowledge about the density of states function. A useful but, of course, arbitrary alternative to this definition of the energy gap is to use that photon energy which corresponds to an absorption coefficient of 10^4 cm^{-1}.

It seems clear, therefore, that the initial shift of the edge to higher photon energies at T_s is increased corresponds to the removal from the pseudogap region of states that would produce low energy optical absorption, presumably as incorporated H more completely compensates dangling bonds. And, that this effect reverses at higher T_s.

Another way of illustrating the same result, shown in Figure 10, is to take a sample prepared at a certain (p_H, T_s) and then anneal it to successively higher temperatures. The hydrogen evolves, and the absorption edge shifts monotonically to lower photon energies.

Similar conclusions may be drawn from transport studies. Figure 11 shows the conductivity vs 1/T for samples prepared at several T_s and p_H = 0. Two things are to be noted. The first is the weak dependence of a relatively high (for amorphous semiconductors) conductivity σ on temperature; the second is the relatively mild (× 5) decrease in the conductivity at any measurement temperature as the substrate temperature T_s is increased. The weak T-dependence of σ on T is attributed to the hopping of carriers between sites in the amorphous material with which are associated localized states. For such a situation, starting with a hopping probability between sites

$$p \propto \exp(-2\alpha r) \exp(-w/kT)$$

Mott[32] has derived an expression for the conductivity

$$\sigma = \frac{Ae^2 nR_o}{6kT} \exp\left(-\frac{T_o}{T}\right)^{1/4}$$

In these expressions, $\exp(-\alpha r)$ describes the rate of decay of the localized wavefunctions, and w the average spacing in energy between the states at sites separated by 2r. A is an attempt frequency, n the density of hopping carriers, and $R_o = [9/8\pi\alpha N(E_f)kT]^{1/4}$ and $T_o = [18\alpha^3/kN(E_f)]$. The central concept of this variable range hopping theory is that the most probable hops will be longer and longer as the temperature is reduced. The weak $T^{-1/4}$ dependence of σ is very often found, when sought, but it should be noted that the

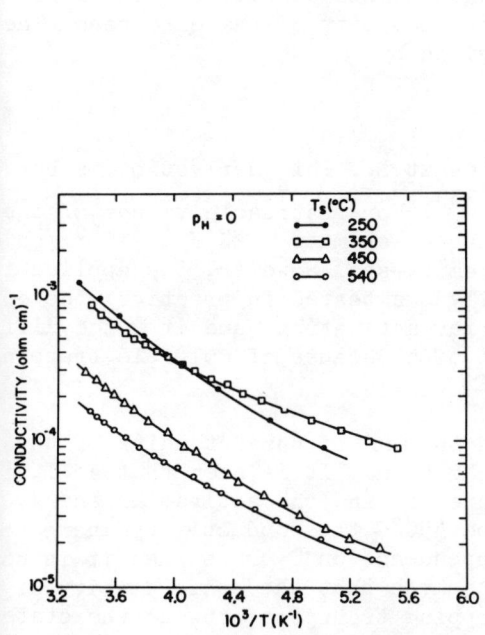
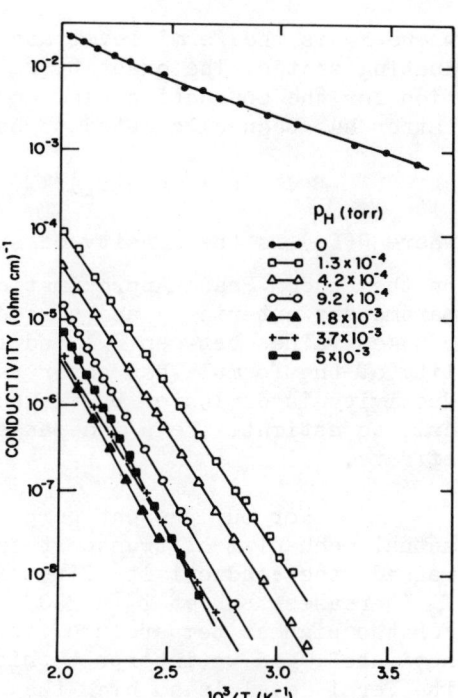

Figure 11. Variation of the conductivity vs temperature relation for unhydrogenated a-Si for samples prepared at several different substrate temperatures. After D.A.Anderson et al., reference 34.

Figure 12. Variation of the conductivity vs temperature relation for a-Si-H alloys prepared at a fixed partial pressure of hydrogen and different substrate temperatures T_s. After D.A.Anderson et al., reference 34.

dependence on T is too weak for sharp tests to be applied[33]. The curves of Figure 11 obey the $T^{-\frac{1}{4}}$ law, and we may assume that the transport is proceeding by hopping between states whose energy is near the Fermi level. The reduction in σ as T_s is raised is then interpreted as due to a reduction in the density of pseudogap states as the defect states are progressively healed.

If the same experiment is carried out at finite p_H, the results shown in Figure 12 are obtained[34]. The absolute magnitude of the conductivity is reduced by several orders, and the transport is now clearly activated. It must be supposed that the dominant transport mechanism results from the activation of electrons from the Fermi level into the extended states of the conduction band (or, at least, states of good mobility). The conductivity may be written

$$\sigma = \sigma_o \exp[-(E_c - E_f)/kT] = \sigma_o \exp[-E_\sigma/kT]$$

where E_f is the Fermi level and E_c the lower energy bound of the conducting states. The quantity σ_0 contains the density of states function for the conduction band and the mobility of the electrons. The latter has been calculated to be given by

$$\mu = A\, N(E_c)/kT$$

where $N(E_c)$ is the density of states at E_c. This derivation is based on the Random Phase Approximation[35]. For reasonable values of the parameters entering A and a reasonable estimate of $N(E_c)$ (10^{20}/cc) μ takes values between 0.1 and 10 cm^2/v-sec. However, the applicability of the formula has never really been tested in practice: the conductivity is dominated by the carrier activation, and it is too difficult to estimate the μ independent of σ because of multiple trapping effects.

For our present purposes the lack of understanding of the actual mechanism of transport can be borne. It is seen that as T_s is raised, the conductivity first decreases and the activation energy E_σ increases. but at T_s = 350°C and 450°C the conductivity increases. For the highest temperature its dependence on T shows that it is not activated, and we interpret this to mean that the state density at the Fermi level is so high that hopping transport between the states there dominates the activated transport. The cause is supposed to be the reduced efficiency of incorporation of H at high T_s.

These results, taken together, show that it is not just that there are fewer defect states for the hydrogen to compensate at higher T_s, but also the H is not efficiently incorporated. The evolution of the H at relatively low annealing temperatures suggests a low activation energy. One might guess that the activation energy is the difference

Net activation energy = (Energy cost of breaking two Si-H bonds)
- (Energy gain from formation of H_2 molecule)
- (Energy gain from reconstructed Si-Si bond)

which is small if you insert bond energies taken from Paulings book. However, this process faces the formidable kinetic difficulty that the Si-H bonds must be close together, which may not be so in Si with a small amount of included H, as well as the problem that an increased spin density is found after anneal. The precise resolution of this problem is not yet to hand.

Next we return to the results of varying p_H at a fixed value of T_s. Figure 13 illustrates the H concentration found from evolution of the gas on heating. Notice two things. First, the maximum concentration of H is above 20%, so that we are really studying Si-H alloys, not Si doped heavily with H. Second, c_H varies non-monotonically with p_H. The precise reason for this is not known - it could be an alteration in the mixture of SiH_n species in the plasma with p_H, or there might be an upper limit to the amount of H the matrix can accommodate.

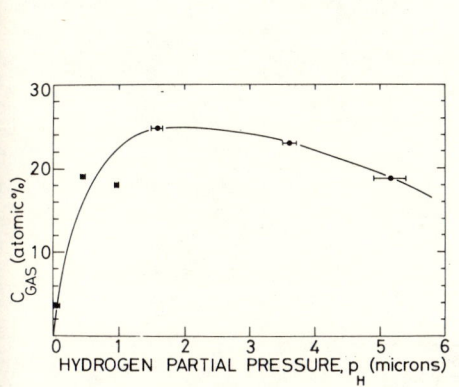

Figure 13. Dependence of the H-concentration in a-Si-H alloys prepared at fixed T_S on the different partial pressures of H in the sputtering gas. After S.Ogus & M.A.Paesler, Bull.Am.Phys.Soc.,23,247 (1978).

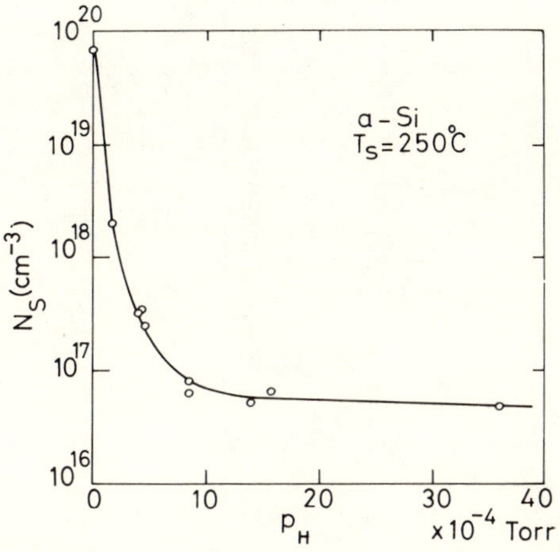

Figure 14. Dependence of the spin density N_s in a-Si-H alloys prepared at fixed T_s on the different partial pressures of H in the sputtering gas. After J.R.Pawlik, unpublished.

In Figure 14 we show the spin density N_s vs p_H. The rapid decrease is interpreted as the removal of single electron states by H-compensation of dangling bonds. We note however, that the starting N_s was 10^{20} /cc or 1/10%, while the incorporated H is several percent. This again illustrates that the H is doing more than just compensating dangling bonds.

Figure 15 illustrates the spectrum of photoluminescence when a sample at 77 K is excited by laser light at 514 nm from an Ar ion laser. This result is typical of the luminescence from $a-Si_{1-x}H_x$ prepared both by r.f. sputtering and by the decomposition of SiH_4. The electronic transition involved has not been finally identified, although the current supposition is that it represents a strongly Stokes shifted emission between tail states of the conduction and valence bands[36]. (Strongly Stokes shifted because, as we shall see, the energy gap for absorption is about 0.5 eV greater than the energy of the center of the emission line.)

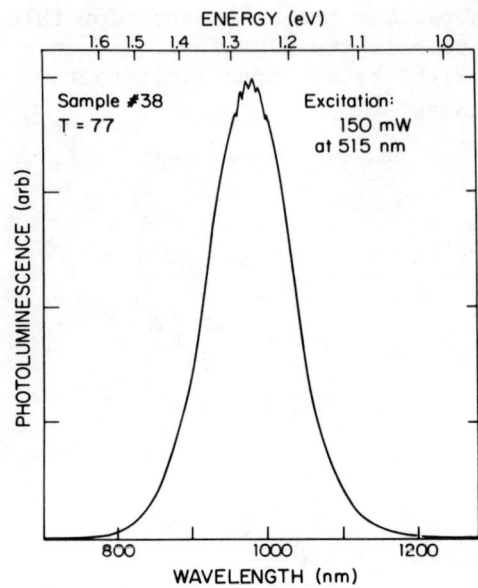

Figure 15. Spectrum of photoluminescence from a sample of a-Si-H alloy excited by 514 nm radiation from an Ar ion laser. After M.A.Paesler and W.Paul, to be published.

Figure 16 and 17 show how the integrated intensity and peak photon energy of this line roughly follow the c_H vs p_H curve of Figure 13. For efficient photoluminescence we have over 10 atomic

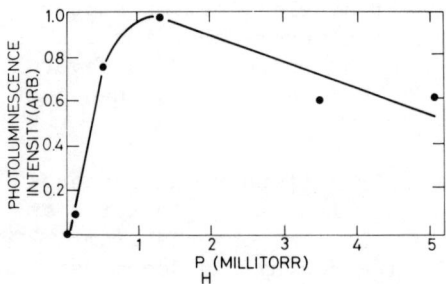

Figure 16. Integrated intensity of the photoluminescence from a-Si-H alloys prepared at fixed substrate temperature on the different partial pressures of hydrogen in the sputtering gas. After M.A.Paesler & W.Paul to be published.

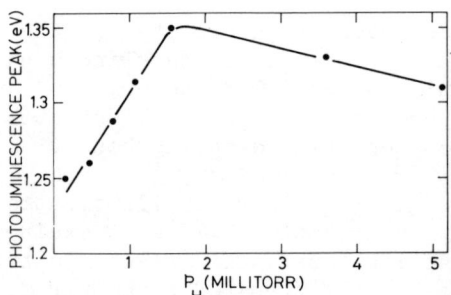

Figure 17. Photon energy of the peak of the photoluminescence from a series of a-Si-H alloys prepared at fixed substrate temperature and different partial pressure of hydrogen in the sputtering gas. After M.A.Paesler & W.Paul, to be published.

% H. No a-Si without H has ever shown photoluminescence! Is the function of the H that of removing dangling bonds and thus potential non-radiative recombination centers, or is the H-center somehow a catalyst for the radiative recombination? This is a question which is not yet settled. Its answer is really linked to other poorly understood properties of the local Si-H environment. However, there is no lack of quantitative data and partly explained phenomena concerning the photoluminescence: an integrated photoluminescence which depends on the laser excitation energy, a peak at about 60 K in the integrated intensity vs measurement temperature and quenching of the photoeffect by the application of a (nonheating) electric field[37].

Next we examine the spectrum of vibrational absorption in the infrared, shown in Figure 18. The modes are identified as a stretching mode near 2000 cm^{-1}, a bending mode caused by an SiH_2 complex near 890 cm^{-1}, and a wagging Si-H mode near 600 cm^{-1}. At lower frequencies we see absorption caused by the a-Si matrix itself. Notice

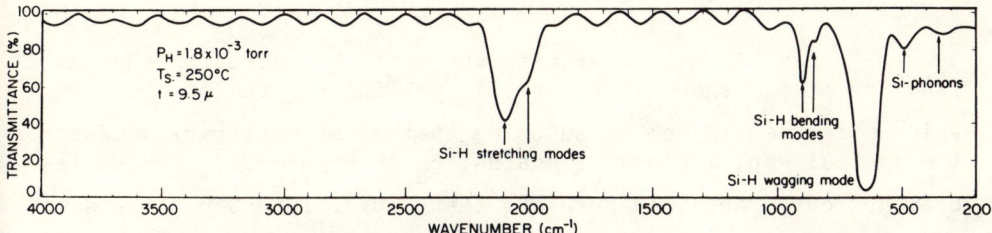

Figure 18. Spectrum of vibrational absorption in the infrared region from a typical a-Si-H alloy. After E.C.Freeman and W.Paul, to be published.

the doublet nature of the stretching mode. There are, in fact, three stretching mode peaks near 2000 cm^{-1} which appear under different preparation conditions. Attributing them and similar richness of feature in the bending modes to different configurations of the Si-H is a subject of some present debate. Here I wish to note only two things. First, there are multiple features, which we interpret as implying the existence of more than one configuration <u>for the single Si-H bond</u>. The question of the existence of multiple sites for the H and the question of what they are, is one of <u>the</u> questions in the field today. For, along with the different vibrational modes of different sites go possibly different electronic states. Second, the area under the whole absorption curve near 2000 cm^{-1} may plausibly be taken as a measure of the relative magnitude of the H-content in the sample. This quantity, A_s, is shown as a function of p_H in Figure 19, and also plotted vs c_{gas} from evolution in Figure 20. The relationship is linear, as it should be, and the non-zero intercept confirms that our sputtered samples contain A.

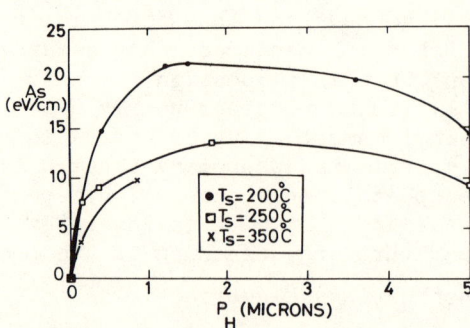
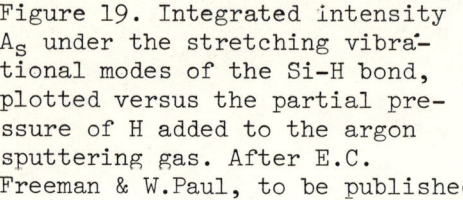
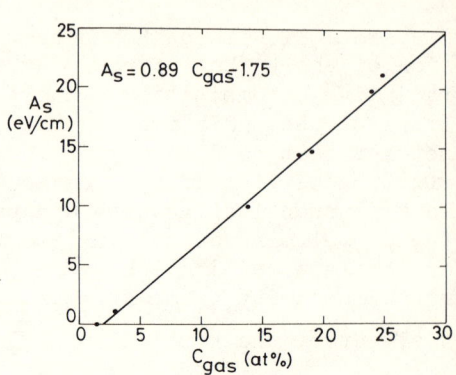

Figure 19. Integrated intensity A_S under the stretching vibrational modes of the Si-H bond, plotted versus the partial pressure of H added to the argon sputtering gas. After E.C. Freeman & W.Paul, to be published.

Figure 20. Integrated intensity A_S of Figure 17, replotted versus the evolved gas shown in Figure 11. After E.C.Freeman & W.Paul to be published.

Next I shall deal with the optical absorption edge as a function of p_H, shown in Figure 21. Calling E_{03} the photon energy at which α reaches 10^3 cm^{-1} and using that as an arbitrary measure of the optical gap, I plot E_{03} against c_H in Figure 22. The double-branched curve obtained suggests that, at different p_H, the same amount of H can be incorporated in different configurations, and these different configurations affect the density of states in the pseudogap differently. This is consistent with the complexity of the vibrational spectra, and it implies that we need to examine the energy levels of the different H-configurations.

Finally, in Figure 23 I illustrate the photoconductivity, which shows the most complex behavior of all, and which I shall not try to rationalize here.

Although I have confined my remarks to the properties of a-Si r.f. sputtered in H so as to give you one coherent case history, there are more similarities than differences between this type of a-Si and that produced by glow discharge decomposition of SiH_4. What I have not said so far is that at present the latter is considerably more photoconductive and photoluminescent. Thus, a-$Si_{1-x}H_x$ made up into a Schottky barrier configuration of the general type illustrated in Figure 24 has given under approximate AM 2 illumination, open circuit photovoltages of 0.8 eV, closed circuit photocurrents of 7.8 mA/cm^2 and overall solar cell efficiency of the order of 6%[38] It would take very little more to make these cheaply made devices

AMORPHOUS SEMICONDUCTORS 333

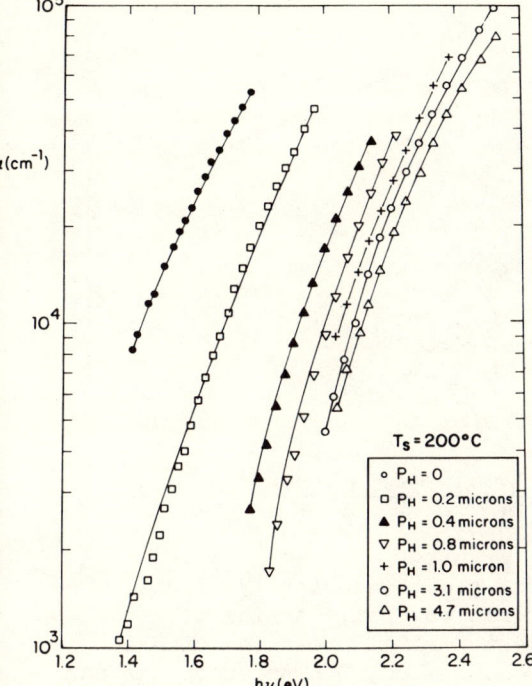

Figure 21. The absorption edges of a series of a-Si-H alloys, prepared at different partial pressures of H at a fixed substrate temperature. After E.C.Freeman and W.Paul, to be published.

Figure 22. Plot of E_{03}, the photon energy at which the absorption coefficient reaches 10^3 cm^{-1}, versus the concentration of H in a series of a-Si-H alloys prepared at different partial pressures of H in the sputtering gas. After E.C.Freeman and W.Paul to be published.

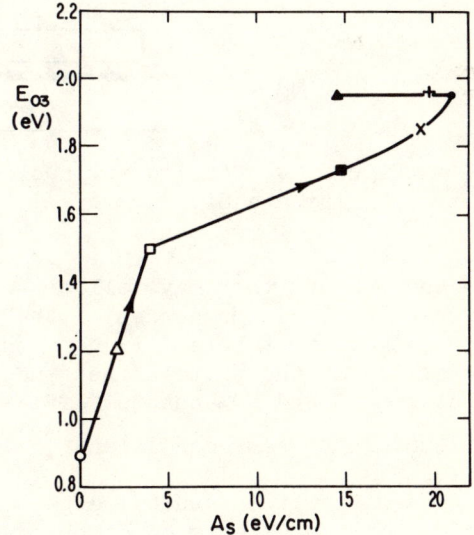

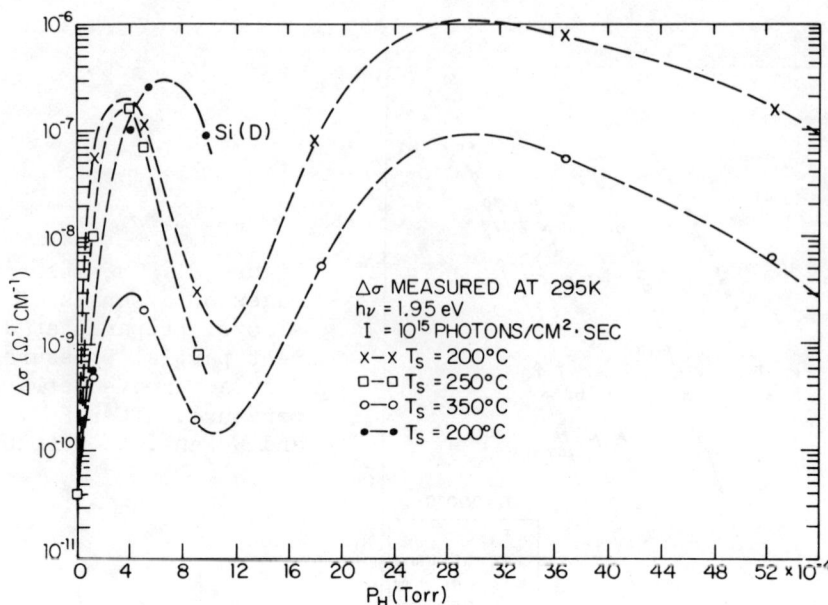

Figure 23. Variation of the photoconductivity signal, under fixed conditions of illumination, for a series of a-Si-H alloys prepared at different partial pressures of H in the sputtering gas. After T.D.Moustakes, D.A.Anderson & W.Paul, Solid State Commun., 23, 155 (1977).

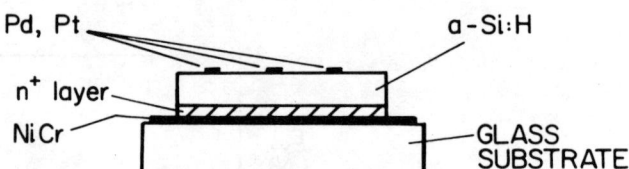

Figure 24. Configuration of a Schottky barrier device utilising a-Si.

economic, and they may well be so already for certain applications. Since the basic processes of the photoconduction are not understood yet, there must be a fair hope that even better performance is possible. As for the photoluminescence, it is commonly supposed to be between 10 and 50% quantum efficiency, higher indeed than crystalline Si under the same conditions[39].

In the future I expect that there will be many investigations of the effects of other parameters of the preparation methods, in an attempt to clear the gap of defect states even more, improve the ease of doping, increase the excess carrier lifetime, and so

AMORPHOUS SEMICONDUCTORS

improve even more the device properties. It is likely that a search will be made for other elements than H with which the natural defects might be compensated. Such other elements might be Li, O or a halogen, and it is conceivable that they could cause fewer complications than the H, and also be more stable at high temperatures. In any event, I expect there will be theoretical work on the spectrum of electronic states for different possible configurations of H in Si, since preliminary considerations suggest that some of these may even lead to new states in the pseudogap of the material.

4. CHALCOGENIDE GLASSES

Finally, I shall discuss the principal properties of chalcogenide materials, such as Se, As_2Se_3 or a four-component alloy of Si-Te-As-Ge (STAG glass). Despite individual variations, it is possible to make some general observations and present some general theory, applicable, mutatis mutandis, to the different systems[40].

4.1 General Experimental Observations

(a) <u>Transport</u>: Figure 25 shows the conductivity vs 1/T behavior of a variety of chalcogenides[41]. The law

$$\sigma = \sigma_o \exp[-E/kT]$$

is generally applicable, i.e., the transport always appears to be

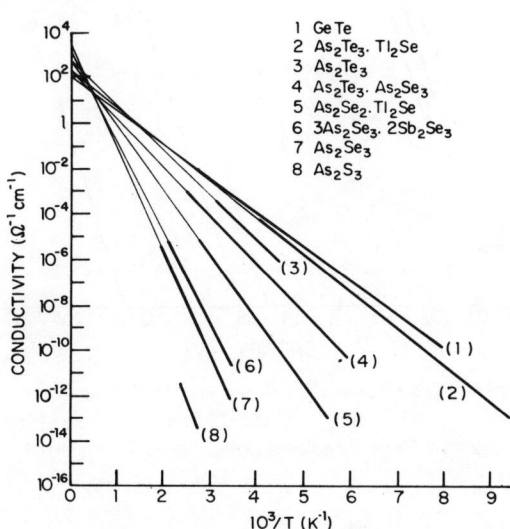

Figure 25. Temperature dependence of the electrical conductivity of various chalogenide glasses. After N.F.Mott and E.A.Davis, reference 41.

activated. The relation

$$\ln \sigma \propto T^{-\frac{1}{4}}$$

which describes transport by hopping among localized states, is not found. Moreover, the E in the above equation is always approximately half of an (admittedly arbitrarily defined) energy gap. Thermoelectric power measurements indicate that the carrier providing the transport is a hole. By assuming a 'reasonable' band density of states, one can deduce a mobility from the conductivity, and this turns out to have a magnitude appropriate for band transport. The conductivity is insensitive to attempts at doping; for example, group 4 and group 7 dopants have no effect on the conductivity in As_2Se_3.

(b) <u>Magnetic properties</u>: The most important observations here are that no spin resonance and no parametric susceptibility are found in chalcogenide glasses cooled from the melt[42].

(c) <u>Optical absorption</u>: Optical absorption edge spectra show that the optical gap is most often very similar to the corresponding crystal. Figure 26 illustrates this. At lower energies than

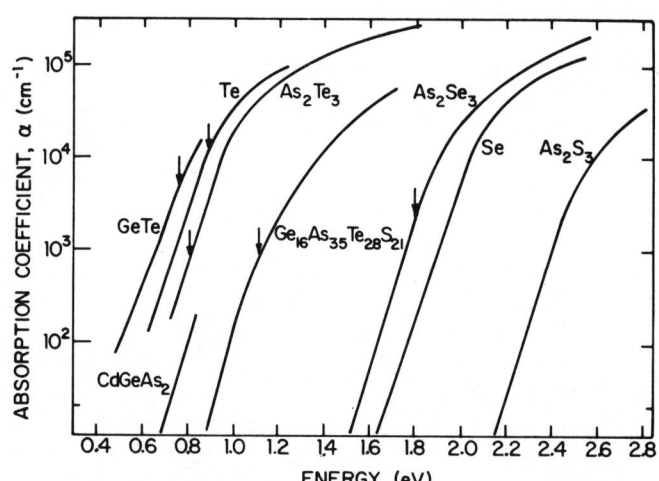

Figure 26. Optical absorption edges of various glasses at room temperature. The arrows indicate twice the conductivity activation energy. For references see N.F.Mott and E.A.Davis, reference 41.

the band gap, the amorphous material is at least as transparent as the crystal. In fact, extreme transparency in the infrared is one of the outstanding observables in these materials.

(d) <u>Photoluminescence</u>: Most of the elemental and binary chalcogenides display very efficient photoluminescence with the photoluminescence spectrum Stokes-shifted from the absorption spectrum to roughly half of the optical band gap energy. As illustration, Figure 27 shows the photoluminescence spectrum, the excitation spectrum and the absorption edge for Se and Figure 28 the corresponding spectra for As_2S_3, As_2Se_3 and $As_2Se_{1.5}Te_{1.5}$.

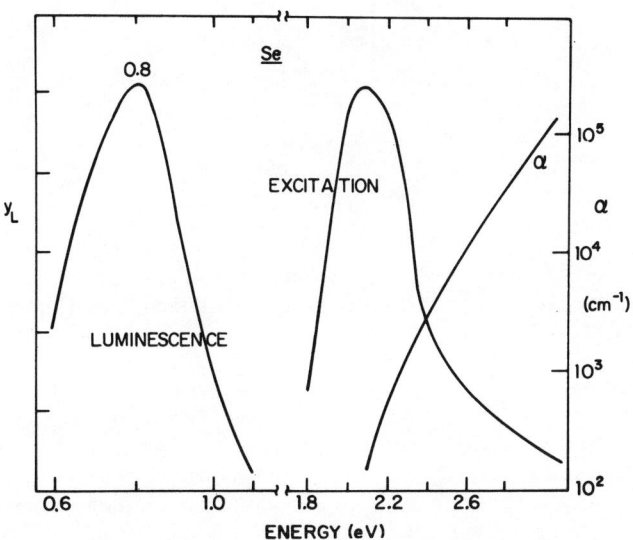

Figure 27. Photoluminescence spectrum, excitation spectrum, and optical absorption (α) for Se.

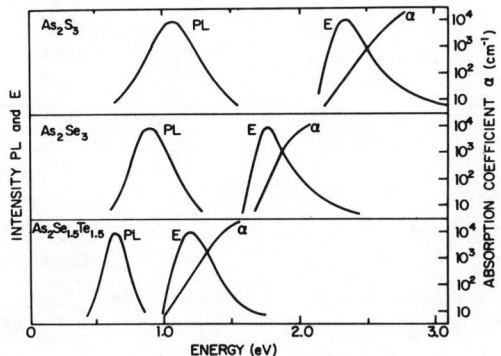

Figure 28. Photoluminescence spectra (PL), excitation spectra (E) and optical absorption (α) for As_2S_3, As_2Se_3 and $As_2Se_{1.5}Te_{1.5}$. After R.A.Street, reference 40.

4.2 Band Structure Ideals

A very simple band structure model, based on the observed two-fold coordination of the chalcogen, is shown in Figure 29. The separation between the atomic s and p levels is some 10 eV; thus,

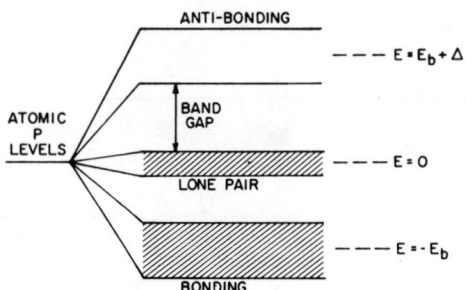

Figure 29. Tight binding scheme for the valence and conduction bands of the chalcogenides, showing bonding, lone-pair and antibonding p-states of the two-fold coordinated chalcogens.

hybridization is not expected. There are two low energy bonding p-states, but the valence band is formed out of lone pair nonbonding p-states. The conduction band is formed out of the antibonding p-states. These expectations are borne out by photoemission experiments. Figure 30 shows the valence band density of states for crystalline and amorphous Se. The structures are remarkably similar and

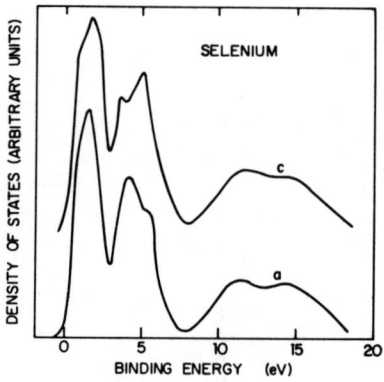

Figure 30. Valence band density of states for amorphous and crystalline selenium. After A.J.Shevchik, M.Cardona and J.Tejeda, Phys. Rev. B $\underline{8}$, 2833 (1973).

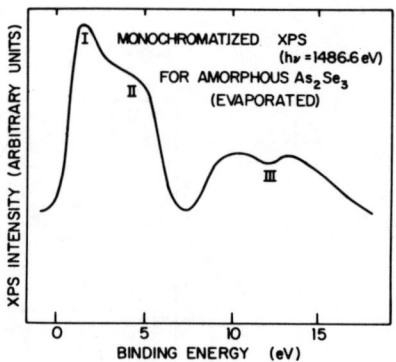

Figure 31. X-ray photoemission spectra of amorphous As_2Se_3. After S.G.Bishop and N.J.Shevchik, Phys. Rev. B $\underline{12}$, 1567 (1975).

AMORPHOUS SEMICONDUCTORS

do correspond to a narrow nonbonding p-band, a wider bonding p-band and an s-band, in that order from the zero of energy at the top of the filled states. Figure 31 illustrates XPS data for As_2Se_3; band I corresponds to the lone pair p-states of the Se, band II to the bonding As-Se states and band III to the s-tates of the As and Se. Thus there appears to be no problem in understanding in this qualitative fashion the main features of the valence band structure.

4.3   States in the Gap

Problems arise, however, when a self-consistent model is sought to explain some of the observed phenomena. The dilemma can be described in terms of arguments for and against the existence of states in the gap between valence and conduction bands. Observations favoring their existence are: (1) the fact that the photoluminescence spectrum is centered on $h\nu = 1/2\ E_g$; (2) pinning of the Fermi level in rectification and tunneling experiments; (3) the absence of field effect, i.e., a shift in Fermi level when charge is added to a surface layer. This implies a large state density at the Fermi level to accommodate the charge. Observations contradicting their existence are: (1) failure to observe any optical absorption at $h\nu = 1/2\ E_g$; (2) absence of conductivity following a $\ln \sigma \propto T^{-\frac{1}{4}}$ law, which would imply carrier hopping in localized states; (3) absence of electron spin resonance and paramagnetic susceptibility; (4) the fact that the photoluminescence spectrum of the amorphous chalcogenides is generally very similar to that of crystal, implying that gap state densities are not involved.

Mott had at one time suggested that any impurities would have no doping effect because they would achieve chemical valence satisfaction. On the other hand, the most visible model, that of Cohen, Fritzsche and Ovshinsky[43], envisaged tails of states from both the valence and conduction bands, overlapping in mid-gap, and giving rise to a high density of + and - states and a high spin density. In summary, there is an apparently clear contradiction between the nonexistence of a spin signal and the pinning of the Fermi level.

4.4   Anderson's Theory (1975)

In 1975 Anderson[44] made the bold hypothesis that the electrons in disordered chalcogenides always made electron pair bonds. The basic idea was not new[45], but its mathematical formulation was. Anderson started from the Hamiltonian

$$H = \sum_{i,\sigma} E_i n_{i\sigma} + U \sum_i n_{i\uparrow} n_{i\downarrow} + \sum_{ij\sigma} T_{ij} c^+_{i\sigma} c_{j\sigma} + \sum_i \left[\frac{1}{2} m x_i^2 + \frac{1}{2} c x_i^2 - \lambda x_i (n_{i\uparrow} + n_{i\downarrow})\right]$$

Here,

- E_i is the electron energy at a 'site' i,
- $n_{i\sigma}$ is the number of electrons (one or zero) at site i, with spin σ,
- U is the correlation energy = $\langle i\sigma_1 | e^2/r_{12} | i\sigma_2 \rangle$ assumed independent of i,
- c^+, c are creation and annihilation operators,
- T_{ij} is a hopping energy between sites i and j,
- x_i is a bond length coordinate at site i,
- m is the electron mass,
- the potential energy at site is given by

$$V = \frac{1}{2} c x_i^2 - \lambda x_i (n_{i\uparrow} + n_{i\downarrow}).$$

For very low frequency processes for which $w \ll (c/m)^{1/2}$, and neglecting the T_{ij} term (44), the Hamiltonian may be rewritten as

$$H_{\text{low frequency}}^{\text{effective}} = \sum_{i,\sigma} E_i^{\text{eff}} n_{i\sigma} + U^{\text{eff}} \sum_i n_{i\uparrow} n_{i\downarrow}$$

where

$$E_i^{\text{eff}} = E_i - \lambda^2/2c \text{ and } U^{\text{eff}} = U - \lambda^2/c.$$

Whereas U > 0, it is postulated that $U^{\text{eff}} < 0$, i.e., that the double occupancy leads to a local distortion that reduces the elastic energy of the system, and that this effect more than offsets the cost in correlation energy. Anderson labelled this circumstance a Negative Effective Correlation Energy. At each site i there are three possible eigenenergies:

$$n_i = 0, \quad E = 0$$
$$n_i = 1, \quad E = E_i^{\text{eff}}$$
$$n_i = 2, \quad E = 2E_i^{\text{eff}} + U^{\text{eff}}$$

At low temperatures all sites will either have $n_i = 0$ or $n_i = 2$:

For $E_i^{\text{eff}} > \frac{1}{2}|U^{\text{eff}}|$, $n_i = 0$

For $E_i^{\text{eff}} < \frac{1}{2}|U^{\text{eff}}|$, $n_i = 2$

It is immediately clear that, if all sites have either zero or two electrons, almost all, if not all, of the problems cited earlier can be explained. There is no spin resonance or paramagnetic susceptibility, since all of the electrons are paired. Ther can be no ($T^{-\frac{1}{4}}$) hopping transport; since energy is required to separate a pair, there are no nearby sites of almost equal energy for an electron to hop to, and electron-pair hops are of lower probability.

Anderson's theory, however, provides for an alternative choice of the totality of sites i, or in other words, a continuous distribution in energy of two-electron states. This explains the 'pinning of the Fermi level' in doping, rectification and tunneling experiments and the absence of a field effect: attempts to dope by adding atoms with more electrons than are required to satisfy the bonds of the existing network will lead to a modified network with only pair bonds. Despite the high density of two-electron states continuously distributed in energy, the energy required to produce an electron hole pair ($|U^{eff}|$) remains finite, which leads to an optical gap and infrared transparency. The photoluminescence at half the energy of the optical absorption energy gap is explained by network distortion, that is large Franck-Condon effect.

4.5 Modification of Anderson's Theory by Mott and Street

Mott and Street[46] sought to modify Anderson's theory through the hypothesis that the concept of a negative U^{eff} applied only to defects like dangling bonds and not to the totality of states of the solid. They assumed

$$2D^\circ \rightarrow D^+ + D^-$$

where D stood for dangling bond, and the superscripts denoted the charge, constituted an exothermic reaction. This is certainly sufficient to explain the observed properties. It is here that a series of (to me, minor) disagreements occur among the various contributors. Anderson considers that it is unnecessary to label D^+ and D^- as **defects**, and that in his perception[47], these are just two of the ways that electrons will produce pair bonds. Of course, Anderson did not explain any more than Mott and Street did why negative U^{eff} were appropriate.

Theory of Kastner, Adler and Fritzsche: Kastner, Adler and Fritzsche[48] (KAF) took a different tack, which amounts to examining the Alternative Coordinations for a chalcogen atom[49] and trying to estimate which of them would give the lowest energy defect (defect is here defined as a departure from the usual two-fold coordination). The result of their examination was to identify one possible source of a negative correlation energy.

In an ideal glass, the coordination number and the bond length are the same in the amorphous material as in the crystal. It is expected that the defect density will be lower in chalcogens prepared from the melt than in tetrahedrally-coordinated material deposited on a cold substrate. The defect density frozen into the nearly ideal CRN at the glass temperature will be

$$N = N_o \exp[-G_f/kT_g]$$

where N_o is the atom density, T_g the glass temperature and G_f the free energy of formation of the defect under consideration. KAF examined the defect likely to have the lowest energy. Their rather simplified theory relies on the fact that, in the formation of bonding and antibonding orbitals, the energy of the antibonding state is raised more with respect to the original orbital than the energy of the bonding orbital is lowered. Thus, in forming bonding (+) and antibonding (-) orbitals $\psi_\pm$ between atoms A and B

$$\psi_\pm = \frac{1}{\sqrt{2}} (\psi_A \pm \psi_B)$$

with energies

$$\psi_- : E_- = \frac{E_o + \beta}{1 - s} \equiv + E_b + \Delta$$

$$\psi_+ : E_+ = \frac{E_o - \beta}{1 + s} \equiv - E_b$$

Where
$E_o \equiv$ onsite matrix element $= \langle \psi_A | H | \psi_A \rangle$
$\beta \equiv$ hopping matrix element $= \langle \psi_A | H | \psi_B \rangle$
$s \equiv$ overlap integral $= \langle \psi_A / \psi_B \rangle$

Then
$$|E_+ - E_o| > |E_- - E_o|.$$

KAF then examined the relative energies of several defects illustrated in Figure 32[52]. (1) The comparison energy used is that of the normal two-coordinated chalcogen atom, labelled C_2^o, i.e., $-2E_b$ below the unbonded p-state.

(2) The single uncharged dangling bond, C_1^o, has an energy $-E_b$. Thus the energy excess over C_2^o is E_b, which is large and argues against the persistence of unmodified dangling bonds.

(3) A three-fold coordinated uncharged chalcogen, C_3^o, has an energy

$$-3E_b + E_b + \Delta = -2E_b + \Delta.$$

Thus, the energy excess over C_2^o is Δ, and this is therefore a much lower defect than C_1^o. This defect, however, is excluded by experiment, since its extra electron would give magnetic effects. Another defect or defects must therefore have lower energy.

AMORPHOUS SEMICONDUCTORS

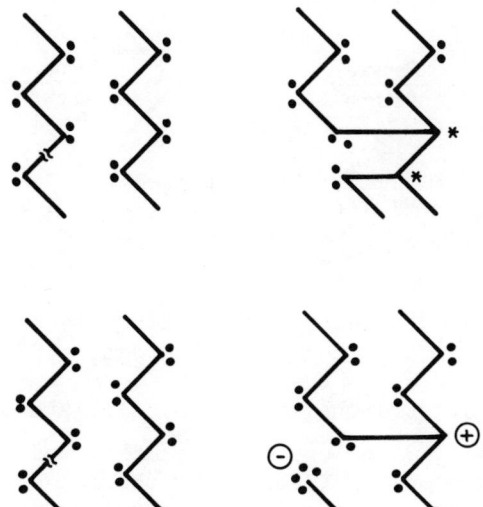

Figure 32. Illustration of formation of two C_3^o defects (top) and a (C_1^-, C_3^+) defect pair (bottom) as a result of normal bond breakage. Based on M. Kastner, reference 52.

(4) Instead of considering here all the possibilities, we describe instead the defect-pair found to have the lowest energy. This is $C_1^- + C_3^+$. Consideration of Figure 32 shows that the pair has energy

$$(-E_b + U) - 3E_b = -4E_b + U$$

The energy cost relative to two C_2^o's is U. The number of bonds is conserved, and all the electrons are paired. The combination (C_1^-, C_3^+) has been called a Valence Alternation Pair. Since experiment does not indicate the existence of spins, we suppose that

$$U < 2\Delta$$

and that
$$2C_3^o \rightarrow C_1^- + C_3^+$$

is exothermic. It would be hard to calculate this a priori, and in this sense the KAF theory - despite claims to the contrary - is based on postulate just as the Anderson and Mott-Street versions were.

The source of the negative correlation energy on the KAF theory may be illustrated as follows[52]. We recall that, in circumstances where the positive correlation energy is dominant, the energy required to remove the first electron from an oribtal is smaller than that to remove the second. Using Figure 33, we can show that this situation is reversed. When an electron is removed from C_1^-, the resulting C_1^o and a nearby C_2^o move closer together to convert the

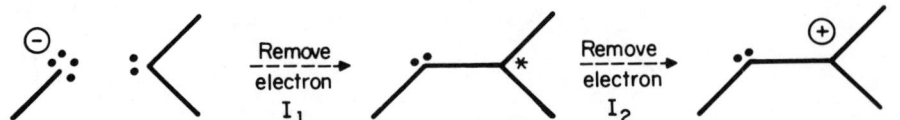

Figure 33. Demonstration that the cost of removing a second electron from a C_1^- defect is less than the cost of the first ionization. After M. Kastner, reference 52.

latter into C_3^0. But C_3^0 has one antibonding electron, which is removable at very little cost in energy to create a C_3^+.

It is now straightforward to calculate the average energy to create electrons and holes both thermally and optically[53]. We shall not elaborate on this, referring the reader to the references, except for an emphasis on the need to consider optical absorption and photoluminescence using configurational coordinate diagrams such as Figure 34. There is nothing different here than in any configurational coordinate diagram, and we note simply that the atom shifts and binding energy W_p are large, in agreement with the model that new bonding arrangements result from electron excitation.

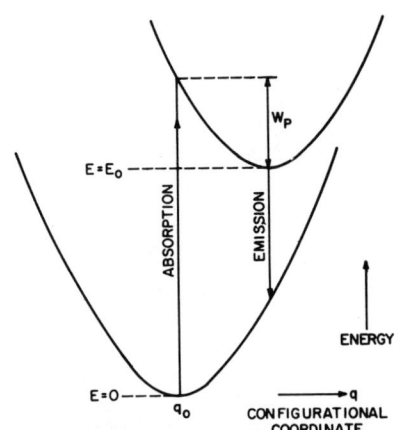

Figure 34. Configurational coordinate diagram for a site showing a strong electron-phonon interaction.

The Street-Mott and KAF models are seen to be identical as presented, despite their different genesis and different notation. Both can explain in a natural way the observable phenomena. However as we said in the introduction, any model that gives electron-pair bonds will do the same, so that more sophisticated tests are required to verify that KAF have indeed identified the lowest energy sites (defects) in the chalcogenide structure. The model can be used as a basis to explain other observed effects and to predict new ones. When some chalcogenides are excited with high energy photons at low temperatures, they show photo-induced spin resonance, a decrease in photoluminescence and a new low energy absorption band[54]. The postulated reason is the creation of electron hole pairs at a $D^-(C_1^-)$ center, followed by the capture of the hole to form D°, with the electron then hopping away. The induced absorption is attributed to transitions between the D° states and states at the band edges. The induced absorption and spin resonance can be removed and the luminescence restored by illumination at a photon energy between $E_g/2$ and E_g. These effects are illustrated in Figure 35.

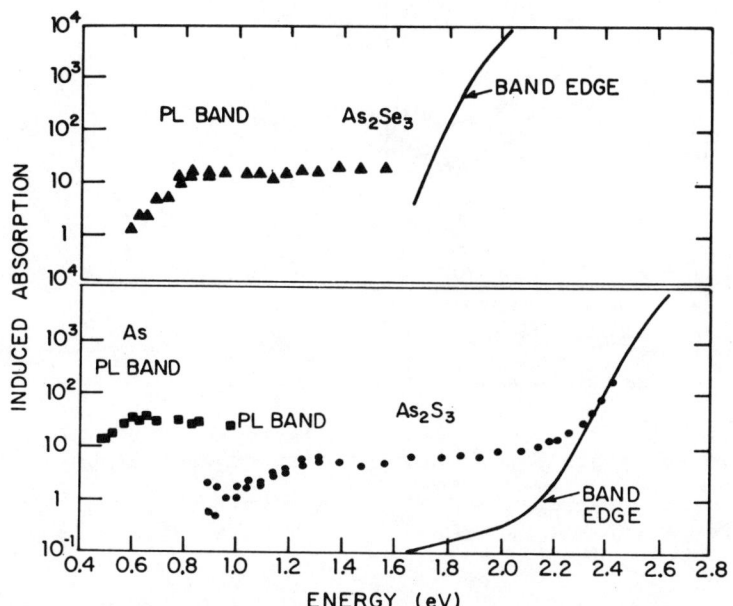

Figure 35. Induced optical absorption as a result of low temperature irradiation of As_2S_3 and As_2Se_3. Shown are the normal band edge absorption, the induced absorption and the energy range of the photoluminescence, After S.G.Bishop et al., reference 54.

4.6 Photostructural Changes

The photo-induced changes in spin resonance and optical properties described above represent only a very small part of a wide variety of photostructural changes in chalcogenides. Illumination with moderate fluxes of high energy photons have been shown[55] to produce changes in thickness, X-ray diffraction spectra, chemical reactivity including oxidation rate, refractive index and optical band gap. Figure 36 shows the effect of prior irradiation on the X-ray diffraction intensity versus angle; without troubling to inver

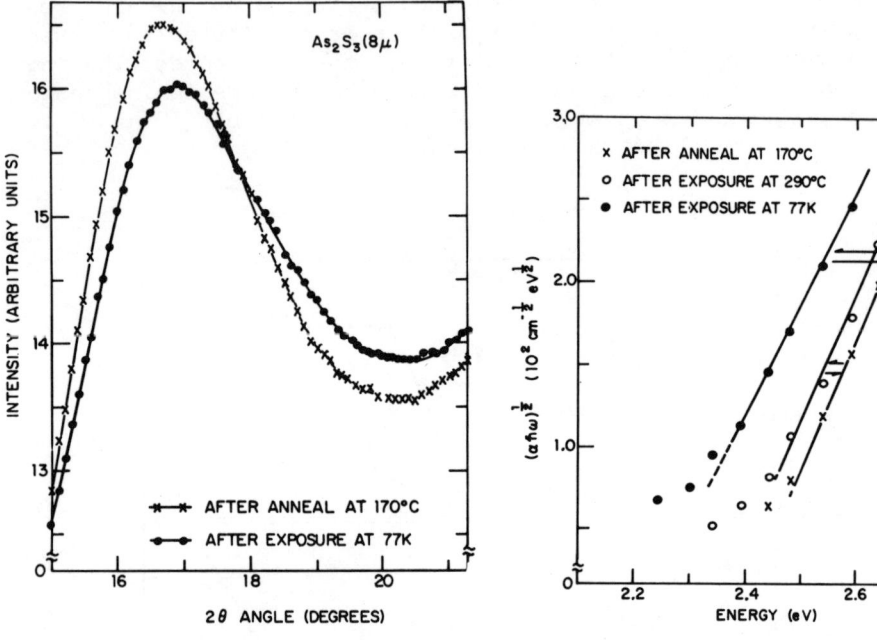

Figure 36. Diffracted X-ray intensity versus angle for As_2S_3, before and after irradiation at 77 K. After K.Tanaka, reference 56.

Figure 37. Shift of the absorption edge spectrum of As_2S_3 as a result of irradiation. After K.Tanaka, reference 57.

this to obtain the changed atomic distribution in real space, it is clear that some redistribution of bonding has occurred. Figure 37 shows the effect on the absorption edge of As_2S_3[57]. The magnitude of the shift increases as the temperature at which the irradiation is performed is decreased. The effect is small for photon energies for which the absorption coefficient is $< 10^3$ cm^{-1}. The curve describing the 'efficiency' versus photon energy for absorption edge shift (photostructural change) is the inverse of that for the excitation of photoluminescence[58]; this is reasonable since the occurance of (geminate) recombination implies no local changes in bonding. Finall annealing at high temperatures can reverse all of the observed changes, consistent with a return to the original metastable glass structure.

4.7 Device Applications

Although considerations of space have forced me to focus on just a few recent developments in the physics of the chalcogenides, no review should omit mention of the striking applications in devices of great potential[59]. Among these applications are (1) xerography, (2) threshold switches and memories, (3) imaging, and (4) multiple copy printing.

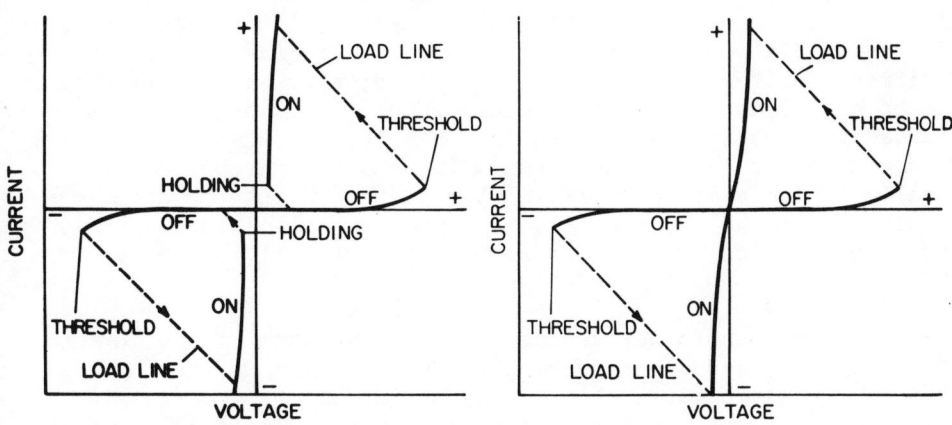

Figure 38. Current-voltage characteristics for (a) the threshold switch (b) the memory switch.

In the threshold switch (see Figure 38) the application of a field greater than a certain threshold changes the conductance of a layer of chalcogenide glass by about a factor of 10^6 in a time of about 10^{-9} seconds. As long as a certain holding current is kept flowing, the switch will stay in the high conductance, or 'on' state. The removal of the small holding voltage permits the switch to revert to the low conductance or off-state. Early controversies between the proponents of a thermal explanation and an electronic explanation for the switching now seem to have been resolved in favor of latter, and the phenomenon now appears as a reproducible and rather special property of these amorphous materials. The memory switch has a current-voltage characteristic that is similar, except that there is no holding voltage. Once switched to the on-state, the device retains a high conductance until a carefully-gauged pulse of current switches it back again to the off-state. The memory effect is connected with local crystallization of material that possesses a small amount of cross-linking. Local memory switching can also be accomplished in large-area devices by laser pulses. In the two memory states the optical properties of the material are different, so that optical transmission (or reflection) may be used to 'read' the state of the material. Mass memories with a capacity of a trillion bits per square meter have been claimed, based on the fact that visible light can be focused down to a spot of the order of 1 micron in diameter. In principle the same ideas may be extended to electron beam writing and reading. The dot size for electron beams may be made much smaller than for visible light. The read process depends on the fact that the emission of secondary electrons produced by an incident beam of electrons is quite different for amorphous material in its crystalline and amorphous phases.

The basis for the imaging applications is again the possibility of an amorphous-crystalline transition. Proper choice of material and radiation can vary the size of the grains produced and the fraction of material that is crystallized; thus, a gradation of tone can be achieved. Since both the amorphous and crystalline phases are stable at room temperature, no 'fixing' in the usual sense of the photographic process is necessary. Information may be added or erase locally from a sheet of film. A photographic print (or many) can be obtained from the exposed film through the application of a diverse set of properties that differ between the crystalline and amorphous phases.

These and other devices all work because of the interplay of easily-produced structural change and changes in the physical properties. The structural changes may be induced by a variety of stimuli such as light, heat, pressure or current. The physical properties that change include, among others, optical transmission, refractive index, electrical conductivity, chemical reactivity and wettability by water, inks and oils. The variety of phenomena here have made amorphous chalcogenide glasses an arena of great challenge both to those interested in the basic physics of solids and to those whose concern is the production of novel and economic devices for technological applications.

REFERENCES

1. Useful general references are the proceedings of biennial conferences on <u>Amorphous and Liquid Semiconductors</u>, the most recent of which have been (1) Ann Arbor (1971), J. Non-Cryst. Solids 8-10 (1972); (2) Garmisch (1973), published as <u>Amorphous and Liquid Semiconductors</u> (Taylor and Francis 1974): (3) Leningrad (1975), published as <u>Structure and Properties of Non-Crystalline Semiconductors</u> (Nauka, Leningrad 1976); (4) Edinburgh (1977), published as <u>Proceedings of the 7th International Conference on Amorphous and Liquid Semiconductors</u>. Also, <u>Amorphous and Liquid Semiconductors</u>, editor J. Tauc, published by Plenum Press, New York (1974).
<u>Physics of Structurally Disordered Solids</u>, editor S.S.Mitra, published by Plenum Press, New York (1976).
<u>Tetrahedrally Bonded Amorphous Semiconductors</u>, edited by Brodsky, Kirkpatrick and Weaire, American Institute of Physics Vol. 20 (1974).

Structure and Excitations of Amorphous Solids, edited by Lucovsky and Galeener, American Institute of Physics, Vol.31 (1976).

2. Several sources give classifications into groups. See, for example, H.Fritzsche in Amorphous and Liquid Semiconductors, editor J.Tauc, reference 1.

3. For further discussion of the glass transition and the conditions for glass formation, see, for example, D.Turnbull, Contemp. Phys., 10, 473 (1969).

4. See, for example, B.E.Warren in X-Ray Diffraction (Addison-Wesley, Reading, Mass., 1969); J.F.Graczyk and P.Chaudhari, J.F.Graczyk and S.R.Herd, in Physics of Structurally Disordered Solids, editor S.S.Mitra (Plenum Press, New York 1974).

5. See, for example, W.Paul and G.A.N.Connell, in Physics of Structurally Disordered Solids, editor S.S.Mitra, reference 1, for a review.

6. See, for example, J.C.Knights, T.M.Hayes and J.C.Mikkelson, Phys. Rev. Lett., 39, 712 (1977).

7. N.J.Shevchik and W.Paul, J. Non-Cryst. Solids, 16, 55 (1974).

8. See, for example, S.A.Solin in Structure and Excitations in Amorphous Solids, reference 1, p. 205; G.Lucovsky and F.L. Galeener, Proceedings of the 1975 Leningrad Conference, reference 1, p. 207.

9. See, for example, G.A.N.Connell and J.R.Pawlik, Phys. Rev., B 13, 787 (1976).

10. See reference 9 for original work plus general references.

11. See reference 4 and also N.J.Shevchik and W.Paul, J. Non-Cryst. Solids, 8-10, 381 (1972); R.J.Temkin, W.Paul and G.A.N.Connell, Advances in Physics, 22, 581 (1973); J.F. Graczyk and P.Chaudhari, Phys. Stat. Solidi, B 58, 163 and 58, 501 (1973).

12. D.E.Polk, J. Non-Cryst. Solids, 5, 365 (1971); D.Turnbull and D.E.Polk, J. Non-Cryst. Solids, 8-10, 19 (1972).

13. G.A.N.Connell and R.J.Temkin, Phys. Rev., B 12, 5323 (1974).

14. N.J.Shevchik and W.Paul, J. Non-Cryst. Solids, 8-10, 381 (1972); M.G.Duffy, D.S.Boudreaux and D.E.Polk, J. Non-Cryst. Solids, 15, 435 (1974); D.Beeman and B.L.Bobbs, Phys. Rev. B 12, 1399 (1975); R.Kaplow, T.A.Rowe and B.L.Averbach, Phys. Rev., 168, 1068 (1968).

15. D.E.Polk and D.S.Boudreaux, Phys. Rev. Lett., 31, 92 (1973).

16. P.Steinhardt, R.Alben and D.Weaire, J. Non-Cryst. Solids, 15, 199 (1974).

17. D.Weaire and M.F.Thorpe, Phys. Rev., B $\underline{4}$, 2508 (1971); Phys. Rev., B $\underline{4}$, 3518 (1971).

18. D.W.Bullett and M.J.Kelly, Solid State Commun., $\underline{12}$, 1379 (1975).

19. M.J.Kelly and D.W.Bullett, Solid State Commun., $\underline{18}$, 393 (1976 J.D.Joannopoulos, Amer. Inst. Phys. Conference Proceedings, $\underline{31}$, 108 (1976).

20. I.B.Ortenburger and D.Henderson, Proceedings of the 11th International Conference on Semiconductors, Warsaw, 1972 (published by Polish Scientific Publishers), p. 465; I.B. Ortenburger, W.E.Rudge and F.Herman, J. Non-Cryst. Solids, $\underline{8-10}$, 653 (1972); J.D.Joannopoulos and M.L.Cohen, Phys. Rev., B $\underline{7}$, 2644 (1973).

21. For a listing, see W.Paul, in Proceedings of the 11th International Conference on the Physics of Semiconductors, Warsaw, 1972 (Polish Academy of Sciences), p. 38.

22. P.W.Anderson, Phys. Rev., $\underline{109}$, 1492 (1958).

23. See, for example, J.C.Phillips, Phys. Stat. Solidi, B $\underline{44}$, K1 (1971).

24. P.W.Anderson, Phys. Rev. Lett., $\underline{34}$, 953 (1975).

25. W.Paul, Bull. Am. Phys. Soc., $\underline{22}$, 358 (1977).

26. G.A.N.Connell and W.Paul, Bull. Am. Phys. Soc., $\underline{22}$, 405 (1977

27. R.A.Street and N.F.Mott, Phys. Rev. Lett., $\underline{35}$, 1293 (1975); N.F.Mott, E.A.Davis and R.A.Street, Phil. Mag. $\underline{32}$, 961 (1975)

28. M.Kastner, D.Adler and H.Fritzsche, Phys. Rev. Lett., $\underline{37}$, 1504 (1976).

29. W.Paul, A.J.Lewis, G.A.N.Connell and T.D.Moustakas, Solid State Commun., $\underline{20}$, 969 (1976).

30. In collaboration with Drs. D.A.Anderson and M.A.Paesler, Messers. G.Moddel, S.Oguz and J.R.Pawlik and Ms. E.Freeman.

31. See, for example, J. Tauc in *Amorphous and Liquid Semiconductors*, reference 1.

32. N.F.Mott, Phil. Mag., $\underline{19}$, 835 (1969); V.Ambegaokar, B.I.Halperin and J.S.Langer, Phys. Rev., B $\underline{4}$, 2612 (1971).

33. For a comprehensive review, see W.Paul, Thin Solid Films, $\underline{33}$, 381 (1976).

34. D.A.Anderson, T.D.Moustakas and W.Paul, Proc. 7th Int. Conf. on Amorphous and Liquid Semiconductors (Edinburgh 1977), p. 334.

35. N.K.Hindley, J. Non-Cryst. Solids, $\underline{5}$, 17 (1970).

36. R.A.Street, Phil. Mag., to be published.

37. M.A.Paesler and W.Paul, to be published.
38. D.E.Carlson, IEEE Trans. on Electron Devices, ED-24, 449 (1977).
39. R.A.Street, private communication.
40. General reviews on this subject are to be found under reference 1. An up-to-date review of luminescence is to be found in R.A.Street, Advances in Physics, 25, 397 (1976). A long review article by G.A.N.Connell and R.A.Street will appear in a forthcoming book to be edited by S.Keller and published by North-Holland Publishing Co.
41. N.F.Mott and E.A.Davis, Electronic Processes in Non-crystalline Materials (Oxford University Press 1971), p. 327.
42. Spin resonance can, however, be found in bulk samples that are quenched at a very rapid rate, or in films deposited onto a cold substrate and kept cold.
43. M.H.Cohen, H.Fritzsche and S.R.Ovshinsky, Phys. Rev. Lett., 22, 1065 (1969).
44. P.W.Anderson, Phys. Rev. Lett., 34, 953 (1975).
45. See, for example, J.C.Phillips, Comments Solid State Physics, 4, 9 (1971); J.A. van Vechten, Solid State Commun., 11, 7 (1972); D.Emin, Phys. Rev. Lett., 28, 604 (1972); D.Emin, C.H.Seager and R.K.Quinn, Phys. Rev. Lett., 28, 813 (1972); N.F.Mott, Phys. Bull., 25, 448 (1974).
46. N.F.Mott, E.A.Davis and R.A.Street, Phil. Mag., 32, 961 (1975); R.A.Street and N.F.Mott, Phys. Rev. Lett., 35, 1293 (1975); N.F.Mott and R.A.Street, Phil. Mag., 36, 33 (1976).
47. P.W.Anderson, J. de Physique, 37, C4-339 (1976).
48. M.Kastner, D.Adler and H.Fritzsche, Phys. Rev. Lett., 28, 355 (1976).
49. The probability that a chalcogen such as Se could have coordinations different from two had been considered before, no doubt several times. The present author first encountered experimental data from EXAFS (extended X-ray absorption fine structure) favoring three-fold coordination[50] when he worked on this problem in 1974[51]. The contribution of Kastner, Adler and Fritzsche was the demonstration that a certain pair of 'defects' constituting one identifiable entity, probably constituted the lowest energy defect state, and that in that state all electrons were paired with none left over. Their model does not allow for a major presence of higher energy defects possessing spin, and they give convincing arguments why the density of such defects will be below the limit of observation for melt-cooled glasses.

50. D.E.Sayers, F.W.Lytle and E.A.Stern, in Amorphous and Liquid Semiconductors (Taylor and Francis, 1974), p. 403.

51. Unpublished.

52. M.Kastner, Proceedings of a Conference on Atomic Scale Structure of Amorphous Solids, IBM Yorktown Heights (1978).

53. See, for example, references 46 and 52.

54. S.G.Bishop, U.Strom and P.C.Taylor, Phys. Rev., B 15, 2278 (1977).

55. J.P. de Neufville, in Optical Properties of Solids - New Developments, editor B.O.Seraphin (North-Holland Publishing Co., 1976), p. 437; J.P. de Neufville, in Amorphous and Liquid Semiconductors (Taylor and Francis, 1974), p. 1351.

56. K.Tanaka, Appl. Phys. Lett., 26, 243 (1975).

57. K.Tanaka, American Institute of Physics Conference Proceedings, 31, 148 (1976).

58. R.A.Street, Advances in Physics, 25, 397 (1976).

59. For a general elementary review of this field, see D.Adler, Scientific American, 236, 36 (1977). See also S.R.Ovshinsky and H.Fritzsche, IEEE Trans. Electron Devices, ED-20, 91 (1973).

SOLAR ENERGY MATERIALS

Lyle H. Slack

Department of Materials Engineering
Virginia Polytechnic Institute & State University
Blacksburg, Virginia 24061, USA

1.   SOLAR/THERMAL MATERIALS

The sun is the engine that drives our universe. It is the source of electromagnetic energy that makes it possible for life to flourish here. The continual presence of solar energy over many eons has facilitated our evolution. Early man recognized the overpowering importance of the sun by worshiping it. The same continual solar flux on earth resulted in the fossil fuels which we are depleting so rapidly.

In order to solve our current and impending energy deficiencies for the industrialization of our countries, man has the sun and the materials of the earth at his disposal. Man's utilization of materials had led to human progress marked by the bronze and brass ages, which are so evident from our present historical perspective. We are now probably in the nuclear age. If man can learn the use of the solar flux for his energy needs, he will be using his ultimate energy source in a much more direct and efficient way than he has in the past, and perhaps man will have entered into the solar age. His entry into this age will depend on his cleverness in using the properties of materials in entrapping, transporting and converting solar energy.

This first section treats some materials science and engineering aspects of solar collectors and other solar thermal devices. The second will focus on the coatings used on architectural windows for the purpose of entrapping solar energy in rooms as heat, or rejecting it so as to keep office workers comfortable and productive in the summer and to minimize air conditioning costs. The third section will review the materials aspects of solar photovoltaic energy conversion.

The entrapment of solar energy in a solar/thermal collector[1], Figure 1.1(b), involves absorbing the energy of the solar spectrum, shown in Figure 1.1(a) for air mass-two* radiation, and keeping the solar heated absorber plate from remitting thermal energy back into space. The solar spectrum[2] is concentrated between 0.3 and 2.0 micrometers. Once this energy is absorbed and heats the absorber plate to about 100°C, the energy attempting to leave is concentrated between 3 and 25 μm as also shown in Figure 1.1(a).

Absorption is usually accomplished using black paint on the thermally conducting absorber plate. Black paint usually consists of oxide pigment particles suspended in a transparent polymer medium. Sometimes metallic droplets or crystallites are dispersed in a polymer medium. In either case, the electromagnetic energy in or near the visible spectrum is converted to infrared (heat) energy. However, this heat must be transferred to the underlying metallic plate through the thermally insulating polymer medium. The paint must be kept very thin to minimize thermal resistance. The maintanence of good thermal conductance from the front face of the absorber layer all the way to thermal storage is necessary in order to keep the absorber surface as cool as possible. If that surface is kept close to room temperature, the amount of infrared radiation lost by reradiation will be minimized. Such a collector surface is considered to be 'solar selective in that it does not emit much energy in the thermal infrared spectrum but does absorb in the visible. In this case, the infrared emissivity is kept low by keeping the surface temperature low.

Using a cover plate helps reduce the loss by infrared reradiation. A glass or plastic cover plate admits much of the solar energy but is opaque and reflecting at infrared wavelengths. The cover plate also serves to reduce the heat loss from the absorber plate by convection and air conduction. Near the equator, a single cover glass is optimum; in the temperate regions two are optimum; and in the colder climates three cover glasses provide optimum solar collection.

Attempts to fabricate surfaces which absorb solar (visible energy but do not re-emit energy in the infrared have been partially successful. Kirchhoff's Law specifies that the absorptance of a surface must equal its emittance (in order to be in equilibrium with its surroundings) at any wavelength. However, the microstructure of the surface can be engineered so that the absorptance at solar wavelengths is high but the absorptance and emittance at the larger infrred wavelengths is low. Such a surface is an electroplated layer of rhenium on a metal surface. A vertical section of the layer is as shown schematically in Figure 1.2. The rhenium crystallites tend to nucleate and crystallize along one crystallographic direction so

---

*AM-2 represents the intensity of solar flux which has passed throug twice as much atmosphere as is encountered in perpendicular incidenc

SOLAR ENERGY MATERIALS

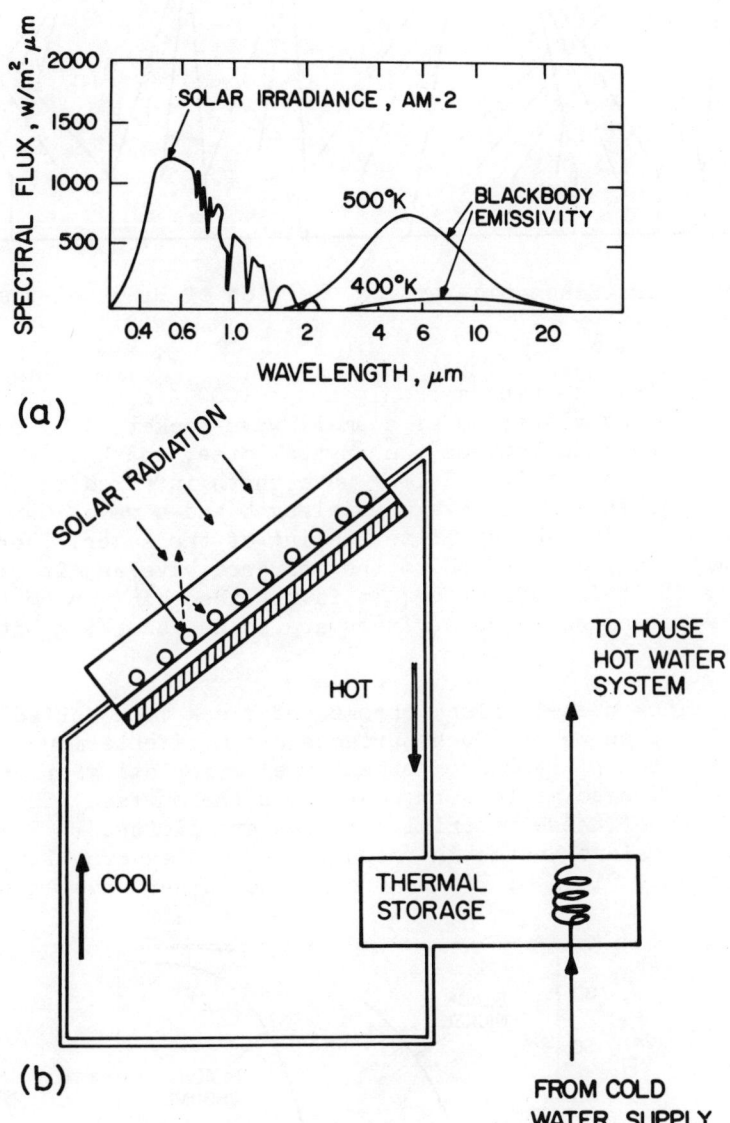

Figure 1.1. (a) Solar and Blackbody Spectra, (b) Solar Collector System Schematic.

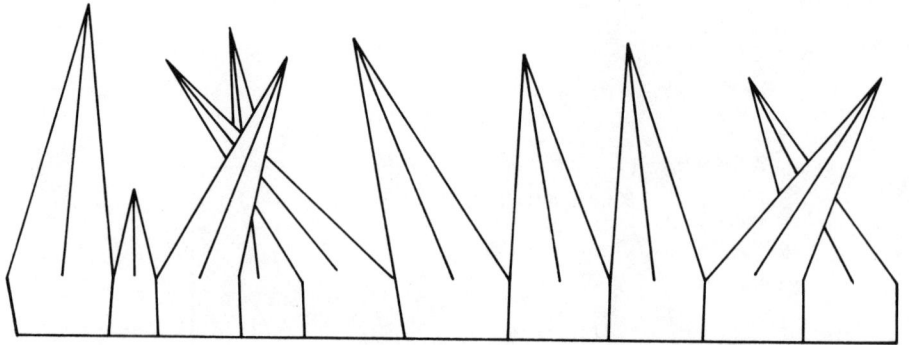

Figure 1.2. Schematic Vertical Section of Electrodeposited Rhenium.

that the crystallite tips are only about 1000 Å apart. A visible photon, which we may visualize as a small wave-packet, will experience multiple reflections between the crystallites, and become completely absorbed into the coating. A long wavelength infrared wave encountering this surface will have a wavelength and a wavefront very large compared to the intercrystallite spacing of the layer. Therefore, the surface will 'appear' smooth to the infrared wave and it will be reflected. If it is reflective, it is not absorbing or emitting at infrared wavelengths. Such surfaces are solar absorbing but not infrared emitting.

Black nickel, black chrome and black iron surfaces also display this phenomenon. Such surfaces are manifestations of the same optical phenomenon. Each are formed by electroplating or chemical vapor deposition so as to form nodules on the surface. The reflectances of two of these material surfaces are plotted in Figure 1.3. As the surfaces are heated in vacuum, the nodules grow larger, and the reflectance rises of Figure 3 shift to larger wavelengths. Such

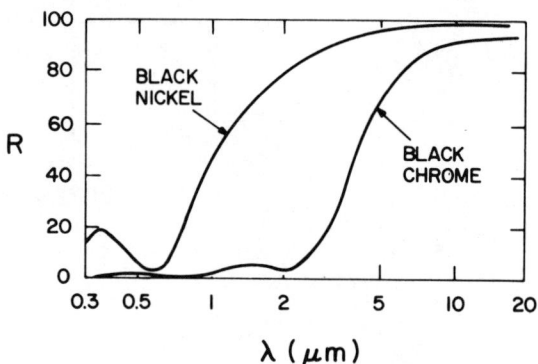

Figure 1.3. Reflectance Edges from Black Nickel and Black Chrome Coatings.

SOLAR ENERGY MATERIALS

surfaces have an advantage over the black paint surfaces mentioned previously in that there is much less heat resistance between the top surface and the base metal. On the other hand, these nodular surfaces have a small component of reflectance not so evident in the black paint surfaces.

The optical properties of any material must obey this relation at any particular wavelength:

Absorptance (A) + Reflectance (R) + Transmittance (T) = 1

The ideal solar selective material would have an absorptance equal to unity with a nonexistant reflectance and transmittance in the solar (visible) spectrum. At the far infrared wavelengths the same ideal material would have no absorptance so that its emittance would be zero. The rhenium, black nickel, black chrome, and black iron coatings tend to approach these idealities (3-6).

Another method of reducing the heat loss from the absorber plate is by removing the air adjacent to it so as to eliminate heat loss by convection. A comparison of the design of a conventional solar collector and an evacuated one is made in Figure 1.4. Goodman et al[7]. demonstrate the importance of vacuum and coatings in redu-

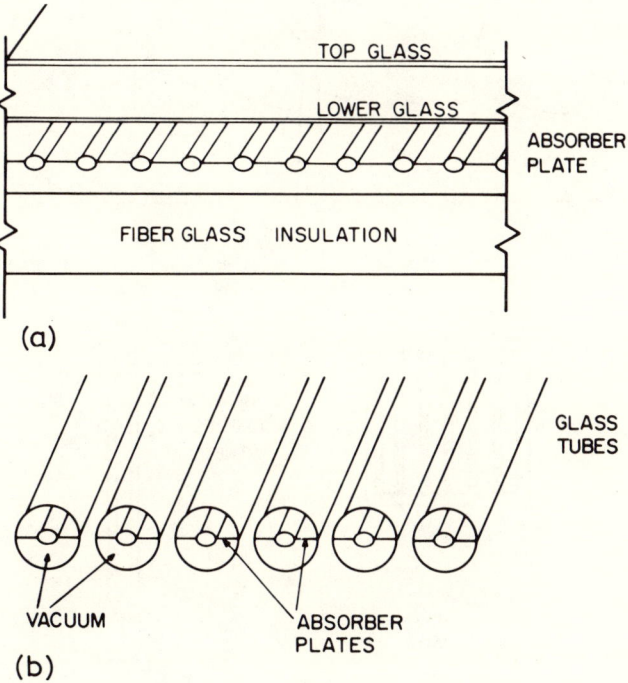

Figure 1.4. Schematic Representation of a (a) Conventional and (b) Evacuated Solar Collectors.

cing thermal losses. The glass tubes of the second collector partially focus the sunlight onto the absorber plate. The lower half of the glass tubes can be silvered to help eliminate the radiative component of heat loss.

Thermal storage in a collector system may be accomplished using the heat capacity of inexpensive material such as gravel, rock or water. More thermal energy can be stored in a material without increasing its temperature by using the heat of fusion of that material. The most common such material is Glauber's salt ($Na_2SO_4 \cdot 10H_2O$). In a typical collector system such a hydrated salt stores about six times as much heat as water, and nearly twenty times as much as gravel. The high degree of hydration of these salts lowers the fusion temperature to near room temperature. Most of these hydrated salts tend to disproportionate during melting which complicates the subsequent freezing process. In order to discourage this microscopic phase separation, various suspension media are used. For example, wood pulp is used to suspend $Na_2S_2O_3 \cdot 5H_2O$, and clay to suspend $Na_2SO_4 \cdot 10H_2O$[8]. An entire conference[9] was devoted to all aspects of storage materials.

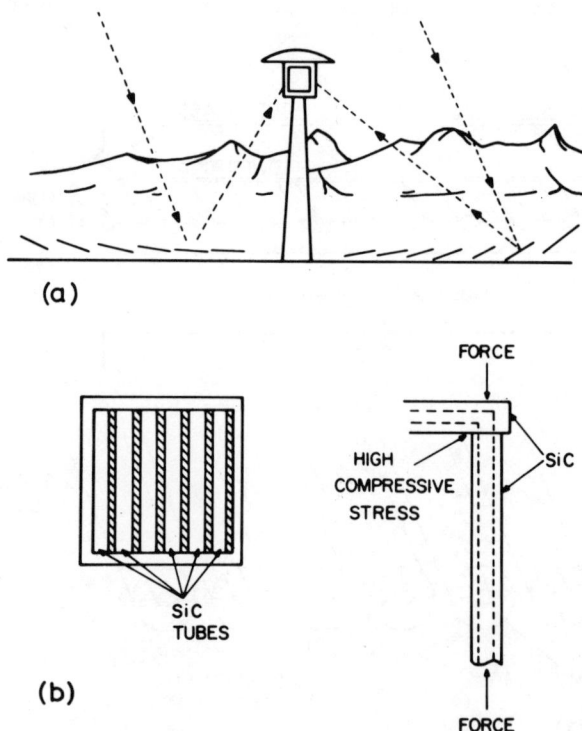

Figure 1.5. Schematics of (a) Solar Tower or Solar Farm Concept, and (b) Details of Silicon Carbide Receptor.

SOLAR ENERGY MATERIALS

The solar tower concept [Figure 1.5(a)] has also presented some interesting materials challenges. The mirrors are made of optically perfect float glass. The mirrors are distorted into a parabolic focusing shape by pulling down on the middle of the glass sheet using a frame as shown in Figure 1.6. The mirrors[10] of the solar tower array[11] are sychronized with the sun by means of semiconducting

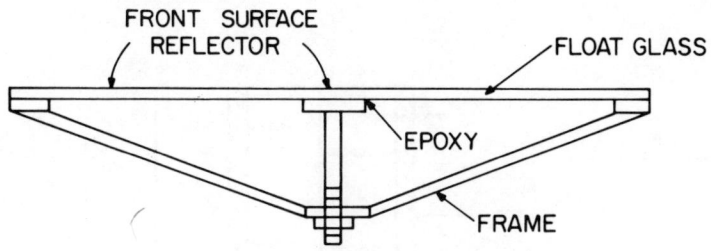

Figure 1.6. Schematic of Solar Mirror.

photosensors placed in the bottom of a tube. The sun shines down the tube onto the sensors. When the tube and the sun became misaligned, a servomechanism realigns the tube (and the mirror) with the sun. One of the most challenging materials problems of the solar tower has been the solar receptor at the top of the tower. It must withstand temperatures in excess of 2000°C and contain high pressure steam. The most successful design to date consists of silicon carbide tubes fitted to silicon carbide endmembers, as shown in Figure 1.5(b). Rather than use threaded tubes or taper-fits, it was found that whenever perfectly flat surfaces were forced together by large compressive forces, leak-free and trouble-free joints could be achieved.

2. SOLAR REVERSIBLE WINDOWS USING OXIDE COATINGS ON GLASS

Expansive areas of glass are used in modern architecture[12]. Such large area windows are often coated with an oxide or metal so as to reduce solar heating of the building's interior. In other cases, the glass is darkened by introducing an oxide colorant, usually a transition metal oxide, to the glass batch. This darkening is done for the same purpose, to reduce solar influx, thus minimizing air conditioning costs. Another purpose for darkening a window, either by films or glass coloration, is to balance the lighting in the room so as to reduce eye strain.

The film or glass coloration accomplish the desired reduction in room solar heating, but it is operative in the cooler winter months when solar heating would reduce the amount of energy needed for space heating. The purpose of our research to be described in this section is to develop a window which permits solar heat to be admitted in one (the 'winter') orientation, and to keep the heat outside in the opposite ('summer') orientation.

The basic concept of the window is demonstrated in Figure 2.1. This represents the summer orientation. Most of the solar radiation would be absorbed in the dark Co_3O_4 film. The solar energy is converted to far infrared which is emitted in both directions. This radiation can be reflected at any of the three glass-air interfaces. Shown in Figure 2.1 is an infrared SnO_2 infrared reflecting film to enhance the probability that the infrared radiant heat will be reflected to the outside.

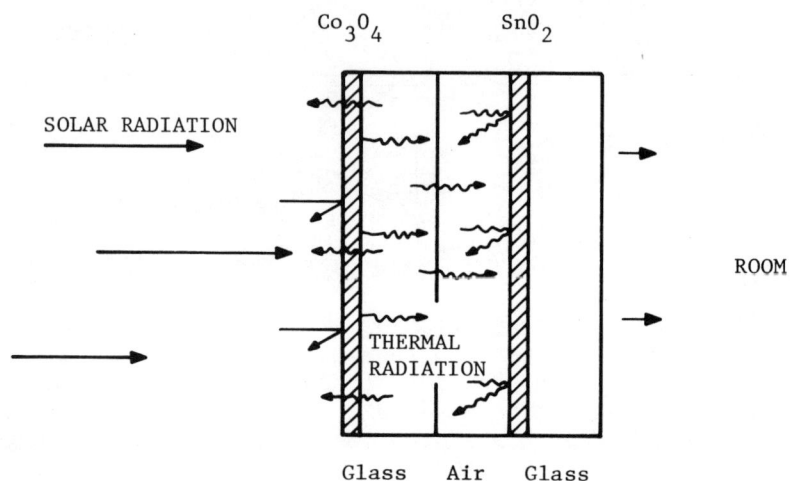

Figure 2.1. Reversible Window Concept, Summer Orientation.

In the opposite orientation, the solar energy would penetrate the optically transparent glass panes and SnO_2 film and once again be predominately absorbed in the dark Co_3O_4 film. This time the Co_3O_4 film is inside the room and its heat is transferred by radiation, convection and conduction. The radiant component that is directed to the outside will mostly be reflected back into the room by the infrared reflecting film or the glass surfaces.

Testing the above concept involved the preparation of 6" × 6" windows coated with Co_3O_4 and SnO_2. Oxide coatings were deposited by the pyrolitic decomposition of Cobalt Acetylacetonate $(Co(C_5H_7O_2)_3)$ in alcohol (0.04 molar concentration) and Stannic Chloride $(SnCl_4 \cdot 5H_2O)$ in alcohol (2.85 molar). These solutions may be sprayed in a conventional sprayer, but the associated velocity of air cools the heated glass substrate so much that it must be preheated to the softening point of the glass. A fogging device[13] was developed in our laboratories which produces a slow-moving fog. This

SOLAR ENERGY MATERIALS 361

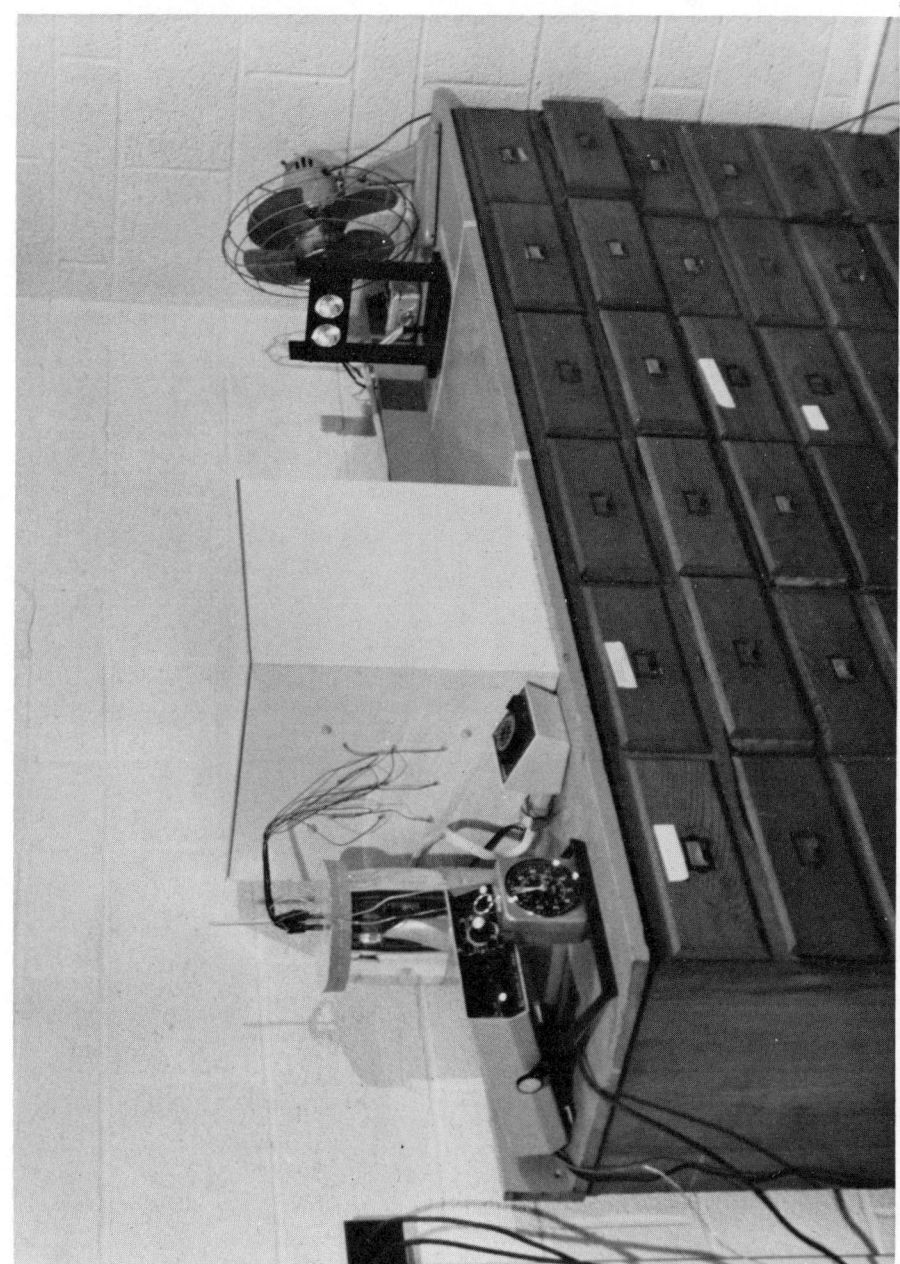

Figure 2.2. General View of Solar-Thermal Test Apparatus.

Figure 2.3. Inside View of Solar-Thermal Test Chamber.

permitted the deposition to occur at a glass temperature of only 350 C. However, that fogging device could only be used to coat 3" × 3" substrates, so an ultrasonic fogging device[14] was used in coating the 6" × 6" × 1¼" windows. The ultrasonic vibrations were produced in a resonance cavity in the same way that sonic vibrations are produced in a whistle. The solution droplets produced by this technique were less than five micrometers in diameter.

The thickness of the thin films were determined by the attenuation of X-rays from the glassy substrate, according to a procedure described by Evans and Fisher[15].

A Beckman model DK-2 spectrophotometer was used to measure the transmittance (T) and spectral reflections (R) of the coated windows. The absorptances (A) were calculated

$$A + T + R = 1.$$

The apparatus for testing the amount of solar heat admitted by a window is shown in Figures 2.2 and 2.3. The solar simulator consisted of two tungsten-halogen projection lamps*. The test box (room) was constructed of three layers of urethane foam sheating** enclosing one cubic foot of interior volume. A 6" by 6" opening in the front of the box accomodated the test windows. Suspended in the rear of the box by four white nylon (thermally insulating) bolts is a 25-mole (678 gram) aluminum plate painted with flat (non-glossy) black paint. Nine thermocouples were inserted into the back of the aluminum plate. These thermocouples, connected in parallel, provided an average reading of the plate temperature.

A standard test procedure of recording the temperature rise during 30 minutes of simulated solar exposure was established. An analysis of the sources of error of this test revealed that the following parameters had to be controlled in order to obtain reproducable (± 1%) results:

1. Laboratory humidity
2. Air movement in the laboratory
3. Glass temperature at the beginning of each test
4. Room temperature.

The temperature rise during the last fifteen minutes of the thirty-minute test was found to be most reproducable, and was adopted as the standard observation time. This simulated solar-induced temperature rise was converted to an equivalent heat gain

---

*Quartzline Lamp, Model ELH, 120V, 300W, General Electric Corp, Cleveland, OH, USA.
**Celotex ½" Thermax Sheathing, Celotex Corp., Tampa, Fla., USA.

using this equation:

$$\text{Heat Gain} = \frac{C_p \times M \times \Delta T}{A}, \text{ in cal/cm}^2\text{-hr.}$$

where, C_p = molar heat capacity of aluminum (5.975 cal/°C-mole)

M = number of moles of aluminum in the absorber plate (25.0)

ΔT = temperature rise per hour ($4 \times \Delta T_{ave}$)

A = area of window (232 cm^2)

1 Cal/cm = 3.69 BTU/ft

Q_w = heat gain (cal/cm^2 hr) when the reversible solar window is in the winter orientation

Q_s = heat gain when the window is in the summer orientation, and

$\frac{Q_w}{Q_s}$ = ratio of winter and summer heat gains, a measure of solar-thermal reversibility, and called the STR Coefficient.

An identification system for different positions of the absorbing and reflecting films on the two glass panes is illustrated in Figure 2.4. For example, a 1A4R array would have an absorbing Co_3O_4 film on the first surface (facing the sun) and a reflecting

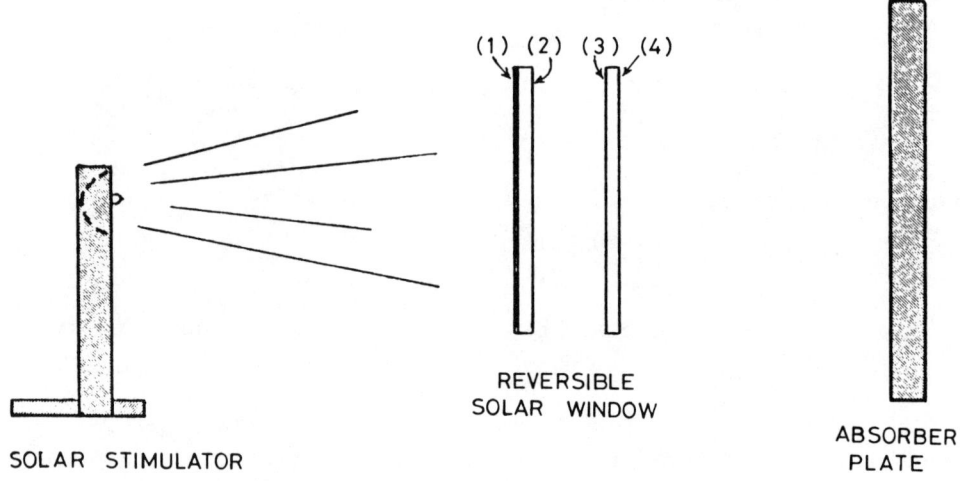

SOLAR STIMULATOR

REVERSIBLE SOLAR WINDOW

ABSORBER PLATE

Figure 2.4. Identification System for Double Light Window Surfaces.
Note: A summer orientation is shown, i.e. Co_3O_4 thin film is on surface(1).

Table 1

Solar Reversibility Tests on Double Pane Windows

Absorber	Orientation	ΔT_{15}	Q	Q_w/Q_s
50Å Co_3O_4	1A	4.73	12.17	1.059
	4A	5.01	12.89	
200Å Co_3O_4	1A	3.95	10.16	1.096
	4A	4.33	11.14	
650Å Co_3O_4	1A	2.64	6.79	1.250
	4A	3.30	8.49	
2025Å Co_3O_4	1A	2.01	5.17	1.726
	4A	3.47	8.93	
Gray Glass, no coating	n.a.	4.25	10.93	1.088
	n.a.	4.63	11.91	
50Å Co_3O_4 on Gray Glass	1A	3.41	8.77	1.217
	4A	4.15	10.68	
160Å Co_3O_4 on Gray Glass	1A	2.28	5.86	1.553
	4A	3.54	9.11	
650Å on Gray Glass	1A	1.99	5.12	1.774
	4A	3.53	9.08	
1700Å on Gray Glass	1A	1.71	4.40	1.988
	4A	3.40	8.75	
1700Å on Gray Glass	2A	1.69	4.32	1.690
	3A	2.84	7.31	
Chrome Coat on Clear Glass	1A	1.58	4.06	1.608
	4A	2.54	6.53	
1700Å Co_3O_4 on Gray with Sunadex	1A	1.84	4.73	1.973
	4A	3.63	9.34	
1700Å Co_3O_4 on Gray with IR Reflecting Glass	1A	1.31	3.37	2.06
	4A	2.70	6.95	
Same with convection	1A	1.17	3.01	2.31
	4A	2.70	6.95	

SnO_2 coating on the fourth surface (facing the room or absorber plate). A 1A array represents a summer window with a Co_3O_4 absorbing layer facing the sun and no other coating used. When this window is rotated into the winter orientation, its designation becomes 4A.

Results

The cobalt oxide films on glass were all polycrystalline, and were the spinel form of Co_3O_4. The tin oxide films were all SnO_2 polycrystalline films with the cassiterite structure.

The results of the solar reversibility tests are listed in Table 1. The solar reversibility coefficient is seen to increase from 1.059 to 1.726 as the Co_3O_4 absorber coating on clear glass increases from 50 Å to 2025 Å. When gray glass was used as the substrate for the Co_3O_4 absorber coating, the coefficient increases from 1.217 to 1.987 as the Co_3O_4 film thickness increased to 1700 Å. This dependence of the coefficient on the Co_3O_4 thickness is plotted in Figure 2.5.

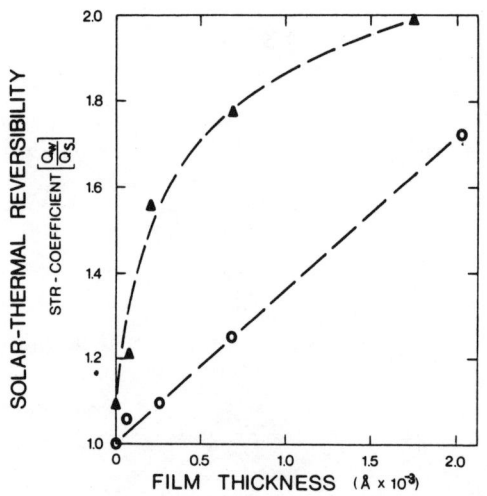

Figure 2.5. Solar-Thermal Reversibility as a Function of Co_3O_4 Film Thickness on Gray Glass (Δ) and Clear Float (O) Substrates. Maximum Heat Gain in 15 minutes.

The optical characteristics of the clear and gray glasses coated with the thickest films of Co_3O_4 are shown in Figures 2.6 and 2.7. From these plots it is seen that the visible absorptance is approximately 80% in each case. The peaks and dips in the transmittance curves are due to interference effects. As the film thickness increases, these features move to longer wavelengths.

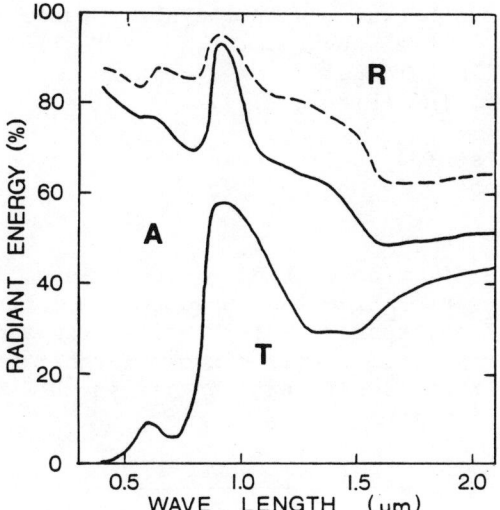

Figure 2.6. Change in Reflectance (R), Absorptance (A), and Transmittance (T) of Light with Wavelength for Co-121 ($\simeq 2000\text{Å}$ Co_3O_4 on CF).

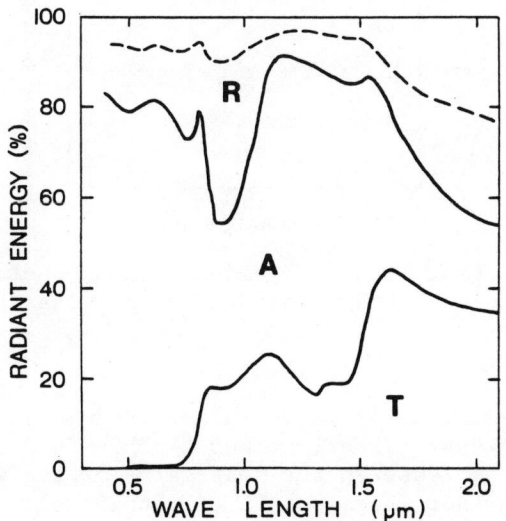

Figure 2.7. Change in Reflectance (R), Absorptance (A) and Transmittance (T) of Light with Wavelength for Co-122 ($\simeq 1700\text{Å}$ Co_3O_4 on Gray).

Discussion of Results

The solar-thermal reversibility coefficient for the double windows increases when the Co_3O_4 absorber film thickness increases. It increases linearly with film thickness when a clear window is used as the Co_3O_4 substrate. When the absorptance is enhanced by using gray glass as the substrate, the coefficient is higher and also increases with film thickness. However, this increase follows a law of diminishing returns (Figure 2.5).

The first few hundred angstroms of Co_3O_4 have the greatest effects on the heat gain observed for the winter and summer orientations.

For the films deposited on clear glass, the heat gain observed for the winter orientation decreases linearly with thickness until ~650 Å is reached. This results from a drop in solar transmittance through the double window. Further increases in thickness do not result in significant changes in the observed heat gains (Figure 2.8).

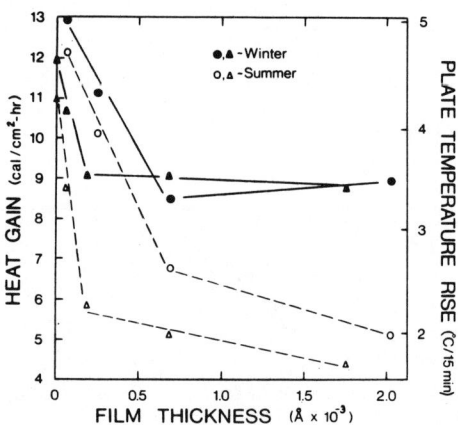

Figure 2.8. Heat Gain (cal/cm²-hr) as a Functon of Co_3O_4 Film Thickness on Gray Glass (Δ,Δ) and Clear Float (O,O) Substrates.

Similar behavior is found for the coefficient for windows made with gray glass. This first 200 Å of Co_3O_4 on gray glass also causes a linear decrease in the heat gain for the winter and summer orientations. Further increases in film thickness do not cause corresponding increases in the coefficient.

In all cases, adding thicker layers of absorbing Co_3O_4 film causes the rate of heat gain to become diminished. For very thin film essentially all of the solar energy is absorbed at the absorber plate in the back of the simulated room. As the thickness of the Co_3O_4 film increases, less energy reaches the absorber plate and the rate of heat gains decreases. When that absorber film is on the outside, the heat is mostly lost to the outside by all three mechanisms of heat transfer, but when it is on the inside (winter configuration), heat from the film is transferred to the absorber plate, again by all three mechanisms. This accounts for the steeper drops in heat gain (vs. film thickness) (Figure 2.8) for the summer window configurations.

After a minimum film thickness has been achieved (200 Å for the gray glass substrate and 650 Å for the clear glass), most of the solar radiation is absorbed in the window. Then heat is transferred to the absorber plate (room) by infrared radiation, convection and air conduction. Further increases in the film thickness has little effect on the rate of heat gain because essentially all of the solar radiation is being absorbed anyway. The slopes for the rate of heat gain vs. Co_3O_4 thickness become very small once this two-step mechanism of heat transfer becomes operative.

The ideal situation from an energy-saving point-of-view would be to have a clear uncoated double window in the winter (providing a heat gain of ~ 14 cal/cm$^2 \cdot$hr) and a coated gray glass on the outside with a clear float glass on the inside for the summer. The heat gain would then be only about 4 cal/cm$^2 \cdot$hr, or even less if convection cooling of the outside of the window was permitted. However, such a window would not provide year-around shading.

It was found early in the research that including an infrared reflecting coating of SnO_2, as shown in Figure 2.1, does little to enhance the reversibility coefficient. Apparently SnO_2 films reflect near-infrared energy readily, but have little reflective ability of infrared radiation beyond that of an uncoated glass surface.

Incorporating a textured glass* of low absorptance in the heat reflection position (3-4 in the summer orientation) also had little effect on the reversibility coefficient. Even though the textured glass may have reflected more heat, it has a lower absorptance than conventional float glass and transmitted more visible solar energy to the room.

3. PHOTOVOLTAIC MATERIALS

The essential components of a photovoltaic solar cell are n-type and p-type semiconductors in contact, and metallic connections to each, as shown in Figure 3.1. In the ideal situation, the top semiconductor would serve as a perfect solar window. That is, it would have a large enough band gap to not absorb any of the solar spectrum. Cadmium sulfide is an example that approaches this ideality. It is transparent with only a slight yellow tint. The window should also have a small lateral resistance so that charge carriers

---

*ASG Sunadex, ASG Industries, Kingsport, TN, USA.

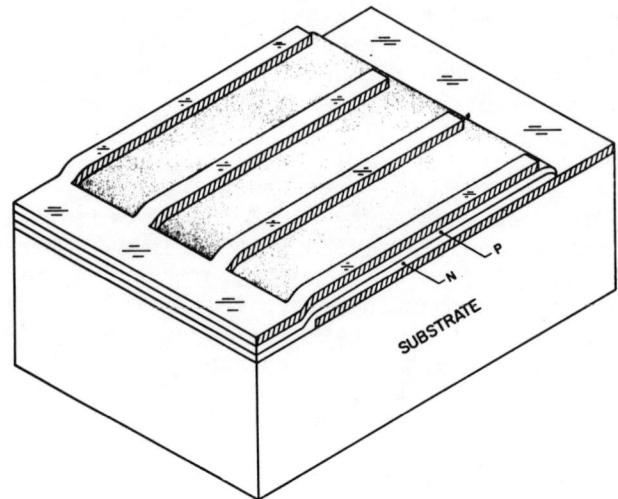

Figure 3.1. Heterojunction Solar Cell Schematic.

can be transported away from the junction to a skeletal or transparent electrode on top of the window. Again, a film of CdS is a good example because its naturally occurring defects tend to make it conducting (n-type) without affecting its solar transparency. CdS may also be made more n-type or p-type by adding impurities[16].

A second ideality is to have the underlying absorber absorb the solar spectrum as completely as possible in a region very close to the junction. If our sun was a monochromatic laser aimed at planet Earth, then an absorber with a direct band gap equal to or slightly greater than the energy of the solar line would absorb all of the sun's energy by creating electron-hole pairs. The solar cell efficiency could be very high, indeed. However, the sun has a spectrum which roughly approximates a 6000°K blackbody radiator (Figure 3.2).

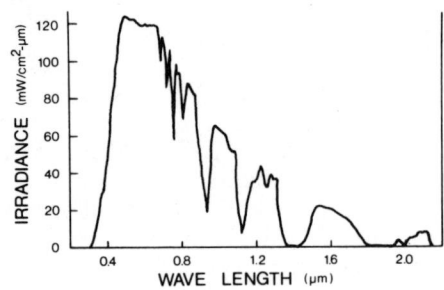

Figure 3.2. Variation of Spectral Irradiater with Wavelength for Air Mass 2.

SOLAR ENERGY MATERIALS

This means that only a very small portion of the solar spectrum matches the absorber's band gap. The portion of the solar spectrum at lower energies is transmitted and the portion at higher energies excites electron-hole pairs but gives these carriers additional translational energy which is lost to the lattice as heat. Therefore, in the real case, much of the solar spectrum cannot be used to create electron-hole pairs.

Once the electron-hole pairs are created, there is no guarantee that the electrons will escape into the n-type semiconductor and the hole to the p-type semiconductor even though these are the lowest energy locations. Crystallographic defects at the junction will be sites for electron hole recombination.

In summary, the ideal heterojunction photovoltaic cell would have the following characteristics:

1. A window with good solar transparency, meaning that its gap is ≥ 2 ev.

2. The window has low resistance, to minimize series resistance losses.

3. The absorber has a very high absorptance, to insure electron hole pair generation near the junction.

4. The absorber is thin to minimize series resistance losses.

5. The window/absorber junction contains few crystallographic defects, in order to minimize electron-hole recombination at the junction.

As indicated previously, CdS is an acceptable window material and can be made to satisfy the first two requirements. CdS is usually used with Cu_2S where the Cu_2S is made by depositing Cu on the top of a CdS layer and thermally diffusing the Cu into the CdS surface to form a thin layer of Cu_2S. If the CdS is deposited onto a transparent substrate and transparent electrode material, it is possible to illuminate the cell from the CdS window side. This type of illumination is called 'backwall illumination' and is shown in Figure 3.3. If the cell is to be used in the backwall configuration, then the top electrode can be a continuous film of conductor. Such a cell could be made by spraying various solutions onto the surface of hot ($\sim 500°C$) glass to deposit all four layers. Such a large area solar cell could be very economic.

The efficiency of CdS/Cu_2S cells can be enhanced by increasing the area of the p/n junction. This may be accomplished by etching the CdS surface as shown in Figure 3.4 to provide a very irregular surface. Then Cu is diffused into the enlarged surface area and an irregular Cu_2S layer results.

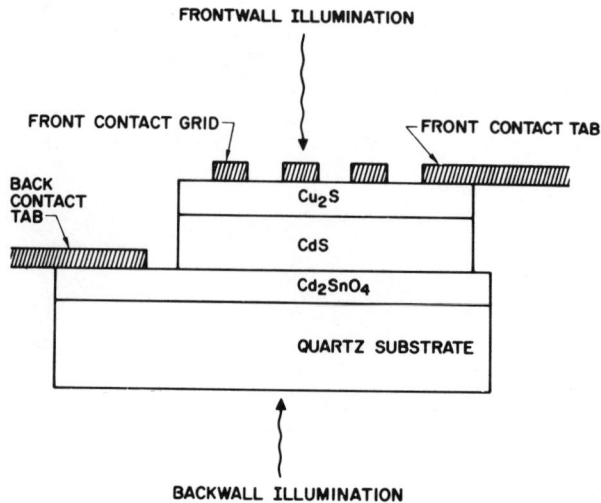

Figure 3.3. Schematic of CdS/Cu_2O Solar Cell for both Back-Wall and Front Wall Configuration.

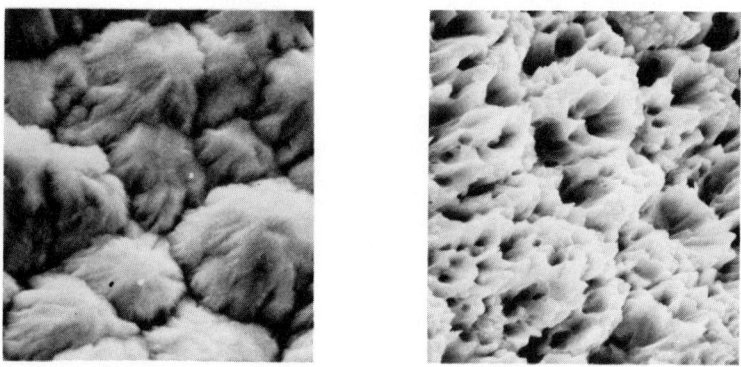

Figure 3.4. CdS thin film on Left, after Etching to Enlarge Surface area on Right.

A number of semiconductors have band gaps in the range of 1.3-1.6 ev. Figure 3.5 shows the theoretical efficiency of homojunction cells as a function of E_g. It is seen that CdTe, GaAs, etc. have the proper band gaps. The thinness of the absorber can be achieved if the band gap is direct so that the absorption rises abruptly with increasing energy. Semiconductors like CdTe and GaAs reasonably satisfy the third and fourth requirements.

The fifth requirement presents a problem. When two materials are put together, a grain boundary results. Here the atoms are

SOLAR ENERGY MATERIALS

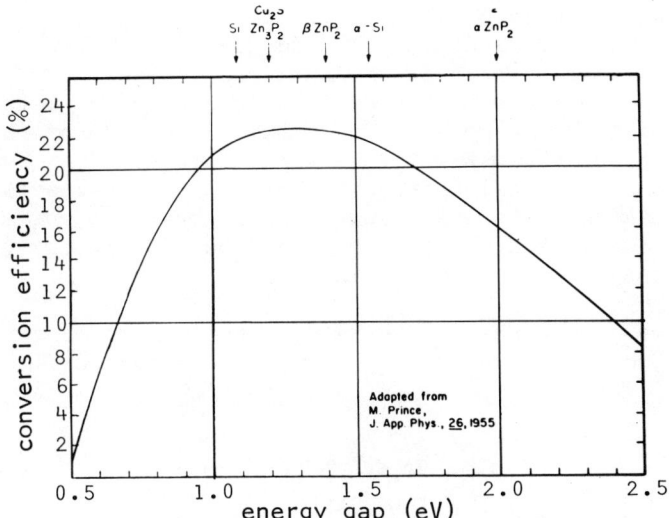

Figure 3.5. Theoretical Efficiency of Homojunction Solar cells, as a function of Semiconductor Band Gap.

in some disarray and the associated defects serve as recombination centers. Then, the formation of electron-hole pairs does not result in external electrical energy. One way to try to eliminate these defects is to match the lattice parameters of the n- and p-type semiconductors.

The chalcopyrites (17-21) are semiconductors which offer the possibility of a lattice match with CdS or other semiconductors. They are II-IV-V compounds, analogs of III-V compounds. Table II

Table II

Data for Several II-IV-V_2 Compounds

Compound	Type	E_g(eV)	A(Å)	x(eV)
$ZnSnP_2$	P	1.66	5.651	4.25
$CdSnP_2$	P,N	1.17	5.900	4.41
$CdGeP_2$	P,N	1.72	5.741	3.98
$CdGeAs_2$	P,N	0.55	5.943	4.60
$CdSiAs_2$	P	1.55	5.884	3.85
CdS	N	2.42	5.851	4.50

E_g = energy gap
A = lattice parameter
x = electron affinity

presents data for some chalcopyrites that we have prepared. Data for CdS is also included. The optimum band gap for an absorber is 1.3 to 1.6 ev, and it can be seen that a number of these phosphide and arsenide chalcopyrites have band gaps in this range. The lattice parameter that is listed here is the length of the side of the cubic unit cell. The lattice parameter listed for CdS is $\sqrt{2}$ times the atomic spacing in its hexagonal unit cell. In both cases, the listed lattice parameters divided by $\sqrt{2}$ represents the interatomic spacings of the hexagonal close-packed layers. The electron affinities (χ) of the chosen chalcopyrites are to be discussed presently.

The dependence of the lattice parameters of various chalcopyrite solid solutions on composition are shown in Figure 3.6. The

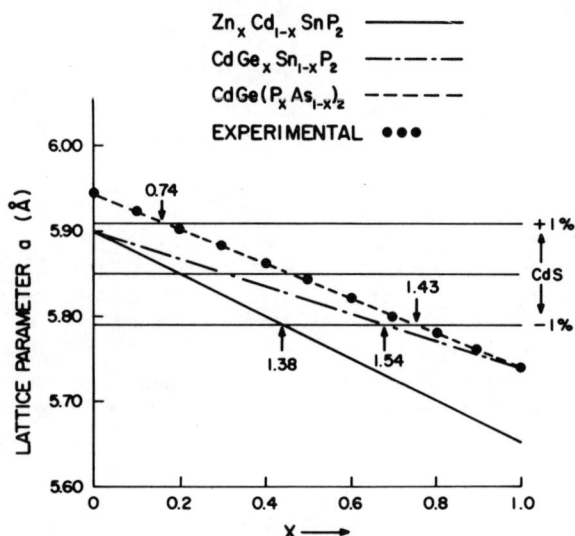

Figure 3.6. Lattice Parameter vs. Composition in chosen Chalcopyrite Solid Solutions.

three horizontal lines here indicate the lattice parameter of CdS and the ±1% deviation from the lattice match condition. Each of the solid solutions represented here can form a precise lattice match with cadmium sulfide. The lattice parameter data for $CdGe(P_xAs_{1-x})$, determined by Mikkelsen[22], indicates a linear relation between lattice parameter and composition. The same linear dependence is assumed for the other two chalcopyrite systems. Both are being studied, but the $CdGe_xSn_{1-x}P_2$ is the focus of this paper. Also, $CdSiAs_2$ has a lattice parameter close (within 1%) to that of CdS, and is being studied here.

A hypothetical band diagram for the $CdSiAs_2$-CdS p-n junction is presented in Figure 3.7. The 0.65 ev step is the difference between the electron affinities of the two materials. Because this

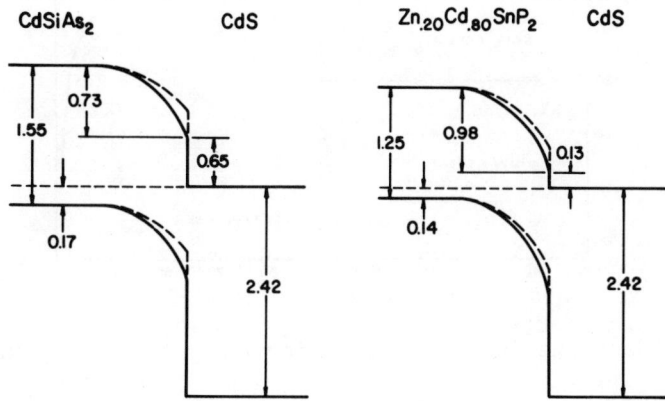

Figure 3.7. Postulated Band Diagram for chosen p-n Junction.

difference is less than the band gap of the p-type material, it is expected that there will not be a spike in the conduction band at the junction. The accuracy of the electron affinity values is not great, but if the step is considerably less than the p-type band gap, as shown, then there will be an internal field in the p-type material. This field is important because it will ensure separating of electrons and holes when the solar photons are absorbed. These considerations make the $CdSiAs_2$ and the $CdGe_xSn_{1-x}P_2$ chalcopyrites appear to be important candidates as absorbers in CdS solar cells.

$CdSiAs_2$ has been synthesized here by the direct fusion technique. There is no phase diagram for this compound but there is a phase diagram for the CdP_2-Sn system containing $CdSnP_2$. This compound is similar to CdSiAs and the CdP_2-Sn, Figure 3.8, was used for guidance in the preparation of $CdSiAs_2$.

The $CdSiAs_2$ composition was sealed into an evacuated thick (2 mm) walled silica ampoule and heated to 950°C. The melt was held at the temperature for 20 hours and quenched to room temperature. It was reheated to 840°C and held for 68 hours. This resulted in homogeneous $CdSiAs_2$ with some porosity but no other crystalline phases. The porosity (Figure 3.9) was caused by the dissolution of a small amount of second phase material which appeared during cooling the melt through the two-phase region. The element diffused into the $CdSiAs_2$ phase during the prolonged heat treatment. No second phase or deviation from stoichiometry was found

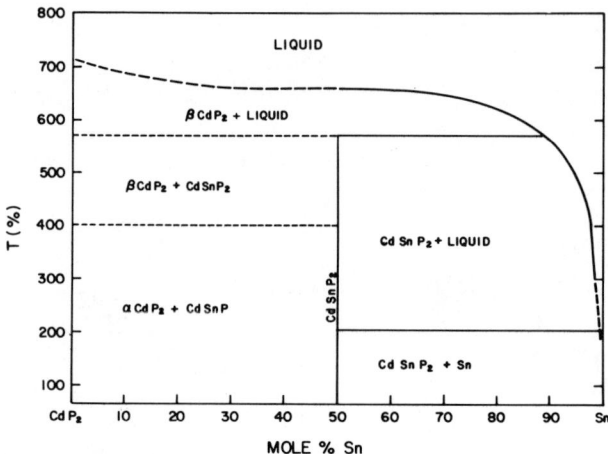

Figure 3.8. The CdP$_2$-Sn Phase Diagram (After Buehler, et al[24]

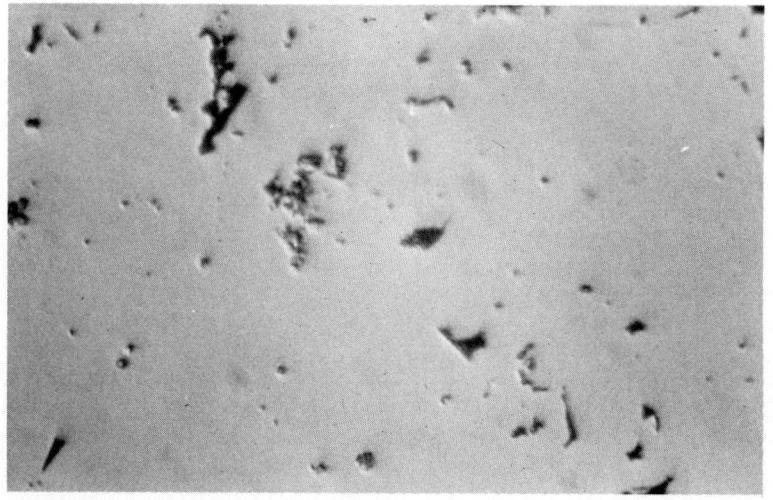

Figure 3.9. CdSiAs$_2$ Microstructure (Length of Micrograph - 200 μ).

using the quantitative microprobe. Also, only the CdSiAs$_2$ chalcopyrite phase was present according to X-ray diffraction.

CdGeP$_2$ was also synthesized by direct fusion. The phase diagram containing this composition is shown in Figure 3.10, and predicts that CdGeP$_2$ is consequently melting. The melt was held at 825°C for 25 hours, and then cooled at 5°C per hour to 750°C. Here

SOLAR ENERGY MATERIALS   377

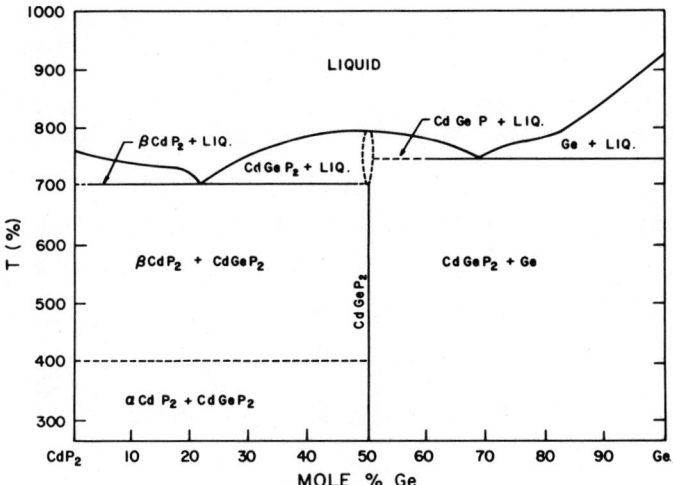

Figure 3.10. The CdP_2-Ge Phase Diagram (After Buehler et al[24]).

it was held at 6 hours before cooling to room temperature by removing from the furnace. Although the resulting $CdGeP_2$ contains porosity, Figure 3.11, it is almost completely single phase. The very minor second phase was identified by the scanning electron microprobe to be CdP_2. A higher magnification microphotograph, Figure 3.11, shows the porosity congregating along grain boundaries. The lighter phase is CdP_2, as determined by the microprobe.

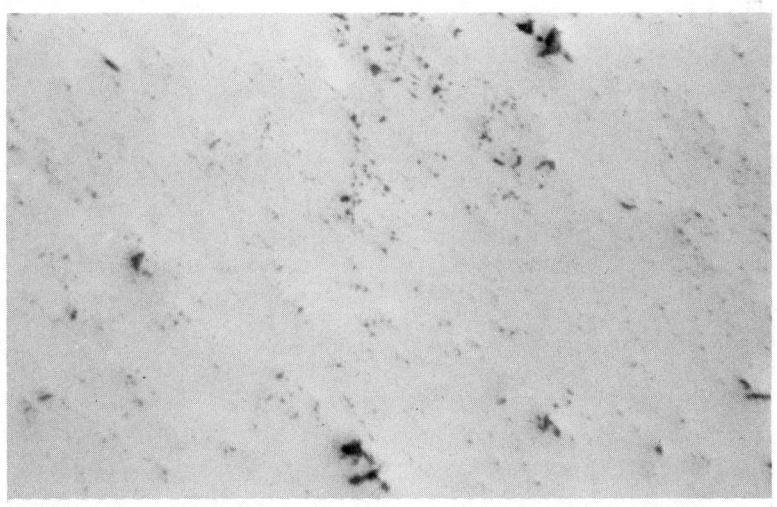

Figure 3.11. $CdGeP_2$ Microstructure (Length of Micrograph = 200 μ).

The X-ray diffraction pattern of $CdGeP_2$ is presented at the top of Figure 3.12. The unlabeled peaks belong to CdP_2 which heretofore had not been reported. It is anticipated that a more rapid cooling rate would preclude the formation of this second phase.

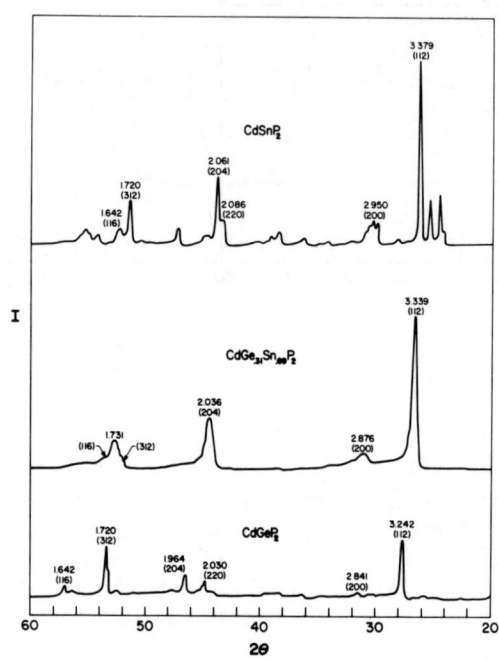

Figure 3.12. X-ray diffraction Pattern of $CdSnP$, $CdGe_{.31}Sn_{.69}P_2$ and $CdGeP_2$.

$CdSnP_2$ has also been prepared by direct fusion. The phase diagram containing this chalcopyrite is shown in Figure 3.8, and predicts that the $CdSnP_2$ liquidus is at $\sim 670°C$. Therefore, this composition was melted at 800°C in a fused silica ampoule and quenched. Eventhough quenching was rapid, Sn and CdP_2 appeared as minor phases Figure 3.13. A subsequent heat treatment at 550°C converted the sample to single phase $CdSnP_2$. The sample was porous, the pores representing the volume previously occupied by the tin crystals.

$CdGe_{.31}Sn_{.69}P_2$ was heated to 805°C in an ampoule and held for three hours before cooling to 790°C at 5°/hr., where it was held for 18 hours.

The X-ray diffraction patterns of $CdSnP_2$ and $CdGe_{.31}Sn_{.69}P$

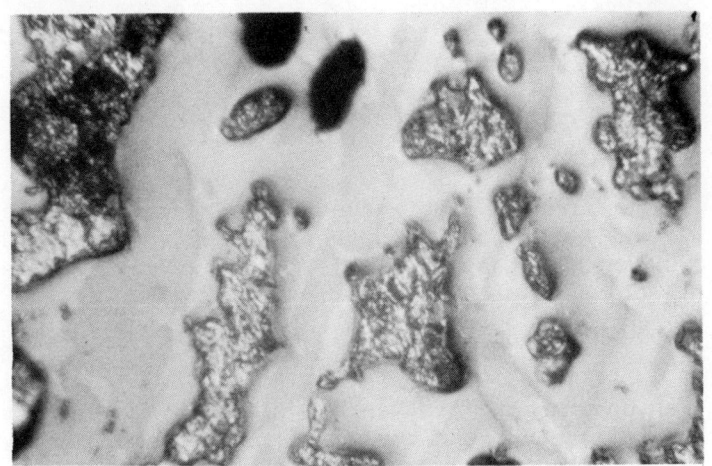

Figure 3.13. CdSnP$_2$ Microstructure, containing Sn (Islands) and CdP$_2$ (surrounding Islands)(Length of Micrograph = 200 μ).

are also shown in Figure 3.12. The CdGe$_{.31}$Sn$_{.69}$P$_2$ has the chalcopyrite structure, as do its end members, CdGeP$_2$ and CdSnP$_2$ with CdS[23] have been prepared and evaluated. Both show the photovoltaic effect but the research is too preliminary to permit conclusions about the benefit of lattice matching.

In conclusion, CdSiAs$_2$ and CdGe$_{.31}$Sn$_{.69}$P$_2$ are relatively easy to prepare in single phase form, and these chalcopyrites have close lattice matches with CdS. Research on the synthesis of these compounds and evaluation of their junction (with CdS) characteristics is continuing.

Acknowledgements

The contributions of graduate students, W.L.Newell and T.R.Vinerito, and the financial support of ASG Industries, made the discovery mentioned in Section 2 possible.

The contributions of Mr.A.F.Carroll, Graduate Research Assistant and Dr.L.C.Burton, Associate Professor of Electrical Engineering in Section 3 are gratefully acknowledged.

References

(1) J.F.Kreider and F.Kreith, Solar Heating and Cooling, Scripta Book Co., Hemisphere Publishing Corp., Washington, and McGraw-Hill Book Co., New York (1975).

(2) J.A.Buffie and W.A.Beckman, Solar Energy Thermal Processes, Wiley (1974).

(3) P.M.Driver, R.W.Jones, C.L.Riddiford and R.J.Simpson, New Chrome Black Selective Absorbing Surface, Sol. Energy, 19, (3) 301-306 (1977).

(4) L.Melamed and G.M.Kaplan, "Survey of Selective Absorber Coating for Solar Energy Technology," J.Energy Dev., 1 (2) 100-107 (1977).

(5) D.P.Grimmer, K.C.Herr and W.J.McCreary, Possible Selective Solar Photothermal Absorber: Ni Dendrites formed on Al Surfaces by the CVD of $Ni(CO)_4$," J.Vac. Sci. Technol., 15 (1) 59-64 (1978).

(6) K.D.Manterson, "Selective Surfaces for Solar-Thermal Conversion," J.Solid State Chem., 22 (1) 41-49 (1977).

(7) R.D.Goodman and A.G.Menke, "Effect of Cover Plate Treatmen on Efficiency of Solar Collectors," Solar Energy, 17, 207-211 (1975).

(8) Proceedings of the Workshop on Solar Energy Storage Subsystems for the Heating and Cooling of Buildings, L.U. Lillelht, J.T.Beard and F.A.Fachetta, Eds., Charlottesvill VA, April 16-18,(1975).

(9) G.Wettermark and J.Kowalewska, "Storage of Low Temperature Heat", Swedish Council for Building Research, Royal Institute of Technology, S-100 44, Stockholm 70, Sweden (1976).

(10) E.J.Lehmann, Optical Coatings for Solar Cells and Solar Collectors (A Bibliography with abstracts) National Technical Information Service, 5285 Port Royal Road, Springfield, Virginia, 22161, USA (1975).

(11) Sharing the Sun, Solar Technology in the Seventies, Conference Proceedings, Published by the American Section of the International Solar Energy Society, 300 State Road 401, Cape Canaveral, Florida 32920, USA (1976).

(12) B.Anderson, *Solar Energy in Building Design*, Arthur D. Little Press (1975) (Available through Total Environmental Action, Church Hill, Harrisville, NH 03450, USA).

(13) T.R.Viverito, E.W.Rilee and L.H.Slack, "Oxide Film Deposition by an Improved Pyrolitic Decomposition Process", Am. Cer. Soc. Bull., 54 (2) 217-8 (1975).

(14) G.Haacke, H.Ando and W.E.Mealmaker, "Spray Depsotion of Cadmium Stannate Films", J.Electrochem. Soc., 124 (12) 1923-26 (1977).

(15) D.L.Evans and G.R.Fisher, "X-Ray Determination of Film Thickness Using the Glassy Halo", Bull. Am. Ceram. Soc., 52 (6) (1973).

(16) Y.Y.Ma, A.L.Fahrenbruck and R.H.Bube, "Photovoltaic Pro-

perties of n-CdS/p-CdTe Heterojunctions Prepared by Spray Pyrolysis", Appl. Phys. Lett., 30, (8) 493-4 (1976).

(17) J.L.Shay and J.H.Wernnick, Ternary Chalcopyrite Semiconductors: Growth, Electronic Properties and Applications, Pergamon Press (1975).

(18) J.L.Shay, S.Wagner and J.C.Phillips, Appl. Phys. Lett., 28, 31 (1976).

(19) A.S.Borsheckevskii, I.A.Mal'tseva, Yu V. Rud' and Yu.K. Undalov, Sov. Phys. Semicond., 10, 655 (1976).

(20) G.A.Medvedkin, K.Ovezov, Yu. V. Rud' and V.I.Sokolova, Sov. Phys. Semicond., 10, 1239 (1976).

(21) K.J.Bachmann, E.Buehler, J.L.Shay and S.Wagner, Zeitschrift für Physikalische Chemie Neue Folge, 98, 365 (1975).

(22) J.C.Mikkelson, Jr. and H.Kildal, Jour. Appl. Phys., 49 426 (1978).

(23) A.G.Milnes and D.L.Feucht, Heterojunctions and Metal-Semiconductor Junctions, Academic Press (1972).

(24) E.Buehler and J.H.Wernick, The CdP_2-Ge System and the Growth of Crystals of $CdGeP_2$, NBS Special Pub. 364, Solid State Chemistry Proceeding of 5th Materials Research Symposium, July (1972).

PART III

COMPUTATIONAL METHODS IN PHYSICS

COMPUTATIONAL METHODS IN PHYSICS

K.V.Roberts

Culham Laboratory, Abingdon,

Oxon, OX14 3DB, England

1.   POWER AND LIMITATIONS OF COMPUTERS

1.1   Introduction

This article discusses some aspects of the application of computers in Physics. Clearly the subject is nowadays very wide: computers are used by physicists in a great variety of ways and the number of different types of application is continually increasing. The article therefore begins by establishing some general principles and then progressively narrows down to one specific field of application, namely the solution of classical partial differential equations in hydrodynamics and plasma physics which is the author's main field of interest. The last section deals with practical programming techniques and considers the scope for international collaboration. It is hoped that in this way both the wood as a whole and some of the more significant and interesting trees will be adequately displayed.

Three good general references that may be mentioned at the outset are 'Computers and their Role in the Physical Sciences'[1], 'Computers as a language of Physics'[2] and 'The Impact of Computers on Physics'[3]. Leading journals are the annual review series 'Methods in Computational Physics'[4] and the periodicals 'Journal of Computational Physics'[5] and 'Computer Physics Communications'[6]. An excellent introductory text on numerical calculations in physics is the book 'Computational Physics by D.E.Potter'[7].

1.2   Relevance of Computers to Physics

Computers are <u>machines for processing information by means of algorithms</u>. Physics like any other science is largely concerned with acquiring, interpreting, storing and disseminating information, and computers will be of use wherever the necessary algorithms exist and the available practical techniques have reached the stage at which they are sufficiently convenient and reliable to use.

Table 1 is a list, probably not complete, of the current applications of computers in a major scientific laboratory such as the UKAEA Culham Laboratory for Plasma Physics and Nuclear Fusion

Table 1

Scientific Applications of Computers

Design of apparatus, (physics, engineering)   Experimental data processing   (acquisition, storage, retrieval, analysis, display)   Control of experiments
Small-scale numerical calculations   Large-scale numerical simulation   Modelling mathematical or scientific problems to gain   intuition   Very accurate or very complex calculations
On-line analytic calculations   Large-scale algebraic manipulation   Interactive graphics
Production of scientific reports and journals, (word-   processing, type-setting)   Production of graphs, slides, microfiche, movies   Information storage and retrieval
Communication between scientists and between laboratories   Communication with remote apparatus
Teaching, formula evaluation, browsing
Management data processing   Project control

Research. The area that will mainly be covered in these lectures is starred, but this is not necessarily the most significant application of computers to thermonuclear research at the present time. With a large scientific experiment such as JET (Joint European Torus), estimated to cost about £120M over the 5-year construction period 1978-1982 and with an operating cost of £25M/year from 1983-1990, it is evident that engineering design calculations, management data processing and project control must be of great importance. Reliability of the data-processing and control computers is also essential since the cost of the project averaged over its working life is about £100K/day, and the cost of any lost time may easily exceed the cost of the data-processing equipment that is at fault. Similar considerations apply to the other major experimental thermonuclear devices being constructed in the USA, USSR and Japan, to the large accelerators used in high-energy physics, and to the equipment used in space research.

Once we are confident about the physical conditions inside the plasma and the equations and numerical techniques to be used, then accurate plasma physics and neutronics simulations will be important in the design and optimization of full-scale fusion reactors, a stage which has already been reached for fission systems. Complex plasma physics simulations are of course being carried out at the present time, but they must to some extent be tentative because the phenomenological transport coefficients, wall interactions, impurity levels and so on which must be included in the codes have not yet been adequately measured by experiment.

All the applications listed in Table 1 are expanding steadily as the available equipment becomes cheaper, more reliable and more versatile. For example, a scientific report cannot conveniently be typed on a card punch which has a restricted upper-case character set, few mathematical symbols, a line width of only 80 columns and only limited facilities for tables and graphics. Punched cards were never designed for this purpose and in fact were not intended for use with electronic computers at all, being an invention of the end of the 19th century which was based on even earlier technology such as that of the Jacquard loom. When electronic computers were introduced in the late 1940's and early 1950's they simply used the mechanical and electromechanical input-output equipment that was already to hand, such as punched-card and paper-tape equipment, Flexowriters, line printers, teletypes and so on. This had a constraining effect not only on the applications of computers but also on programming languages themselves, the restricted character set of Fortran being based on IBM punched-card practice and that of Algol 60 on the Flexowriter.

Now that on-line terminals, displays such as the TEKTRONIX 4014 and printer-plotters such as the VERSATEC are available with a more extended character set including lower case, characters of several sizes, special mathematical symbols, a full line width and

the ability to include graphics within the text, one must expect that
scientific reports and papers will increasingly be typed and edited
on-line by the author or secretary using word-processing techniques,
stored in digital form so giving immediate access by information-
retrieval systems, and transmitted from one laboratory to another
via international communication links. This will inevitably have a
substantial influence on the way in which scientific information is
exchanged over the next few decades, although not necessarily for
the better since printed books and journals are pleasant and convenient to handle. There is also the important question of permanence
and reliability. Information printed on paper in multiple copies
which are then widely distributed through the normal publication
mechanism represents a very safe form of storage, and such information is almost impossible to destroy completely or to corrupt. By
contrast, information which is stored in one location in digital
form on magnetic or semi-conducting media is highly volatile and
elaborate precautions must be taken to protect it. Such precautions
are of course being developed by banks, insurance companies and large
international firms and can no doubt be taken over by scientists in
due course.

The introduction of international information banks represents a challenge to developing countries and to bodies such as
UNESCO. Provided that the necessary satellite links, efficient internal telephone services and local peripheral equipment can be provided,
then a wide variety of commercial and scientific information which
will (and to a large extent already does) exist in countries such as
the USA will automatically become available world-wide. Distance
limitations and the need to purchase large numbers of expensive books
and journals will be avoided. But without such ancillary equipment
much of the world commercial and scientific data traffic will become
inaccessible.

Evidently computers are having and will continue to have
a major impact on the way in which physicists conduct their day-to-
day business, but one may also ask whether they will have a <u>fundamen-
tal</u> impact on physics research in the way that new theoretical and
experimental techniques frequently do; for example the discovery of
Schrödinger's equation, the invention of the telescope and micro-
scope, or the more recent widening of the electromagnetic spectrum
available to astronomers to include the range from γ-rays to radio
waves. Such an impact has hardly happened yet, and perhaps it never
will. The point is that computers require <u>algorithms</u> to drive them,
and it is usually much easier to do something than to devise an algorithm for doing it. Einstein could develop the theory of relativity,
but it is hard to imagine anyone devising an algorithm for generating such theories mechanically. Until this happens computers are
bound to play a subsidiary role in physics but one which may nevertheless be of great importance; for example they can carry out large
amounts of routine analysis or algebra quickly and have helped in

the recent proof of the 4-colour theorem[8]; they can assist in the design of new diagnostic equipment and the reduction and display of experimental results on a massive scale which is totally impracticable for humans. They can also provide intelligence in environments which humans cannot reach.

1.3   Digital Information

Computers are mainly concerned with <u>digital</u> information. An advantage is that once any necessary analogue-to-digital conversion has been made (which will involve some degree of uncertainty or approximation) the remaining processes can be determined to any degree of reliability required. For example a digital document can be handed down from generation to generation in the presence of random 'noise' with a chance of error 2^{-N} where N (the number of protection bits) can be made arbitrarily large. A telephone call sent by digital transmission can be equally uncorrupted however far the distance, and the use of digital telephony may be expected to increase. Another advantage is that digital information is readily converted into analogue form but not necessarily vice-versa; it is quite easy to make a computer talk but much more difficult to teach it to understand the human voice although advances have been made in this direction. Therefore there is a progressive tendency to store, process and transmit all information in a 'universal' digital form, and to use other information formats only for transitory display to human beings.

Of course analogue storage and processing techniques do have advantages and can sometimes be simpler and faster for example the holographic reconstruction of a 3D image by optical Fourier transform methods[9].

1.4   Advantages and Disadvantages of Computers

The advantages of computers are well known (Table 2). Performance factors are increasing all the time and are concerned with the relatively sophisticated state of the hardware, which with the advent of microprocessors is month-by-month becoming physically smaller, faster, cheaper, more reliable and easier to assemble.

The disadvantages are severe and stem from the fact that one starts from the 'primitive' state in which not only do very few algorithms exist, but there is also no fully adequate language in which to express them, either for discussion between people or for presentation to the computer[10]. A comparison with standard mathematical formalism emphasizes the fact already mentioned in sub-section 1.2 that compared to mathematics, existing programming languages have an artificially limited character set (mainly upper-case letters and a few special symbols) as well as an awkward syntax. Analysis

Table 2

Advantages and Disadvantages of Computers

ADVANTAGES

    Speed

    Low cost

    Ability to store and process large quantities of data

    Can accept and display data in any chosen form

    Links to communication channels

    Reliability (potential)

DISADVANTAGES

    Limited character set

    Inadequate programming languages

    Lack of algorithms and software

    Time required to write programs

    Undue complexity

    Preoccupation with inessential details

    Incompatibility between systems

    General inconvenience

    Unreliability (actual)

of the vector formula $\underline{V} \times \underline{V} \times \underline{B}$ carried out some years ago by the author and his colleagues showed that it required 100 times as many symbols to express in a 3D Fortran difference scheme as in the original mathematical notation[11]. A recent 'Introduction to Programming and Problem Solving'[12] recommends that algorithms should initially be worked out in a form of English rather than in any of the current programming languages. These disadvantages are concerned with the relatively unsophisticated state of the software which is developing quite slowly compared to the hardware; the main languages Fortran[13], Algol 60 and Cobol have hardly altered since their intorduction some 20 years ago and have not yet been successfully superseded, although an updated version of Fortran[14] with some useful extensions has now been agreed.

       The introduction of a new algorithmic or programming language involves subtle difficulties. The requirements are partly intellectual but practical simplicity and convenience are as important as theoretical power and elegance, and although formalized languages such as Algol 60, Algol 68 and Pascal[12] have been popular with the Computer Science departments of the universities it is the more loosely-defined Fortran and Basic that have caught on with the physicists and engineers. I do not think that the physicists and engineers are wrong in this. Simple problems should be handled by simple methods and one does not want a massive superstructure of obviously redundant declarations and other formalism for every small computing job. My own preference would be to pursue a course similar to that which used to be adopted in the teaching of school mathematics; first arithmetic, then algebra and plane Euclidean geometry, followed by elementary calculus and coordinate geometry and finally the more advanced branches of the subject. In teaching computing one would accordingly start with Basic at late primary or early secondary school level as soon as algebra was available, taking advantage of the cheap home computers that will shortly become as pervasive as pocket calculators are now. Basic can be used for a wide variety of small calculations of the scientific engineering and business type, and with minor extensions can be used to control experimental equipment, to analyse data and to drive graphical display units and plotters. It would therefore remain useful throughout most of the school and much of the university syllabus. It does however become awkward when programs exceed two or three pages in length, and one would use this fact to demonstrate the advantages of a more powerful language such as Pascal by analysing a number of problems solved by both methods, in much the same way that the more advanced mathematical techniques begin to show their advantages once problems reach a certain stage of complexity.

Fortran falls between the two extremes of an elementary language such as Basic and a more advanced language such as Pascal and no doubt will be discarded eventually even by physicists, but this will not happen until there is a rival language which is at least as universal, well-implemented and practically convenient as Fortran is at present. For example, Algol 60 failed to penetrate for practical rather than theoretical reasons; it was not well-adapted to the card punches and line printers that were used at the time, the hardware representation was not standardized and was generally inconvenient, there was no standard input-output system, complex numbers were not part of the language, and there was no adequate way of dividing the very large programs used by physicists into segments which could be independently compiled.

My own feeling is that it would be wrong to standardize on a new scientific language until a wider character set is generally adopted. Mathematics has made great use of the various Latin and Greek fonts that became available when printing was first invented and programming languages ought to be given a similar opportunity. Even the obvious use of lower-case characters for comments can make a big difference to the readability of a code (Section 5).

1.5   Complexity

Amongst the disadvantages of computers for the physicist are their undue complexity, apparent preoccupation with inessentials, mutual incompatability and general inconvenience for the user. This is a growing trend: where once a simple Fortran job could be put straight on to the machine as a deck of cards with no preliminaries one then had control cards, followed by an untidy Job Control Language (JCL) with obscure catalogue procedures, and now an elaborate System Control Language (SCL) with a syntax of its own. At the same time there is a proliferation of incompatible word lengths, alternative character codes such as EBCDIC and ASCII, and different magnetic tape formats, hardware interfaces, modems, communication protocols, local subroutine and procedure libraries, editing and file manipulation facilities and so on. Mathematics, physics, numerical analysis, computer science and the various high-level programming languages are almost universal: they can be taught as academic disciplines and mastered once-for-all, but this proliferating undergrowth of the computing world is badly documented, ever changing and varies from one computer system to the next; it distracts attention from the real problems of physics.

Probably most computer professionals and users do not realise how simple and straight-forward the use of computers can actuall be made by the application of general principles which are commonplace in other fields. For example, TV sets, automatic telephones and automobiles are highly complex mechanisms but they are carefully designed so that any member of the public can use them. Computers ought to be the same and there are enough examples to show that the desired result can be achieved, especially in areas where the competition is very intense, for example pocket calculators and commercial on-line computing systems which would lose business if they were made too difficult to use. Some of the ways in which computing can be simplified will be discussed in Section 5.

1.6   Reliability

The question of reliability deserves serious consideration and quantitative planning (Table 3). Computers are at present notoriously unreliable as every newspaper reader is aware, and their performance is not optimized. A reliability or availability of 96% for a central laboratory computer system is regarded as good or even high, but such a level would not be tolerated in other services such as electricity, water, gas or telephones and certainly not for the library. It is not optimized because the costs of the people or equipment using the computer (about £12.50/man-hour at Culham and £100K/day for JET) are greater than those of the computer itself, and their work may be seriously disrupted when the computer system fails.

Table 3

Reliability of Computers

FACTORS FAVOURING UNRELIABILITY

    Hardware complexity

    Software complexity

    Inadequate documentation

    Communication links

    Different manufacturers' equipment

    Time taken to locate and correct errors

    New errors introduced by changes

    Transitory nature of data storage media

    Concentration of data in one place

    Lack of data protection

    Serialism

FACTORS FAVOURING RELIABILITY

    Parallelism

    Automatic error-detecting and correcting codes

    Self or mutual checking and reporting

    Dispersed multiple copies of data

    Built-in checks (e.g. energy conservation)

    Redundancy

    Protection mechanisms

    Importance of many vital applications

It is actually more scientific to talk about the <u>unreliability</u> or unavailability of a computer system than its reliability or availability. This is because a great number of processes have to work simultaneously so that their unreliabilities are (very nearly) additive, like resistances in a series electrical circuit. A reliability of 96% is equivalent to an unreliability of 4%, and is only 1/4 as good as a reliability of 99% which corresponds to an unreliability of 1%.

If n subsystems with low unreliability x are placed in <u>series,</u> (i.e. if they all have to work together), then the overall unreliability of the complete system is increased to

$$u_s = 1 - (1 - x)^n \simeq n x \qquad (1)$$

This illustrates the dangers inherent in the enthusiastic development of complex systems involving local microcomputers, data links, switching equipment, front-end minicomputers and a large central computer; such a system is certain to be less reliable than a single computer unless careful precautions are taken. Unfortunately a quantitative reliability analysis is very often not made when such systems are proposed and installed, a procedure which is equivalent to an electrical engineer failing to take into account the resistance and other characteristics of his transmission lines. The problems are compounde when equipment supplied by different manufacturers is coupled togethe so that each can blame the others for any trouble that occurs, and when it is difficult to tell whether the hardware or the software is at fault. Since faults will undoubtedly occur, complex systems requir error detection equipment and correction procedures to be built in at the design stage, so that troubles can be detected, diagnosed and corrected quickly and in a positive way.

Von Neumann[15] was one of the first to point out that a system with arbitrarily high reliability can be built from basically unreliable components by proper design procecures. The fundamental idea is to use parallelism: if two devices with unreliability $x \ll 1$ are in 'series' so that both must work together then the total unreliability is nearly doubled as Eqn.(1) shows, but if they are in parallel so that only one of the two has to work then the unreliability is decreased to x^2, and in general for n devices in parallel

$$u_p = x^n \qquad (2)$$

By this method any degree of hardware reliability may be obtained, at a cost proportional to

$$n = \log u_p / \log x \qquad (3)$$

which increases only logarithmically with the performance required and with the unreliability of the individual device.

It is evident that many complex computer systems need to operate with a high degree of reliability, for example those used by banks, insurance companies large department stores, airline booking, air, sea and ground traffic control, telephone networks, space research and the military. The fact that such systems can only rarely be allowed to fail guarantees that the necessary reliability theory and techniques will be developed; physicists should then try to make sure that these improvements are then incorporated into their own laboratory computer systems which are of course less vital.

Some of the snags may not be obvious. For example if a computer which is operating a single batch stream breaks down, then probably only the job that is currently in progress will fail, and in many cases the operators will be able to rerun this so that although there may be some delay the users need not know. The effective down time is however increased by half the average length of job and this ought to be taken into account in reporting the performance of the system; moreover it is clear that several short breaks are worse than one long one because they cause a greater loss of time.

When there is a failure in a complex time-sharing and multiprogramming system which has many-on-line users and several concurrent batch streams, then the activities of all the users will be disrupted and some of their work may be lost together with all the batch jobs which are currently in progress. Therefore the loss of time

$$t_{lost} = t_{down} + \frac{1}{2} f \bar{t}_{job} \qquad (4)$$

is increased by a <u>concurrency factor</u> f which is the average number of jobs resident in the system, t_{down} being the time for which the machine is actually down and $\bar{t}_{job}$ the average length of the individual lost jobs.

Furthermore unless the users are notified promptly as soon as the machine goes down they will waste time in sitting at inactive consoles and by fruitlessly trying to log in again or by telephoning the operators for information, and unless they are notified as soon as service has been restored part of the available power of the system will be lost because the number of on-line users will increase only slowly to its full level. One obvious way to avoid such inefficiencies is to have a signal lamp on each console which is automatically illuminated whenever the system is available for use.

Another important requirement for increasing the reliability of a complex multiprogramming system is the automatic checkpointing of batch jobs and on-line user interactions, so that a dump of the current state of each item of work is taken at frequent intervals and copied to permanent storage. Then the loss of time due to a break is decreased since $\bar{t}_{job}$ in (4) is replaced by the mean time $\bar{t}_{dump}$ between dumps.

Rather little work has been published so far on the reliability of complex multiprogramming systems such as those used in physics laboratories and universities, and this would appear to be a fruitful topic for research.

Clearly as computers are used more and more widely a discipline which might be called <u>reliability engineering</u> will need to develop and perhaps is already doing so. The physicist will need to take account of this. If he jots down something in pencil in a notebook it will be there 100 years later, but if he writes it to magnetic tape it may get lost, or overwritten, or the tape damaged, or the character code forgotten in a much shorter space of time.

1.7   Paradoxes

As already remarked in 1.3, digital techniques exist by which information can be handed down from generation to generation virtually uncorrupted, i.e. with a chance of undetected error 1 in 2^N where N is the number of protection bits. The trick is to associate with the character string to be preserved an <u>authentication code</u> ($0 \leq A \leq 2^{N-1}$), in such a way that any change in the character string, however small, produces a <u>random</u> change in the number A. If the final user is given the character string, the authentication code A and the algorithm by which it was produced, then he can readily recalculate A for himself to check whether any corruption has taken place. The number of protection bits required is quite small and increases only logarithmically with the degree of protection required for example 30 bits provide a chance of undetected error approximately 10^{-9}, 60 bits 10^{-18} and so on.

Techniques of this kind are used in the transmission of digital information via noisy communication lines but they can also be used for historical transmission, including manual copying from a printed record. If the ancients had used such techniques then scribes could have copied documents endlessly and negligible textual corruption would have occurred. On the other hand if they had simply stored their information on magnetic tape <u>without</u> such protection then probably everything would have disappeared by now. Or worse, there might have been one single copy, stored on-line in a data centre at Alexandria.

One wonders whether protection bits are used in the genetic code? This could protect against mutations with any degree of reliability required - perhaps even against cancer? But if the genetic code were too well protected against mutations then probably no evolution could have occurred. At any rate we do not want random mutations in our computer programs or in our scientific literature or experimental data.

Another interesting paradox pointed out by Schrödinger[16] is that it is <u>quantum mechanics</u> that preserves the information contained in the genetic code by the use of bound states with very long lifetimes. Classical systems are much less reliable by comparison, despite Heisenberg's principle of uncertainty which might lead one to take a contrary view.

As an example of a classical system, consider a hard sphere gas contained in a cubical box of side L = 100 m consisting of N rigid spherical 'particles' of mass m = 1gm, radius r_o = 1 cm, average thermal velocity v_o = 100 cm/sec and free path ℓ = 100 cm.

The phase-space volume is

COMPUTATIONAL METHODS IN PHYSICS

$$\left[\frac{4\pi}{3}(mv_o)^3 L^3\right]^N = \left[4.2 \times 10^{18}\right]^N \quad (5)$$

In order to think of this system as an information store we must take into account the maximum accuracy with which the position of the representative point in phase space may be measured, corresponding to a quantum-mechanical cell of size

$$h^{3N} = (2.91 \times 10^{-79})^N \quad (6)$$

where N is Planck's constant. In these units the phase-space volume is

$$W = (1.44 \times 10^{97})^N \quad (7)$$

As usual $S = k\log W$ is the entropy of the system, where k is Boltzmann's constant, but we may equally well think of it as a computer memory of size $\log W$ bits.

As a computer memory this system is very inefficient because collisions between the particles rapidly destroy the information that it contains, each small volume element in phase-space being drawn out into a longer and longer 'filament' as time goes on. The destruction of information occurs surprisingly quickly. The solid angle of the target at each collision is

$$4\pi \left[\frac{r_o}{\ell}\right]^2 = 4\pi \times 10^{-4} \quad (8)$$

so that there is an increase of coarsed-grained phase-space volume by a factor 10^2/particle pair at each collision due to amplication of small deviations of the approaching particles. The total coarse-grained volume therefore increases by a factor 10^N/sec and in a relaxation time of less than 2 minutes all the information has been lost. Moreover although it has been necessary to use a quantum-mechanical argument to fix the size of the unit phase-space cell, the time needed to destroy all the information depends only logarithmically on Planck's constant h.

1.8   Computer Hardware

As is well known, computer hardware is nowadays produced by computerized techniques and we are in the middle of an explosive development in which more and more processing and storage capability is being packed on to tiny silicon chips. Because of mass production these can cost only $1 or so to make and are entering the domestic market and even the market for home computers, toys and games. These microprocessor chips and storage devices are highly complex on the small scale but their complexity is essentially only that of the photograph and these may readily be reproduced. Moreover the original of the 'photograph' which has in the past been designed by engineers and drawn by artists at high cost can increasingly be designed and

drawn by computer. This also applies to the 'wiring' which connects
the chips together. Thus the actual design and mass production of
computers is becoming an automated, largely unskilled activity.

How far can the development of hardware go? We are limited
by the velocity of light (30 cm/nanosecond) so in order to obtain
adequate speed with devices of ever-increasing storage capacity must
continue to make everything physically smaller. At present this means
etching with high-energy electron beams to avoid the wave length
limitations of visible light. A considerable heat production must be
dissipated as sizes are reduced. The recently-introduced CRAY-1
computer produces 115 Kw of heat within a cylinder of radius 4' 8"
and height 6' 3" and removes this heat by Freon cooling. It is not
only that the computer must be prevented from melting; the temperature
tolerances are quite tight for the electronics itself. Chips are also
electrically sensitive and must for example be protected against
destruction by static electricity during assembly.

The ultimate limit is presumably to use individual quantum-
mechanical bound states or molecular structure for storage, and the
example of the genetic code shows that this can be done. Perhaps we
might store 1 bit/100 atoms. This would correspond to 10^{20} bits/gm,
a long way ahead of the present CRAY-1 store of 10^8 bits in $4\frac{1}{2}$ tons.
The heat dissipation is also several orders of magnitude less in
biological systems.

1.9   Storage and Speed Requirements

The solid state physicist as computer hardware engineer is
of course playing a major part in this development. The physicist as
computer <u>user</u> can probably assume that computers will go on getting
larger, faster and cheaper for some time ahead with limits that are
determined by quantum-mechanics and the speed of light and fore-
shadowed by biological systems. However, his appetite is fairly insa-
tiable. Consider for example a straightforward 3D time-dependent
calculation in hydrodynamics or magnetohydrodynamics (MHD). Suppose
that there are N mesh points in each direction of a cubical box. The
simplest method of calculation is the <u>explicit</u> method, described in
Section 2, in which the new values at each mesh point P at the end
of a timestep t + Δt depend on the values at the point P and its
immediate neighbours at time t. Because of a numerical stability
restriction it is necessary to limit the timestep to

$$\Delta t \leq \Delta/v_c \qquad (9)$$

where Δ is the spacestep and v_c is the maximum characteristic velocity of the system. Then if the numerical accuracy of the calculation
is improved by increasing N, the storage required increases as N^3
and the amount of computer time needed to model a given physical
process increases as N^4.

With $N = 100$, and 1000 timesteps, we need 10^6 words and 10^9 space-time operations for each physical variable, (e.g. 8 variables $\rho, T, \underline{v}, \underline{B}$ in MHD). The CRAY-1 at present has a maximum of 10^6 words of main storage and can therefore only handle such a problem with the help of backing store. On the other hand 100 mesh points in each space direction may not be sufficient to model the physics accurately since it only corresponds to 10 Fourier modes with 10 points for the shortest-wavelength mode.

When one takes into account the fact that <u>implicit</u> calculations (Section 2) on a longer physical timescale require more computing time/step, that many classical calculations such as those in plasma physics involve 6-dimensional phase space, and that a full quantum-mechanical simulation of an n-particle system would require a 3n-dimensional configuration space, it is easy to see that most physical problems can <u>never</u> be accurately solved by putting them <u>directly</u> on to any conceivable computer. They have too great a range of space- and time-scales, that is, more bits of information (or degrees of freedom) in the real problem in the store of the computer. It is necessary first to develop a suitable model.

1.10   Development of a simulation model

The stages by which a numerical simulation model and its associated computer program are developed are illustrated in Figure 1 and will be further examined later in this article using plasma physics, hydrodynamics and MHD as examples.

We start from a set of <u>primitive equations</u>; for example in plasma physics these might be Newton's equations of motion or the corresponding relativistic equations for the particles together with Maxwell's equations for the electromagnetic field; in quantum chemistry the primitive equations would be Schrödinger's equation for the electrons and nucleus together with a radiation interaction term. There are many such sets of equations in classical and quantum mechanics which can be regarded as exactly known[10]; furthermore mathematical theorems guarantee that solutions exist, and numerical techniques are available which can be proved to approach these solutions with any desired degree of accuracy in a finite number of steps n_s.

In most cases however n_s is of purely abstract interest because it is many orders of magnitude too large for actual computer solution and it is necessary first to develop a simplified theoretical model with a much smaller number of degrees of freedom. Such models were of course developed long before computers were introduced, examples being the replacement of the co-ordinates and momenta of all the molecules of a gas by the macroscopic thermodynamic variables density, velocity, temperature and pressure or by the Boltzmann distribution function, and the use of Vlasov's equation in plasma physics. Frequently it is necessary to introduce phenomenological transport coefficients in such a model, and these may either be estimated

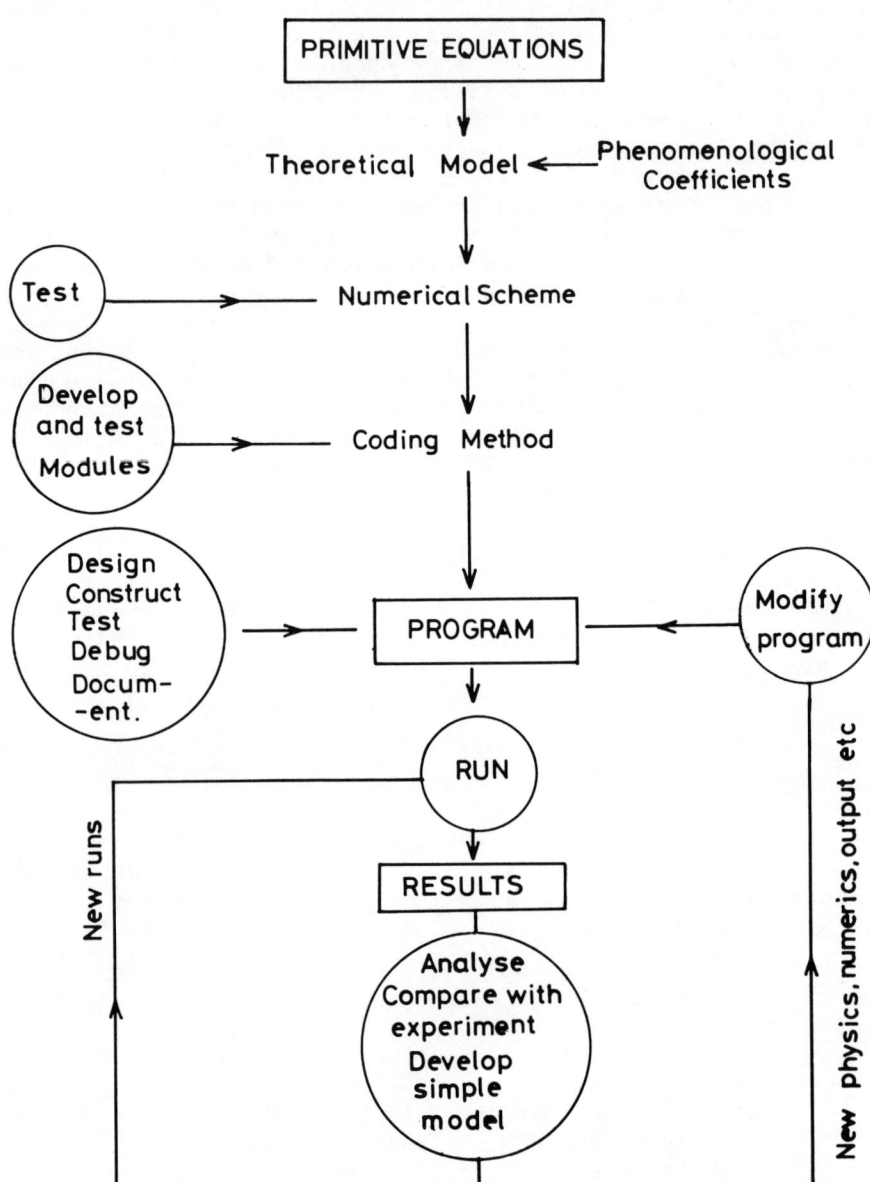

Figure 1. Development of a Simulation Model.

theoretically or determined by experimental measurement. Examples are the turbulent transport coefficients that must be used in plasma physics and meteorology to simulate the effect of motions whose space-time scales are too fine to be accommodated on the computer.

Although some people have felt in the past that theoretical physics might be 'taken over' by computer so that analytic theory would no longer be required, we can now see this to be an illusion and in fact many leading theoreticians who at first criticised computing are becoming progressively more enthusiastic about it*. A considerable amount of analytic or intuitive theory will always be needed to transform the primitive equations into a model which a real computer can handle, and subsequently to interpret the results and to check their validity.

Sometimes however it turns out that a more direct approach is useful in the preliminary formulation of a model. For example, although a real plasma may contain 10^{20} particles, a relatively crude simulation with 10^6 or even 10^3 particles may bring out unexpected qualitative phenomena which can then be subjected to a more refined theoretical analysis. Such calculations are by analogy often called computer experiments; their role is to guide theory rather than to provide the accurate quantitative results needed in nuclear reactor simulation or in numerical weather prediction.

When a practically-computable model has been found, the next stage (Figure 1) is to develop a useable numerical scheme and here the aid of the numerical analyst must be sought. Once more the emphasis must be on practicality. Abstract numerical analysis is concerned with whether or not a given process will converge to within a distance ε, however small, of the correct result in a finite number of steps n_s, and most of the theorems are proved in the asymptotic limit $\varepsilon \to 0$, $n_s \to \infty$. The computational physicist is interested in obtaining 'acceptable' accuracy (difficult to define) in the strictly limited number of steps that he can afford, and the asymptotic theorems often seem inappropriate to his problem. Numerical theory must therefore be supplemented by numerical experiment in order to test the validity of each scheme that is proposed. It is efficient to carry out such tests in a controlled way, usually on sections of the numerical scheme rather on the complete scheme itself, on individual modules or specially-written test modules rather than on the complete computer program, and often in the first instance on simplified equations which emphasize the essential points.

Further steps in the development of a simulation model will be discussed in Section 5.

*Some of the greatest names in mathematics and theoretical physics have been concerned in the development of computers and their applications including Leibnitz, Turing, Von Neumann, Birkhoff and Fermi so that enthusiasm was never lacking.

1.11   Information Engineering

Although I have stressed that analytic <u>theoretical physics</u> is unlikely to be supplanted by the computer, there is a considerable likelihood of individual <u>physicists</u> being caught up by computing either in electronics hardware, in data processing or in programming. I have noticed this many times in job interviews; people originally trained for example in high energy physics or in radio astronomy find after a few years that they have become specialists in particular branches of computing, where there is such a quantity of obscure but fascinating detail that the real physics (which is much more difficult) tends to get forgotten.

The solution perhaps is to recognize <u>information engineering</u> as a profession separate from physics, and to remove the obscurities, introduce general principles and raise the general standards. To take an analogy, during the 1930's and 1940's particle accelerators were developed and built by nuclear physicists, and research students (including the author) often found themselves helping to construct cyclotrons. Now they are developed by applied physicists and accelerator engineers and largely built by industry and work more reliably and efficiently. Another example would be the growth of the profession of nuclear reactor engineer.

1.12   Computer Software

Although computer hardware is developing rapidly (section 1.9) it is widely recognised that the software is still relatively unsophisticated. This is an area where physicists and mathematicians may be able to play a part (Table 4) and work in this direction would be vital for applications in industry, commerce and government. Two analogies are useful: firstly a computer program can be thought of as an engineering mechanism which is intended to carry out a specific task, and secondly it may be thought of as a document to be read and understood by humans.

Using the first analogy the experimental physicist or engineer can help by introducing the concepts of quantitative efficiency; of the modular construction of programs out of units which may be prefabricated and standardised; of improved design and construction techniques including automation; and of techniques for program testing and operation.

Using the second analogy the theoretical physicist or mathematician can introduce the concepts of universality, abstraction, rigour; the construction of software according to general principles; the use of more powerful symbolism; conciseness and clarity; the open publication of selected programs to serve as copybook examples; collaboration between organisations with similar

Table 4

Development of Software

<u>Analogy 1.</u>   Software as engineering mechanism
<u>Analogy 2.</u>   Software as document to be read and understood.

<u>Contribution from experimental physicist, engineer:</u>
- Quantitative efficiency
- Modularity, prefabrication
- Standardization
- Design and construction
- Testing and operation

<u>Contribution from theoretical physicist, mathematician:</u>
- Universality, abstraction, general principles
- Rigour
- Symbolism, conciseness, clarity
- Open publication, collaboration
- Common notation
- Refereeing, teaching

problems; a common notation; the refereeing of published programs and their use in teaching.

These points are examined further in Section 5.

1.13   The Man-Computer analogy

To illustrate the relation between hardware and software we may compare (Table 5) computer hardware to the mechanism of the human brain, and computer software to all the apparatus of civilization: language, recorded literature, mathematics, science, technology. The 'hardware' of the brain is believed to have reached its present stage some 10^5 years ago, but at that time there was very little 'software', and even though the brain of some individual early man may well have been that of an Einstein it would have been inconceivable for him to have developed relativity or quantum theory. At present our mental 'hardware' appears to have been static for a long while but in some directions the 'software' is growing rapidly: hence the exponential progress in mathematics and science which has taken place throughout the last few centuries with a growth period of perhaps 10 years.

This analogy is meant to suggest that we cannot judge the

Table 5

Man Computer Analogy

	MAN	COMPUTER
Hardware	BRAIN (Apparently static)	* ELECTRONICS (Growth period 1-2 years)
Software	* CIVILIZATION (Growth period $\sim$ 10 years)	SYMBOLISM, LANGUAGES PROGRAMS (Volume increasing but techniques static)

*Present growth areas

ultimate significance of computers in physics by what has occurred in the first 30 years. Even if comptuer hardware were to remain static, (which is far from true at present since the exponential growth time in power and versatility is only a year or so), there is always the possibility of an explosive development of the software leading to a much wider range of applications for the same type of equipment.

Until this happens we must expect that computers will only touch the intellectual fringe of physics: there are no algorithms for thinking up really new ideas or for developing theories so that the deepest level remains unaltered. On the other hand a computer experiment can sometimes exhibit an enexpected phenomena which will suggest a new idea, a good example being the soliton.

Two fascinating constraints are imposed by the theorems of Turing and Gödel. According to Turing's theorem all 'universal' computers are equivalent. This is a limitation because it implies that no fundamentally-new hardware can ever be produced: on the other hand it is a challenge because it implies that any problem that can be solved by any conceivable computer can, in principle, be solved by those that already exist, so why not start to tackle it now? According to Gödel's theorem there are true mathematical theorems that no computer can ever prove. Does this also represent a restriction on the human brain? And is the brain a universal computer in the Turing sence?

Further developments are the automatic construction of software and possibly also algorithms, and the hardware and software symbiosis of man and computer on which experiments are already in progress, for example the use of nerve impulses to control artificial

COMPUTATIONAL METHODS IN PHYSICS

limbs and the implantation within the human body of electrical equipment to assist the deaf and the blind.

1.14   Collaboration between computers, mathematic and theoretical physics

In contrast to such advanced developments, a possibility that already exists but could be much more fully exploited is that computers should do routine work for the theorist, freeing him for more intellectual activities. For example they can evaluate functions, plot graphs, do elementary algebra, evaluate analytic or contour integrals, look up and assemble references and even help to write papers. The main limitation here is practical; if the overheads are too high in cost and especially in general inconvenience it will be more efficient to use the older techniques, e.g. to look up an integral in a table or to visit the library instead of using an information retrieval service.

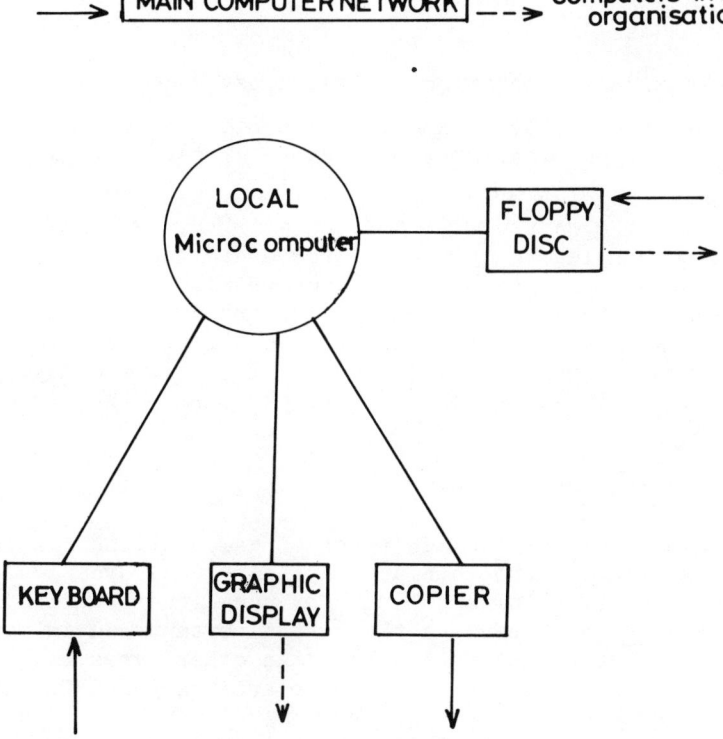

Figure 2. Intelligent office terminal.

Figure 2 illustrates in diagrammatic form the type of intelligent terminal that one can expect to see in theoretician's offices within the next decade or so, consisting of a keyboard with provision for upper and lower case Roman and Greek letters and appropriate mathematical symbols, a character and graphics display, a unit to make permanent copies, and a floppy disc or other peripheral device to provide demountable backing storage. The terminal would be supported by a local microprocessor and by the main laboratory computer network, and connected through this to the computers of other collaborating organisations. The hardware for such a terminal could already be installed at a cost of perhaps £20,000 but the cost must be reduced by perhaps a factor 10 before it becomes economic to provide one for every theoretician, and since the terminal is intended to support concentrated intellectual work it can only be used efficiently in the quiet of an individual office rather than in a communal computer users' room. Much of the necessary software also exists in protype form but this again has to be made more versatile, more convenient to use, better documented and more widely disseminated before it can come into general use.

2. CLASSICAL COMPUTATIONAL PHYSICS

2.1 Definition

Classical computational physics may be defined as

"The use of controlled numerical experiments to simulate the behaviour of physical systems obeying classical laws."

There are many examples of classical systems that one would like to be able to simulate on the computer. Usually this will be because they cannot be handled analytically, for example because they are complicated or strongly non-linear since weakly non-linear systems can often best be treated by perturbation theory. Other uses for the computer in classical physics are to evaluate analytic formulae, to plot or display graphs, to make movie films of time-dependent processes and to elucidate complex structures with the aid of interactive graphics.

2.2 Evolutionary calculations

Amongst the more common types of calculations (Table 6) are those dealing with the time evolution of a system, dynamical steady states, static equilibrium and linear or non-linear stability. Of these <u>time evolution</u> is perhaps the most important and intuitively the easiest to picture and furthermore the other types may be reduced to this; for instance we start from some arbitrary initial condition, follow the time development of the system, and seek for a dynamical steady state or static equilibrium which emerges when the intial perturbations have died down. This technique is of course only appro-

Table 6

Type of Calculations

Time evolution

Steady state

Equilibrium

Stability

    Linear (eigenfunctions, eigenvalues)

    Non-linear

        Limiting amplitude

        Mode coupling

        Turbulence

priate for steady states or equilibria which are stable. If on the other hand a steady state or equilibrium has been found its stability may be tested by linearizing the equations, making an arbitrary perturbation, and following the time-development of the linearized equations to see whether the perturbation increases in amplitude. By following the development for sufficiently long the growth rate of the fastest-growing mode of an unstable system may be determined. Non-linear stability can also be examined by evolutionary methods to study saturation amplitudes, mode-coupling and turbulence.

In practice one does not always use 'physical' evolutionary methods. The elliptic equations which describe electrostatic equilibria or thermal steady states may have been derived from the time-dependent Maxwell (hyperbolic) or diffusion (parabolic) equations respectively, but for numerical solution it is often preferable to disregard their physical origin and to solve them by techniques which have been specifically developed for elliptic problems, for example by iteration or direct matrix inversion. Even so, in understanding the method of solution it may well be helpful to the computational physicist to think of the iteration counter as a 'pseudo-physical' time variable, and to analyse from a physical point of view the effect of any error terms that have become added to the original equation during the discretization process. For example the successive over-relaxation (SOR) method replaces the elliptic equation by a damped hyperbolic equation which describes the linear decay of the residual error as it propagates across the mesh[17], and such understanding may be of practical advantage in improving the rate of convergence.

2.3   Fields and particles

The basic models of classical theoretical physics are fields and fluids (continous) and particles (discrete). These two basic models recur separately or in combination in countless different ways throughout the subject. Fields and fluids may be in real space (1-3 dimensions) or in phase space (2-6 dimensions). Particles may interact either with a fixed field (ordinary differential equations) or with each other (coupled ordinary differential equations). They may also interact with each other via an intermediary field (electrostatics), or they may interact dynamically with a field which has its own degrees of freedom (electromagnetic theory). The situation is further complicated by the fact that in developing a theoretical model (Figure 1) one type of system is frequently approximated by another; for example a liquid composed of discrete molecules is treated as a continuous fluid.

Classical computational physics inherits all these theoretical models and approximations and introduces further models and approximations of its own, replacing for example a continuous fluid by an assembly of fluid elements or simulation particles as in the PIC (particle-in-cell) method, or representing the large number ($\sim 10^{20}$) of point ions and electrons in a plasma by a much smaller number ($10^3 - 10^6$) of super-particles of specific size and shape.

Table 7

Some Methods of Calculations

FIELDS, FLUIDS	MESH	
	Eulerian*	
		MAC (Marker and cell)
	Lagrangian*	
	FINITE ELEMENT	
	PIC (particle in cell)	
PARTICLES	1 - particle	
	N^2 interactions	
	Particle-mesh*	
	P^3M	(particle-particle/particle mesh)
	Monte-Carlo	

*to be discussed in this section.

Table 7 indicates some of the methods that are used, those discussed in this article being starred. Mesh techniques are treated in Section 2.3 and particle-mesh techniques in Section 2.4.

2.4   A fundamental problem

Any classical physical system contains an infinite amount of information because each variable can in principle be measured or specified exactly. This is true whether the number of independent variables is finite (particles) or continuously infinite (fields and fluids). Theoretically it follows from the classical limit in which Planck's constant $h \to 0$ but there is no need in practice to bring in quantum theory; for example in a thermonuclear reactor plasma the number of modes of oscillation is far greater than can be handled numerically. Consider a plasma with temperature 10 Kev, volume $10^3 m^3$, density 10^{20} ion pairs/m^3 and a lifetime of 100s. The Debye length is 7×10^{-5} m and the plasma frequency is 9×10^{10} Hz.

The number of independent electrostatic modes of oscillation is of order 10^{15} and the product modes x periods is of order 10^{28}. The number of bits needed to represent the time-evolution of the plasma depends on the numerical method that is used. The simplest but expensive technique is to use a Eulerian space-time mesh with space-step Δ and timestep ΔT. If we assume that 10 points/wavelength are needed in each of the 4 directions of the mesh for the shortest-wavelength, highest-frequency, mode, than 10^{32} mesh-points are required altogether, and with 10^2 bits/point for accuracy this requires the calculation of 10^{34} bits.

Any digital computer can only generate a finite (and relatively small) amount of information in a finite time, e.g. the very fast CDC STAR-100 can generate about 6×10^9 bits/second or 10^{13} bits in a ½-hour run, so that such a calculation can only simulate 1 in 10^{21} of the effective degrees of freedom in the reactor plasma. A similar restriction applies to a global weather calculation which typically has a mesh size of 100 Km and 15 vertical layers; here all the smaller-scale motions down to the minimum dimension imposed by molecular viscosity can only be taken into account by their effect on phenomenological transport coefficients as has already been pointed out in Section 1.

Any numerical solution can therefore only be approximate and we may pose as a fundamental problem:

'For which types of classical system can an acceptable numerical solution be obtained?'

This problem is not new - it occurred already in analytic theory: Thus to get a useful model for the behaviour of a gas it is necessary for many purposes to replace the 6N co-ordinates and momenta of the molecules by a limited number of thermodynamic variables such as density, velocity, pressure and temperature. Nevertheless the use of computers allows us to express the problem in a sharper quantitative form by comparing the rate of generation of information (bits) in the physical system with the corresponding rate of generation in the computer.

2.5 Optimum representation

Since only a small fraction of the total number of degrees of freedom can be handled on the computer an important aim must be to get an <u>optimum representation</u> in which the storage and information handling capacity of the computer are used in the most effective way. The best choice of representation must depend on the physics and will differ from case to case but some examples may be given (Table 8).

Table 8

Choice of Representation

Problem	Optimum representation	Non-optimum representation
Newtonian gravitation	$\underline{x}_1 \cdots \underline{x}_N$	Gravitational field $\Phi(x,y,z)$
2D inviscid incompressible hydrodynamics	Positions of sun and planets $\underline{x}_1 \cdots \underline{x}_n$ Boundaries $\underline{x}_\alpha(s)$ of finite-area vortex regions	Stream-function $\psi(x,y)$ Velocity field $\underline{u}(x,y)$ Vorticity distribution $\zeta(x,y)$
1D shock	Shock position, strength	Density $\rho(x)$ Pressure $p(x)$ Velocity $u(x)$ Temperature $T(x)$
Step-function distribution	Boundary curves or surfaces	Function $f(\underline{x})$
Fourier component	Wave-number Amplitude Phase	Function $f(\underline{x})$

Firstly it is clear that the use of an Eulerian space mesh in the plasma physics example of section 2.4 is relatively wasteful in storage since it uses 1000 mesh points for the shortest-wavelength modes whereas only two numbers ought to be needed to describe the mode amplitude and phase. It is also wasteful in the number of time-steps by perhaps a factor 1000, since we have assumed 10 steps/minimum period whereas in fact each individual mode might well be coherent over (say) 100 periods. It would be preferable to use a Fourier representation (equivalent to the interaction representation in

quantum mechanics) in which only the changes in the coefficients of
the linearized modes are computed. However we do not at present know
how to do this efficiently since to calculate the non-linear interaction of every mode with every other mode would require even more
computation than does the Eulerian method.

The examples shown in Table 8 are relatively obvious. The
configuration of the solar system might be defined by specifying the
gravitational potential $\Phi(x,y,z)$ at every point, but clearly it is
more economical to give the co-ordinates $\underline{x}_i$ of the sun and planets.
Similarly for an electrostatic system of point charges. In 2D inviscid,
incompressible hydrodynamics this reduction of information content
can be turned to good account by the use of point vortex methods. The
fluid flow may be described in three equivalent ways by the stream
function ψ, the velocity $\underline{u}$ or the vorticity ζ, satisfying

$$\underline{u} = \text{Curl } \psi, \quad \text{div } \underline{u} = 0 \qquad (10a,b)$$

$$\zeta = \text{Curl } \underline{u} \qquad (11)$$

$$\nabla^2 \psi = -\zeta \qquad (12)$$

$$\frac{D\zeta}{Dt} \equiv \frac{\partial \zeta}{\partial t} + \underline{u}\cdot\nabla\zeta = 0 \qquad (13)$$

The fluid is assumed to have uniform constant density ρ_o and to move
in the (x,y)-plane, so that ψ, ζ have only their z-components non-zero
and may be treated as scalars. Equation (13) shows that the vorticity
is attached to the fluid and moves with it (Helmholtz theorem). Then
several methods of calculation suggest themselves:

1. Define ζ at points of an Eulerian mesh and solve a set of
   2D partial differential equations.
2. Define u_x, u_y at points of an Eulerian mesh and solve
   a set of 2D partial differential equations.
3. Approximate the voriticity distribution $\zeta(x,y)$ by a finite
   number of point vortices $\underline{x}_i$ interacting via Green's functions.
4. Approximate the vorticity distribution by a finite number
   of point vortices $\underline{x}_i$ interacting via their potential $\tilde{\psi}$.
5. Represent a finite-area vortex region (FAVR) of uniform
   constant vorticity by its boundary curve and compute the
   function $\zeta(x,y)$ geometrically.
6. Represent an FAVR of uniform constant vorticity ζ_o by its
   boundary curve on which Curl $\zeta \neq 0$ and solve the equation
   $$\nabla^2 \underline{u} = -\text{Curl } \zeta$$
   obtained from (13) by taking the curl.

7. Represent an FAVR of uniform constant vorticity ζ_o by its boundary curve, each segment of the curve interacting with itself and every other segment via a Green's function.

Most of these methods have been tried successfully. The point I wish to emphasize is that while method (1) represents the instantaneous state of the fluid by a function $\psi(x,y)$ of two independent variables x,y, and method (2) by two such functions $\underline{u}(x,y)$, methods (5-7) represent the state by one or more boundary curves $\underline{x}_\alpha(s)$, i.e. by one or more functions of a single parameter s, and methods (3) and (4) by a finite number of point co-ordinates. These latter methods are therefore more economical in information content and in some sense are better adapted to the physics of the problem.

Method (1) is the well-known stream-function and vorticity method[18] which proceeds as follows:

(a) Given $\zeta(t)$ at time t, solve Poisson's equation (12) for $\psi(t)$.

(b) Compute $\underline{u}(t)$ from ψ by differentiation.

(c) Compute $\zeta(t + \Delta t)$ by solving the advection equation (13).

This works best for systems with rigid boundaries on which the normal velocity $u_\perp = 0$, i.e. ψ = constant. For a system with a free boundary at which the pressure p is constant or zero it may be better to use method (2) and to solve the equation of motion

$$\rho_o \frac{D\underline{u}}{Dt} = - \text{grad } p \qquad (14)$$

directly. To step the velocity on from time t to t + Δt it is necessary to know the pressure distribution at time t + $\Delta\tau/2$. One can get a Poisson equation for p by taking the divergence of (14):

$$\nabla^2 p = - \rho_o \text{ div} \left(\frac{D\underline{u}}{Dt} \right) \qquad (15)$$

The right-hand side of (15) contains a term non-linear in $\underline{u}$ (which should be iterated) and a linear term

$$- \rho_o \frac{\partial}{\partial t} (\text{div } \underline{u}) \qquad (16)$$

In the physics problem div $\underline{u}$ = 0 is an identity, but numerical errors may prevent this from being exactly achieved. We therefore try to adjust div $\underline{u}$ to zero at the new time t + Δt by expressing (16) as a correction difference term:

$$\rho_o \text{ div } \underline{u}(t)/\Delta t \qquad (17)$$

This is the basis of the successful MAC (Marker and Cell) method[19].

Physically we may think of the vorticity distribution as a moving, self-interacting imcompressible 'fluid'. A point vortex of strength ζ at the origin of an infinite medium induces a rotational velocity field

$$u_x = -\frac{\zeta_u}{r^2} , \quad u_y = \frac{\zeta_x}{r^2} \tag{18}$$

(corresponding to the magnetic field of a line current). For a vortex in a finite region with <u>rigid</u> boundaries the Green's function is approximately modified. Therefore one way (method 3) of solving a 2D incompressible inviscid flow problem is to represent the instantaneous state of the fluid by N point vortices, and to compute the velocity of each vortex i by summing the N-1 contributions from all the other vortices $j \ne i$ (together with a self-interaction from i itself if images exist). This was done several years ago by Abernathy and Kronauer[20] who computed non-linear wave problems such as the formation and stability of the Von Karman vortex street by following the motion of only 24 point vortices. The interaction of hurricanes can be calculated in a similar way.

If the finite size of the vortices is important it may be better to use an alternative representation in which $\zeta(x,y)$ is treated as a step-function or 'waterbag' distribution consisting of regions λ of uniform, constant vorticity ζ separated by boundary curves. The Helmholtz theorem (13) together with the incompressibility condition (10b) guarantee that the area of each region remains constant during the fluid motion, that the topology of the set of curves is preserved, and that the vorticity values ζ remain invariant. The instantaneous state of the fluid is therefore defined by the configurations of the boundary curves and it is not necessary to follow the motion of the fluid as a whole but only that of the boundaries. The simplest configuration is a set of one or more regions of vorticity $\pm \zeta_o$ within a background of zero vorticity.

Christiansen, Roberts and Zabusky[21] calculated the evolution of such waterbag systems using method (4) by packing the interior of each waterbag region with a set of uniformly-distributed 'point' vortices. The stream-function ψ was then calculated from ζ by solving Poisson's equation (12) with a Fast-Fourier Transform method[22], the velocity distribution $\underline{u}$ determined from (10a), and then the motion of each individual 'point' vortex computed by solving its equations of motion

$$\frac{dx}{dt} = \underline{u}(\underline{x}) \tag{19}$$

(Actually an individual vortex was not strictly a point but an element of finite area equal to that of a single mesh cell).

This makes for a simple efficient code, but physically it is evident that one does not need to follow the motion of points within the boundaries but only that of the boundary curves themselves in order to compare the distribution ζ by a geometrical technique (method 5) similar to that used by Roberts and Berk[23] in the 2-stream instability problem in plasma physics.

One can go further and notice (method 6) that instead of solving Poisson's equation (12) for the stream-function ψ from the voriticity ζ, one can take the curl of ζ and obtain a distribution which is non-zero only on the boundaries. A vector Poisson equation (14) is then solved for the velocity y directly. This gives us a new physical picture in which it is not the vortices that mutually interact with one another but the vorticity gradients, (which in this case are δ functions located on the moving boundary curves).

This picture has been exploited by Zabusky, Roberts, Hughes and Tappert[24] using method (7) in which each boundary segment interacts with every other segment (and with itself) according to a Biot-Savart law.

Models of this kind are of considerable help in understanding how a physical system 'works' and may suggest new concepts as well as facilitating more economical computer calculations. For example an incompressible, irrotational water wave can have non-zero vorticity only on the fluid boundaries namely on the free surface and on the bottom. In the case of a flat bottom a simple image of the surface vorticity may be used. Therefore the motion of a water wave may be represented as the mutual interaction of a surface vorticity sheet with itself and with its image. Using this picture Zaroodny and Greenberg[25] have computed the motion of a solitary wave by a Biot-Savart technique.

Non-linear problems frequently exhibit physical processes that appear to reduce the amount of information in the optimum representation, making analytic or computer calculations more tractable than might otherwise be expected. Examples are the concentration of pressure or velocity gradients into one or more shocks, the formation of thin boundary layers and other localized vortex regions including hurricanes, the concentration of magnetic flux into sunspots and the absorption of mass by black holes. Not all of these processes are understood at present. One way to make progress is to search for qualitatively new phenomena by means of a relatively crude calculation using for example an Eulerian mesh with a coarse grid. If interesting phenomena appear then the analytic or computational model can be further refined. Solitons were first recognised in early computer calculations[26] and more recently a simple 3DMHD Eulerian calculation has been used by Sykes and Wesson[27] to study the formation of reversed field pinch configurations by a dynamo process.

2.6 Choice of representation

One can pose two general questions:

Size: How many bits are needed to define the essential state of the system?

Speed: At what rate (bits/second) is this information changing?

The answers to these questions should determine the size and speed of the computer required but at present they can only be asked and answered in a loose way.

One way in which the representation of the state of the system may not be optimum is that all the information which it contains may not be independent as already discussed in section 2.5. This also applies to the history or evolution of the system, two examples being a linear problem in which the phases of the independent eigenmodes change in a known way, and the problem of advection in a fixed velocity field[28].

Consider a fixed velocity field $\underline{u}(x,y)$ and a function f which is advected with the fluid according to

$$\frac{Df}{Dt} \equiv \frac{\partial f}{\partial t} + \underline{u} \cdot \nabla f = 0 \qquad (20)$$

The solution can be obtained in principle by solving the characteristic equations

$$\frac{d\underline{x}}{dt} = \underline{u}(\underline{x}) \qquad (21)$$

to give the transformation

$$\underline{x}(0) \to \underline{x}'(t) \qquad (22)$$

for which

$$f(\underline{x}',t) = f(\underline{x},0) \qquad (23)$$

This is actually done in the Lagrangian method of calculation (Section 3) which works well in 1 dimension. In several dimensions all but the simplest types of fluid motion cause the mesh to become distorted due to velocity shear, and in fact the transformation (20) can often be ergodic in certain regions of the flow and laminar in others[29]. Therefore the Lagrangian method is usually impracticable and it is necessary to solve the advective equation (20) by finite difference methods on a fixed Eulerian mesh. Curiously enough this simple equation gives rise to major numerical problems to be examined in the next section.

A fruitful approach may be to divide the information that represents the instantaneous state into two parts:

$$\text{STATE} = \text{ESSENTIAL} + \text{INESSENTIAL} \tag{24}$$

(persistent information)   (information that is rapidly lost in the physical system)

Examples are shown in Table 9 which indicates that detailed knowledge of molecular positions, small-scale turbulent eddies and short-wavelength temperature modes is rapidly lost and should not affect the future state of macroscopic variables such as the thermodynamic functions, the distribution function, the mean flow and the long-wavelength temperature modes. If this assumption is physically correct then it may not matter that such inessential information is either not represented at all in the numerical solution or is computed inaccurately.

Table 9

Essential and Inessential Information

Problem area	Essential information	Inessential information	Prescription
High-density gases	Thermodynamic functions	Molecular positions and velocities	Use hydrodynamic and thermodynamic equations
Low-density gases or plasmas	1-particle distribution function	Molecular positions and velocities Correlations	Use Vlasov, Fokker-Planck or Boltzmann equations
Hydrodynamics	Mean flow	Turbulent eddies	Use turbulent transport coefficients?
Heat diffusion	Long-wavelength modes	Short-wavelength modes	Ignore short wavelengths

As a problem which is further studied in Section 3, consider the 1D diffusion equation

$$\frac{\partial T}{\partial t} = D \frac{\partial^2 T}{\partial x^2} \tag{25}$$

solved on an Eulerian mesh of uniform step-size Δ with timestep Δt by the fully implicit method. Physically each Fourier mode of wavenumber k should decay according to:

$$T(k,t) = T(k,0) \, e^{-Dk^2 t} \tag{26}$$

Numerically the following errors arise:

(a) Modes with wavenumber $k > k_{max} \equiv \pi/\Delta$ cannot be represented.
(b) The derivative $-\partial^2/\partial x^2$ is represented by $k_x^2 \equiv 2(1-\cos k \Delta)/\Delta^2$ instead of by k^2.
(c) The exponential in (26) is represented by

$$\left(\frac{1}{1 + Dk_*^2 \Delta t} \right)^{T/\Delta t} \tag{27}$$

Clearly if $Dk_*^2 \Delta t \geq 1$ the short-wavelength modes are damped much less in the numerical solution than they are in the physical problem. However they are damped sufficiently to cause them to disappear and this may well be all that is required; furthermore the very short wavelength modes which cannot be represented in the solution at all would disappear physically even more rapidly and it is therefore reasonable to ignore them. An accurate solution to the diffusion equation (25) may therefore be anticipated. The situation is different for example in 3D turbulence where energy is known to be transported towards the short-wavelength modes; here the omission of modes beyond the cutoff $k_{max} = \pi/\Delta$ and the inaccurate treatment of modes near the cuttoff requires a justification which has not yet been adequately given.

3. MESH TECHNIQUES

3.1 Uniform Eulerian mesh

Consider a 1D time-dependent problem involving a partial differential equation for a single function f containing derivatives $\partial f/\partial t$, $\partial f/\partial x$, $\partial^2 f/\partial x^2$ etc, and use a simple fixed Eulerian mesh with uniform mesh intervals Δ, Δt to represent the derivatives by finite differences (Figure 3). Any function and its derivatives can be

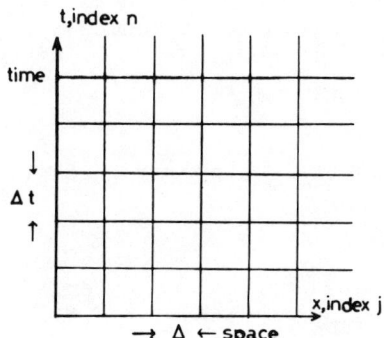

Figure 3. Uniform 1D Eulerian Mesh.

represented in this way but it is necessary to consider the truncation errors caused by dropping the higher-order terms in the Taylor expansions, e.g.

$$f(x + \Delta) = f(x) + \Delta f'(x) + \frac{\Delta^2}{2!} f''(x) + \frac{\Delta^3}{3!} f'''(x) + \cdots \tag{28}$$

The first space derivative may for example be represented by a second-order-accurate formula

$$\left(\frac{\partial f}{\partial x}\right) \rightarrow \frac{f_{j+1} - f_{j-1}}{2\Delta} \quad \left(-\frac{\Delta^2}{3!} f'''_j \cdots \right) \tag{29}$$

and the second derivative by

$$\left(\frac{\partial^2 f}{\partial x^2}\right) \rightarrow \frac{f_{j+1} - 2f_j + f_{j-1}}{\Delta^2} \tag{30}$$

Similarly for the time derivative, either

$$\left(\frac{\partial f}{\partial t}\right)^n \rightarrow \frac{f^{n+1} - f^{n-1}}{2\Delta t} \tag{31}$$

or

$$\left(\frac{\partial f}{\partial t}\right)^{n+\frac{1}{2}} \rightarrow \frac{f^{n+1} - f^n}{\Delta t} \tag{32}$$

Here j labels the space mesh and n labels the time mesh as indicated in Figure 3.

3.2 Linearization, Fourier modes

A fruitful method for studying truncation errors is to linearize the difference equations and to examine the behaviour of small perturbations about a given solution treating them as if they were physical modes. To get a simple model it is convenient to consider a linear problem with constant coefficients solved on a uniform space-time mesh so that the eigenmodes exp i(kx - wt) are independent. The differential equation

$$\frac{\partial f}{\partial t} = L\left(\frac{\partial}{\partial x}\right) f \tag{33}$$

corresponds to the <u>algebraic</u> dispersion relation

$$\omega = G(k) \tag{34}$$

with $G = iL(ik)$. In the numerical solution (34) is replaced by a <u>trigonometric</u> dispersion relation

$$\omega = \Gamma(k) \tag{35}$$

which results from representations of the type

$$\frac{\partial}{\partial t} \to \frac{e^{-i\omega\Delta t} - e^{i\omega\Delta t}}{2\Delta t}, \quad \omega \to \omega\left(\frac{\sin \omega\Delta t}{\omega\Delta t}\right) \quad (36)$$

$$\frac{\partial}{\partial x} \to \frac{e^{ik\Delta} - e^{-ik\Delta}}{2\Delta}, \quad k \to k\left(\frac{\sin k\Delta}{k\Delta}\right) \quad (37)$$

$$\frac{\partial^2}{\partial x^2} \to \frac{e^{ik\Delta} - 2 + e^{-ik\Delta}}{\Delta^2}, \quad k^2 \to \frac{2}{\Delta^2}(1 - \cos k\Delta) \quad (38)$$

This replacement causes the roots of true dispersion relation (34) to be displaced in the complex ω-plane, leading to pseudo-physical effects which appear as numerical instability, damping and dispersion [28]. Another possibility is additional roots (computational modes) due to the transcendental form of (35).

3.3   Advective equation

The advective equation

$$\frac{Df}{Dt} \equiv \frac{\partial f}{\partial t} + u\frac{\partial f}{\partial x} = 0 \quad (39)$$

has already been mentioned in Section 2. The importance of (39) is that a Lagrangian operator of the type D/Dt appears on the left-hand side of all hydrodynamic equations, so that numerical errors in the treatment of (39) are prevalent in all Eulerian hydrodynamic and MHD difference schemes. Such errors can most easily be seen if we choose u = constant and take the limit $\Delta t \to 0$, when the algebraic relation (34):

$$\omega = uk \quad (40)$$

is replaced by the trigonometric relation (35):

$$\omega = uk\left(\frac{\sin k\Delta}{k\Delta}\right) \quad (41)$$

Clearly (40) corresponds to a disturbance of any shape propagating unchanged in form with constant velocity u; there is no dispersion. The phase velocity and group velocity are both constant and equal to u. Furthermore (40) preserves positivity which is important for the propagation of positive-definite quantities such as density and temperature.

The numerical relation (41) introduces dispersion, the phase and group velocities respectively being

$$u_p \equiv \frac{\omega}{k} = u\left(\frac{\sin k\Delta}{k\Delta}\right) \quad (42)$$

$$u_g \equiv \frac{d\omega}{dk} = u \cos k\Delta \quad (43)$$

Modes with $k > \pi/\Delta$ cannot be represented on the mesh at all, but in addition with $\pi/\Delta > k > \pi/2\Delta$ are non-physical because they have a negative group velocity. For $k = \pi/4\Delta$ the phase velocity has fallen to 0.900 u and the group velocity to 0.707 u. If 5% accuracy in the group velocity is required then only modes with $k < \pi/10\Delta$ are admissible.

The propagation of a localized disturbance leads to a change in shape which may cause non-physical ripples to appear, and if these travel into a space-time region where f is small they may also cause f to change sign since unlike the physical differential equations this second-order-accurate numerical scheme does not preserve positivity. Usually it is necessary to take precautions to prevent f from becoming negative, either by <u>ad hoc</u> methods such as a cavalier Fortran statement.

$$F(J) = \text{Max } [F(J), \text{FMIN}], \qquad (44)$$

by using a less accurate different scheme that preserves positivity by damping out the short-wavelength modes, or by one of the more sophisticated approaches[30,31] that have been developed in recent years.

3.4 Method of the modified differential equation

Although dangerous numerical instabilities frequently appear first for large k, it is often convenient to analyse errors such as damping and dispersion by a small-k expansion. Higher-order terms in the Taylor expansion (28) which are wrongly treated by approximations such as (29)-(32) appear as additional terms in an <u>equivalent partial differential equation</u> which the difference scheme 'really' solves. This is called the 'method of the modified differential equation'[32]. For example the second-order accurate representation of (39) using (37) corresponds to the linear dispersive equation

$$\frac{\partial f}{\partial t} + u \frac{\partial f}{\partial t} + \frac{u\Delta^2}{6} \frac{\partial^3 f}{\partial x^3} + \cdots = 0 \qquad (45)$$

There is no damping because all even-order derivatives cancel.

A less accurate method for solving (39) is the one-sided Lelevier method[33] in which $\partial f/\partial x$ is represented by

$$\left(\frac{\partial f}{\partial x}\right)_j \rightarrow \begin{cases} \dfrac{f_j - f_{j-1}}{\Delta} & (u \geq 0) \qquad (46a) \\ \dfrac{f_{j+1} - f_j}{\Delta} & (u < 0) \qquad (46b) \end{cases}$$

This corresponds (for case (a)) to a linear interpolation

$$f_j^{n+1} = f_j^n\left(1 - \frac{u\Delta t}{\Delta}\right) + f_{j+1}^n\left(\frac{u\Delta t}{\Delta}\right) \qquad (47)$$

so that a positive-definite function f will remain positive definite provided that the stability condition on the timestep Δt

$$\frac{u\Delta t}{\Delta} \le 1 \qquad (48)$$

is obeyed. On the other hand compared to the second-order-accurate expression (30) there is an additional error term

$$\frac{u\Delta}{2}\left(f_{j+1} - 2f_j + f_{j-1}\right) \qquad (49)$$

so that the equation being solved is not (39) but the modified differential equation

$$\frac{Df}{Dt} \equiv \frac{\partial f}{\partial t} + u\frac{\partial f}{\partial x} = D\frac{\partial^2 f}{\partial x^2} \qquad (50)$$

with an additional numerical diffusion coefficient $D = u\Delta/2$. It is just this extra diffusion that damps the short-wavelength modes before ripples can lead to non-positivity.

3.5   k-space

Evidently only a small fraction of the modes are accurately treated with Eulerian mesh. For example if positive group velocity is taken as the criterion only 1/8 of the modes are acceptable in 3D, but if one requires the group velocity to be accurate to 5% then only 0.1% of the modes meet this condition. This is a serious problem, largely unsolved. In addition to <u>linear dispersion</u> there is also <u>non-linear aliasing</u>[34] illustrated in Figure 4 for the 2D case: if the resultant vector

$$\underline{k} = \underline{k}_1 + \underline{k}_2 \qquad (51)$$

corresponding to the combination of two modes $\underline{k}_1$, $\underline{k}_2$ lies outside the allowed region

$$(-k_{max} \le k_x, k_y \le k_{max}) \qquad (52)$$

then the energy of the physical mode (51) reappears under an 'alias' in a mode that does lie within the region (52). The usual solution to both these problems is to damp the short-wavelength modes before they can cause too much trouble, so that in multi dimensional Eulerian calculations most of k-space is treated as a kind of 'guard-region' which protects the inner physical domain. Evidently this occupies a considerable amount of storage space and computation time. An alternative solution (spectral method[35]) is to Fourier-analyse the

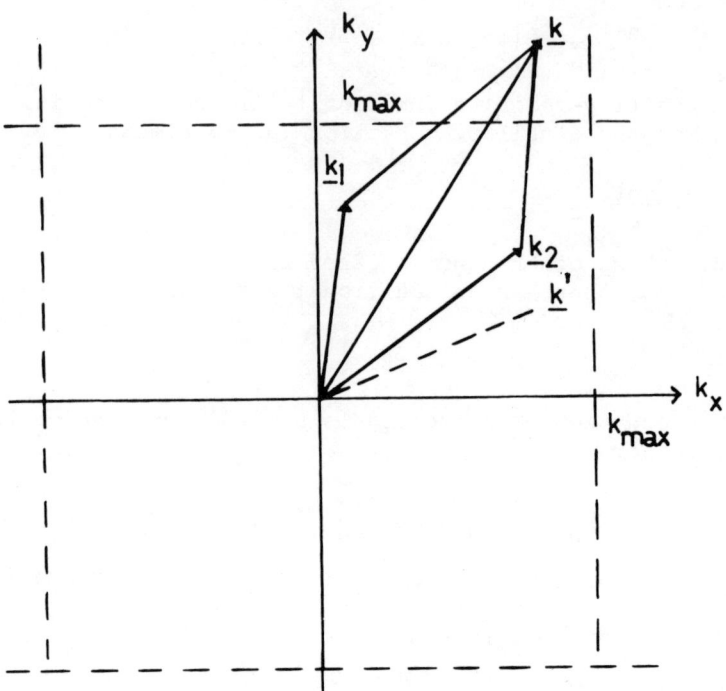

Figure 4. Aliasing. Two allowed modes $\underline{k}_1$, $\underline{k}_2$ combine non-linearly to generate a mode $\underline{k} = \underline{k}_1 + \underline{k}_2$ which is outside the allowed region $(|k_x|, |k_y|) \leq k_{max}$. This is then incorrectly represented by a mode $\underline{k}'$.

differential equations and treat the physical modes (say M in number) more accurately, but the $M^2, M^3, \cdots$ separate calculations needed to evaluate the non-linear interaction terms may then also be prohibitive in computation time.

3.6   Amplification factor

Instead of working with the angular frequency ω it is convenient to use the amplification factor $G = f^{n+1}/f^n$. Questions of stability and damping are then investigated by examining the value of the modulus:

$$|G| \begin{array}{ll} > 1 & \text{instability} \\ < 1 & \text{damping} \end{array} \qquad (53)$$

In the following we consider a single mode, (e.g. by taking a Fourier transform in space), so that the partial differential equation (33) becomes an ordinary differential equation

$$\frac{df}{dt} = \lambda(k)f \tag{54}$$

with λ simply a complex number for given k, e.g. for $k \to 0$

$$\begin{aligned} \lambda &= -iku_p & \text{(wave)} \\ \lambda &= -Dk^2 & \text{(diffusion)} \end{aligned} \tag{54}$$

where u_p is the phase velocity of the wave and D is the diffusion coefficient.

The difference equation for a 2-level scheme can then be written as

$$\frac{G-1}{\Delta t} = \lambda \left[\Theta G + (1-\Theta) \right] \tag{55}$$

where the value of f on the right-hand side is interpolated between the values f^n, f^{n+1} at the beginning and end of the timestep according to

$$f = (1 - \Theta)f^n + \Theta f^{n+1} \tag{56}$$

with $0 \leq \Theta \leq 1$. From (55) we obtain

$$G = \frac{1 - \alpha(1-\Theta)}{1 + \alpha\Theta} \tag{57}$$

with $\alpha = -\lambda \Delta t$.

This resolution into individual modes has a similar basic significance to that of the resolution of a classical or quantized field into harmonic oscillators. All the well-known 2-level difference schemes can be considered as special cases of equation (57) as Figure 5 illustrates. Each case corresponds to a different rational approximation (57) to the exponential $\exp(-\alpha)$.

The formulae in Figure 5 are valid in the limit $k \to 0$. For finite k the trigonometric replacements

$$\begin{aligned} k \to k^1 &= k \left(\frac{\text{Sin } k\Delta}{k\Delta} \right) \\ k^2 \to (k^2)^1 &= \frac{2(1 - \text{Cos } k\Delta)}{\Delta^2} \end{aligned} \tag{58}$$

should be made.

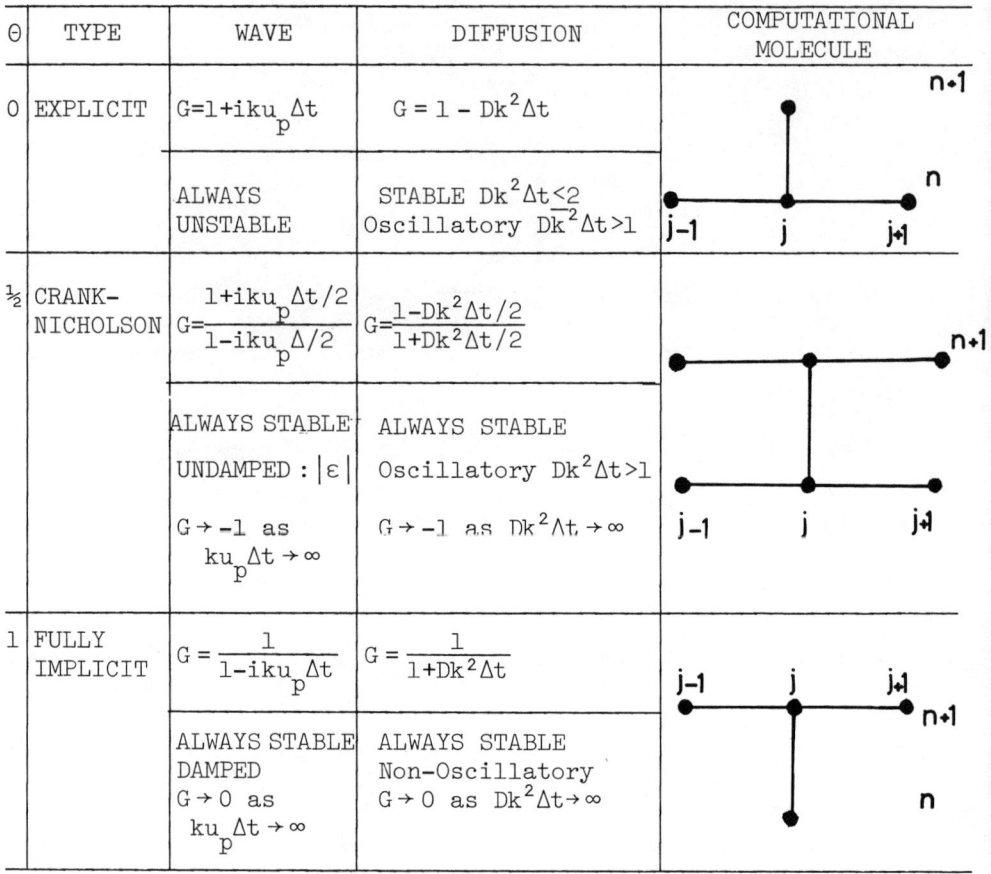

Figure 5. 2-level difference schemes.

3.7 Properties of 2-level schemes

Although the time-centred Crank-Nicholson scheme is often preferred because it is second-order accurate for small α, this preference is to some extent misguided and there is no universally 'best' scheme. The numerical analyst is in a position similar to that of the designer of an electronic circuit or a compound lens system, and the individual schemes in Figure 5 may be thought of as <u>modules</u>, analogous to electronic filters or lens components, which are to be combined together to produce the desired effect. Even unstable schemes can be used in combination with damped schemes to provide a compound scheme which is stable overall.

Taking the diffusion equation as an example, we see that the explicit scheme is unstable for

$$\alpha = D(k^2)_{max} \Delta t > 2 \quad \text{where} \quad (k^2)_{max} = \frac{4}{\Delta^2} \tag{59}$$

i.e. for timesteps Δt larger than the stability limit

$$(\Delta t)_{max} = \frac{\Delta^2}{2D} \tag{60}$$

The Crank-Nicholson scheme is stable for all Δt but it becomes quite inaccurate for Δt comparable with (60), and when $\alpha > 1$ it represents a damped physical mode by a numerical mode that oscillates in sign. It may therefore be preferable to use the fully implicit scheme since although this has a lower order of accuracy as $k \to 0$ all modes remain damped for arbitrarily large Δt. The damping for large α is not as strong as that given by the correct physical exponential but it is nevertheless strong enough to make the unwanted high-k modes disappear from the solution.

The advantage of implicit schemes with $\theta \geq \frac{1}{2}$ is that there is no stability restriction on the timestep. Our goal is to use a Δt which is limited only by the accuracy of the solution, e.g., by a restriction that there should (say) be not more than $\varepsilon = 0.05$ relative change in important physical quantities during one step. This is quite difficult to do successfully, and often one has to use a much smaller Δt than one feels should really be required. The oscillations introduced by the Crank-Nicholson scheme are independent of the timestep as $\Delta t \to \infty$, and once the accuracy limit ε has been exceeded a reduction in Δt may produce little effect until a value comparable with the explicit timestep (60) has been reached. An accuracy control on the timestep can therefore sometimes cause Δt to fall catastrophically.

The answer for diffusion equations is probably to use the fully implicit scheme. For wave equations it may be best to choose θ somewhat greater than $\frac{1}{2}$ so that a measure of damping is introduced, thus reducing the amplitude of the high-k modes without too much error in the wanted low-k part of the solution.

Explicit schemes have the advantage that they are easy to code in any number of dimensions and that the individual timesteps are fast. The timestep is however limited by the maximum physical propagation speed u_{max} according to the stability condition

$$\Delta t \leq \Delta x / u_{max} \tag{61}$$

Implicit schemes have no such stability restriction on Δt but they require a solution algorithm since the formula for f^{n+1} at one mesh point involves the unknown values of f^{n+1} at adjacent points. Such algorithm is available in 1 dimension and is quite fast and straight forward to code. In 2 and 3 dimensions the solution is less fast and considerably more complex but nevertheless may have advantages over the explicit method.

All schemes limit the mode frequency since the period cannot be less than $2\Delta t$. Physical modes with frequency ω_{phys} such that $\omega_{phys}\Delta t \gtrsim 1$ cannot therefore be properly represented.

3.8 Conservation laws

So far we have adapted a mathematical approach and considered point functions and Taylor expansions. In many ways it is more convenient to use a physical approach and to think of discrete masses, energies, magnetic fluxes etc. enclosed in finite boxes, as illustrated in Figure 6. Exact physical conservation laws can then often be represented by exact difference identities, within the round-off accuracy of the computer, and provide a powerful check on the validity of the algebra, programming, key-punching and so on. (Note that with 32-bit computers it is essential for this purpose to work in double length).

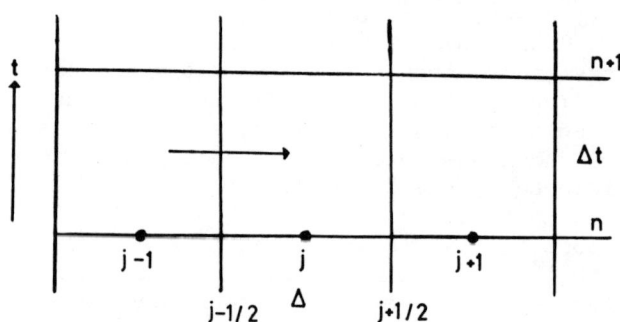

Figure 6. Conservation of heat. In solving the heat diffusion equation $\partial T/\partial t = \kappa\, \partial^2 T/\partial x^2$ the same difference expression $\kappa_{j-\frac{1}{2}}(T_{j-1} - T_j)\Delta t/\Delta$ is used for both the loss of heat by cell $j-1$ in time Δt, and the gain of heat by cell j. This guarantees heat conservation as a difference identity.

Conservation laws that involve quadratic or higher-order expressions cannot usually be expressed in exact difference form, although sometimes this can be done. However, such laws have their uses since quantities that are identically conserved by the physics but not by the difference scheme provide a convenient check on the numerical accuracy of the solution. Ideally therefore a code should probably include both types of conservation law.

Figure 6 shows how exact conservation is achieved for the heat diffusion equation by expressing the loss of heat by box $j-1$ across wall $j-\frac{1}{2}$ in the same form as the gain of heat by box j across the same wall.

COMPUTATIONAL METHODS IN PHYSICS

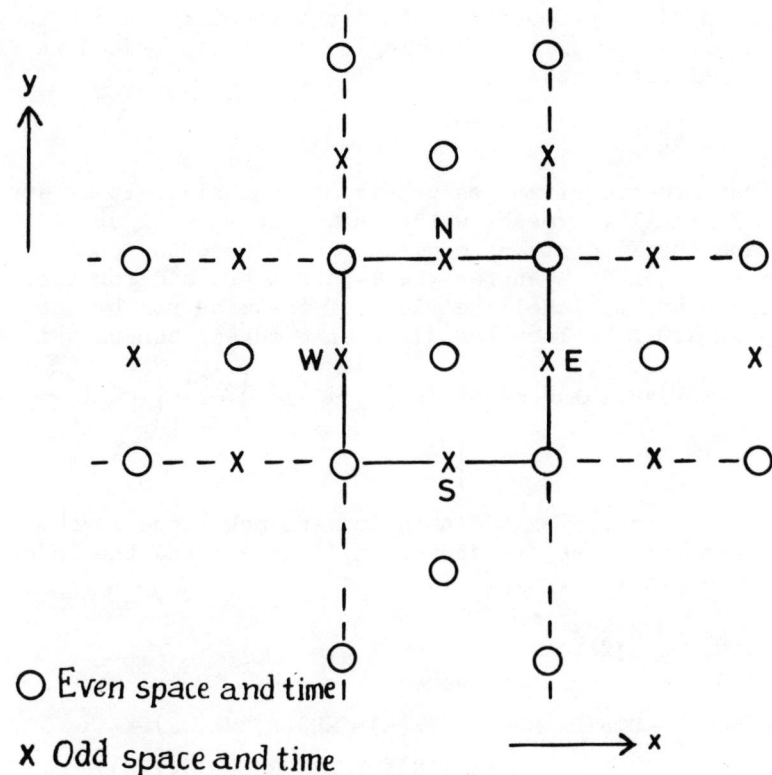

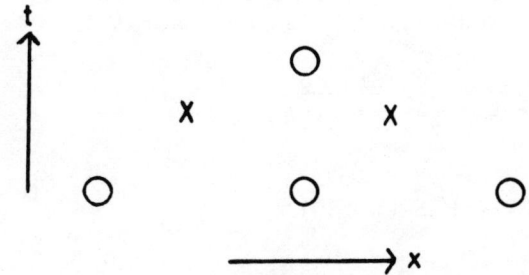

Figure 7. 2D explicit leapfrog scheme. Points are staggered in space and time (NaCl lattice) so that all differences are entered.

3.9 2D Explicit leapfrog scheme

A physically-appealing conservative difference scheme in 2 dimensions is the explicit leapfrog scheme, illustrated in Figure 7 for the continuity equation

$$\frac{\partial \rho}{\partial t} = - \text{div} (\rho \underline{u}) \tag{62}$$

Since first-order derivatives appear on both sides it is appropriate to use a 3 time-level mesh, with the O points in Figure 7 at even level n and the X points at odd n. If (j,k) are the space indexes we see also that j+k is even for the O-points and odd for the X-points. It is convenient to label the sides of the mesh box by the points of the compass N,S,E,W. Then the difference scheme can be written

$$(\rho^{n+1} - \rho^{n-1}) 4\Delta \times \Delta y = -2\Delta t \left\{ \frac{\{[\rho V_x]_E - [\rho V_x]_W\} 2\Delta y - \{[\rho V_y]_N - [\rho V_y]_S\}}{2\Delta x} \right\}^n \tag{63}$$

in FORTRAN it simplifies the notation and makes the code more efficient if the variables are stored in 1D arrays and the indexes are calculated explicitly, e.g.

```
    INTEGER    C,N,S,E,W
    --------   calculate indexes   --------
RHO(C) = RHO(C) - DTBYDX(RHO(E)*VX(E) - RHO(W)*VX(W)) -
                  DTBYDY(RHO(N)*VY(N) - RHO(S)*VY(S))            (64)
```

3.10 Leapfrog stability analysis

An elementary leapfrog stability analysis is that of the 1D advective equation (39). Because of the double timestep the equation for the amplication factor G is now quadratic:

$$\frac{G^2 - 1}{2\Delta t} = \frac{u}{i} \left(\frac{\sin k\Delta}{\Delta} \right) G \tag{65}$$

with solution

$$G = -i\beta \pm (1 - \beta^2)^{\frac{1}{2}} \tag{66}$$

where

$$\beta = \frac{u\Delta t}{\Delta} \sin k\Delta \tag{67}$$

The scheme is therefore stable with an amplification factor modulus 1 if $\beta \leq 1$ for all k, and since the maximum value of $\sin k\Delta$ is 1 we obtain the CFL stability condition $u\Delta t/\Delta \leq 1$. This is to be expected since $\Delta/\Delta t$ is the maximum speed of propagation of information on the mesh.

3.11 3D Maxwell equations

Figure 8 shows an interesting leapfrog scheme for the Maxwell equations

$$\frac{\partial \underline{B}}{\partial t} = -\text{Curl } \underline{E}, \quad \frac{\partial \underline{E}}{\partial t} = \text{Curl } \underline{B} \qquad (68)$$

Examination of (68) shows that it is sufficient to define $\underline{E}$ only at even times (O-points), and $\underline{B}$ only at odd times (X-points). Furthermore all 6 components are to be defined at different space-points. In Figure 8 (E_x, E_y, E_z) are defined respectively at the centres of

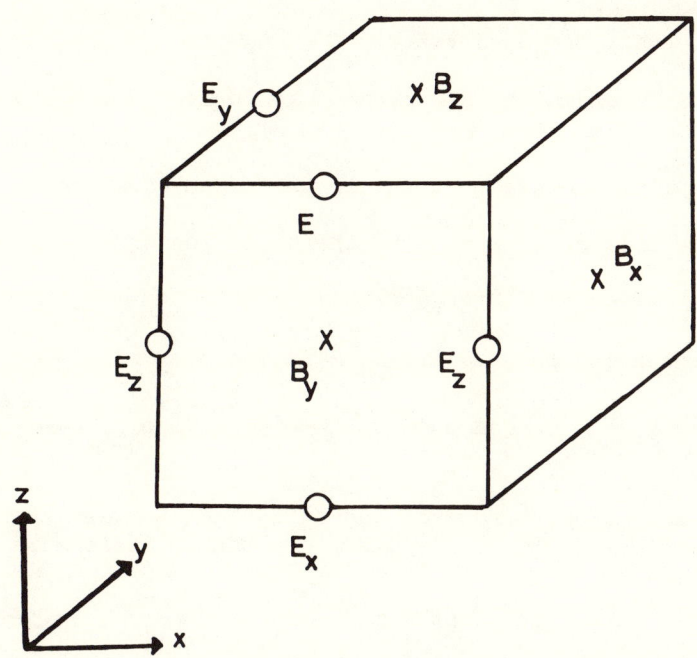

Figure 8. Leapfrog Scheme for 3D Maxwell Equations. Each field component is defined on a different lattice. $\underline{E}$ is at even times and $\underline{B}$ at odd times.

the (x,y,z) edges of a cube, while (B_x, B_y, B_z) are defined respectively at the centres of the (yz, zx, xy) faces. However, the relation between $\underline{E}$ and $\underline{B}$ is entirely symmetrical apart from the $-$ sign in (68). The difference equation for the B_x component has the form

$$(B_x^{n+1} - B_x^{n-1})_{i+1,j,k} \Delta y \Delta z = -\Delta t \left(\{ E_{z,i+1,j,k+1} - E_{z,i+1,j,k-1} \} \Delta z \right.$$
$$\left. + \Delta t \{ E_{y,i+1,j+1,k} - E_{y,i+1,j-1,k} \} \Delta y \right) \qquad (65)$$

i.e.

$$\text{(Change of } \underline{B} \text{ flux)} = -(\text{line integral of } \underline{E}) \tag{66}$$

It may be verified that the difference equivalents of

$$\frac{\partial}{\partial t}(\text{div } \underline{B}) = 0, \quad \frac{\partial}{\partial t}(\text{div } \underline{E}) = 0 \tag{67}$$

(obtained by taking the divergence of (68)) are satisfied as difference identities to within the rounding error of the machine. Thus it is only necessary to satisfy the difference equivalents of the remaining Maxwell equations

$$\text{div } \underline{B} = 0, \quad \text{div } \underline{E} = 0 \tag{68}$$

as an initial condition, and perhaps to correct for rounding errors from time to time.

3.12   Explicit solution of the 1D diffusion equation

The explicit solution of the diffusion equation

$$\frac{\partial f}{\partial t} = \text{div } (D \text{ grad } f) \tag{69}$$

can be written in conservative form in 1 dimension as

$$(f_j^{n+1} - f_j^n)\Delta = \left[D_{j+\frac{1}{2}} \frac{(f_{j+1} - f_j)}{\Delta} - D_{j-\frac{1}{2}} \frac{(f_j - f_{j-1})}{\Delta} \right] \Delta t \tag{70}$$

The right hand side represents the difference between the fluxes across the sides $j\pm\frac{1}{2}$ of cell j. For constant D it is more convenient to write (70) as

$$f_j^{n+1} - f_j^n = \frac{D\Delta t}{\Delta^2} \left(f_{j+1}^n - 2f_j^n + f_{j-1}^n \right) \tag{71}$$

Stability analysis gives

$$G = 1 - \frac{2D\Delta t}{\Delta^2}(1 - \cos k\Delta) \tag{72}$$

The worst case occurs when $\cos k\Delta = -1$, when the condition $|G| \leq 1$ leads to

$$\frac{2D\Delta t}{\Delta^2} \leq 1 \tag{73}$$

Corresponding to the maximum speed of a drunkards-walk propagation across the mess.

3.13   Implicit Crank-Nicholson scheme, Gauss tridiagonal algorithm

To avoid the stability restriction we can use an implicit

scheme with $\Theta = \frac{1}{2}$ i.e. the Crank-Nicholson scheme. Then for constant D

$$f_j^{n+1} = f_j^n + \frac{D\Delta t}{2\Delta^2}\left(f_{j+1}^{n+1} - 2f_j^{n+1} - f_{j-1}^{n+1} + f_{j+1}^n - 2f_j^n + f_{j+1}^n\right) \quad (74)$$

which can be written as

$$-A_j f_{j+1}^{n+1} + B_j f_j^{n+1} - C_j f_{j-1}^{n+1} = D_j \quad (75)$$

with

$$\begin{aligned}
A_j &= C_j = D\Delta t/2\Delta^2 \\
B_j &= 1 + A_j + C_j \\
D_j &= A_j f_{j+1}^n + (2 - B_j)f_j^n + C_j f_{j-1}^n
\end{aligned} \quad (76)$$

The general solution to (75), (which applies to non-uniform D as well as to other 1D second-order difference equations) is

$$f_j^{n+1} = E_j f_{j+1}^{n+1} + F_j \quad (77)$$

with

$$\begin{aligned}
E_j &= \frac{A_j}{B_j - C_j E_{j-1}} \\
F_j &= \frac{D_j + C_j F_{j-1}}{B_j - C_j E_{j-1}}
\end{aligned} \quad (78)$$

which is a particular case of the Gauss tridiagonal algorithm[36]. Starting from the boundary conditions at x_{min} which define the initial values of E and F, equations (78) are solved for increasing x. Then the boundary conditions at x_{max} are used to fix an initial value for equation (77) which is solved for decreasing x. Minor variants of the algorithm enable different boundary conditions of f to be handled, e.g. prescribed f, $\partial f/\partial x$ or $f+\lambda \partial f/\partial x$.

3.14   Plane isentropic Lagrangian scheme

In one dimension, the difficulty in handling the advective term may be avoided by using a Lagrangian space mesh in which the points or cells are attached to the fluid and move with it. This is illustrated in Figure 9 for the isentropic equations

$$\frac{\partial \rho}{\partial t} + \text{div}(\rho \underline{u}) = 0 \quad (79)$$

$$\rho \frac{\partial \underline{u}}{\partial t} + \underline{U} \cdot \nabla \underline{u} = -\text{grad } p \quad (80)$$

$$p = \rho^\gamma \quad (81)$$

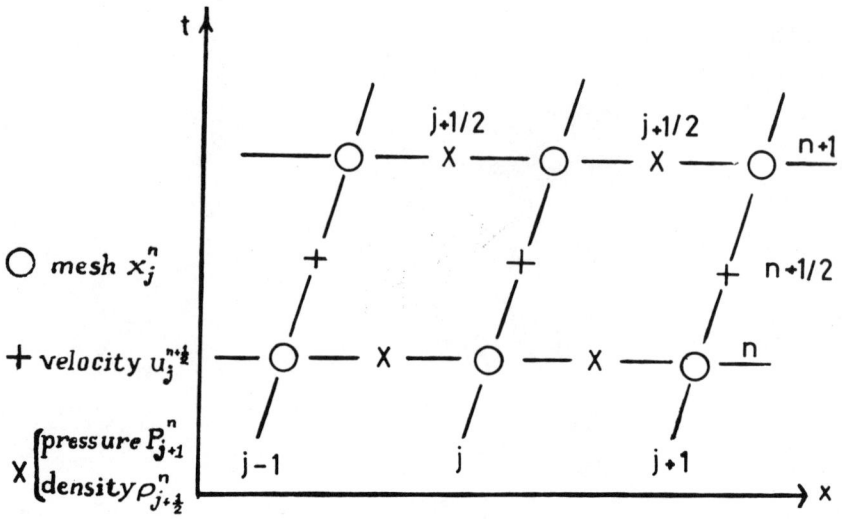

Figure 9. 1D Plane Isentropic Lagrangian Scheme.

where p is the fluid pressure and γ is the adiabatic exponent. It is convenient to think of each cell enclosed between its boundary points x_j, x_{j+1} as a moving element with fixed mass $m_{j+\frac{1}{2}}$ and variable 'volume' $V_{j+\frac{1}{2}} = x_{j+1} - x_j$. In computing the motion of these boundary points we attribute half the mass of a cell to each of its boundaries:

$$m_j = \tfrac{1}{2}(m_{j-\frac{1}{2}} + m_{j+\frac{1}{2}}) \tag{82}$$

The difference equations are:

Mesh motion
$$x_j^{n-1} = x_j^n + u_j^{n+\frac{1}{2}} \Delta t \tag{83}$$

Pressure
$$p_{j+\frac{1}{2}}^{n+1} = p_{j+\frac{1}{2}}^n \left(\frac{V_{j+\frac{1}{2}}^n}{V_{j-\frac{1}{2}}^{n+1}}\right)^\gamma \tag{84}$$

Force
$$F_j^n = -(p_{j+\frac{1}{2}}^n - p_{j-\frac{1}{2}}^n) \tag{85}$$

Equation of motion $\quad m_j(V_j^{n+\frac{1}{2}} - V_j^{n-\frac{1}{2}}) = F_j^n \Delta t \tag{86}$

The indices show how each quantity is located in space and time. If the density is required it is computed from

$$\rho_{j+\frac{1}{2}}^n = m_{j+\frac{1}{2}}/V_{j+\frac{1}{2}}^n \tag{87}$$

A published example of 1D Lagrangian code is the MEDUSA laser fusion

COMPUTATIONAL METHODS IN PHYSICS

code[37] which is based on these difference equations but includes a variety of physics and may be run in plane, cylindrical or spherical geometry.

4. PLASMA SIMULATION

4.1 Introduction

The aim of simulation methods in plasma physics is to solve a set of Vlasov equations

$$\frac{\partial f_i}{\partial t} = [H_i, f_i], \quad (i = 1, 2, \cdots, N), \tag{88}$$

one for each species of charged particle, coupled with Maxwell's equations which describe the behaviour of their collective electromagnetic fields. (Individual particle encounters are neglected in this approximation). Here $f_i(\underline{q}, \underline{p}, t)$ is the distribution function in n_i-dimensional phase space for the ith particle species, $\underline{q}, \underline{p}$ are the n_i coordinates and momenta, and H_i is the particle Hamiltonian. The Poisson bracket of two dynamical variables A, B is defined by

$$[A, B] \equiv \frac{\partial A}{\partial \underline{q}} \frac{\partial B}{\partial \underline{p}} - \frac{\partial B}{\partial \underline{q}} \frac{\partial A}{\partial \underline{p}}. \tag{89}$$

Recalling the Hamiltonian equations of motion

$$\dot{\underline{q}} = \frac{\partial H}{\partial \underline{p}}, \quad \dot{\underline{p}} = -\frac{\partial H}{\partial \underline{q}}, \tag{90}$$

we see that (88) is equivalent to Liouville's equation for a 1-particle system:

$$\frac{df_i}{dt} \equiv \frac{\partial f_i}{\partial t} + \dot{\underline{q}} \frac{\partial f_i}{\partial \underline{q}} + \dot{\underline{p}} \frac{\partial f_i}{\partial \underline{p}} = 0. \tag{91}$$

In general the Hamiltonian H_i will depend on the local electromagnetic fields through the scalar and vector potentials $\phi, \underline{A}$, and since the charge and current functions $\rho, \underline{j}$ are to be computed as moments of the f_j, we see that each H_i must be regarded a <u>functional</u> of all the f_j, $(j = 1, 2, \cdots, N)$.

The time-derivatives $\dot{\underline{q}}, \dot{\underline{p}}$ may according to (91) be regarded as the components of a velocity in phase space, if we write this equation in the advective form

$$\frac{\partial f_i}{\partial \underline{\xi}} + \underline{v}_\xi \cdot \underline{\nabla}_\xi f_i = 0, \tag{92}$$

and from (90) we see that

$$\frac{\partial \dot{\underline{q}}}{\partial \underline{q}} + \frac{\partial \dot{\underline{p}}}{\partial \underline{p}} = 0 \,, \text{ or } \underline{\nabla}_\xi \cdot \underline{v}_\xi = 0 \,. \tag{93}$$

An illuminating picture is therefore obtained by thinking of $f_i(\underline{q},\underline{p},t)$ as the local density of an incompressible fluid which moves in the n_i-dimensional phase space with local velocity $(\dot{\underline{q}},\dot{\underline{p}})$. Because this velocity depends on the f_j through H_i we speak of a set of <u>incompressible self- and mutually-interacting phase fluids, one for each species of particle</u>.

Mathematically we are faced with the initial-value problem of solving the set of N coupled partial differential equations (91) and (92) each in a space of n_i dimensions. The equations appear very simple but a number of difficult numerical problems arise, in order to obtain a solution it is necessary to examine the specific feature of the equations and to introduce a number of special numerical techniques, amongst which is particle simulation.

The fluid picture is particularly clear[23] for the case of a single species of particles (electrons) in one space dimension x, interacting via their collective electrostatic field $E = -\partial\phi/\partial x$ where ϕ is the potential. The Hamiltonian is

$$H = \frac{mv^2}{2} + e\phi \tag{94}$$

where m is the electronic mass and e the charge (regarded as negative). For simplicity we use the velocity coordinate v in phase space rather than the momentum $p = mv$. The equations of motion (90) then become

$$\dot{x} = v, \quad \dot{v} \equiv \alpha = \frac{e}{m} E(x,t) \tag{95}$$

while the Vlasov-Liouville equation (91) is

$$\frac{df}{dt} \equiv \frac{\partial f}{\partial t} + v \frac{\partial f}{\partial x} + \frac{e}{m} E \frac{\partial f}{\partial v} = 0 \,. \tag{96}$$

The electric field is to be computed at each stage by solving Poisson's equation

$$\nabla^2 \phi = -4\pi\rho \tag{97}$$

where ρ is the charge density, which can be written as

$$d/dx \left(\frac{eE}{m}\right) = \omega_p^2 \left(\int_{-\infty}^{\infty} f \, dv - 1\right) \,. \tag{98}$$

Here $\omega_p^2 = (4\pi n_0 e^2)/m$ is the square of the electron plasma frequency and n_0 is the mean density of the electrons which we assume to be neutralized by a uniform background of fixed positive charge. ω_p determines the characteristic timescale of the problem, which the numerical solution has to follow. The charge density of the electrons at the point x at time t is obtained by a velocity integral:

$$\rho(x,t) = n_o e \int_{-\infty}^{\infty} f(x,v,t) \, dv \quad , \tag{99}$$

the space average of the integral being normalized to unity to ensure charge neutrality.

A continuous function f(x,v,t) can be represented graphically by contours of constant f in the 2-dimensional (x,v) phase plane (Figure 10). Since df/dt = 0 the contours C_k move with the fluid and

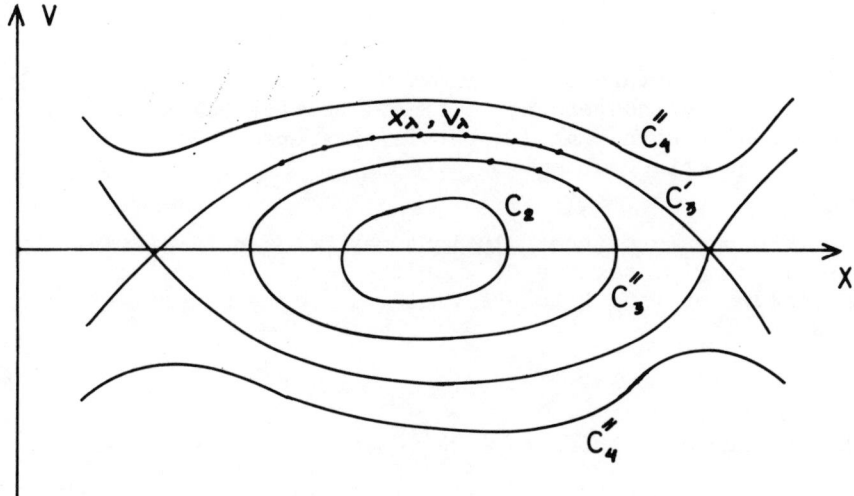

Figure 10. Representation of a continuous function by contours. The contours C_k move with the incompressible phase fluid and preserve topology and area. They may be represented by a set of points (x_λ, v_λ), joined by straight-line segments and satisfying the particle equations of motion (13).

preserve their topology as the system evolves, so that no crossing or smearing occurs, and furthermore, the area enclosed between any two contours is a constant of the motion. Each point (x_λ, v_λ) on a contour obeys the 'particle' equations

$$\frac{dx_\lambda}{dt} = v_\lambda \quad , \quad \frac{dv_\lambda}{dt} = \alpha(x_\lambda, t) \quad . \tag{100}$$

This motion is of a very special kind, being a combination of a constant horizontal shear (Figure 11a) with a vertical shear which depends on position and time (Figure 11b).

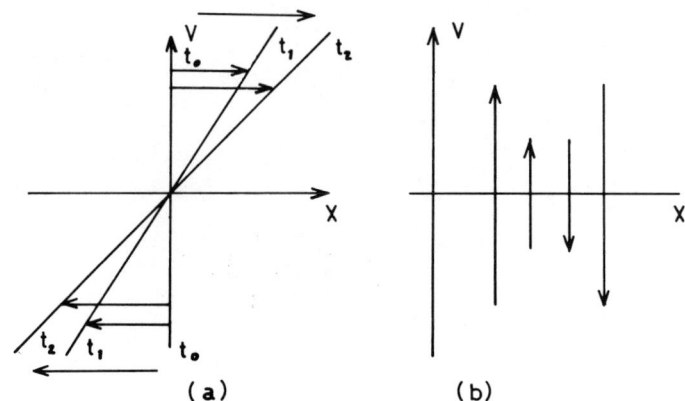

Figure 11. Nature of the flow. The horizontal shear (a) is uniform and independent of time. The vertical shear (b) depends on both x and t, but is independent of v for an electrostatic problem.

The nature of the solutions may be seen by examining the 'free-streaming' case, e = 0, when the particles are uncharged and so no longer interact with the field. Then the general solution of (96) is

$$f(x,v,t) = f\left(x - v(t - t_o), v, t_o\right) \quad (101)$$

so that a Fourier component $g(v)e^{ikx}$ at time t_o is translated into

$$g(v)e^{ik\left(x-v(t-t_o)\right)} \quad (102)$$

at any later (or earlier) time t. This means that modes which have unlimited 'wave-number' $k(t - t_o)$ in the v-direction will tend to develop as time goes on, making it difficult to solve the Vlasov equation on a computer. Figure 12 illustrates the time evolution of an initially sinusoidal contour in phase space.

4.2 Eulerian difference method

Since (91) is simply the advective equation (96) in a generalized ξ-space, it is natural to think of covering this space with a discrete Eulerian mesh, and solving the equation by one of the standard methods, e.g. the leapfrog scheme which for (96) would be

$$f_{\mu\nu}^{n+1} - f_{\mu\nu}^{n-1} = -\frac{v_\nu \Delta t}{\Delta x}\left(f_{\mu+1,\nu}^n - f_{\mu-1,\nu}^n\right) - \frac{\alpha_\mu \Delta t}{\Delta v}\left(f_{\mu,\nu+1}^n - f_{\mu,\nu-1}^n\right) \quad (103)$$

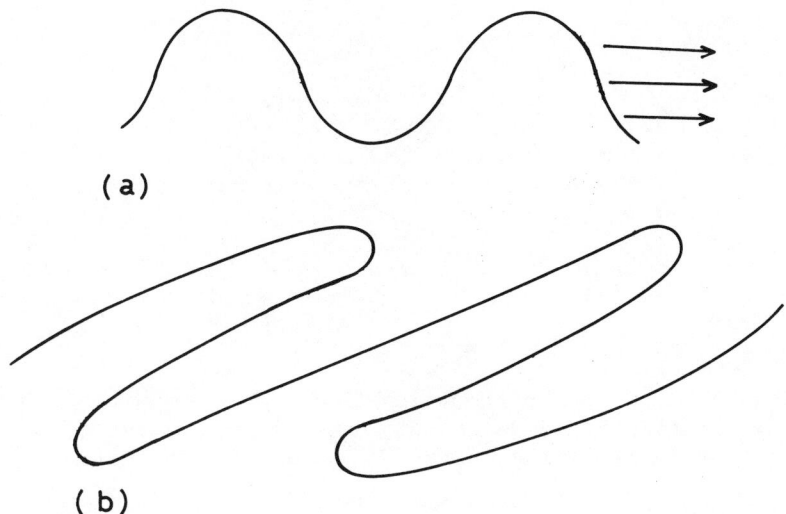

Figure 12. Free-streaming evolution of a sinusoidal contour. The effect of free-streaming on a contour such as (a) is to draw it out into ever thinner and longer filaments, as shown in (b).

where $x = \mu \Delta x$, $y = \nu \Delta y$, $t = n \Delta t$. There are however serious problems:

(a) The machine time and storage space needed for problems in 2 and 3 space dimensions, (4- and 6-dimensional phase space respectively), plus time are usually prohibitive.

(b) Difference techniques for solving the advective equation are not very accurate, since they distort the high wave-number modes.

(c) The high wave-number components in the v-direction, rapidly go beyond the range of any reasonable mesh size.

Other methods of calculation must therefore be sought, and a good account of these is given in the review volume ref. 38 with a further detailed description in the conference proceedings ref. 39.

4.3   Waterbag method

One of the simplest is the waterbag method[23] in which the distribution function $f(x,v)$ is chosen to be a step-function, so that the phase fluid consists of regions of uniform density f_ρ bounded by a finite set of curves C_k as illustrated in Figure 13. The state of the system is then completely defined by the position of the curves at any time, and it is unnecessary to calculate the motion of the phase fluid in the interior of the regions. (The general

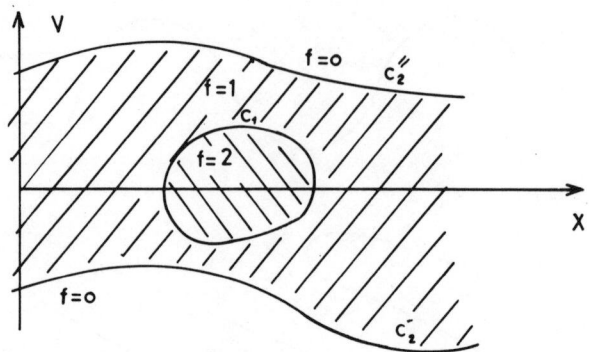

Figure 13. Step-function distribution. A step-function distribution of phase density is completely defined by the positions of the curves C_k, which move with the fluid and preserve area and topology as in Figure 1.

problem of studying the interaction of such curves is referred to as contour dynamics[40]). Figure 14 show the evolution of an unstable initial distribution

$$f(v) = \begin{cases} 1/v_o & (v_o/2 < |v| < v_o) \\ 0 & \text{(elsewhere)} \end{cases} \quad (104)$$

on which a small perturbation has been imposed. This is the well-known 2-stream instability and the calculation shows how it evolves from the linear into the non-linear regime.

The waterbag model represents the curves C_k by a set of Lagrangian points x_λ, v_λ satisfying (100) and for the purpose of display and of calculating the charge integral (91) these are assumed to be joined by straight segments. The curves become longer and more contorted as time proceeds as seen in Figure 14 and we therefore add extra points in order to preserve numerical accuracy. Eventually more points are needed than the computer can handle and the calculation has to stop, unless some way can be found to simplify the curves[23].

4.4   Lagrangian particle method : coarse-graining

Returning to more general distribution functions, we see that there is no difficulty of principle in solving (91) or (96) by a Lagrangian method provided that the functions $(\dot{q},\dot{p})$ or (v,α) are known, which would be true for independent particles moving in a prescribed <u>external</u> field. The distribution function could be repre

COMPUTATIONAL METHODS IN PHYSICS

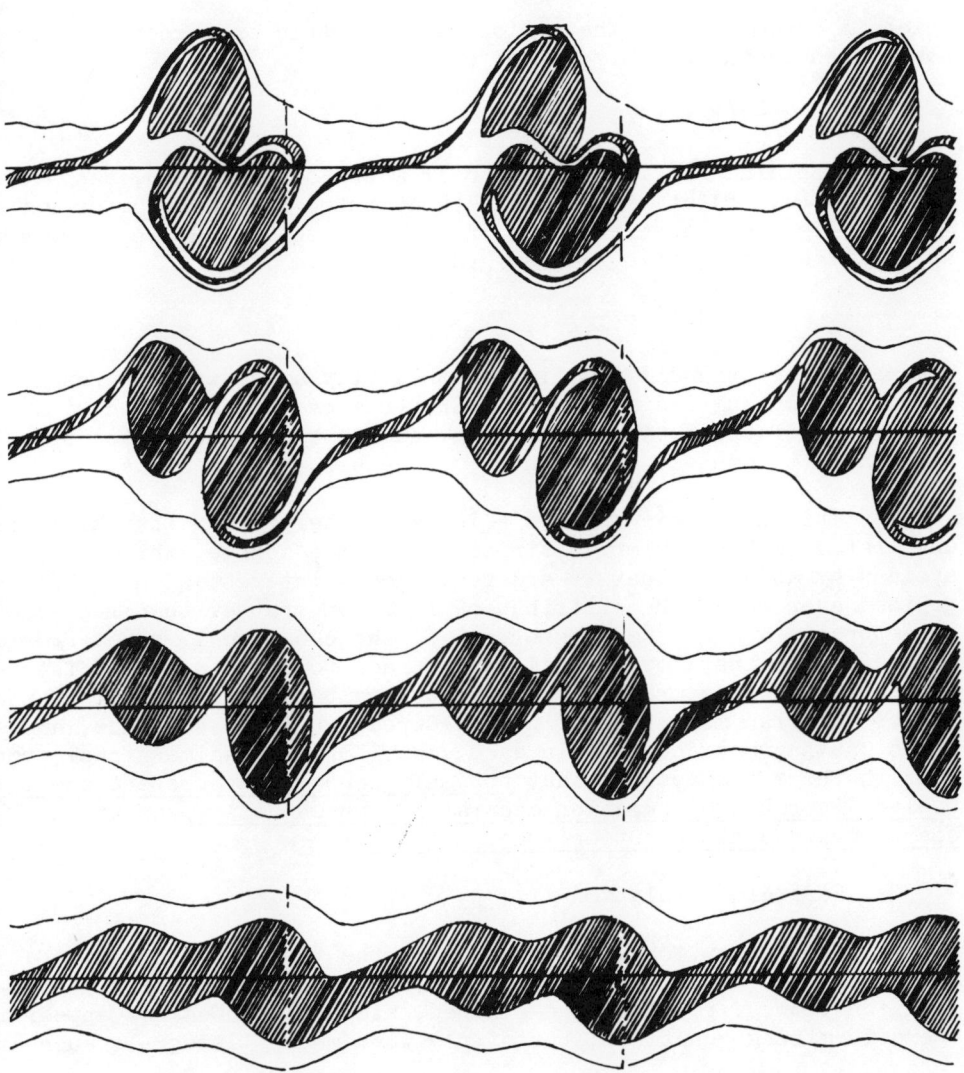

Figure 14. Two-stream instability. Evolution of the 2-stream instability in the phase plane, starting from the initial state (17) and calculated by the waterbag method[1]. The 'hole' region between the upper and lower electron beams has been shaded for clarity.

sented by a discrete set of points P_λ moving according to the equations of motion (90) or (100), and each carrying its own value f_λ which is a constant of the motion. The electric or magnetic fields to be used in the Lorentz force term

$$\frac{e}{m}(\underline{E} + \underline{v} \times \underline{B}) \qquad (105)$$

would be obtained by interpolation from an Eulerian mesh, and the value of $f(\underline{q},\underline{p},t)$ at time t would be computed by averaging over the m particles that happened to be in a finite cell $\Delta\underline{q}\,\Delta\underline{p}$ in phase space at that instant of time (coarse-graining):

$$\bar{f}(\underline{q};\underline{p},t) = \frac{1}{m}\sum_m f_\lambda \quad . \qquad (106)$$

The characteristic difficulties that usually arise with Lagrangian techniques in multi-dimensional hydrodynamics are concerned with the distortion of the moving mesh cells, and in the present case there is no need to connect the points P_λ together to form cells.

This Lagrangian method eliminates the problem (b) which was associated with an Eulerian finite difference scheme, while (c) is avoided by assuming that we are really only interested in the coarse grained average of the distribution function, rather than its fine structure on a scale less than that of the chosen cell size. Problem (a) still remains, however, since in order to represent the distribution function accurately it is necessary to have a sufficient number of Eulerian mesh points in each direction of phase space, and at least one Lagrangian point in each cell. For multi-dimensional problems it therefore appears that <u>not only cannot we represent the fine scaled distribution function accurately; we cannot represent the coarse-grained distribution function either.</u>

4.5 Particle simulation

The assumption which is made in the particle simulation method is that actually we need not be concerned with the distribution function, fine or coarse-grained, since this does not appear in the determining equations, but simply with moment integrals such as

$$\int_{-\infty}^{\infty} f\,dv \quad , \quad \int_{-\infty}^{\infty} vf\,dv \qquad (107)$$

which define the charge and current densities $\rho,\underline{j}$ to be used in calculating the collective electric and magnetic fields associated with a self-interacting system. These are functions only of the space-time coordinates $(\underline{x},t)$ and therefore can be represented with reasonable accuracy on a fixed Eulerian mesh. Because the integration over velocity space may be expected to produce a considerable smoothing,

COMPUTATIONAL METHODS IN PHYSICS

this allows us to reduce the number of velocity points to a reasonable number provided that a suitable algorithm can be found for evaluating the integrals (20) from the current positions and velocities of a set of discrete Lagrangian points.

4.6 The NGP method for electrostatic systems

We now make a significant change in the meaning of the Lagrangian points P_λ. Instead of thinking of them as mathematical points within a continuous fluid of smoothly-varying density f, we discretize the distribution function so that it becomes a set of δ-functions:

$$f(\underline{q},\underline{p},t) = \sum_\lambda w_\lambda \, \delta(\underline{q} - \underline{q}_\lambda(t)) \, \delta(\underline{p} - \underline{p}_\lambda(t)) \,, \quad (108)$$

where w_λ is an appropriate weighting factor. With each point is associated a mass $w_\lambda m$ and a charge $w_\lambda e$ so that it behaves like a physical point particle. The weight is often chosen to be the same for each species of 'simulation particles' and is frequently much larger than unity, so that the system behaves like a set of physical particles with greatly enhanced collision rates. These pseudo-physical collisions lead to numerical errors which must be carefully examined and minimized if an accurate solution is to be obtained, since the approximation made in deriving the Vlasov equation is that the collision rate is zero.

The nearest grid point (NGP) method[41] calculates the charge density simply by summing the charges in each Eulerian cell:

$$\rho = \sum_{\text{cell}} w_\lambda e \,. \quad (109)$$

When the value of ρ associated with each Eulerian grid point has been determined (Figure 15) the potential ϕ is obtained by solving Poisson's equation

$$\nabla^2 \phi = -4\pi\rho \quad (110)$$

and then the electric field $\underline{E}$ is obtained by differencing

$$E(x) = - \frac{(\phi(x+\Delta x) - \phi(x-\Delta x))}{2\Delta x} \,. \quad (111)$$

The field is assumed to be uniform within each cell and is used to update the velocity according to

$$v_\lambda^{n+\frac{1}{2}} = v_\lambda^{n-\frac{1}{2}} + \frac{e}{m} \Delta t \, E(x_\lambda^n) \quad (112)$$

after which the particles can be moved one step using

$$x_\lambda^{n+1} = x_\lambda^n + \Delta t \, v_\lambda^{n+\frac{1}{2}} \quad (113)$$

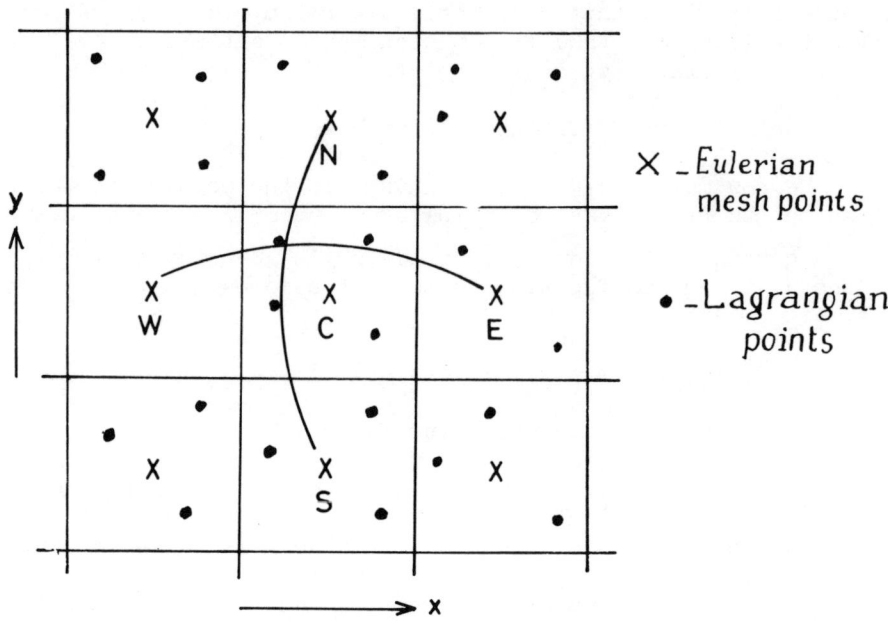

Figure 15. NGP method. (i) All charges in the cell C are credited to the centre of the cell. (ii) This generates a function ρ, defined at the X-points. (iii) Solution of Poisson's equation yields a function ϕ at the X-points. (iv) The electric fields E_x, E_y, uniform thoughout cell C, are obtained from $\phi_W - \phi_E$, $\phi_S - \phi_N$ respectively.

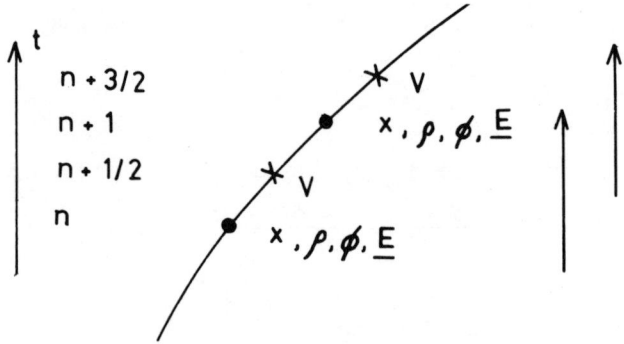

Figure 16. Leapfrog scheme. Equations (25) and (26) are solved by the leap-frog method. All Eulerian quantities ρ, ϕ, E are constructed and used at the same (integral) set of times as the coordinates x_λ.

COMPUTATIONAL METHODS IN PHYSICS   443

This results in a rather neat leapfrog scheme (Figure 16) in which
the velocities are defined at half-integral time, and all other quantities at integral times.

4.7   Area-weighting

Although the electric field used in the NGP method is influenced by the actual positions of the simulation particles, and is
therefore subject to statistical fluctuations as the particles pass
from cell to cell, the model does not include the full effect of
physical point-particle encounters. For example, two simulation particles within the same cell exert no force on one another by symmetry,
since each is assumed to be located at the cell centre. Nevertheless,
the high-k modes in the electric field do appear with excessive magnitude and random phase, and this can cause non-physical diffusion
of the particles in both coordinate and velocity space in certain
problems. To reduce these numerical errors the area-weighting or
Cloud-in-Cell (CIC) method is often used[42]. This may best be illustrated for the case of a rectangular Cartesian space mesh in 2 dimensions, with square cells of side Δ (Figure 17). Instead of using
point simulation particles as in the NGP method, we now assume them

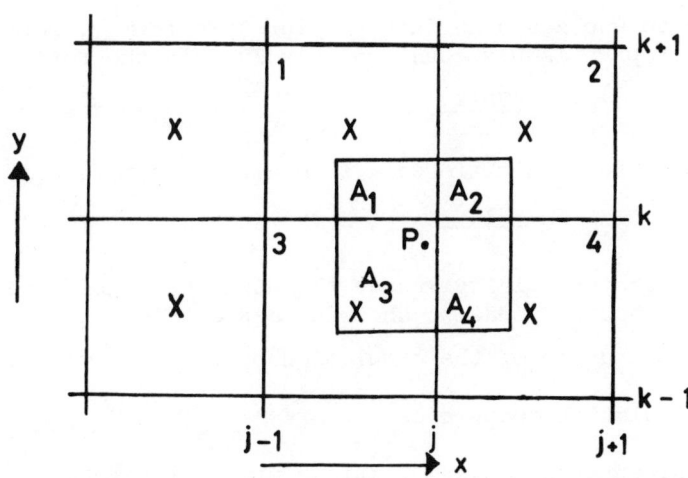

Figure 17. Area-weighting or CIC method. The charge associated
with the point P is allocated to the centres of the cells 1,2,
3,4 in proportion to the areas of overlap A_1, A_2, A_3, A_4.

to be squares of dimension equal to the Eulerian cell size. Each
'particle' will then in general overlap 4 cells (1,2,3,4) as shown
in the diagram, and the increments in the charges at the cell centres

are chosen proportional to the areas of overlap,

$$\delta\rho_i \propto A_i, \qquad (i = 1,2,3,4). \tag{114}$$

This ensures that the charges vary continuously as the particles move from cell to cell. The electric fields which act on the particles are linearly interpolated in a similar way. There have been several detailed numerical comparisons[43] of the relative accuracy of the NGP and CIC methods, and of the amount of machine time that they require.

4.8   Poisson's equation

One of the main problems in particle simulation lies in the solution of Poisson's equation (110). There is no difficulty in 1 dimension since (98) can be solved directly, but in 2 or 3 dimensions it is customary to involve a fast direct technique based on Fast Fourier Transforms[44] or Recursive Cyclic Reduction[45], so achieving an order-of-magnitude saving in computer time over iterative methods. This means that the spatial geometry must be simplified, for example to a rectangular region with periodic boundary conditions.

Although not quite the fastest method, an attractive possibility is to use a double fast Fourier transform[46]. This means that at every step Poisson's equation is solved in the form

$$\phi(\underline{k}) = \frac{4\pi\rho(\underline{k})}{k^2} \tag{115}$$

and it is quite straightforward to modify this to some form[47]

$$\phi(\underline{k}) = \frac{4\pi f(\underline{k})}{k^2} \rho(\underline{k})$$

which will preferentially reduce the unwanted high-k modes. Such a modification corresponds to choosing a specific 'shape' for the charge distribution of the simulation particle, e.g. a Gaussian[48].

4.9   Simulation techniques

The straightforward particle simulation method can only represent a microscopic region of physical space and time, especially for problems involving 2 or 3 space dimensions. The reason for this is that the plasma is subject to electron plasma oscillations on the scale of the inverse plasma frequency $\omega_p^{-1} = m/4\pi n_o e^2$ and the Debye length $\lambda_D = (kT/4\pi n_o e^2)^{\frac{1}{2}}$ which the numerical calculation has to follow. For a plasma with density $n_o = 10^{13} \text{cm}^{-3}$ and temperature $kT = 1$ keV we find $\lambda_D \simeq 10^{-2}$ cm, $\omega_p^{-1} \simeq 5 \times 10^{-12}$ secs. It is usually only practicable to follow only a few hundred periods of oscillation at the most, and not more than (say) 50 Fourier modes in each space direction, e.g. on a 256 × 256 mesh. In order for the system of charged simulation particles to behave like a plasma it is necessary for $n_o \lambda_D^{-m}$ to be large compared to unity (where m is the number of space dimensions), so that for a 2-dimensional problem the total number of particles should be at least several tens of thousands. Finally, if ions as well as electrons are to be followed the time-scales of the calculation may need to be increased because ion velocities are smaller by a factor $\sqrt{m_e/m_i}$, the square root of the mass ratio, (~ 40 for protons).

All these factors make it necessary for particle simulation codes to be as efficient as possible and many sophisticated programming techniques have been developed. There are also numerical prescriptions for improving the accuracy of the calculation, and physical approximations for eliminating unwanted modes, e.g. to remove the high-frequency electron plasma oscillations in a problem in which only the low frequency ion waves are of interest.

4.10 Optimization techniques for particle codes

Some of the techniques that are available are described in articles by Boris and Roberts[46] and by Orens et al[47]. A balance should be sought between flexibility, development time, core requirements and computational speed. Codes should be flexible, so that minor conceptual modifications do not require significant revisions. Usually this means constructing them in a modular fashion. A considerable amount of computer time can be saved by programming the inner loops in assembly language, both for the 'particle pusher' which solves equations (112) and (113) and also for the Poisson-solver, but elsewhere the code should be written so far as possible in higher-level language in order to ensure machine-independence and good reliability. There are several advantages in requiring the number of grid points in each direction to be a power of 2, since not only is this convenient for the Fast-Fourier-Transform technique[44] but it also enables many of the arithmetic operations to be carried out by shifts and masking instead of by the slower instructions such as addition, multiplication and division. On some computers it is desirable to perform all the particle calculations in fixed-point integer arithmetic rather than floating point. Consideration should also be given to eliminating any binary digits that do not materially enhance the accuracy of the solution, so reducing both the machine time and storage space needed. For many problems it is necessary to keep the particle coordinates and velocities on backing storage (disc or tape), and to transfer them in and out of the core for processing as required. This again is an area where a considerable degree of optimization can be achieved.

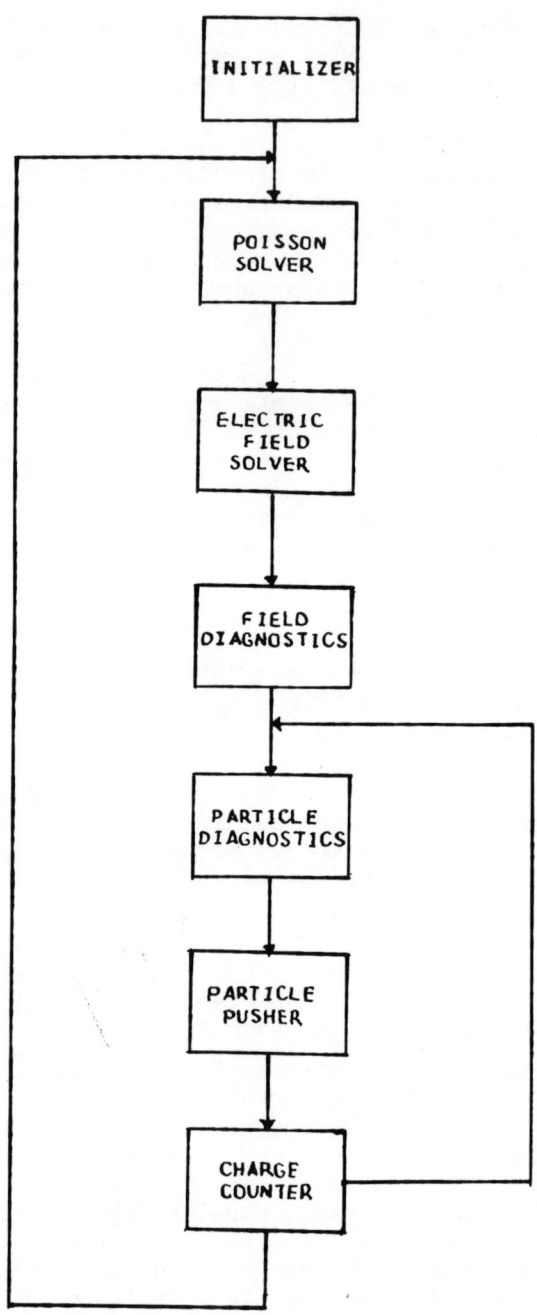

Figure 18. Flow chart of a particle simulation code. This diagram shows a flow chart of the P POWER simulation code discussed in ref. 11 and § 2.1.

COMPUTATIONAL METHODS IN PHYSICS

Figure 18 indicates the structure of the P Power simulation code[47]. Each block represents a modularized segment which is designed to be replaced easily. This allows for independence of each major section and flexibility in modifications. The code is divided into 7 major segments:

1. <u>Initializer</u>. All parameters are read in, the particle tables initialized, and important constants calculated. The particles are divided into classes in which all members have the same charge, mass etc.

2. <u>Poisson solver</u>. Equation (110) is solved by Fast-Fourier-Transform techniques. Assembly-language programming is used, the high-k modes are truncated, and a Gaussian form factor applied.

3. <u>Electric-field solver</u>. The equation $\underline{E} = -\nabla\phi$ is solved using a higher-order (6-point) difference scheme.

4. <u>Field diagnostics</u>. Modes of the electric field $\underline{E}(k)$ are available, and the electric field energy is calculated both in k-space and x-space.

5. <u>Particle diagnostics</u>. The momenta, energies and temperatures of each individual type of particle are calculated.

6. <u>Particle pusher</u>. The particle velocities and positions are updated according to a desired scheme (e.g. classical or relativistic dynamics, NGP or other method).

7. <u>Charge counter</u>. The charge density matrix ρ is calculated by a scheme according to the type of particle pusher used.

The code is designed so that the main loop for diagnosing, updating and counting the particles requires only one pass per timestep through the particle tables, which is important when these are stored on an external device.

4.11 Noise suppression techniques

Unwanted 'noise' due to the finite number of simulation particles can be eliminated or reduced by a method introduced by Byers and Grewal[49,50] and known as the 'quiet start'. We are really interested in a solution of the Vlasov equation for an absolutely collisionless system, discretized though it may be. Therefore instead of thinking of the limited number of particles as an extremely poor plasma (very low $n_o \lambda_D^{-m}$) with all the attendant enhanced kinetic properties such as fluctuations and collisions, we emphasize that the particles are to be considered as discretized elements of a continuous Vlasov phase fluid. The goal is to obtain, with a manageable number of particles, a model which approaches as closely as possible a truly continuous system.

The quiet start technique assumes that the calculation begins in the neighbourhood of an equilibrium state, presumably as in Figure 14. Accordingly it loads the initial values of the particles' phase space positions uniformly along the particle orbits and, in addition, smooths out all wavelengths equal to and smaller than the repetition length, defined as the distance over which the velocity distribution repeats, so giving an ordered regular pattern in phase space. In the simplest 1-dimensional homogeneous system where the equilibrium trajectories are simple straight-line orbits, the quiet start loading produces a number of beams. For computational efficiency, the smoothing is applied to the total charge density, rather than to each individual beam. It is clear that the unperturbed motion of a continuous beam will result in no fluctuation, and in practice the codes give proof of the success of this idea by producing fluctuation levels of an undisturbed system corresponding to machine round-off levels. The real frequencies and growth rates predicted by linear Vlasov theory are reproduced within the precision of measurements. Care must however be taken to avoid pseudo-physical instabilities due to the discrete number of particle beams.

It is also useful from a noise reduction point of view to use particles of different weight w_λ, so that the resolution can be focussed on that part of $f(v)$ which is the most active. In many calculations the regions of phase space where $f(v)$ is largest behave in a smooth laminar fashion and can be represented by a small number of simulation particles, leaving the major part to describe turbulent regions where $f(v)$ may be quite small.

4.12   Pseudo-physical oscillations, fluctuations and collisions

The use of a discrete Eulerian space mesh, a finite timestep, a finite number of particles, and the beam loading employed in the quiet start technique has a subtle effect on the physics which has been examined in detail by Langdon[51,52]. The plasma is found to interact coherently with the periodicity of the spatial grid on which the electromagnetic fields are defined and with the periodicity of the finite-difference time integration. Various parametric instabilities are introduced, which may be weak or strong and may be difficult to distinguish from real instabilities. There is also high-frequency noise associated with the rate at which particles cross the spatial grid cells. The theory developed by Langdon[51] can be checked against numerical experiments performed on the computer[52,53], and can then be used to optimize the parameters used in the solution.

4.13 Applications

Many plasma problems have been attacked by particle simulation methods. Natural extensions of the electrostatic, non-relativistic, homogenous model discussed in this course include:

(a) A fixed external magnetic field $\underline{B}$ which causes the particles to move in helices:

$$m \frac{d\underline{v}}{dt} = e(\underline{E} + \underline{v} \times \underline{B}_{ext}) . \qquad (117)$$

(b) A self-consistent magnetic field generated by the particles themselves.

(c) Interaction with electromagnetic waves.

(d) Relativistic particle motion.

(e) Generalized spatial geometry.

Electron plasma oscillations can be eliminated by assuming that the electron density satisfies a Boltzmann distribution

$$n_e = n_o e^{-e\phi/kT_e} , \qquad (118)$$

while the rapid Larmor precession of the charged particles can be removed by working with the equations for the drift motion of the guiding centres. It remains true however that the short space- and time-scales are still a problem in all these Vlasov calculations.

The following partial list of problems tackled by particle simulation has been given by Dawson[18]

(1) Landau damping.

(2) Nonlinear Landau damping (associated with two waves).

(3) Landau damping of finite amplitude waves.

(4) Collisional emission and absorption of plasma oscillations.

(5) Linear and nonlinear development of numerous instabilities;
   (a) two-stream instability,
   (b) cyclotron instability (Harris type),
   (c) verification of the quasi-linear theory for the weak two-stream instability with fixed ions (for mobile ions, demonstration that parametric instability may follow the two-stream instability in which case quasi-linear theory fails),
   (d) drift-cyclotron instability,
   (e) ion-acoustic instability and associated anomalous dc resistivity,

(f) side band instability due to electrons trapped in a large amplitude plasma oscillation.

(6) Development of finite-amplitude ion-acoustic waves.

(7) Parametric instabilities associated with high-frequency oscillating electric fields and the associated anomalous absorption of radiation.

(8) Diffusion of a plasma across a magnetic field.

(9) Quasi-resonant mode coupling and the damping of plasma oscillations caused by finite amplitude ion waves. The possibility of laser action by plasma.

(10) Investigations of collisionless shocks and the numerous nonlinear and turbulent dissipative mechanisms which may be involved.

5. THE OLYMPUS FORTRAN IV PROGRAMMING SYSTEM

5.1 Introduction

Two striking features of the CTR (controlled thermonuclear research) project are first, the complexity of the plasma physics and engineering, and second, the very considerable scale and therefore cost and construction time of the minimum-sized apparatus needed to reach ignition. In most other areas of technology it is possible to try out basic ideas rather quickly and empirically on prototypes which are relatively small and cheap to construct. Detailed theoretical analysis, diagnostic measurements and computer optimization come in later as the scale is gradually increased. In CTR we wish to be able to predict, with sufficient accuracy, the working of a very large and complex experimental fusion device or prototype reactor 'from the drawing board' before it is built, so enabling many variations to be evaluated and discarded before the optimum design is chosen. This requires a substantial and increasing effort on computing as well as on theoretical and experimental physics, plasma diagnostics and technology.

Such a quantitative increase in the scale and also the complexity of computing compared to that hitherto needed for other scientific and technological projects means that a qualitatively different method is needed for organising the programs. The scientist or engineer should be free to concentrate on the physical working of the device itself, and for this he should in the first place be able to call up and use a large selection of previously-constructed and tested programs from a library without having to spend unnecessary time and effort in finding out for himself how each individual program operates or in designing and programming his own version. He must however be able to establish without difficulty precisely which scientific and engineering assumptions have been adopted in any program that he uses, and also to vary or extend them where necessary, since many of the details of the underlying CTR model will remain in a state of flux for some time and new effects will be found to be important and so included. He or his colleagues must be able to understand the numerical schemes that have been used in a program, and to re-optimize or modify them to suit some previously untested regime of operation, as well as to add any extra types of output or program control that are needed.

These requirements imply that fusion library programs should have a clear standard structure, and that they should be easy to use, well-documented and flexible. Because CTR is a long-term international project we can expect that computational plasma physicists will be increasingly concerned in helping to design, construct, use and extend internationally-available suites of programs rather than in writing and using their own individual programs as hitherto. Major programs should therefore be planned to operate without essential modification on any existing or future computer and to be transferable without difficulty from one laboratory to another.

The OLYMPUS programming system[55] which has been designed and implemented by the Computational Physics Group at the UKAEA Culham Laboratory is intended to meet these and other specifications which are listed in Table 10. Both the specifications and their implementation will be discussed in more detail in the following parts of this section.

Table 10

OLYMPUS Specifications

Portability	Documentation
Immutability	Control
Flexibility	Diagnostics
Architecture	Construction
Notation	Testing
Layout	Automatic Program Generation

An important class of fusion computer calculations is the simulation of plasma behaviour as a function of time. This means solving an initial-value or evolutionary problem

$$\frac{\partial f}{\partial t} + G(f) = 0 \qquad (119)$$

where f is the solution vector and G is a linear or non-linear operator. The details of f and G as well as the initial boundary conditions depend on the calculations to be undertaken and on the choice of plasma model. Many interesting problems in hydrodynamics, astrophysics and other areas of classical physics as well as in CTR can be expressed in this form, and it seemed worthwhile to establish a common programming strategy for dealing with them. Once the OLYMPUS system had been developed, however, it was found to be equally useful with only minor changes for a much larger class of computing work, including the equilibrium and stability calculations used in CTR and programs which have no connection with physics at all, e.g. text manipulation.

OLYMPUS programs are written in (ANSI) Standard FORTRAN[56] because this is the only language subset available at virtually all computer installations and the hardware representation is reasonably well-defined so that tapes or cards can be interchanged. The basic ideas of OLYMPUS could however readily be adapted to programs written in other languages such as PL/I, Algol 60, Algol 68, or in assembler code. The present FORTRAN version has so far been successfully operated on at least 30 computer installations of 7 different types in several countries, and versions of the package have been published for the ICL 4/70[3], IBM 360/370[4] CDC SCOPE[5] systems.

5.2 Portability

Each OLYMPUS physics program consists in reality of a set of subprograms written in accordance with the rules of standard FORTRAN[56] These normally run under the supervision of a universal main program and associated control routine COTROL which form part of a Standard Control and Utility Package[57] it is stored in binary form in a local system library and brought into use by appropriate control statements. A separate version of the package is provided for each different type of computer on which the system has been implemented[57,58,59] The different versions are written mainly in FORTRAN and are similar to one another; however they need to take account of system-dependent features such as word and byte length, channel conventions and supervisor calls. Segregation of these system-dependent features within the OLYMPUS package makes it possible to transfer large physics programs from one computer to another with only minor changes, mostly to control statements, provided that the recommended conventions[55,56] have been followed.

5.3 Immutability and Flexibility

Because of the complexity of a typical CTR physics program it is necessary for the original programming team, the current user and those who are interpreting the results to be certain that no <u>unknown</u> changes have been made to the original version which would destroy the validity of the tests and documentation, and yet at the same time it must be possible for people to introduce <u>well-defined</u> modifications to meet any new requirements that arise. The scheme

COMPUTATIONAL METHODS IN PHYSICS

that has been planned (though not yet fully implemented) for this purpose is indicated in Figure 19.

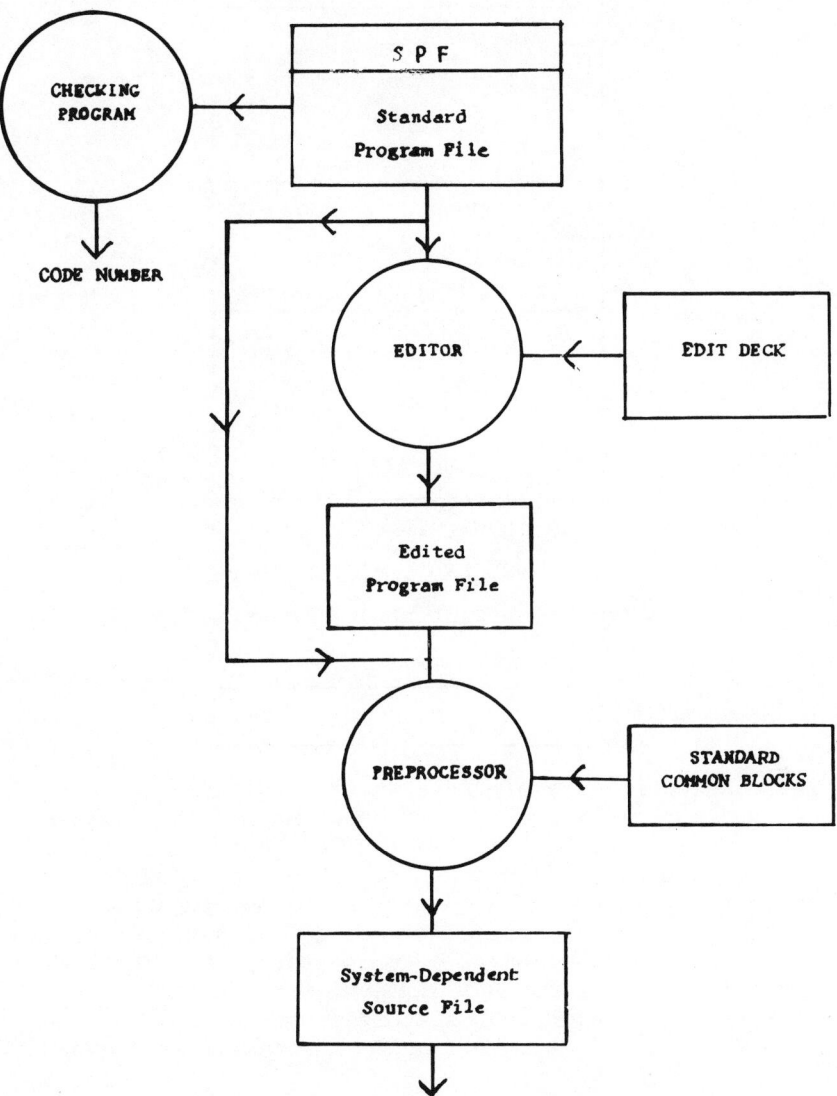

Figure 19. Portable, Immutable but Flexible Programs. OLYMPUS programs will be issued in a standard system-independent form (SPF) and a Checking Program will be provided to verify the validity of the complete character string. A Preprocessor will convert the SPF into the format required for a particular computer system. Changes may, if required, be introduced in a controlled way by using a system-independent Editor.

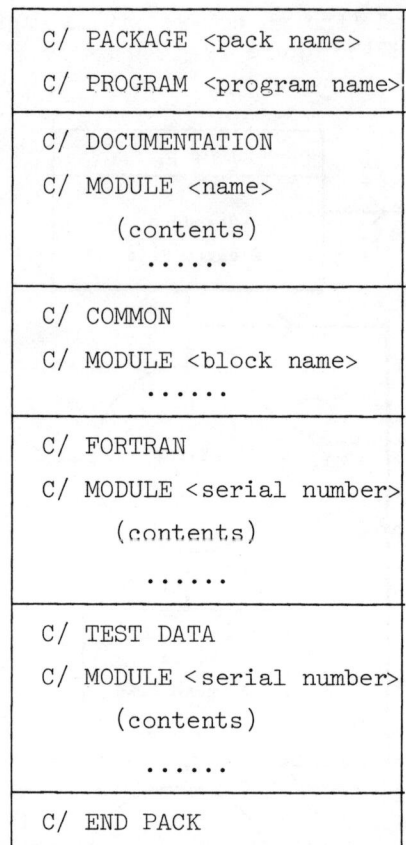

Figure 20. Structure of the Standard Program File.
The SPF is divided into 4 main sections concerned with
documentation, COMMON blocks, FORTRAN subprograms and
test data. Each section cintains one or more modules
headed by an identifying control statement beginning
'C/ MODULE'. The control combination 'C/' has been chosen
for the system-independent control statements since, as
it defines a legal FORTRAN comment, it ought not to
conflict with the control statements of any local system,
and OLYMPUS does not itself use comments with this format.

Each OLYMPUS program will in future be issued as a Standard
Program File (SPF) containing documentation, COMMON blocks, source-
language, test data and system-independent control statements (Figure
20). As part of a new edition of the OLYMPUS package, a Preprocessor
will be provided to convert the SPF into the form needed for a particular type of computer so that it does not need to be altered by
hand. This mainly involves generating a limited number of control
statements. To ensure that no undetected alterations have been made,

COMPUTATIONAL METHODS IN PHYSICS

a Checking Program, called the AUTHENTICATOR, will be provided which uses a system-independent random number generator to form a Code Number from the entire SPF character string by a hash-addressing technique. Finally, a system-independent Editor will be available to carry out any necessary modifications to the SPF after it has been checked and before it is preprocessed. The exact program version that is being run is then defined in a system-independent way by the Code Number and the Edit Deck. Provided that the edit deck is not too large and has been well laid out and commented, the modification should be sufficiently clear and the calculations can readily be checked or extended by another group of physicists using the same edit deck on their own computer.

5.4   Architecture

Understanding is facilitated by adopting the same basic structure for all programs of a similar kind and by laying this out as clearly as possible with a well-organized decimal numbering scheme. Each program can be thought of as consisting of two main parts:

INSTRUCTIONS

DATA STRUCTURE

The instructions are organized into a (universal) main program together with a set of subprograms, while the data is organized into labelled COMMON blocks. It may be useful to picture these COMMON blocks as operands and the subprogram as operators which act on them and change the values of the data that they contain.

So far as possible each subprogram and each block should have a well-defined purpose, and it is useful to adopt the broad classification indicated in Table 11. The subprograms are divided into <u>classes</u> m with decimal labels <m,n> while the COMMON blocks are divided into <u>groups</u> r with labels [Cr.s]. Only one copy of each

Table 11

Program Architecture

0 Control	4 Epilogue
1 Prologue	5 Diagnostics
2 Calculation	U Utilities
3 Output	

DATA: Groups of labelled COMMON blocks

1 General OLYMPUS data	4 Housekeeping
2 Physical problem	5 Input/output & diagnostics
3 Numerical scheme	6 Text manipulation

labelled block is used throughout the entire program; it is stored on-line and brought into use where required by means of a preprocessor control statement which in the system-independent version will have the form

$$C/INSERT \text{ (blockname)} \tag{120}$$

Table 12

CRONUS subprograms and COMMON blocks

Name	No.	Title
Subprogram		
Class 0. Main control		
(MAIN)	0.0	FORTRAN main program
BASIC	0.1	Initialize basic control data
MODIFY	0.2	Modify basic data if required
CONTROL	0.3	Control the run
EXPERT	0.4	Modify standard operation of program
Class 1. Prologue		
LABRUN	1.1	Label the run
CLEAR	1.2	Clear variables and arrays
PRESET	1.3	Set default values
DATA	1.4	Define data specific to run
AUXVAL	1.5	Set auxiliary values
INITAL	1.6	Define physical initial conditions
RESUME	1.7	Resume run from previous record
START	1.8	Start the calculation
Class 2. Calculation		
STEPON	2.1	Step on calculation
Class 3. Output		
OUTPUT	3.1	Control the output
Class 4. Epilogue		
TESEND	4.1	Test for completion of run
ENDRUN	4.2	Terminate the run
Class 5. Diagnostics		
REPORT	5.1	Control the diagnostics
CLIST	5.2	Print COMMON variables
ARRAYS	5.3	Print COMMON arrays
COMMON blocks		
Group 1.		General OLYMPUS data
COMBAS C1.1		Basic system parameters
COMDDP C1.9		Development & diagnostic parameters

COMPUTATIONAL METHODS IN PHYSICS

This makes the source code shorter and the data structure easier to understand.

To help fix the architecture, a standard library program called CRONUS is installed on-line in binary form as part of the OLYMPUS system. It contains the subprograms and COMMON blocks listed in Table 12, and although it does not itself perform any computation, it sets the basic structure for all programs in the OLYMPUS family.

This is achieved in the following way. Subroutine CONTROL <0.3> is a standard control subprogram which is used for all OLYMPUS programs developed so far and which calls lower-level subprograms in Classes 1-4 as well as EXPERT <0.4>. In CRONUS these are simply dummies which perform no actual work; in any 'real' program they are replaced by programmer-supplied versions with the same names and decimal numbers which automatically supplant the dummy library versions. Thus the programmer has to supply a subroutine STEPON <2.1> which organizes the calculation step and calls in other subprograms <2.2>,<2.3> to do the actual work; a subroutine OUTPUT (K) <3.1> to organize the output, and so on.

The group 1 COMMON blocks [C1.1]-[C1.9] are intended to be standard library versions which are part of the OLYMPUS package and are available to all programs. So far only two have been established containing channel numbers, step counter, timing information, restart details, labels and diagnostic control parameters. Others will deal with fundamental physical constants, character codes, input-output control parameters and other general-purpose information.

5.5 Notation, layout and documentation

Standardization of the architecture, notation and layout makes the SPF neater and easier to read, and once a new programmer has mastered the rules he can in practice develop and debug codes more quickly, partly because he is relieved from the need to make ad hoc decisions and partly because he and his colleagues can readily find their way about the listing and hence locate errors.

Tables 13 and 14 indicate some of the elements that have now been standardized. Decimal numbering has been found particularly useful, giving the SPF broadly the same structure as that of a well-planned mathematical textbook consistent with the limitations of FORTRAN. In addition to the numbering of subprograms and COMMON blocks we also divide an individual subprogram where appropriate into sections and subsections, and correlate these with FORTRAN statement numbers (and also line numbers, when working on-line). The initial-letter conventions defined in Table 14 allow the reader to see at a glance which variables and arrays are in COMMON and which are local, thus avoiding possible mistakes when introducing extensions. The methodical use of symbolic rather than arithmetic notation enables parameters to be readily updated should the need arise, as well as making their meaning clearer.

Table 13

Notation and layout

(a) Decimal numbering scheme

    Subprograms, e.g. <2.1> STEPON
    Common blocks, e.g.[C3.2]CONTIM
    Division of subprograms into sections and subsections
    Statement numbers correlated with sections
    Line numbers correlated with sections

(b) Notation for variables and arrays

    Initial letters are used to distinguish between:

    Common/internal
    Real/integer/logical
    Variable/formal parameter/index

(c) Layout

    Standard columns for comments, statements, headings, declarations etc.

(d) Symbolic notation

    Channel numbers   Table sizes
    Character codes   Constants
    Dimensions

(e) Standardization

    Control variables   File names
    Subprograms   Fundamental constants
    Common blocks   NAMELIST data input

Table 14

Initial letters and array names

	Real, complex	Integer (and Hollerith)	Logical
Subprogram dummy arguments	P	K	KL
Common variable and array names	A-H,O,Q-Y	L,M,N	LL,ML,NL
Local variable and array names	Z	I	IL
Loop indexes		J	

```
C              SUBROUTINE COTROL
C
C 0.3 Control the run
C
C      Version 2a         1/8/73      kvr/jpc       culham
C
//SUBSTITUTE COMBAS
//SUBSTITUTE COMDDP
C-----------------------------------------------------------------
C
       DATA ICLASS,ISUB/0.3/
       CALLMESAGE(48H    0.3 ENTER RUN CONTROL
C
C-----------------------------------------------------------------
CL              1.         Prologue
C
       IF(NLRES) GO TO 170
C
C              A.         New Run
C
CL              1.1        Label the run
   110     CALL LABRUN
                              CALL EXPERT(ICLASS,ISUB,1 )
C
CL              1.2        Clear variables and arrays
   120     CALL CLEAR
                              CALL EXPERT(ICLASS,ISUB,2 )
C
CL              1.3        Set default values
   130     CALL PRESET
                              CALL EXPERT(ICLASS,ISUB,3 )
C
CL              1.4        Define data specific to run
   140     CALL DATA
                              CALL EXPERT(ICLASS,ISUB,4 )
C
CL              1.5        Set auxiliary values
   150     CALL AUXVAL
                              CALL EXPERT(ICLASS,ISUB,5 )
C
CL              1.6        Define physical initial conditions
   160     CALL INITAL
                              CALL EXPERT(ICLASS,ISUB,6 )
           GO TO 180
C
C              B.         Resume a previous run
C
CL              1.7        Pick up record, modify required parameters
   170     CONTINUE
                              CALL EXPERT(ICLASS,ISUB,7 )
C          Label the continuation run
           CALL LABRUN
                              CALL EXPERT(ICLASS,ISUB,8 )
C          Clear variables and arrays
           CALL CLEAR
                              CALL EXPERT(ICLASS,ISUB,9 )
```

(continued)

```
C         Pick up record and print details
          CALL RESUME
                                    CALL EXPERT(ICLASS,ISUB,10)
C         Read any new data needed
          CALL DATA
                                    CALL EXPERT(ICLASS,ISUB,11)
C         Modify auxiliary variables as required
          CALL AUXVAL
                                    CALL EXPERT(ICLASS,ISUB,12)
C
C                C.         Preliminary operations
C
CL               1.8        Start or restart the run
  180 CALL START
                                    CALL EXPERT(ICLASS,ISUB,13)
C         Initial output
          CALL OUTPUT(1)
                                    CALL EXPERT(ICLASS,ISUB,14)
C
C
C----------------------------------------------------------------
CL               2.         Calculation
C
CL               2.1        Step on the calculation
  210 CALL STEPON
                                    CALL EXPERT(ICLASS,ISUB,15)
C
C----------------------------------------------------------------
CL               3.         Output
C
CL               3.1        Periodic production of output
  310 CALL OUTPUT(2)
                                    CALL EXPERT(ICLASS,ISUB,16)
C
C----------------------------------------------------------------
CL               4.         Epilogue
C
CL               4.1        Test for completion of run
  410 CALL TESEND
                                    CALL EXPERT(ICLASS,ISUB,17)
      IF(.NOT.NLEND) GO TO 210
C
C         Final output
          CALL OUTPUT(3)
                                    CALL EXPERT(ICLASS,ISUB,18)
C
CL               4.2        Terminate the run
  420     CALL ENDRUN
C
          RETURN
          END
```

Figure 21. Subroutine COTROL <0.3>. This subroutine controls the operation of all programs in the OLYMPUS family. Comments have been converted to lower case to make the listing more readable. EXPERT calls enable ad hoc sections of code to be inserted where needed.

```
C
              SUBROUTINE STEPON
C
C 2.1.  Advance one time step
C
//  SUBSTITUTE COMBAS
//  SUBSTITUTE COMDDP
//  SUBSTITUTE COMCRV
//  SUBSTITUTE COMINT
//  SUBSTITUTE COMAX
//  SUBSTITUTE COMCON
C-----------------------------------------------------------------
      DATA ICLASS,ISUB/2.1/
C
          NSTEP=NSTEP+1
          IF(NLOMT2(ISUB)) RETURN
C
C-----------------------------------------------------------------
CL        1.              Update variables
C
C     Change parity
          MIDDLE=NMAXH-MIDDLE
          MJUMP=NMAXH-MJUMP
C                                  CALL EXPERT(ICLASS,ISUB,1)
C
C-----------------------------------------------------------------
CL        2.              Calculate charge
C
C     Calculate charge within eulerian intervals
          CALL SIGMA          CALL EXPERT(ICLASS,ISUB,2)
C
C-----------------------------------------------------------------
CL        3.              Calculate field
C
C     Calculate field at eulerian points
          CALL FIELDE
C                                  CALL EXPERT(ICLASS,ISUB,3)
C     Find particle charge distribution
          IF(NUMBRT.NE.0) CALL SIGMAP
C                                  CALL EXPERT(ICLASS,ISUB,4)
C     Calculate field at lagrangian points
          CALL FIELDP
C
C                                  CALL EXPERT(ICLASS,ISUB,5)
C
C-----------------------------------------------------------------
```

(continued)

```
CL                4.         Insert and remove points
C
C      Check if we insert and remove points this step
            IF(MOD(NSTEP,NADD).NE.0)GO TO 500
C
C      Adjust points
            CALL ADJUST
C                                      CALL EXPERT(ICLASS,ISUB,6)
C      Check if table is full
            IF(.NOT.NLFULL) GO TO 500
C
C      Table is full , print a warning
            CALL IVAR(8HNSTEP   ,NSTEP)
            CALLMESAGE(48H *** TABLE FULL - HALVE CALLED ***
C
C      Halve number of points and re-adjust
            CALL HALVE
            NLFULL=.FALSE.
            CALL ADJUST
C
C                                      CALL EXPERT(ICLASS,ISUB,7)
C      Stop the run if table still full
            IF(.NOT.NLFULL)GO TO 500
            CALLMESAGE(48H *** RUN TERMINATED ***
C
            CALL ENDRUN
C
C-------------------------------------------------------------------
CL                5.         Advance contours
C
   500      CONTINUE
C
C      Advance co-ordinates and velocities of points
            CALL MOVE(MX,MVDT)
            CALL MOVE(MVDT,MADTSQ)
C                                      CALL EXPERT(ICLASS,ISUB,8)
C
            RETURN
            END
```

Figure 22. Subroutine STEPON <2.1> of the VLASOV code. Each OLYMPUS program has its own version of subroutine STEPON, which controls the physics calculation and calls in other subroutines of Class 2 to do the actual work. The figure also illustrates the use of the message facility [56].

Figures 21 and 22 show two examples of subprograms. The comments have been automatically converted to lower case to make them stand out more clearly and printed on a General Electric TermiNet 300. It is intended to reproduce all future OLYMPUS listings in this form.

Although most existing programs are extremely hard to understand, there is in fact no real difficulty in making a program intelligible if the need for this is foreseen from the outset. Table 15 lists some standard documentation tools. In OLYMPUS programs each numbered section or subsection is preceded by an explanatory heading, and other comments are used to explain the purpose of individual details of the program. Headings, comments, and statements begin in different standard columns in order to distinguish them more clearly. Indexes of variables and arrays and of subprograms are provided. A useful technique is to cross-reference the listing with a program commentary or write-up so that the two can be ready together; the listing refers to the equation numbers in the write-up, whilst the write-up refers to the subprogram and section numbers of the program. Automatic documentation tools are being developed for use with OLYMPUS and other programs and will be made available in due course.

Table 15

Standard Documentation Tools

Titles and headings	Program commentry
Decimal numbering	Equation numbers and references
Indexes	Cross-references
Comments	Program map or flowchart

5.6 Control

Many large physics programs are in a continual state of development since ad hoc modifications have to be introduced to enable particular numerical experiments to be carried out. This is especially true of the programs used in fusion research. Unless precautions are taken such modifications can lead to untidy coding and multiple copies of source and object files, together with some confusion as to which version of the program has been used for a particular run. Some thought has therefore been given to the way in which OLYMPUS calculations should be controlled and ad hoc modifications made (Table 16). A program is thought of as an engineering 'apparatus' and the control techniques are standardized, so making it straightforward for users to transfer their attention from one program to another as the physics problem demands:

1. All OLYMPUS programs share the same main program <0.0> and the same master control subprogram CONTROL <0.3> which are called automatically from a binary library. All such programs therefore have a similar basic structure.

Table 16

Control of Calculations

1. Standard main program <0.0> and CONTROL <0.3>
2. Default values optimally overwritten by NAMELIST data
3. Arbitrary functions with adjustable parameters
4. Logical switches set by NAMELIST data
5. EXPERT <0.4> facility for ad hoc variations
6. System-independent Editor for changes to code.

2. A simple standard method is used for initializating the data structure, illustrated in Figure 23. First, all COMMON variables and arrays which can be independently set are assigned reasonable

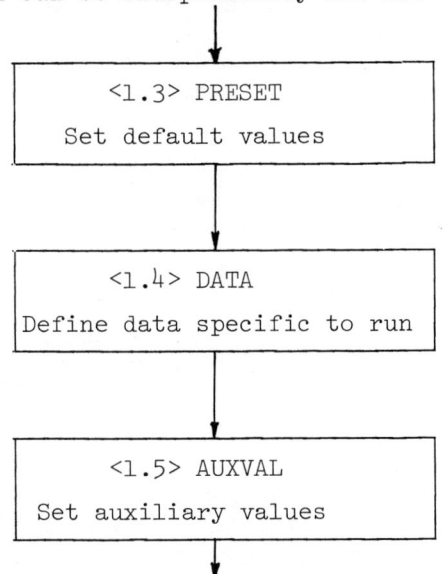

Figure 23. Initialization of the Data Structure. The data structure of OLYMPUS programs is always initialized in the same way. COMMON variables and arrays are first cleared to 0.0,0. FLASE. or blank (as appropriate) in CLEAR <1.2> , and those that can be independently defined are then given suitable default values in PRESET <1.3>. The user is subsequently able to overwrite any or all of the standard settings in NAMELIST NEWRUN which is ready by DATA<1.4>. The input deck thus has a convenient symbolic free format and is generally quite short; even if the user chooses to submit no data at all the program will generally execute a short test run for instructional purposes. AUXVAL <1.5> finally sets the values of auxiliary parameters that depend on the input data.

default values in subprogram PRESET <1.3>, so that subsequently the user only has to supply data values for parameters that he wishes to alter: in fact an OLYMPUS program will usually carry out a simple test run with no data at all other than the 4 cards which are required to label the calculation. Next, subprogram DATA <1.4> uses the NAMELIST input facility to overwrite the values of any of these variables or array elements as required: this is achieved (for a new run) by a <u>single problem-independent input statement</u>.

```
                    READ(NREAD,NEWRUN)
```

where NEWRUN is a NAMELIST name and NREAD is the symbolic input channel. Similarly for a restart. Finally, subprogram AUXVAL <1.5> constructs the values of any auxiliary variables which depend on the input data and cannot therefore be independently specified.

```
            T E S T     3   1 6 / 1 1 / 1 9 7 3
            S T A N D A R D    T E S T    3
            I S E N T R O P I C     C O M P R E S S I O N
            I N    P L A N E    G E O M E T R Y
              &NEWRUN
                    NCASE=3,
                    NGEOM=1,
                    NLABS=F,
                    NLBRMS=F,
                    NLBURN=F,
                    NLCRI1=F,
                    NLDEPO=F,
                    NLDUMP=F,
                    NLECON=F,
                    NLFILM=F,
                    NLFUSE=F,
                    NLHCPY=F,
                    NLICON=F,
                    NLX=F,
                    NPRNT=200,
                    NRUN=200,
                    PIQ(3)=1.3466E-7,
                    PIQ(4)=-1.25,
                    PIQ(6)=1.0E5,
              &End
```

Figure 24. OLYMPUS Input File. An OLYMPUS input file consists of 4 cards which are used to label the run, followed by a NAMELIST which has a convenient free-format symbolic structure. Only those data values that are to be changed from their preset values need be specifically mentioned and the order is immatrial. This example illustrates the use of switches to turn off most of the physics in order to test a section of the code.

This technique provides for symbolic freeformat input (Figure 24) and avoids the many complicated formatted and ordered input statements and data cards that are used by the majority of FORTRAN programs. NAMELIST data input and output are not unfortunately part of Standard FORTRAN[56] although they certainly should be; they are however available with only minor variations on all the computers on which OLYMPUS has been implemented so far.

3. Any arbitrary functions (e.g. initial conditions) are defined by function subprograms or statement functions containing adjustable parameters which are placed in COMMON. If a user wishes to alter the analytic form of a function he recompiles that small section of the program; otherwise he specifies the appropriate parameter changes in the NAMELIST DATA.

4. Logical switches are provided (again in NAMELIST data) for turning program facilities on or off; for example subprograms, physical effects, output options, diagnostics and so on.

5. Ad hoc additions or modifications to sections of the code can in many cases be localized within a single subprogram EXPERT <0.4> which is called from many points throughout the program (outside inner loops) with 3 arguments defining the location from which the call is made. Normally a standard library version of EXPERT is loaded which inserts no extra code; however the user can provide his own version, containing the additional or modified sections of code and controlled by IF or GO TO statements. Interaction with the body of the program is that the ad hoc alterations to the standard version are fully defined by the compilation listing of EXPERT, which can of course be laid out in a readable way with numbered comments, whilst the original program remains unchanged in object module form. This enables several people to use a program at the same time for independent calculations without interfering with one another, and allows the alterations to be recorded in a precise and convenient fashion.

6. The system-independent Editor which has been mentioned earlier will provide a similar facility for those alterations to the program that cannot conveniently be made using EXPERT, and in this case the EDIT Deck will be used to record the changes that have been made.

5.7 Diagnostics

A fairly elaborate set of standard diagnostics tools has been developed to enable OLYMPUS programs to be checked out as quickly as possible. Each program is 'instrumented' like any other engineering apparatus whilst it is being written, i.e. with built-in calls of the form

$$\text{CALL EXPERT (ICLASS,ISUB,IPOINT)} \qquad (122)$$

already mentioned together with certain other diagnostic statements, and the diagnostics can then be switched on or off at will either by including coded data in the NAMELIST input deck or by inserting appropriate control statements in the program. Facilities are available to output messages, to trace the flow of the program, to print the name and values of individual real, integer, logical or Hollerith variables or arrays, to print out COMMON blocks in alpha-numeric order, to time sections of the program[59], and to switch individual subprograms off if they are found to contain catastrophic errors so that other tests can continue.

5.8 Construction and testing

The OLYMPUS system enables a program to be constructed and tested in a methodical way. It is recommended that the data structure should be checked out first, using the CRONUS program with its dummy STEPON <2.1> as a testbed and the diagnostic utility routines for temporary output. Once the data initialization has been checked the program can next be run for one or two timesteps. All these initial tests are carried out with small array sizes in order to limit the output, particularly when working on-line. Because the program is modular its various parts can often be developed and checked out separately, again using CRONUS as a testbed; this applied in particular to the output routines which are logically independent of the physics calculation. When a program has been tested on one computer it is good practice to try it on another of different type in order to bring to light any unintended dependence on the local system.

5.9 Automatic program generation

Because OLYMPUS programs have a well-defined standard structure, there is no reason why many of the 'housekeeping' sections of any new program should not be constructed automatically by the computer itself, thus relieving the programmer of unnecessary work. The scheme which has been implemented for this purpose at the Culham Laboratory by Dr. M.H.Hughes is indicated in Figure 25. Information about the program and its data structure, i.e., the COMMON variables and arrays, is contained in a Master Index which is used by the Generator to construct subsidiary indexes, COMMON blocks, many of the Class 1 subprograms such as LABRUN, CLEAR, PRESET and AUXVAL, and the Class 5 diagnostic subprograms CLIST <5.2> and ARRAYS <5.3> which are used to produce annotated dumps. COMMON variables and arrays can be added, removed and reorganised, or array dimensions changed, simply by updating the MASTER Index and then regenerating these parts of the program.

Another facility provided by a word-processing program called

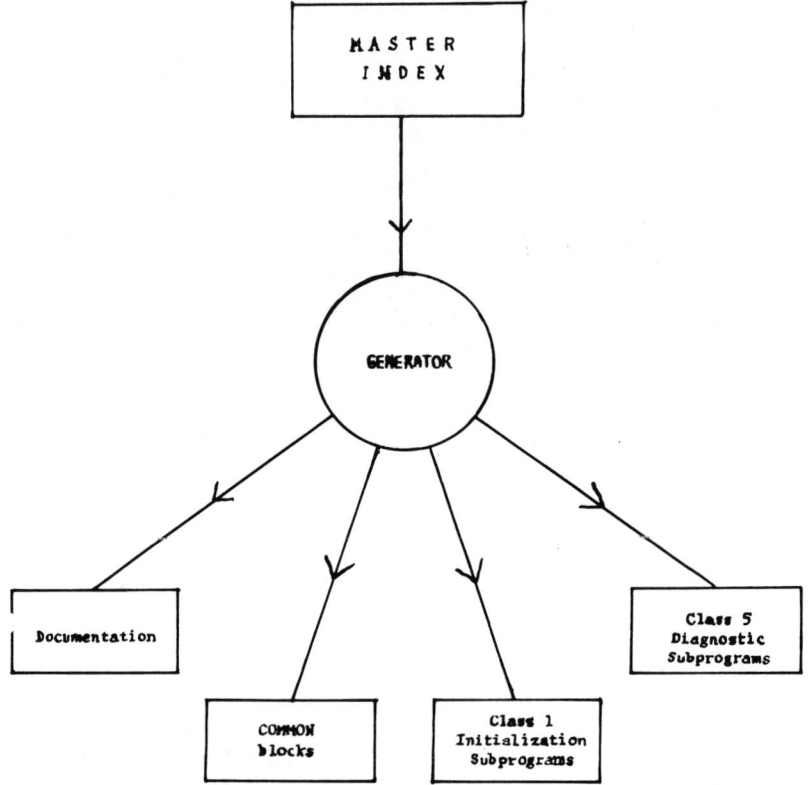

Figure 25. Automatic Program Generaion. To minimise the amount housekeeping work to be done by the programmer, he is simply asked to maintain a Master Index which contains enough information to define the data structure. This is then used by the Generator to construct in a suitably neat form those sections of the program that depend on this information.

the COMPOSITOR is for the programmer to be able to write his code in a free-format shorthand designed for maximum input convenience, and then for the Generator to convert this into the standard OLYMPUS format which is designed for maximum readability; the two requirements are not of course identical.

Once the Generator has been finalized and made available as a system-independent package, it is planned to exchange programs in a more concise format than that of Figure 20, containing the Master Index, since this is shorter and more convenient for the reader than those sections of the SPF that it generates.

5.10   Concluding remarks

Although considerable difficulty has hitherto been encountered in developing, testing, operating, modifying, exchanging and understanding scientific computer programs and in transferring them from one type of machine to another, it does appear that most of the technical problems can in fact be solved by a methodical and consistent approach such as that adopted by the OLYMPUS programming system. Standard Fortran is certainly not an ideal vehicle for this purpose but it is adequate for much of the work and it is the only suitable language subset that is generally available. Once the exchange and publication of scientific programs has become more widely accepted it is however to be hoped that some more powerful language or an improved version of FORTRAN will in due course be able to supplement it.

References

1. 'Computers and their Role in the Physical Sciences', ed Fernbach and Taub, Gordon & Breach, New York (1970).

2. 'Computers as a Language of Physics', Seminar Course held at the International Centre for Theoretical Physics, Trieste, August 1971. IAEA Vienna (1972).

3. 'The Impact of Computers on Physics', Comp. Phys. Commun. Vol.3 Suppl. (1972).

4. 'Methods in Computational Physics', ed. Alder, Fernbach & Killeen (and previously by Rotenberg), Vol.1 (1963) - Vol. 17 (1977), Academic Press, New York.

5. 'Journal of Computational Physics', ed. Alder, Fernbach & Killeen (and previously by Rotenberg), Vol.1 (1966) - Vol. 30 (1979), Academic Press, New York.

6. 'Computer Physics Communications', ed. Burke, Vol. 1 (1969) - Vol. 15 (1978), North Holland, and the 'International Physics Program Library, Queens University, Belfast.

7. D.E.Potter, 'Computational Physics', Wiley, London (1973).

8. K.Appel & W.Haken, 'The Solutions of the Four-Colour-Map Problem' Scientific American Vol. 237, No. 4, 108 Oct. (1977).

9. W.E.Kock, 'Engineering Applications of Lasers and Holography', Plenum Press, London & New York (1975).

10. K.V.Roberts, 'Trends in Computational Physics', in Computational Methods in Classical and Quantum Physics, (ed. M.B.Hooper, Advance Publications Ltd., (1976).

11. K.V.Roberts & J.P.Boris, 'The Solution of Partial Differential Equations using a Symbolic Style of Alogl', J. Comp. Phys., 8, 83 (1971); M.Petravic, G.Kuo Petravic & K.V.Roberts, 'Automatic Optimization of Symbolic Algol Programs, I', General Principles, J. Comp. Phys., 10, 503 (1972).

12. G.M.Schneider, S.W.Weingart & D.M.Perlman, 'An Introduction to Programming and Problem Solving in PASCAL', Wiley (1978).

13. K.V.Roberts, 'FORTRAN', in 'Software Engineering' (ed. R.H. Perrott), Academic Press (1977).

14. Draft Proposed ANS FORTRAN, Sigplan Notices, 11, No. 3 (March 1976).

15. J. von Neumann, Probabilistic Logics and the Synthesis of Reliable Organisms from Unreliable Components', Automata Studies, Princeton University Press (1956).

16. E.Schrödinger, 'What is Life? The Physical Aspect of the Living Cell', Cambridge University Press (1945).

17. K.V.Roberts & D.E.Potter, 'MHD Calculations', Methods in Computational Physics, 9, 340 (1970).

18. J.E.Fromm, Methods in Computational Physics, 3, 345 (1964).

19. F.H.Harlow, 'Numerical Methods for Fluid Dynamics, an Annotated Bibliography, Los Alamos Report LA-4281 (1970).

20. F.H.Abernathy & R.E.Kronauer, J.Fluid Mech., 13, 1 (1962).

21. J.P.Christiansen, J. Comp. Phys., 13, 363 (1973); K.V.Roberts & J.P.Christiansen, Comp. Phys. Commun., 3 (Suppl.) 14 (1972); J.P.Christiansen & N.J.Zabusky, J. Fluid Mech., 61, 219 (1973).

22. R.W.Hockney, J. Assoc. Comput. Mach., 12, 95 (1965).

23. K.V.Roberts & H.L.Berk, Phys. Rev. Lett., 19, 297 (1967); H.L. Berk & K.V.Roberts, Phys. Fluids, 10, 1595 (1967), Methods in Computational Physics, 9, 88 (1970).

24. G.S.Deem & N.J.Zabusky, Phys. Rev. Lett., 40, 859 (1978).

25. S.J.Zaroodney & M.D.Greenberg, J. Comp. Phys., 11, 440 (1973).

26. N.J.Zabusky, Comp. Phys. Commun., 5, 1 (1973).

27. See, C.W.Gowers et al, Proceedings of the sixth Conference on Plasma Physics and Controlled Nuclear Fusion Research, 1976, Vol. II, P.429. IAEA Vienna (1977).

28. K.V.Roberts & N.O.Weiss, Math. Comp., 20, 272 (1966).

29. K.J.Whiteman, Rep. Prog. Phys., 40, 1033 (1977).

30. J.P.Boris, D.L.Book & K.Hain, J. Comp. Phys., 18, 248 (1975).

31. G.A.Sod, J. Comp. Phys., 27, 1 (1978).

32. R.F.Warming & B.J.Hyett, J. Comp. Phys., 14, 159 (1974).

33. R.D.Richtmyer & K.W.Morton, 'Difference Methods for Initial Value Problems', 2nd Ed. p. 292, Interscience (1967).

34. N.A.Phillips, 'An Example of Non-Linear Computational Instability in 'The Atmosphere and the Sea in Motion', Rockefeller Institute Press, New York (1959).

35. S.A.Orszag, Stud. in Appl. Math., 50, 293 (1971).

36. loc.cit.ref. 33 p. 198.

37. J.P.Christiansen, D.E.T.F.Ashby & K.V.Roberts, Comp. Phys. Commun. 7, 271 (1974).

38. A survey of numerical methods in plasma physics is contained in the review volume 'Methods in Computational Physics', Vol. 9, 'Plasma Physics', ed. B.Alder, S.Fernbach & M.Rotenberg, Academic Press, New York (1970).

39. Work on plasma simulation is covered by 'Proceedings of the Fourth Conference on Numerical Simulation of Plasmas', Naval Research Laboratory, Washington DC, November 1970, Office of Naval Research (1971).

40. F.H.Harlow, 'Contour Dynamics for Numerical Fluid Flow Calculations', J. Comp. Phys., 8, 214 (1971); N.J.Zabusky, M.H.Hughes & K.V.Roberts, 'Contour Dynamics for the Euler Equations in Two Dimensions', J. Comp. Phys, to be published, see ref. 23.

41. R.W.Hockney, Phys. Fluids, 9, 1826 (1966); P.Burger, D.A.Dunn & A.S.Halstead, Phys. Fluids, 8, 2263 (1965); O.Buneman, J. Comp. Phys., 1, 517 (1967).

42. R.L.Morse, 'Multidimensional Plasma Simulation by the Particle-in-Cell Method', loc.cit. ref 2, pp 213-239; C.K.Birdsall, A.B.Langdon and H.Okuda, loc.cit. ref. 2, pp 241-258.

43. R.W.Hockney, 'Particle Models in Plasma Physics', paper C, Vol.1, IPPS Conference on Computational Physics, Culham Laboratory, July 1969, Proceedings published as Culham Report CLM-CP (1969), (2 vols) HMSO London.

44. For references see R.W.Hockney, 'The Potential Calculation and Some Applications', Methods in Computational Physics, Vol.9, 'Plasma Physics', pp 135-211.

45. R.W.Hockney, Journ. ACM., 12, 95 (1965).

46. J.P.Boris & K.V.Roberts, 'The Optimization of Particle Codes in 2 & 3 Dimensions', J. Comp. Phys., 4, 552 (1969).

47. J.H.Orens, J.P.Boris & I.Haber, 'Optimization Techniques for Particle Codes', loc.cit. ref. 3, p. 526.

48. B.Rosen, W.L.Kruer & J.W.Dawson, 'A New Version of the Dipole Expansion Scheme', loc.cit. Ref. 3, p. 561.

49. J.A.Byers & M.S.Grewal, Phys. Fluids, 13, 1819 (1970).

50. J.A.Byers, 'Noise Suppression Techniques in Macroparticle Models of Collisionless Plasmas', loc.cit. ref. 3, p. 496.

51. A.B.Langdon, 'Non-physical modifications to oscillations, fluctuations and collisions due to space-time differencing', loc.cit ref. 3, p. 467.

52. A.B.Langdon, J. Comp. Phys., <u>6</u>, 247 (1970).

53. H.Okuda, 'Non-physical noises and instabilities in plasma simulation due to a spatial grid', J. Comp. Phys., <u>10</u>, 475 (1972).

54. J.M.Dawson, 'Contributions of computer simulation to plasma theory', Comp. Phys. Comm., Suppl. <u>3</u>, 79 (1972).

55. K.V.Roberts, Computer Phys. Commun., <u>7</u>, 237 (1974).

56. 'Standard Fortran Programming Manual', Computer Standards Series National Computer Centre Ltd, Manchester, England (1970).

57. J.P.Christiansen and K.V.Roberts, Computer Phys. Commun., <u>7</u>, 245 (1974).

58. M.H.Hughes, K.V.Roberts and P.D.Roberts, Computer Phys. Commun., <u>9</u>, 51 (1975).

59. M.H.Hughes, G.G.Lister and K.V.Roberts, Computer Phys. Commun., <u>10</u>, 167 (1975).

60. M.H.Hughes and A.P.V.Roberts, Computer Phys. Commun., <u>8</u>, 118 (1974).

PART IV

PHYSICS AND FRONTIERS OF KNOWLEDGE

OBSERVATIONAL TRAITS OF BLACK HOLES IN THE OPTICAL BAND

P.L. Bernacca, L. Bianchi, and R. Turolla

Asiago Astrophysical Observatory

The University of Padova

1. INTRODUCTION

A black hole is, still today, only a consequence of the theory of General Relativity. Till the discovery of Pulsars also neutron stars were a theoretical product, in this case of stellar evolution, and a consequence of the stability of superdense bodies. Thirty four years are elapsed between early predictions on their existence made by Zwicky in 1934 and a serious proof of their existence in 1968, when pulsars were discovered. In the case of black holes a convincing evidence, in terms of unambiguous interpretations of observations with a view to current predictions of their observational features, is still missing. The scene is not however discouraging, and at least in one case, Cyg x-1, the basic requirements for the existence of a black hole are satisfied. They are[1]

1. Emission of X-Rays as a result of the interaction of a strong gravitational field and matter around the hole.

2. Short term variability of the X-Ray yield denoting a small object.

3. Mass larger than about 3 $M_\odot$ which would exlude a neutron star.

What is a black hole? It is what is left behind after an object has undergone complete gravitational collapse. Ruffini and Wheeler[2] write: "Spacetime is so strongly curved that no light can come out, no matter can be ejected and no measuring rod can ever survive being put in". The reader is referred to the above authors for an introduction to black hole physics and the related literature. For the purpose of this lecture we adopt the following operative definition: A black hole is the product of the collapse of one or several stars with total mass larger than about 3 $M_\odot$, such that no radiation whatsoever can escape and reach a remote observer. Radiation however may be produced by interaction of the hole with the surrounding medium.

In the following we restrict this presentation to the observable characteristics of Black Holes in binary systems (Section 2), in the cores of Globular Clusters (Section 3) and of isolated Holes in the general interstellar medium (Section 4). Problems like the formation of X-Ray binaries or the coalescence of stars inside Globular Clusters are beyond the purpose of this lecture note.

2. BLACK HOLES IN BINARY SYSTEMS

Zeldovich and Novikov[3] were the first to suggest that the accretion of matter lost from the primary onto a neutron star or black hole in a close binary system can produce x-Ray emission by conversion of gravitational energy into radiation. Based on this idea early models of x-Ray sources have been suggested by several authors including Sholokovsky[4], Cameron and Mock[5], Prendergast and Burbidge[6]. The first observational evidence for such systems came with the identification of Sco x-1[7] and of Cyg x-2[8] with stars showing spectral similarities to old novae: these stars are well known to be close binaries with mass transfer from the star to a white dwarf companion[9].

To-day, it seems that all the observed Compact x-Ray source other than SNR can be accounted for by the model of accretion in close binary systems; the efficiency of this process is such that the observed accretion rate from early type stars can produce luminosity and temperature characteristic of galactic x-Ray sources.

It is now clear that the study of Compact x-Ray sources has a fundamental importance in the search for collapsed objects. In fact, a black hole does not radiate gravitational nor electromagnetic waves[10], and therefore it can be detected only through its gravitational interaction with a circumstant dense gas medium (the gas would accrete the black hole, releasing energy, see, e.g. ref. 11 and 12), or with a neighbouring near star. Searching for black holes in the optical band rests on the photometric and spectroscopic features of this accretion process. Before discussing these features it is convenient to summarize the general characteristics of the x-Ray sources themselves.

2.1   Compact x-Ray Sources: General Observed Characteristics

We pointed out the great importance that the compact x-Ray

sources have for the search of collapsed star. Really, all the observed galactic x-Ray sources other than SNR may be accounted for by considering an accretion model in a close binary system containing a compact star. This statement is based on the following main reasons:

(1) Many galactic x-Ray source are firmly established to be in binary systems (from evidence of x-Ray eclipses, or association with binary stars, or regular Doppler variations in the pulsation period induced by the orbital motion, e.g., Cen x-3, Her x-1).

(2) the 'flat' spectrum and the short time scale variation observed require that such emission originates from a compact region.

(3) Mass accretion onto a collapsed object from a mass companion is an extremely efficient mechanism to produce x-Rays (the other known events which imply high energies are cosmic Rays and Supernovae). Presently about 12 galactic x-Ray sources are identified as SNR; if we consider that the stars with a mass greater than a fairly low limit may evolve up to the last stages, and that at least the 50% of the stars are double[13], it is possible that an appreciable number of binary systems, will not be destroyed by the collapse of one component and the stars will remain bound. In this connection, contemporaneously, as the beginning of the x-Ray observational astronomy, but in an apparently correlated activity, astronomers became concerned how the stellar evolution is affected by membership of a star in a close binary system, in which mass transfer may become important, (as it seems to be the case of the 'old novae').

(4) the spectra of many sources show large and variable 'low energy cut-offs', that may be interpreted as due to gas streaming, i.e., to the presence of a great (and variable, along the line of sight) amount of matter in the system.

The observed characteristics of the compact x-Ray sources are here briefly summarized:

(1) Most of them have a 'flat' spectrum in the range 1-10 Kev, which indicates temperatures from 50 to $500 \cdot 10^6$ °K if the radiation is produced by hot gas.

None has a flat spectrum for energies > 30 Kev but many of them show 'non-thermal tails' extending beyond 100 Kev. Most of the radiated power is in the range 2-10 Kev.

(2) the total number of the Galctic x-Ray sources is not very high, it is $\sim$ 100.

(3) Spatial distribution: galactic x-Ray sources are concentrated along the galactic plane, and particularly at low galactic longitudes. There are two populations of sources: the sources in the outer regions of the Galaxy are of extreme Pop.I and are found in spiral arms[14,15], while the strong sources in the central region

are distributed as ordinary stars and are not of extreme Pop I.

(4) Although distance estimates are difficult to obtain for x-Ray stars, it seems secure that the intrinsic liminosity has a maximum of 10^{38} erg/s and a minimum value of around 10^{36} - 10^{35} erg/s.

(5) Temporal variations are observed on every time scale, from millis. to years; variability may be casual or periodic.

2.2   Optical Identification of Compact X-ray Sources

Recent progress in identifying optical counterparts of galactic X-Ray sources has been reviewed by Hutchings[16]. Since we deal only with possible black holes, we simply mention only those binary systems with secondaries having a mass larger than 3.2 $M_\odot$, the accepted upper limit for the mass of a neutron star. These are very few and are listed in Table 1.

Table 1

Black hole candidates

X-Ray Source	Optical Object	M_p	M_x	S_p	Sep(R_p)	$\dot{M}_p$
Cyg X-1	HDE 226868	25	15	09.7 Iab	1	$2.6 \cdot 10^{-6}$
0352 + 30	X Per	20	40	B0Ve	50	10^{-7}
0053 + 60	Cas	20	6	B0Ve		10^{-7}
0531 + 05	OriA	22	13	09V	36	$6 \cdot 10^{-8}$

Columns 3 to 8 give respectively: the mass of the visible star (primary), the mass of the unseen companion thought to produce X-rays, the spectral type, the separation of the two components in units of the radius of the primary and the mass loss rate from it. The case of OriA should be considered still uncertain because it is not definitly established as optical counterpart of the source. Bernacca and Bianchi[17] have studied this object in the far UV by means of the satellite Copernicus and found a mass loss rate $\dot{M} = 6.4 \cdot 10^{-8}$ which can account for the observed X-Ray luminosity $L_x = 10^{33}$ erg/sec by accretion onto the hypothetical compact secondary.

2.3   X-Ray Yield by Mass Accretion

The basic mechanism of x-Ray production by accretion onto a black hole is the following: a fraction of the mass lost from the normal star (essentially by radiation pressure or x-Ray heating driven stellar wind or by Rochelobe overflow, (see Figure 1) falls into the sphere of influence of the gravitational field of the black hole. Due to the angular momentum conservation, this matter cannot reach

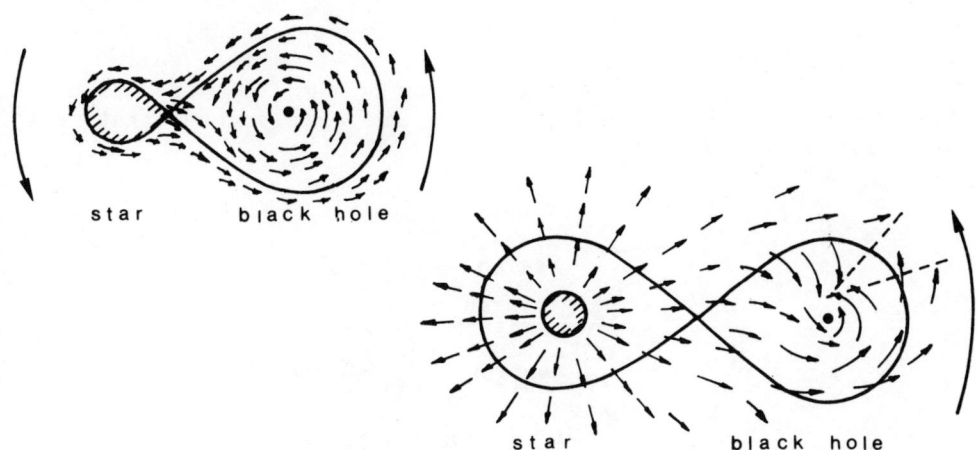

Figure 1. Two regimes of matter capture by a collapsar: (a) a normal companion fills up its Roche lobe, and the outflow goes in the main through the inner lagrangian point:(b) the companion's size is much less than Roche lobe the outflow is connected with a stellar wind. The matter loses part of its kinetic energy in the shock wave and thereafter, gravitational capture of accreting matter becomes possible.

directly the surface of the collapsed star (free radial infall), it rather spiralizes around this star, forming an accretion disk[18]. The angular momentum is dissipated, or transferred outwards by turbulence or by the magnetic field of the matter flowing into the disk[18] so that the gas finally accretes the black hole, releasing gravitational potential energy. Part of this energy increases the rotation energy, part turns into thermal energy, that heats the inner few hundred km of the disk to temperatures of the order of $10^7 - 10^8 °K$. Such a hot gas will emit x-Rays.

In the case of a strong magnetized neutron star (Figure 2) the magnetic field prevents the formation of the inner few thousands km of the disk, and matter only flows in along the open field lines above the magnetic poles, where two accretion columns form. The matter is funneled into two x-Ray emitting hot columns (of diameter ~ 1 km) above the stellar surface. Magnetic and rotation axes being not aligned, the rotation of the neutron star causes the apparent pulsation of the x-Ray emission (cf. 20 and 21). Examples are Her x-1 and Cen x-3. Thus, one of the observational properties that enables to discriminate between neutron stars and black holes is the temporal feature of the emission. Matter which falls from the normal star toward the collapsed object will reach the surface (or Schwarzschild radius) with a velocity V comparable with the escape velocity of the campact object (see Table 3).

$$V \sim (G\ Mx/Rx)^{\frac{1}{2}} \qquad (1)$$

(M_x is the mass of the accreting object, R_x its radius, G the gravitational constant). This is exactly true if the matter falls radially (V free fall ~ V escape) but the results of the following calculations do not change appreciably if we take into account that matter spiralizes in the accretion disk.

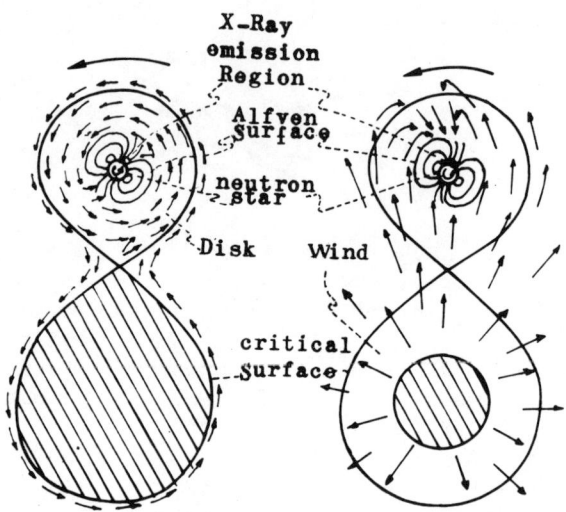

Figure 2. Schematic representation of the rotating neutron star model for pulsating X-ray stars. Both accretion disk and stellar winds cases are shown.

Therefore the gravitational energy released per gram will be of the order of ~ GM_x/R_x. If $\dot{M}$ is the accretion rate, the total gravitational potential energy released per second is:

$$L \sim \dot{M} V^2 \sim G \dot{M} M_x/R_x \simeq 10^{33} \left(\frac{M_x}{M_\odot}\right)\left(\frac{R_\odot}{R_x}\right)\left(\frac{\dot{M}}{10^{-8} M_\odot yr^{-1}}\right) \quad (2)$$

For main sequence stars, that are sustained by gas and radiation pressure, the mass-radius ratio (in solar units) is not greater than 3; for dwarfs sustained by pressure of degenerate gas, $M_x/R_x \sim 100$ (the degeneracy pressure of the electrons halts the collapse of the star at a radius $R \sim 10^{-2} R_\odot (M_\odot/M_x)^{1/3}$ so $(M_x/M_\odot)(R_\odot/R_x) \sim 100 (M_x/M_\odot)^{4/3}$), for neutron stars or black holes is about 10^5 (Table 2).

Thus we see from Eq.(2) that an accretion rate, e.g. of the order of $10^{-8} M_\odot yr^{-1}$ would release, onto a degenerate dwarf, a total power of 10^{35} erg/s, onto a neutron star or black hole it would produce a luminosity of the order of 10^{38} erg/s.

Table 2

Energetics of mass accretion

Object	V_{esc} (km/s) from surface or horizon	Kin.Energy	M/R (solar units)	$\overset{\circ}{M}$ required to produce $\sim 10^{38}$ erg/s
Sun	620	1 KeV	1 m.s.stars ≤ 3 red giants < 1	10^{-3} $M_\odot$/yr
W.D.	6200	100 KeV	$100(M/M_\odot)^{4/3}$	$10^{-5} M_\odot$/yr
N.S.	$1.4\ 10^5$	100 MeV	$10^5(M/M_\odot)^{4/3}$	$10^{-8} M_\odot$/yr
B.H.	c	300 MeV	$10^5(M_\odot/M)$	$10^{-8} - 10^{-9}$ $M_\odot$/yr

Already before reaching the surface, collisions will thermalize the gas to temperatures corresponding to the velocity V.

$$T \sim E/k \sim \alpha\, m_p\, G M_x/k R_x \sim 10^7\, \alpha \left(\frac{M_x}{M_\odot}\right)\left(\frac{R_\odot}{R_x}\right)\ °K \quad (3)$$

where m_p is the proton mass, k is the Boltzmann constant, α is the efficiency factor which depends on how the gas is heated, i.e. how the angular momentum is transferred outward[19]. For adiabatic heating in a strong shock wave, $\alpha \sim 0.1$[22]. For slower heating processes (e.g. the viscosity), radiative losses become important, and so the temperature is determined by the balance between radiative losses and heating. In this case α is smaller, $\sim 10^{-5} - 10^{-6}$[18,19].

It follows therefore from Eq.(3) that the accretion will produce temperature of the order of 10^8 °K if the mass-radius ratio of the collapsed object, in solar units, is:

$$M_x/R_x \geq 10/\alpha \geq 100 \quad (4)$$

Thus, such high temperatures, for the degenerate dwarfs, can be produced only by shock wave heating, while in the case of neutron stars or black holes they can be produced even by viscous heating.

Eq.(2) gives the total power released in the accretion process. The resulting x-Ray luminosity is:

$$L_x \simeq \eta\, L \simeq \eta\, \overset{\circ}{M}\, G\, M_x/R_x \quad (5)$$

η is the efficiency factor for converting potential energy into radiation; $\eta \simeq 1$ for white dwarfs and neutron stars, while $\eta \simeq 0.06$

for a non-rotating black hole (Schwarzschild metric) and η ∼ 0.42 for a rotating black hole (kerr black hole).

This value η < 1 is due to the fact that the energy released by a particle which falls into a black hole is limited by the gravitational redshift (the maximum energy released is ∼ 0.1 mc² and ∼ 0.42 mc², in the Schwarzschild and Kerr metrics, respectively). This limits the effective radius of the black hole to a value of the order of the Schwarzschild or gravitational radius, $R_g = 2\,G\,M_x/c^2$, so that

$$\left(\frac{M}{M_\odot}\right)\left(\frac{R_\odot}{R_x}\right) \sim 10^5\,(M_\odot/M_x)$$

which is the value given in Table 3.

Table 3

Accretion rates and timescales for Roche-lobe overflow as a function of the primary star mass

$M_p/M_\odot$	1	1.3	2.1	4.0	8.0	15.0	20.0	30.0	45
$\dot M(M_\odot/yr)$	3×10^{-8}	10^{-7}	10^{-6}	10^{-5}	10^{-4}	7×10^{-4}	10^{-3}	$2.5\,10^{-3}$	10^{-2}
Approximate duration (yrs)	3×10^7	10^7	2×10^6		10^5	2×10^4	$1.5\,10^4$	$7\,10^3$	$3\,10^3$

From Eq.(5) we obtain that the accretion rate required to produce an x-Ray luminosity of the order of 10^{38} erg/s (see Table column 5) is ∼ 10^{-8} $M_\odot$/yr for black holes or neutron stars, while it is very high (10^{-3} $M_\odot$/yr) for white dwarfs. It can be shown[23] that, in the case of a white dwarf, the high amount of matter required to supply the x-Ray emission would prevent the x-Rays from escaping directly and they would be degraded to radiation of lower energy. In the case of neutron star or black hole, the amount of matter above the level of T ∼ 10^8 °K seems to have an x-Ray optical depth less than the unity.

2.4 Critical Luminosity and Critical Accretion Rate

The total energy released by the accretion process (Eq.(2)) and the spectrum of the outgoing radiation[19] essentially depend on the accretion rate $\dot M$, i.e.. due amount of matter which flows in the disk per unit time. In a fully ionized gas, the dominant x-Ray opacity source is the electron scattering. This process is independent by the wavelength; the cross-section for each electron is $\sigma_e = 6.6\,10^{-25}$ cm², the radiation pressure force per proton mass

m_p at a distance r from an isotropically emitting source with luminosity L is

$$F_{rad} = \frac{L}{4\pi r^2} \frac{\sigma_e}{c} \qquad (6)$$

and the resulting outward acceleration of the protons is

$$g_{rad} = \frac{L}{4\pi r^2 m_p} \frac{\sigma_e T}{c} \qquad (7)$$

the gravitational inward acceleration, is

$$g_g = G M_x/r^2 \qquad (8)$$

so the effective acceleration becomes positive (outward), when the luminosity L becomes greater than the critical Eddington luminosity L_{crit}

$$L_{crit} = \frac{4\pi G M_x m_p c}{\sigma_e} = 1.2 \; 10^{38} \left(\frac{M_x}{M_\odot}\right) \; erg/s \qquad (9)$$

Then, using Eq.(5), we may define a critical accretion rate, (for which L_x become equal to L_{crit})

$$\overset{\circ}{M}_{crit} = \frac{4\pi m_p R_x c}{\eta \sigma_e} \; (c.g.s.) \qquad (10)$$

The radiation pressure will prevent accretion rates larger than this value, e.g. for a neutron star with R_x = 10 km, $\overset{\circ}{M}_{crit} = (1.5/\eta)10^{-8}$ $M_\odot/yr$.

If the main opacity source would be the photoinisation L_{crit} should may be smaller up to a factor 100[24].

More detailed calculation can be made, for non spherically symmetric accretion. In the case of an accretion disk around a black hole[19]

$$\overset{\circ}{M}_{crit} = 3 \; 10^{-8} \; \frac{0.06}{\eta} \left(\frac{M_x}{M_\odot}\right) \; M_\odot/yr \qquad (11)$$

Therefore dealing with the theory of accretion disk, we should consider two separate cases: subcritical regime ($\overset{\circ}{M} < \overset{\circ}{M}_{crit}$) and supercritical regime ($\overset{\circ}{M} > \overset{\circ}{M}_{crit}$).

2.5 Accretion Disks Around Black Holes

The structure and characteristics of the accretion disk has been studied in detailed models by Pringle and Rees[18], Shakura and Sunyaev[19], Shakura[25] and Novikov and Thorne[26].

We have seen in the previous sections that, in order for radial infall to occur, angular momentum within the disk must be dissipated, through both turbulent and magnetic viscosity[19] and by the Poynting-Robertson effect[27]. In either case the infall velocity increases inward. The <u>viscosity parameter</u> α is defined as the efficiency of the mechanism of angular momentum transport through the disk ($\alpha \leq 1$). In the inner parts of the disk, the main causes of viscosity are expected to be turbulence and magnetic fields, and thus $\alpha \simeq 0.01 - 1.0$.

The emitted radiation from accretion disk, however, depends more strongly on the accretiom rate than on the viscosity (Fgirure 3) since a smaller viscosity would imply a smaller local emissivity, but longer infall times. The local emissivity of the disk can be estimated by considering the total energy balance, including the energy dissipated by viscosity[18,19].

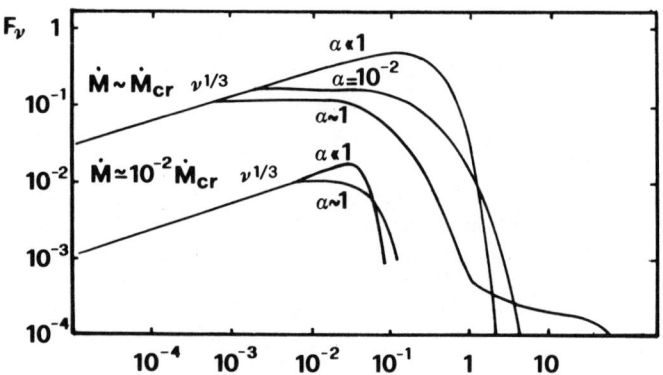

Figure 3. The integral radiation spectrum of the disk for different $\dot{M}$ and α (Ref. 19).

The main mechanism for x-Ray emission is bremsstrahlung, but if the density is high we have mainly black-body radiation, while when a strong magnetic field ($\sim 10^8$G) is carried along the gas, cyclotron radiation dominates. If electron scattering is important, a Wien spectrum is produced by Compton-scattered cyclotron emission from high harmonics[28,29]. The resulting radiation spectrum can appear thermal with a high-energy tail due to the Compton-scattering. Characteristic local spectra are shown in Figure 4.

Integrating over the entire disk, we obtain a radiation spectrum that resembles a non-thermal one. In Figure 3, the integrated radiation spectrum calculated by Shakura and Sunyaev, is shown

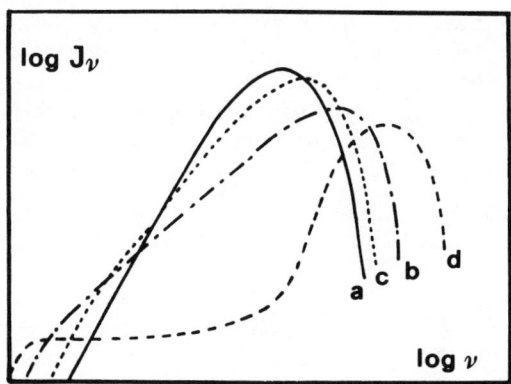

Figure 4. Local spectra of radiation formed in the disk.
a.- black body. b.- isothermic medium with opacity due to scattering. c.- isothermal exponential atmosphere. d.- spectrum formed in the presence of Comptonization (Ref. 19).

for the two cases of subcritical regime ($\overset{\circ}{M} \sim 10^{-2} \overset{\circ}{M}_{crit}$) and critical regime ($\overset{\circ}{M} \sim \overset{\circ}{M}_{crit}$). It can be seen that the spectrum peaks at somewhat higher energies than those expected for a black-body spectrum, due to the electron scattering of radiation (a black-body spectrum would peaks at $h\nu_{max}$ (2.44 · 10^4 eV)(Ts/10^8°K) where T_s is the temperature of an emitting surface[30]. Figure 3 also shows that for $\overset{\circ}{M} < 10^{-2} \overset{\circ}{M}_{crit}$ the energy distribution peaks in the UV or soft x-Rays. Only for $\overset{\circ}{M} \sim \overset{\circ}{M}_{crit}$ hard x-Rays are produced.

The total luminosity of the disk appears to increase quite linearly with $\overset{\circ}{M}$ (Figure 5), until it reaches the critical value L_{crit} (Eq.(9)); in the supercritical regime, the expansion of the disk leads to an outward flow of matter above the plane of the disk. This matter can leave the system through the Lagrangian point L_2.

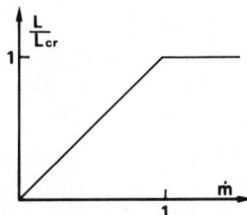

Figure 5. Luminosity of the disk as a function of inflowing matter (Ref. 19).

The thickness Z_o of the disk appears to increase weakly with $\overset{\circ}{M}$ ($Z_o \sim \overset{\circ}{M}^{1/5}$ in the innter part of the disk). If $\overset{\circ}{M}$ is very large, Z_o

increases so much that the disk turns into a more or less spherical configuration, which is opaque the x-Rays generated from its interior. Then, the maximum effective surface temperature decreases, (Figure 6), and the frequency of the emission peak shifts towards UV and visible part of the spectrum, as can be seen from Figure 3. E.g. for an accretion rate $\dot{M} \approx 10^2 \dot{M}_{crit}$ ($M \approx 10^{-6} M_\odot/yr$) the maximum effective temperature is below 10^6 K, so that no more x-Ray emission is expected

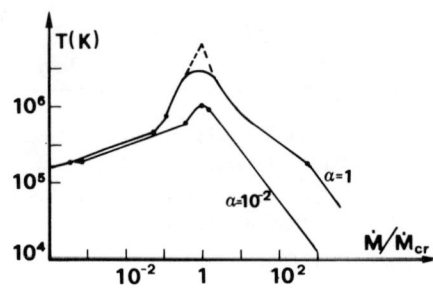

Figure 6. Maximum effective temperature of the disk as a function of inflowing matter (Ref. 19).

to be observable. We point out the difference about this aspect, with the case of a neutron star. For strongly non-spherical accretion onto a such object, in fact $\dot{M}$ can also exceed $\dot{M}_{crit}$ by a factor 10-100 before the x-Ray emission is estinguished (without extintion of the x-Ray emmission).

As far as concernes the temporal structure of the x-Ray emission from a disk accreting a black hole, the observational characteristics are different from the regular pulsations seen in the case of a neutron star. Chaotic rapid fluctuations of the observed radiation flux and of its spectrum may be expected, the strongest occurring for the hard radiation. This irregular activity would be due both to the turbulence in the disk and to its chaotic magnetic field[19]. In disks surrounding black holes, in fact, hot spots can be formed through reconnection of magnetic fields lines of force (as occurs, e.g., in solar flares), and this may lead to fluctuations or pulsations in brightness, on orbital time scales ($10^{-4} - 10^{-1}$s;[31]. Qusai pulsations in emissivity or time scales of 0.1 - 1 s can also be produced by Alfven waves[32].

2.6   Types of Accretion in x-Ray Binaries

The primary star can transfer matter onto a compact companion by means of the following mechanisms:

2.6.1 Roche Lobe Overflow

The primary star fills up its Roche-lobe and the matter that exceeds the critical surface flows through the inner lagrangian point, coming into the influence region of the compact star. In this case the matter leaves the stellar surface with low ($\sim$ thermal) velocities, and most of it will be captured by the companion. Thus the accretion rate $\mathring{M}$ will be of the same order as the mass loss rate $\mathring{M}_p$ from the primary.

We can estimate the mass transfer rate for Roche-lobe overflow in the case of a type B binary system[30], i.e., a system in which the mass transfer occurs when the primary is in the hydrogen shell-burning stage. In this case almost all the hydrogen-rich envelope is rapidly transferred, so we obtain the mass loss rate by dividing the mass of this envelope (about $0.8\ M_p$) by the thermal (Kelvin-Helmholtz) timescale of the primary (assuming that it is the more massive companion. This time scale for Roche lobe overflow is[33]

$$\tau_{KH} \simeq 3.1\ 10^7\ M_p/RL \text{ years}$$

where M_p is the primary star mass, R and L its radius and luminosity, in solar values. The mass transfer rate will be

$$\mathring{M} \simeq \mathring{M}_p \simeq 0.8\ M_p/\tau_{KH} = (0.8/3.1\ 10^7)\ \frac{RL}{M_p}\ [M_\odot/yr] \qquad (12)$$

Taking, for example, R, L values of the models of Iben and Simpson[34], just after the star has left the main sequence, we obtain the following values for M and the timescales[30] (Table 3):

We see from these values that the Roche lobe overflow is a process that occurs only at particular stages of the stars evolution, and its duration is very short, if compared with characteristics timescales (nuclear burning timescales) of an even very massive star life. Thus, this mechanism for mass transfer, in spite of its high efficiency and its importance in the evolution of close binary stars (see for instance, Ref. 35), is completely negligible from an observational point of view.

2.6.2 Stellar wind

If the primary star is very massive, it can have a considerably high stellar wind. In this case, matter will be lost by the entire system, and only a fraction of the wind will be captured by the companion. The stellar wind is one of the main properties of the stars, and the rate of mass loss varies from $2 \times 10^{-14}\ M_\odot/yr$ from the Sun upto $10^{-5}\ M_\odot/yr$ for very hot supergiants or Wolf-Rayet stars. This constitute a link between the x-Ray and UV astronomy.

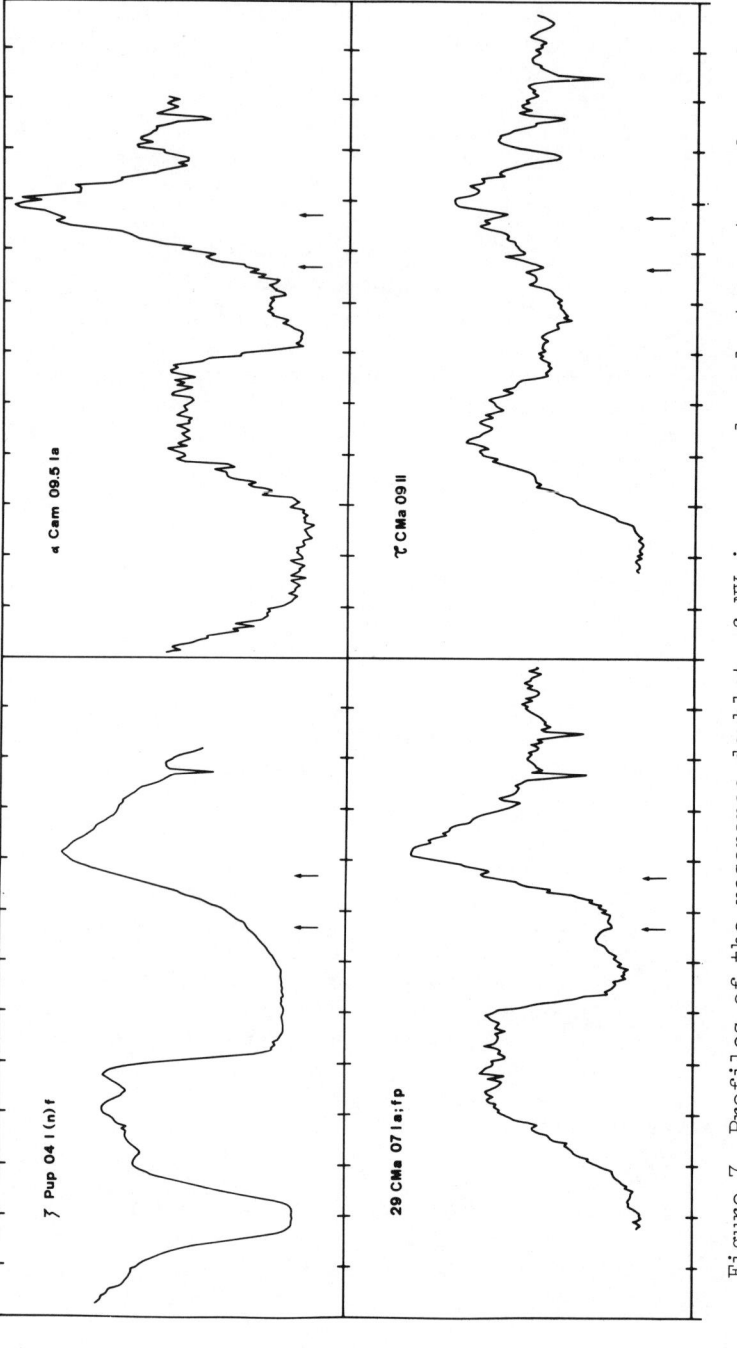

Figure 7. Profiles of the resonance doublet of NV in several early type stars observed by the satellite Copernicus, showing mass loss effects. (Figure arranged on the basis of original tracings kindly made available by Princeton Observatory).

TRAITS OF BLACK HOLES IN OPTICAL BAND

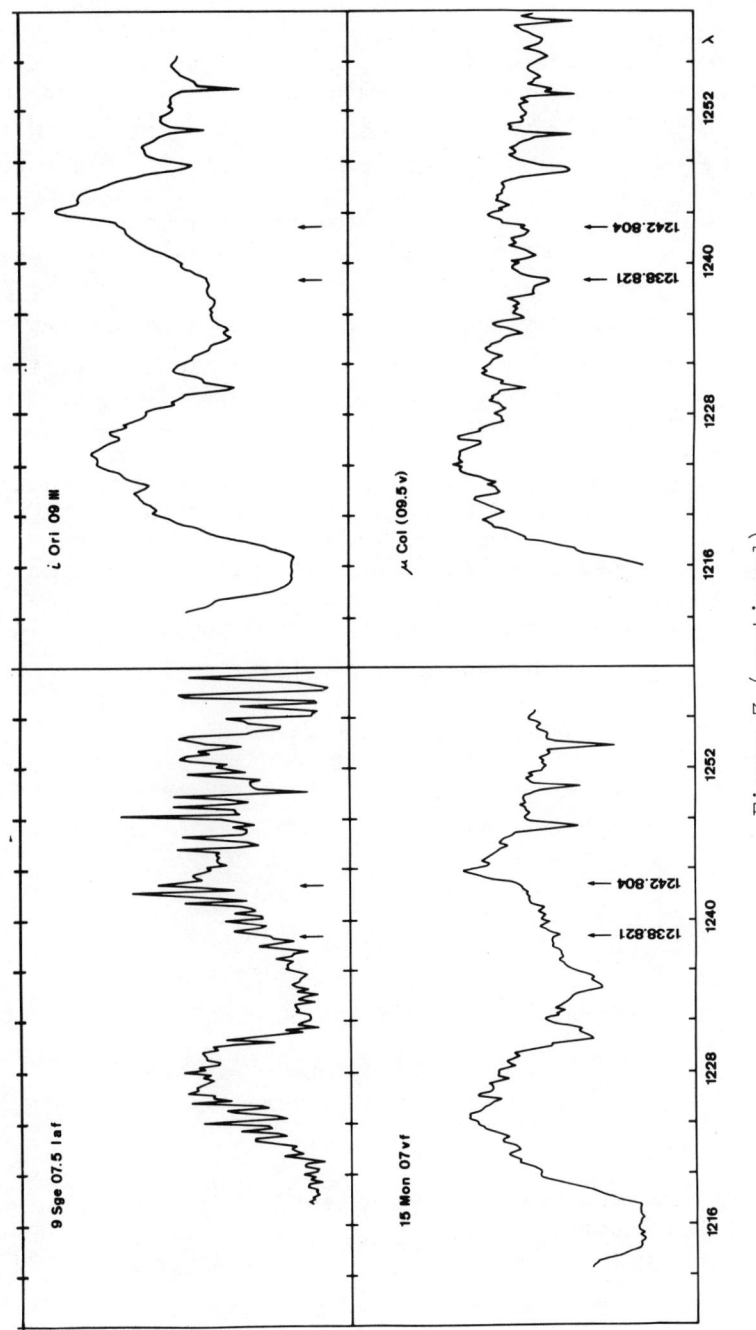

Figure 7 (continued).

In fact, observations of mass loss effects, expanding extended envelopes, stellar chromospheres and coronae, and mass loss rate estimates, are greatly facilitated by the availability of UV data, because early-type stars emit most of their light in this spectral region, and because of the large number of resonance lines and other low excitation transitions in the far UV. That early type stars are losing mass continuously was clear since the first[36] explanation of the so-called P-Cygni profile, consisting of an emission contribution and an asymmetric, violet-shifted absorption component: from the profile of this component and from the outflow velocity that we observe the mass loss rate M_p can be deduced[17,37,38,39,40]. Figure 7 shows a set of absorption profiles of the resonance doublet of NV in a number of early type stars observed by Copernicus satellite[41]. The steep edge of the shortwavelength side of the profile shows that these stars are loosing mass.

In the case of supersonically outflowing wind, the compact star will only capture the fraction of matter which passes it very closely. A shock front forms around the compact star, where the matter is accelerated to subsonic velocities, by dissipation of the velocity component perpendicular to the front. The accretion radius r_a (i.e., the radius of the cylinder, around the compact star, from which matter is captured) is approximately given by[42]

$$r_a = 2 G M_x/(V_{rel}^2 + C_s^2)$$

C_s is the sound velocity, that can be neglected as it is always much smaller than V_{rel}, the velocity of the wind relative to the collapsed object.

$$V_{rel}^2 = V_W(a) + \left(V_{orb} \frac{R_p}{a} V_{eq} \right)^2 \qquad (13)$$

where $V_w(a)$ is the wind velocity, a is the separation between the components, V_{eq} the equatorial rotational velocity of the primary ($V_{eq} = 0$ for a non-rotating primary, $V_{eq} = (R_p/a) V_{orb}$ for a co-rotating primary[30]). V_{orb} is the orbital velocity of the secondary (for circular orbits $V_{orb}^2 = G(M_p + M_x/a)$).

Finally, the amount of wind captured by the collapsed star (for a radially outflowing wind) is

$$\overset{\circ}{M} = \overset{\circ}{M}_p \pi r_a^2/4\pi a^2 \qquad (14)$$

$\overset{\circ}{M}_p$ being the mass loss rate from the primary.

Another interesting observational feature of the stellar

wind is the fact that radial mass loss from the system will increase the binary period. On the contrary, note that both mass transfer from the more massive to the less massive component, and mass loss from the system though the external lagrangian point L_2 in a Roche model tend to decrease the period[43].

An example for this may be Cen X-3, where the period is observed to change in both directions (the slow-down effect of the radial stellar wind may at times predominate over the speed-up effects).

2.6.3   X-Ray heating of the normal stars

A strong X-Ray source can heat the outer layers of the primary to produce a wind. This case is similar to the previous one, but the wind is driven by the X-Ray emission (no more radiation pressure driven wind o coronal wind).

In some cases the X-Ray induced wind is unstable above a threshold X-Ray luminosity, resulting in a total luminosity near L_{crit}[44,45].

The optical appearance of a close binary system, when a significant part of the X-Ray radiation hits the surface of the normal star and is reradiated by its atmosphere[4] can be unusual. An observed example of this effect is the system Her X-1 - HZ Her, in which the emisphere that faces the X-Ray source is brighter than the opposite one by a factor 3 and has a different spectral class[46].

2.7   Observational Evidence of Accretion Wakes

2.7.1   Optical lines

Variation with phase of the profile are observed in the system HD 77581-3U 0900-40[47,48,49] and in Cygnus X-1[16,50] and of the HeI5876 line in HD 153919-3U 1700-37[51].

The profile of these lines has an unshifted emission component (probably formed in the expanding envelope of the star) and a violet-shifted absorption component, formed in the outflowing gas along the line of sight. In the above three examples the radial velocity of this blue component varies with phase, the largest short-wavelength shift occurring between phases 0.6 and 0.7, i.e. where the compact star just has passed in front of the primary. This can be accounted for assuming that, at this phase, we are observing or a stream outflowing through the Lagrangian point L_2, or a trailing wake of the stellar wind behind the compact star. A possible model for a such accretion wake is shown in Figure 8[51] in the case of 3U 1700-37. At phase 0.6 - 0.7 the velocity component along the line of sight, in the wake, is small, in spite of the large outflow radial velocity in the wind. The part of the wake that we see in H_α seems to be the broad one, where there is a high density due to the cooling behind the shock.

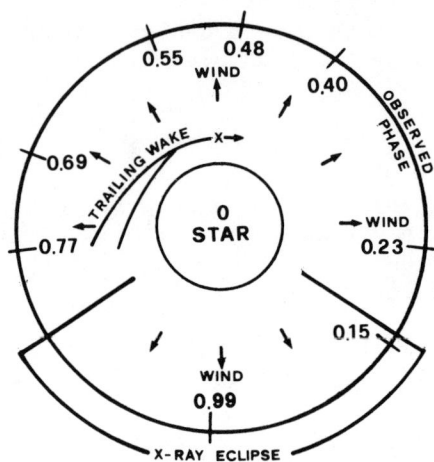

Figure 8. Model of accretion wake in HD 153919 (RU 1700-37) and in HD 77581 (3U 0900-40) (Ref. 51).

Detailed calculations of line profile variations due to gas streaming in a binary system have been made by Sima[52]. In Cen X-3, turn-ons have been observed by both UHURU and UK 5 (Ariel) satellites[53,54,55]. UK 5 observations show at phase 0.6 a strong absorption dip during turn-on; such a dip may be interpreted[56] as due to an accretion wake passing between us and the X-Ray source, the wake is derived to make an angle of about 45° with the line joining the centres of the stars.

Similar effects of X-Ray absorption due to a wake are observed at phases around 0.6 - 0.7 in the Vela X-1 (3U 0900-40)[57].

3. BLACK HOLES IN GLOBULAR CLUSTERS

Seven among the most luminous X-ray sources are located in globular clusters (Table 4): NGC 6624, NGC 6440, NGC 6441, NGC 1851, NGC 6712, NGC 7078 (M 15)[58,59,60]. The X-ray emission from these objects could be explained taking into account accretion onto a collapsed object, (either a neutron star or a black hole) from a normal companion in a binary system as discussed in Section 2. However, in the case of globular clusters the following fact must be pointed out: the number of known globular clusters is about 120 containing 2.10^7 stars in comparison with 10^{11} of the whole galaxy, therefore while the relative number of X-ray sources among stars in globular clusters is $\geq 10^{-7}$, that one for the stars in the whole galaxy is only $\sim 10^{-9}$ [61]. This suggests that X-ray sources associated with globular clusters could be intrinsically different from the other sources elsewhere in the galaxy and that one would need a completely different model to explain their nature. Many authors have suggested that in the centre of the globular X-ray clusters could lurke a massive black hole of $\sim 10^3 M_\odot$ and that the X-ray emission could be produced by matter falling onto such a collapsed object[62,63,64]. Accretion would be due either to the gas in the cluster (some of them are bound enough to retain the gas lost by the stars during their evolution) or to debris arising from tidal disruption of stars which cross near the central black hole[62]. The spherical symmetry of globular clusters and relativistic calculations on the dynamic of accretion let one think that the space-time around the hole would differ very slightly by the one described by the Schwarzchild metric once a stationary condition is obtained[65] and therefore an efficiency η of converting ingoing mass ($\overset{\circ}{M}$) into outgoing radiation of ~ 0.1 may not be far from truth.

A rough calculation of the X-ray luminosity (2-10 Kev band) by means of (5) here rewritten as

$$L_X = \eta(\overset{\circ}{M}_{acc}/10^{-9} \, M_\odot \, yr^{-1}) \times 10^{37} \, erg \, s^{-1} \qquad (15)$$

with typical values of η and $\overset{\circ}{M}$ can account for the observed X-ray luminosity in globular clusters[62]. The stellar distribution function in the nuclear region of X-ray globular clusters, and thus in the vicinity of the hypothetical hole, is an important observational feature.

This question was first raised by Wyller[66] and was extensively investigated by Peebles[67] and by Bahcall and Wolf[68,69]. Let us briefly outline the theory; assuming $M_{bh} \ll M_c$ (M_{bh} is the mass of the black hole and M_c is the mass of the core of the cluster) the effect of the presence of the hole on the stellar distribution will be relevant only not far from the centre; more exactly the black hole will strongly perturb the stellar distribution only for $r < r_h$ where r_h is the radius of influence given by $GM_{bh}/\langle\Delta v^2\rangle$ where $\langle\Delta v^2\rangle$ is the dispersion of the velocity of the stars in the central region. Moreover at a certain distance r_t tidal forces will overcome the binding internal energy disrupting the infalling stars.

There are therefore two boundary conditions for the perturbed function: for $r > r_h$ it must coincide with the unperturbed function which is assumed to be maxwellian while for $r < r_t$ it must vanish. Under suitable assumptions one can show that when $r_h < r < r_t$ the stellar distribution function $f(E,t)$ depends only on the total energy $E = G M_{bh}/r - v^2/2$ and on time t; imposing the further condition $\partial f/\partial t = 0$ the solution is $f = \text{const. } E^p$ with $p = 1/4$.

The exponential law is not however the only stationary solution but this choice can be justified considering the behaviour of dynamical solutions. Numerical computations have shown that the distribution function changes substantially in a time τ which is of the order of half of the relaxation time t_{rel} while for $t > \tau$ the solution is quasistationary and in the energy range corresponding to $r_h < r < r_t$ it is of the order of $2 \times \text{constant } E^{\frac{1}{4}}$. All the previous results were obtained assuming the stars in the cluster to have the same mass M_* ($M_* \ll M_{bh}$). More sophisticated models which take into account also stars of unequal masses do not give substantially differences with observable quantities[69].

Once the distribution function is known, it is straightforward to obtain the surface stellar density $\sigma_{tot}(S)$ as a function of the apparent distance S from the centre.

$$\sigma_{tot}(S) \simeq \text{cost.} \left[(1 + S^2/r_c^2)^{-1} + 1.57 \ (S/r_c)(r_h/S)^{7/4} \right] \quad (16)$$

$\sigma_{tot}(S)$ gives the number of stars per arcmin^{-2} projected on the sky at an apparent distance S from the centre and holds for $r_{tot} > r > r_t$. Here r_{tot} and r_c are respectively the total radius and the radius of the core of the cluster.

In Eq.(16) the quantity $(1 + S^2/r_c^2)$ describes the contribution of the stars which are not bound to the central black hole, while the second term describes the contribution of those which are gravitationally. Unlike the curves given by King[70] whose first derivatives vanish in the limit of $S = 0$, Eq.(16) shows a sharp density cusp for $S = r_t$. The observations which have been carried out on some of the X-ray clusters (NGC 6624, NGC 6440, NGC 6441, M15) have not given a definite proof either for or against the existence of a massive central black hole.

Using the 1m telescope of Wise Observatory in Israel, John and Neta Bahcall and their collaborators have taken during 1973-76 several exposures of 4 X-ray clusters[71,72,73]. Using the count method suggested by King et al[74], they were able to obtain surface density curves up to $S \sim 0".05$ ($0".125$ for M15), a limit beyond which overcrowding effects become important at the available plate resolution limited by a seeing of about 2".

Unfortunately to be able to show the existence of a density cusp it is necessary to make precise counts or photometric measure up to a distance $\leq 0.1 r_c$ and therefore an angular resolution of about $0".15$ is needed.

Very accurate photometric measures were carried out in 1976-77 by Canizares et al[75] with the 4 m telescope of Cerro Tololo on NGC 6624. They were able to draw a surface luminosity curve up to a distance at $0.5\ r_c$, but they could not detect any luminosity cusp. They have also obtained a certain number of spectra of the nucleus of NGC 6624 and of the innermost stars; the nucleus was classified as G2 and perhaps the only pecularity in the strong absorption corresponding to H_α and H_β : the spectra of the nucleus are somehow noisy around H_α and this fact could make one assume the presence of gas in the cluster core.

More recent observations of M15 carried out by Newell, Da Costa and Norris[76], show a certain evidence for the presence of a massive black hole in the cluster core. The surface brightness profile obtained by electrographic methods goes up to $0".05$ and for $r < r_c \sim 0".25$ the luminosity curve differs considerably from the King's one, showing a continuous increase for decreasing S. The best fit with the experimental data points to the existence of a central collapsed object of about 800 $M_\odot$; this value is consistent with the upper limit for a central black hole in M15 of $\sim 10^4\ M_\odot$ as extimated by Bahcall[72].

A fundamental contribution to the solution of this relevant problem will be given by orbital instruments; the Faint Object Camera on board of the 2.4 m NASA Space Telescope with an expected resolution of $0".03$ will be able to perform extremely precise photometric measurements even in the central regions of globular clusters showing or disproving the presence of collapsed objects of mass even less than 500 $M_\odot$[68].

4. ISOLATED BLACK-HOLES

Let us assume that the collapsed object is a Schwarzschild hole of stellar mass at rest in the interstellar medium and that the angular momentum per unit mass of the captured gas is less than $\sqrt{3}\ (2GM/c^2)c$. This ensures a spherically symmetric flow rather than the formation of a disk.

In a hydrodynamic description the form of the flow is governed by two fundamental equations: the conservation of rest mass and Bernoulli equation (the flow is assumed to be adiabatic). It is possible to solve this set of coupled equations, suitably rewritten, for the accretion rate $\dot{M}$ obtaining:

$$\dot{M} = \alpha\, r_g^2\, c\, \rho_\infty (m_p\, c^2/kT_\infty)^{3/2} \tag{17}$$

or

$$\dot{M} = (1 \times 10^{11}\ \text{gs}^{-1})(M/M_\odot)\, (\rho_\infty/10^{-24}\text{g cm}^{-3})(T_\infty/10^4\text{K})^{-3/2} \tag{18}$$

Here α is a constant of order unity, r_g is the gravitational radius of the hole, ρ_∞ and T_∞ are respectively the density and the temperature of matter at infinity.

One immediately realizes that $\dot{M}$ and consequently the luminosity L of such an object strongly depends on the conditions of the interstellar matter far from the hole.

As the gas moves inward it is heated up by pdV work and by dissipative heating and consequently it starts to radiate. At low temperature, $T \sim 10^4$ K, the principal cooling mechanism is radiative recombination, then, as T increases, electron-proton and electron-electron bremasstrahlung become more and more important; at very high temperature, $T \sim 10^{10}$ K reached at $r \sim 10\, r_g$, the cooling mechanisms, namely synchotron radiation and especially inverse Compton scattering, are so efficient that cooling exceeds heating and temperature starts to decrease. This picture may describe the accretion of a hole embedded in a dense H II region near galactic nucleus with $\rho_\infty \sim 10^6 - 10^7$ cm^{-3}. As far as interstellar gas is concerned with $\rho_\infty \sim 1$ cm^{-3} $T_\infty \sim 10^4$ K, cooling never exceeds heating and the temperature increases monotonically up to $T \sim 10^{12}$ K near r_g. One can derive from this and other considerations the continuous emission spectrum expected from such a source, (see Figure 9).

These spectra are characterized by a sharp peak due to thermal synchotron radiation; the frequency at the peak is given by

$$\nu_{peak} = (7 \times 10^{14}\ \text{Hz})(\rho_\infty/10^{-24}\text{g cm}^{-3})^{\frac{1}{2}}(T_\infty/10^4\ \text{K})^{-3/4} \tag{19}$$

The peak at 1.9×10^{20} Hz shown in Figure 9, can be explained taking into account the accretion of high energy cosmic ray particles, namely electrons, as proposed by Meszaros[77].

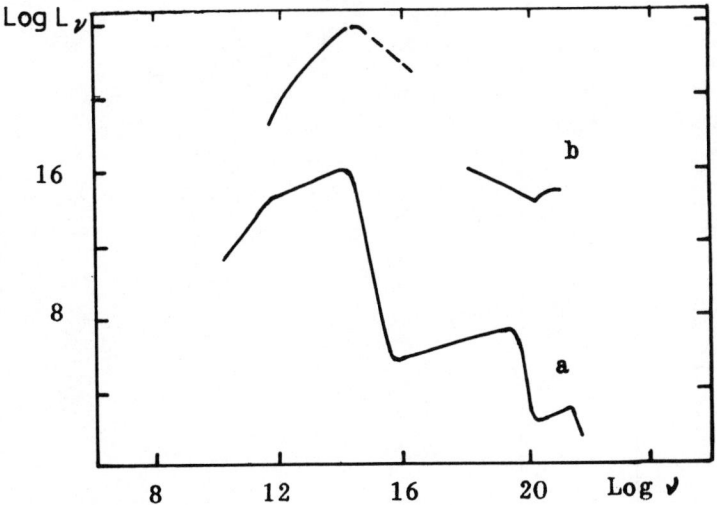

Figure 9. Luminosity per unit frequency in the case of normal HII regions (a) and in the case of dense nuclear regions (b). (Ref. 77).

For a black hole of 10 $M_\odot$ accreting in the general interstellar medium ($\rho_\infty \sim 1$ cm^{-3}, $T_\infty \sim 10$ K) the expected luminosity will be L=10^{31} erg s^{-1} [77,78]. At distance of 1 Kpc this would give a flux of $\sim 10^{-28}$ erg cm^{-2} s^{-1} H2^{-1} in the 2-10 KeV band and of $\sim 4 \times 10^{-19}$ erg cm^{-2} s^{-1}, in the infrared.

Swartsman[79] remarks that such a spectrum and luminosity resembles in the visible those of DC white dwarf stars (white dwarfs with continuous, line free-sepectra); it is possible that objects till now classificated as DC stars are actually black holes accreting in the interstellar medium.

Novikov and Thorne[26] point out that possible instabilities in the flow may produce luminosity fluctuations whose timescale is of the order of the gas travel time from 10 r_g to 2r_g that is the principal radiating region:

$$\Delta t_{fluctuations} \sim (10^{-3} \text{ to } 10^{-4} \text{ s})(M/M_\odot) \qquad (20)$$

typically 10^{-3} s for a 10 $M_\odot$ object.

Unfortunately it will be very difficult to carry out such observations on objects so faint (m $\sim$ 18 assuming a distance of 1 Kpc and L $\sim 10^{31}$ erg s^{-1}) by ground based instruments. Time resolved photometry of DC white dwarfs, that are heretofore the best optical candidates for isolated holes, will be on the contrary within reach for ST and a survey of optical properties of DC stars will be an interesting program for space instruments.

Table 4

X-Ray sources in Globular Clusters

Name	α δ	L_x × 10^{36} erg/s	Name	Apparent Diameter (arcmin)	ρ_c × $10^4 M_\odot/pc$	v_{esc} (km/s)	t_{rel} (10^7 yr)	r_{core} (pc)	Distance (kpc)
MX 0513-40	5 13 / -40 3	4.7	NCG 1851	11	8.2	40	3.6	0.37	10
MX 1746-20	17 57 / -17 29	75	NGC 6440	—	≥ 30	≤ 48	≤ 3	≤ 0.3	(10)
3U 1746-37	17 46 / -37 0	48	NGC 6441	—	≥ 9	≤ 48	≤ 10	≤ 0.55	9.3
3U 1820-30	18 20 / -30 23	25	NGC 6624	4.2	5	28	2.8	0.4	4.7
3U 2131-11	21 29 / +12 0	3.7	NGC 7078	9.4	9	37	3.6	0.4	10
4U 1850-08	18 50 / -8.41	1.2	NGC 6712	~ 10	~ 0.07	—	~ 17	1.6	6.8

References

1. H. Gursky and R. Ruffini, in 'Neutron stars, B.H. and binary X-ray sources', Ed. D. Reidel P.C., Dordredit, Holland (1975).
2. R. Ruffini and J.A. Wheeler, Cosmology from space platforms, Ed. H.V. Hardy, ESRO SP52, Paris (1971).
3. B.Ya Zeldovich and I.D. Novikov, Doklady Akad. Nauk, USSR, 158, 811 (1964).
4. I. Sholokovsky, Ap. J., 148, L1 (1967).
5. A. Cameron, M. Mock, Nature, 215, 464 (1967).
6. K. Prendergast and G. Burbidge, Ap. J., 151, L83. (1968).
7. A. Sandage, et al., Ap. J. 146, 316 (1966).
8. G. Burbidge, C. Lynds and A. Stockton, Ap. J., 150, L95 (1967).
9. R. Kraft, Ap. J., 139, 457 (1964).
10. B.Ya Zeldovich and I.D. Novikov, The theory of the gravitation and stars evolution, Nauka, Moscow (1971).
11. E. Salpeter, Ap. J., 140, 796 (1964).
12. B.Ya. Zeldovich, Doklady Akad. Sci., USSR, 155, 67 (1964).
13. D.Ya. Martynov, Course of General Astrophys, Nauka, Moscow.
14. H. Gursky, P. Gorenstein and R. Giacconi, Ap. J., 150, L75 (1967).
15. E. Salpeter, in 'X-and Gamma-rays Astronomy', IAU Symp. No. 55, 135, Eds. H. Bradt and R. Giacconi (1973).
16. J.B. Hutchings, Ap. J., 192, 677 (1974).
17. P.L. Bernacca and L. Bianchi, Astron. Astrophys (in press).
18. Lynden-Bell, Nature, 223, 690 (1969);
    J. Pringle and M. Rees, Astr. Ap., 21, 1 (1972);
    K.H. Prendergast, Ap. J., 132, 162 (1960);
    N.I. Shakura, Astron. Zh., 49, 625 and 921 (1972).
19. N.I. Shakura and R.A. Sunyaev, Astr. Ap., 24, 337 (1973).
20. K. Davidson and J.P. Ostriker, Ap. J., 179, 585 (1973).
21. F.K. Lamb, C. Pethick and D. Pines, Ap. J., 184, 271 (1973).
22. L. Landau and E. Lifshitz, Fluid Mechanics, Reading, Mass. Addison-Wesley (1959).
23. B.A. Peterson, Proc. Astron. Soc., Australia, 2, 1978 (1973).
24. R. McCray, Nature Phys. Sci., 243, 94 (1973).
25. N.I. Shakura, Soviet Astronomy, 16, 756 (1973).

26. I.D. Novikov and K.S. Thorne, in 'Black holes', Eds. de Witt and de Witt, Gordon and Breach, p. 345 (1973).
27. G.R. Blumenthal, Ap. J., **188**, 121 (1974).
28. Y. Gnedin and R. Sunyaev, M.N.R.A.S., **162**, 53 (1973).
29. J. Felten and M. Rees, Astr. Ap., **17**, 226 (1972).
30. E.P.J. van den Henvel, Mem. S.A.It., **47**, no. 34 (1975).
31. R.S. Sunyaev, Sov. Astron. A.J., **16**, 941 (1973).
32. G.R. Blumenthal and W.H. Tucker, Nature, **235**, 97 (1972).
33. B. Paczynski, Ann. Rev. Astron. & Ap., **9**, 183 (1971).
34. E.E. Simpson, App. J., **165**, 295 (1971).
35. C. Chiosi, E. Nasi and S.R. Sreenivasan, Astr. & Ap., **63**, 103, (1978).
36. C.S. Beals, MNRSA, 90, 202 (1929).
37. D.C. Morton, Ap. J., **147**, 1017 (1967).
38. H.J.G.L.M. Laurers and D.C. Morton, Ap. J. Suppl., **32**, 715 (1976).
39. H.J.G.L.M. Laurers and J.B.Jr. Rogerson, Astr. Ap., **66**, 417 (1978).
40. M.J. Barlow and M. Cohen, Ap. J., **213**, 737 (1977).
41. T.P. Snow and D.C. Morton, Astrophys. J. Supp., **32**, 429 (1976).
42. H. Bondi and F. Hoyle, MNRAS, **104**, 273 (1944).
43. A. Kruszewski, 'Advances in Astronomy and Astrophysics', IV, Academic Press, New York, p. 233 (1966).
44. J. Arons, Ap. J., **184**, 539 (1973).
45. M. Basko and R. Sunyaev, Ap. Space Sci., **23**, 117 (1973).
46. A.M. Cherepashchuk, N.Yu, Efremov, N.E. Kurochikiu, Shakura, R.A.NI Sunyaev, Inform. Bull. Var. Stars, n 720 (1972)
47. E.J. Zuiderhijk, E.P.J. van den Henvel and G.H. Henseberge, Astr. and Ap., **35**, 353 (1974).
48. G. Wallerstein, Ap. J., **194**, 451 (1974).
49. M.S. Bessel, N.V. Vidal and D.T. Wikromasinghe, Ap. J., **195**, L117 (1975).
50. R.J. Brucato and R.R. Zappala, Ap. J., **189**, L71 (1974).
51. P.S. Conti and A.P. Cowley, Ap. J., **200**, 133 (1975).
52. Z. Šíma, Ap. and Space Sci., **24**, 421 (1973).
53. H. Gursky and T. Shereier, in 'Neutron Stars, Black Holes and Binary X-Ray Sources', Eds. H. Gursky and R. Ruffini, Dordrecht Reidel Publishing Company, p.175 (1975).

54. R.Giacconi, Proc. 7th Texas Symp. on Relativ. Astrophys., (1976).
55. K.A.Ponnds, B.A.Cooke, M.J.Ricketts, M.J.Turner and M.Elvis, M.N.R.A.S., 172, 473 (1975).
56. J.C.Jackson, MNRAS, 172, 463 (1975).
57. P.A.Charles, et al., NASA preprint X-660-75-285, p.335 (1975).
58. C.R.Canizares, Ap. J., 201, 589 (1975).
59. C.R.Canizares and J.E.Neighbours, Ap. J., 199, L77 (1975).
60. J.Grindlay, H.Gursky and H.Schnopper, Ap. J., 205, L127 (1976).
61. G.W.Clark, Lecture given at the E.Fermi Summer School for Physics and Astrophysics of Neutron stars and Black Holes, Varenna, Italy (1975).
62. J.B.Bahcall and J.P.Ostriker, Nature, 256, 23 (1975).
63. J.Silk and J.Arons, Ap. J., 200, L131 (1975).
64. J.Frank and M.J.Rees, M.N.R.A.S., 176, 633 (1976).
65. P.J.Young, Ap. J., 212, 227 (1977).
66. A.A.Wyller, Ap. J., 160, 443 (1970).
67. P.J.E.Peebles, Gen. Rel. and Grav., 3, 63 (1972).
68. J.N.Bahcall and R.A.Wolt, Ap. J., 209, 214 (1976).
69. J.N.Bahcall and R.A.Wolt, Ap. J., 216, 883 (1977).
70. I.King, Ap.J., 71, 64 (1966).
71. J.N.Bahcall, N.A.Bahcall and D.Weistrop, Astrophys. Lett., 16, 159 (1975).
72. N.A.Bahcall, Ap. J., 204, L83 (1976).
73. N.A.Bahcall and M.A.Hausman, Ap.J., 207, L181 (1976).
74. I.King, E.Jr.Heidmann and S.M.Hodge, A.J., 73, 456 (1968).
75. C.R.Canivares, J.E.Grindlay, W.A.Hiltner, W.Liller and J.E.McChintock, Astrophys. J. (in press)(1976).
76. B.Newell, G.S.Da Costa and J.Norris, Ap. J., 208, L55 (1976).
77. P.Meszaros, Astron. and Astrophys., 44, 59 (1975).
78. S.L.Shapiro, Ap. J., 180, 531 (1973).
79. V.F.Swartsman, Soviet Astron., 15, 37 (1971).

THE ROLE OF POLARIZATION IN MICROSCOPIC PHYSICS

Michael J. Moravcsik

Institute of Theoretical Science

University of Oregon, Eugene, Oregon 97403, USA

In this lecture I want to give a brief summary, accessible to physicists in all fields of research, of the essential role polarizations play in microscopic (that is, atomic and subatomic) physics. I will outline the essential features and illustrate them on a simple example that is familiar to everybody from studies of elementary quantum mechanics.

To describe the overall role of polarization experiments, I want to use an analogy. If we want to explore the (three-dimensional) shape of an unknown object some distance from us, we will take pictures of it with a camera. If we restrict ourselves to taking pictures from only one location fixed with respect to the object(i.e., from just one given direction), no improvement in the quality of the camera or film will help us beyond a certain point in determining the shape of the object. If, however, we also take pictures from other directions (in this case, three orthogonal directions suffice) then we get full information.

Using only differential cross section data corresponds to taking pictures only from one fixed direction: No matter how these data are refined and extended in energy and angle, they give us only a certain limited type of information. Various polarization experiments correspond to taking pictures from other directions, and they furnish <u>qualitatively</u> different information.

What kind of information do we seek? In general, polarization data can be used for the following purposes:

(1) Experimental determination of the reaction matrix. If a sufficient number of experiments are carried out, the amplitudes

in the reaction matrix can be unambiguously determined, which is the most an experimentalist can do.

(2) The validity of conservation laws can be checked. As we will see, polarization experiments allow this even in the absence of knowledge about the forces that act between the particles.

(3) Theoretical models can be tested. Mere differential cross section experiments are a very poor way to test such theories, because such experiments are sensitive only to some gross features of these theories. On the other hand, polarization experiments can probe in the details of such models.

Each of these general areas have various subramifications. For example, even in the absence of a set of experiments that determines the reaction matrix completely, one can get information out of a partial set of experiments and can also plan further polarization experiments which give new information, independent of that provided by previous experiments.

We will start with writing down the structure of the reaction matrix. The richness of this structure is provided by the spins of the particles participating in the reaction. The strucutre can be described in terms of many different (but basically equivalent) formalisms. In these lectures I will use one that is familiar to both nuclear and particle physicists.

The illustrative reaction I want to use is one in which the initial two particles have spin 0 and ½, respectively, and the same is true for the final two particles (which may or may not be the same two particles). To begin with, I will assume only rotation (or Lorentz) invariance, since such an assumption seems to be a very good one in all areas of subatomic physics, and since the discussion without it would be very cumbersome.

Under this assumption, the reaction matrix M can be written as

$$M = b_0 1 + b_1 \vec{\sigma}\cdot\vec{q}_1 + b_2 \vec{\sigma}\cdot\vec{q}_1 \times \vec{q}_2 + b_3 \vec{\sigma}\cdot\vec{q}_2 \equiv \sum_i b_i \phi_i \quad (1)$$

where $\vec{\sigma}$ is the usual Pauli spin matrix, 1 is the 2×2 unit matrix and the b's are amplitudes, complex numbers which depend on the kinematic variables (angle and energy). The numerical values of the b's depend on the dynamics (i.e., the forces) between the particles. The rest of the structure, however, is independent of such forces.

The quantities q_1 and q_2 in the above equation are two noncollinear momenta that describes the kinematics of the reaction. (For example, they may be the initial and final center of mass momenta). There are an infinite number of ways to describe this kinematics. It will be to our advantage to choose these momenta in such a way that the structure of the polarization experiments is simple.

Experimental observables are always bilinear in amplitudes. In particular, we will define our experimental observables as

$$L(\phi_I, \phi_F) = \text{Tr}(M\phi_I M^\dagger \phi_F) \qquad (2)$$

where ϕ_I and ϕ_F describe the initial and final states in the particular experiment. This expression originated from the fact that the expectation value of an operator in the final state is the trace of the product of that operator with the final state density matrix, and the latter can be written as

$$\rho_F = M \rho_I M^\dagger \qquad (3)$$

Substituting into (2) the form of (1), we get

$$L(\phi_I, \phi_F) = \sum_i \sum_j b_i b_j^* \text{Tr}(\phi_i \phi_I \phi_j^\dagger \phi_F) \qquad (4)$$

As we expected, the observable quantities are bilinear in the amplitudes, and the coefficients are traces of the products of four spin-momentum tensors. The latter are independent of the dynamics and depend only on the values of the spins of the particles. Hence they can be evaluated and tabulated once and for all for the various values of the particle spins.

There are 16 bilinear combinations of amplitudes, and also sixteen different observables, since each of the two indices in the observables can take on four different values. Thus the relationship between bilinear products and observables in general involves 256 coefficients, and the relationship appears to be cumbersome even in this very simple example. It is evident that for higher values of spins the relationship would be even more complicated.

Fortunately in reality the situation is very much simpler. Depending on the skill with which one selects spin tensors and momenta, most of the coefficients in these relationships can be made to vanish. For example, let us choose our momenta as follows:

$$\hat{\ell} \equiv \frac{\vec{q}_1 - \vec{q}_2}{|\vec{q}_1 - \vec{q}_2|} \qquad \hat{m} \equiv \frac{\vec{q}_2 \times \vec{q}_1}{|\vec{q}_2 \times \vec{q}_1|} \qquad \hat{n} \equiv \hat{\ell} \times \hat{m} \qquad (5)$$

In that case, the table of relationships between the observables and the amplitude-products is given on next page.

We see that all but 40 of the coefficients vanish, and that the large problem is subdivided into many independent small problems. This has very significant benefits for all the purposes to which polarization experiments can be utilized.

		$\|b_0\|^2$	$\|b_1\|^2$	$\|b_2\|^2$	$\|b_3\|^2$	Re b_0b_2	Im b_1b_3	Im b_0b_2	Re b_1b_3	Re b_0b_1	Im b_2b_3	Im b_0b_3	Re b_2b_1	Im b_0b_1	Re b_2b_3	Im b_0b_3	Re b_2b_1
L	(0,0;0,0)	1	1	1	1												
L	(0,l;0,l)	1	1	−1	−1												
L	(0,l;0,m)	1	−1	1	−1												
L	(0,n;0,n)	1	−1	−1	1												
L	(0,m;0,0)					1	−1										
L	(0,0;0,m)					1	1										
L	(0,l;0,n)							−1	1								
L	(0,n;0,l)							1	1								
L	(0,l;0,0)									1	−1						
L	(0,0;0,l)									1	1						
L	(0,n;0,0)											1	−1				
L	(0,0;0,n)											1	1				
L	(0,n;0,m)													−1	1		
L	(0,m;0,n)													1	1		
L	(0,l;0,m)															1	1
L	(0,m;0,l)															−1	1

Let us now illustrate what happens if conservation laws additional to mere rotation (or Lorentz) invariance are imposed. For example, consider parity conservation. If this is imposed, and if the product of the four <u>intrinsic</u> parities of the particles in the reaction is $+1$, then the reaction matrix in (1) must also be invariant under reflection. On the right hand side, the amplitudes are always scalars (and not pseudoscalars) because they are functions of the rank-zero tensors that can be composed out of the momenta that characterize the reaction, and for a four-particle reaction those are all scalars (q_1^2, q_2^2, $\vec{q}_1 \cdot \vec{q}_2$). Concerning the spin-momentum tensors, the 1 is a scalar and the $\vec{\sigma}$ a pseudovector, so that if the reaction matrix is to be a scalar, only the first and third terms can appear. Thus in this case $b_1 = b_3 = 0$ for all angles and energies independent of the forces. In the reverse case, when the product of the four intrinsic parities is -1, the reaction matrix (1) must also be a pseudoscalar, and hence only the second and fourth terms can survive and $b_0 = b_2 = 0$.

The vanishing of some amplitudes will, however, have an effect on our table relating the amplitude-products and the observables. While the number of different observables will of course remain the same, the number of non-zero amplitude products is cut down to 4. This will then create 12 <u>linear</u> relationships among the 16 observables, and these relationships are tests of parity conservation because they did not exist when we did not assume this conservation.

In fact some of these relationships merely test parity conservation but do not give information on the product of the intrinsic parities of the four particles, while others also provide the latter.

It should be emphasized that these tests are independent of dynamics, and that they hold at all energies and angles, and hence also in a range of energies and angles.

Let me now turn to the question of determining the amplitudes of the reaction matrix. If one performs all 16 experiments at a given energy and angle, naturally one can determine the 4 complex amplitudes uniquely. But it is also clear that doing all 16 experiments is unnecessary. We have to determine only 4 complex numbers, and with one overall phase being arbitrary and unmeasurable, there are only 7 real numbers we want to fix. Thus seven experiments may be enough. Since, however, the relationship between experiments and amplitudes is a bilinear one, there might also be some discrete ambiguities which require some additional experiments before they are resolved.

Furthermore, it is also evident that not <u>any</u> set of seven experiments will do the job. If, for example, one already measured $|b_i|^2$, $|b_j|^2$, and Re $a_i a_j^*$, an additional experiment measuring Im $a_i a_j^*$ is redundant even though we are still far short of the magic number of 7.

Thus the problem of determining which set of experiments fix the amplitudes uniquely is not a trivial one. In fact, this problem has not been solved completely. It has two parts: The <u>linear</u> problem between observables and amplitude products, and the <u>non-linear</u> problem of the relationship between amplitude products and the real and imaginary parts of the amplitudes. Ironically, the second part, which appears the more difficult one, has been solved completely by constructing a geometrical, diagrammatic analog to this algebraic problem and then giving a set of quite simple rules to tell which diagram is 'good' from the point of view of giving a unique determination. The linear part of the problem, however, has not been solved completely, and the remaining problem pertains to situations when 'accidental' cancellations occur in the determinants relating the observables and the amplitude products.

Work has also been done on getting particle information out of a partial set of experiments, and on treating situations when one or a few amplitudes can be expected to play a dominant part and hence a successive iteration procedure can be used to fix the amplitude.

Perhaps the most extensive use of these results have been made in pion-nucleon and nucleon-nucleon scattering. The latter in the range up to 10 GeV is the target of an intensive experimental effort at Argonne National Laboratory, where techniques for making polarized beams, polarized targets, and for measuring polarizations are strong, and hence already an extensive set of nucleon-nucleon scattering polarization quantities have been measured.

There are, however, many other situations when polarization experiments have made a major impact. Very recently, for example, the polarization in electron scattering has been measured to see if there is any other particle beside a photon exchanged in such reactions. The tiny but non-zero result reported to have been found in that experiment in turn is claimed to provide support to one of the presently popular theoretical models aiming at the unification of electromagnetic and weak interactions.

But results on what polarization experiments can provide are of relevance in a much broader range of physical problems. In atomic, nuclear, and particle physics experimental techniques of performing such polarization experiments have been greatly improved in recent years, so that the possibility is open for a much broader use of such experiments.

In this lecture I could only touch upon the most important conceptual features of the polarization structure of reactions. There is a considerable amount of detailed work that needs to be done when one applies these ideas to specific situations. In this respect, I would like to refer to the already existing articles in the literature. A summary of the results of the work of my own plus of my collaborators was given in "The Polarization Structure of Nuclear and Particle Reactions", notes that I prepared in April 1976 for a set of lectures at the Troisieme Cycle de la Physique en Suisse Romande. These notes are available by writing to Professor R.Mermod, Department de Physique Nucleaire et Corpusculaire, 32 boulevard d"Yvoy, CH-1211 Geneva 4, Switzerland.

References

1. P.L.Csonka, M.J.Moravcsik and M.D.Scadron, Non-existence of Parity Experiments in Multiparticle Reactions. Phys. Rev. Letters, $\underline{14}$, 861 (1965). Proves that in a reaction involving more than four particles, even if the parity of all but one particle is known, there are no experiments which can determine model-independently the parity of the remaining particle.

2. P.L.Csonka, M.J.Moravcsik and M.D.Scadron, Parity Determination in Particle Physics. Phys. Letters, $\underline{15}$, 353 (1965). Seven theorems about parity determinations, regarding the type of experimental information one must have.

3. P.L.Csonka, M.J.Moravcsik and M.D.Scadron, The Non-Dynamical Structure of the He^3 (d,p) He^4 Reaction. Phys. Rev., $\underline{143}$, 775 (1966). A specific nuclear application of the early results of the spin-structure studies and a demonstration of the kind of information one can obtain from some partial experimental results.

4. P.L.Csonka, M.J.Moravcsik and M.D.Scadron, General Non-dynamical Formalism for Reactions with particles of Arbitrary Spin: Composite Reactions and Partial Wave Expansions. Phys. Rev., $\underline{143}$, 1324 (1966). Mathematical Formalism for composits reactions and also for the partial wave expansions of the spin structure amplitudes.

5. P.L.Csonka and M.J.Moravcsik, Non-dynamical Formalism and Tests of Time Reversal Invariance. Phys. Rev., $\underline{152}$, 1310 (1966). A general discussion of the constraints of time reversal invariance and the tests thereof in the non-dynamical formalism.

6. P.L.Csonka, M.J.Moravcsik and M.D.Scadron, General Phenomenological Formalism for Reactions with Particles of Arbitrary Spin. I. Rotation Invariance. Annals of Physics, $\underline{40}$, 100 (1966). The first comprehensive article on the non-dynamical spin formalism, with mathematical details. It is a shortened version of UCRL 14222

report, which, however is probably no longer available. The report has all the mathematical details of the formalism.

7. P.L.Csonka, M.J.Moravcsik and M.D.Scadron, Non-Dynamical Methods to Determine Parities and to Test Invariance Principles. Nuovo Cimento, 42, 743 (1966). Four theorems about parity determination and other invariance principles, concerning the type of information needed.

8. P.L.Csonka, M.J.Moravcsik and M.D.Scadron, Non-Dynamical Spin and Parity Experiments. Phys. Letters, 20, 67 (1966). General Formula for the linear combination of observables which vanish.

9. P.L.Csonka and M.J.Moravcsik, Productset Populations of Reactions Involving Particles of Arbitrary Spin. Journal of Natural Science and Math. (Pakistan), 6, (1966). Tables for the number of bilinear products of amplitudes in the non-dynamical formalism under various conditions.

10. P.L.Csonka, M.J.Moravcsik and M.D.Scadon, General Non-Dynamical Formalism for Reactions with Particles and Arbitrary Spin:Number of Invariants, Parity Conservation. Annals of Physics, 41, 1 (1967). Detailed exposition of the formalism when parity in conserved, and a discussion of how to count amplitudes.

11. P.L.Csonka, M.J.Moravcsik and M.D.Scadron, The Non-dynamical Structure of Photoproduction Processes. Rev. Mod. Physics, 39, 178 (1967). An application of spin-structure studies to various photoproduction processes. Amplitude-observable tables, something about amplitude determination.

12. P.L.Csonka and M.J.Moravcsik, A Maximum Complexity Theorem for Non-Dynamical Parity Tests. Nuovo Cimento, 50A, 422 (1967). To determine parities it is sufficient to measure multipole moments only up to half the maximum allowed.

13. P.L.Csonka and M.J.Moravcsik, Mirror Relations as Non-Dynamical Tests of Conservation Laws. Phys. Rev., 167, 1516 (1968). A general discussion of the origin and role of mirror relations in various tests of conservation laws.

14. P.L.Csonka and M.J.Moravcsik, Non-Dynamical Tests of the CPT Theorem and Related Symmetries. Phys. Rev., 165, 1915 (1968). A general discussion of the constraints and tests of CPT symmetry and other combinations of C,P and T.

15. P.L.Csonka and M.J.Moravcsik, Non-Dynamical Structure of Particle Reactions Containing Identical Particles, Phys. Rev., D1, 1821 (1970). Constraints in the non-dynamical spin structure when one has two or more identical particles.

16. P.L.Csonka and M.J.Moravcsik, Non-dynamical Properties of Particle Reactions. Párticles and Nuclei, 1, 337 (1971). This is a simple, pragmatic summary of the non-dynamical formalism, written for experimentalists. Uses $1 + 0 \rightarrow 1 + 0$ as an example.

17. P.L.Csonka, T.Haratani and M.J.Moravcsik, An introduction to X-coefficients and a Tabulation of their Values. Nuclear Data, 9, 235 (1971). An exposition and evaluation of the coefficients in the orthonormal l-m-n coordinate system which connect the observables and the bilinear products of amplitudes.

18. P.L.Csonka and M.J.Moravcsik, Identical Particles, Crossing, and CPT. Phys. Rev., D5, 2546 (1972). If crossing is a valid operation, and if identical particles are really indistinguishable, then $A + B \to A + B$ is CPT invariant.

19. G.R.Goldstein, M.J.Moravcsik, J.F.Owens III and J.P.Rutherford, Spin Correlation Measurements in Pseudoscalar Meson Photoproduction. Nucl. Phys., B80, 164 (1974). For spin-0 photoproduction, amplitude-observable are constructed, and bounds on observables are worked out. Experimental data are used to demonstrate the results.

20. G.R.Goldstein, M.J.Moravcsik and D.Bregman, Determination of Reaction Amplitudes from Experimental Observables. Nuovo Cimento Letters, 11, 137 (1974). Gives a general prescription, using a geometrical analogue, for selecting sets of observables which completely determine the amplitudes. After publication it was discovered that the prescription is not absolutely complete.

21. G.R.Goldstein and M.J.Moravcsik, Determination of Amplitudes from Observables in Quantum Mechanical Systems. J. of Math. Phys. (to be published) (1976). Presents a detailed proof of the prescription to select sets of observables which completely determine the amplitudes. The prescription later turned out to be not absolutely complete.

22. G.R.Goldstein and M.J.Moravcsik, Optimally simple Connection between the Reaction Matrix and the Observables. Annals of Phys., (to be published)(1976). Determines the type of amplitudes and observables for which the connecting matrix is a diagonal as possible. Works out $0 + \frac{1}{2} \to 0 + \frac{1}{2}$ as an example.

23. G.R.Goldstein and M.J.Moravcsik, A No-Go Theorem for Polarization Structure, Journ. Math. Physics (in press, 1978). Shows that it is impossible to construct a polarization formalism in which the differential cross section is the absolute value squared of a single amplitude and the simplest type of polarization quantities are the function of two amplitudes only.

24. L.J.Gutay, J.E.Lanutti, P.L.Csonka, M.J.Moravcsik and M.D.Scadron, Final State Enhancement at 1920 MeV in Inelastic Scattering. Physics Letters, 16, 343 (1965). Claims Treiman-Yang test also holds for spin-$\frac{1}{2}$ exchange.

25. G.L.Kane and M.J.Moravcsik, Can the Parity of the Ω^- be measured? Phys. Rev., 176, 1733 (1968). A cute application of the parity test theorems of the non-dynamical formalism to show that the determination of the parity of the Ω^- is virtually impossible.

26. K.C.Lam and M.J.Moravcsik, The Spin Structure of Exchange Processes, Phys. Rev., D8, No. 11 (1973). Shows that the spin-structures of a given reaction with various one-particle and one-reggeon exchanges can be reduced to a few very simple cases which helps in devising polarization experiments checking such dynamical models.

27. M.J.Moravcsik, Second Resonance in Pion Photoproduction. Phys. Rev. Letters, 2, 171 (1959). A pioneer paper in using polarized photons to resolve dynamical problems, in this case the identification of the multipole state of the second π-N resonance.

28. M.J.Moravcsik, Boson Photoproduction Experiments. Phys. Rev., 125, 1088 (1962). An early paper, discussing spin-0 photoproduction from spin ½ particles, with parity conservation, giving spin structure, relationship between amplitudes and experiments, and a bit about 'complete' sets of experiments. Uses somewhat ad hoc methods.

29. M.J.Moravcsik, The Non-Dynamical Structure of Particle Reactions. In 'Recent Developments in Particle Physics', ed. M.J.Moravcsik (Proceedings of the First Pacific International Summer School in Physics, Hawaii 1965), Gordon & Breach, New York, 1966, pp. 197-263. An early summary of spin-structure studies, in considerable detail, but lacking some of the simplifying insight gained later. Covers the mathematical details of trace calculations well.

30. M.J.Moravcsik, The Spin Structure of Nucleon-Nucleon Scattering. Proceedings of the International Conference on Polarization Phenomena of Nucleons, 1965, Burkhauser Verlag, Basel 1966. An exposition of the four-spin ½ reaction using still somewhat ad hoc methods.

31. M.J.Moravcsik, Non-Dynamical Structure of Particle Reactions. Proceedings of the Conference on Intermediate Energy Physics, Williamsburg, 1966, p. 517. An exposition in simple terms of the spin-structure studies, and a detailed discussion of the $0 + 1 \to 0 + 1$ process as an example.

32. M.J.Moravcsik, Experimental Restrictions in the Determination of Invariant Amplitudes. Phys. Rev., 170, 1440 (1968). A partial contribution to the task of determining amplitudes from observables.

33. M.J.Moravcsik and Wing-Yin Yu, Determination of Invariant Amplitudes from Experimental Observables. J. of Math. Phys., 10, 925 (1969). Introduces graph-methodology and gives some necessary and sufficient conditions for amplitude determination.

34. M.J.Moravcsik, Optimal Coordinate Representation for Particle Reactions. Nucl. Phys., A160, 569 (1971). Investigates the problem of selecting the simplest coordinate system in describing polarization, depending on the criteria used.

35. M.J.Moravcsik, Non-dynamical Structure of Collinear Processes, Phys. Rev., D5, 836 (1972). The constraints on the spin structure for forward or backward reactions.

36. M.J.Moravcsik and P.Ghosh, Deuteron Wave Function at Small Distances, Phys. Rev. Letters, 32, 321 (1974). Gives an illustration of the power of polarization experiments in yielding new dynamical information by showing that the tensor polarization in elastic.e-d scattering at a few GeV is very sensitive to the small distance part of the deuteron wave function.

37. G.Ohlson, Polarization Transfer and Spin Correlation Experiments in Nuclear Physics, Rep. Prog. Phys., 35, 717 (1972). General review of polarization formalism, using the most straightforward methods. Extensive tables, elaborate formulae, collection of results for low spin reactions.

38. R.Reid and M.J.Moravcsik, The Non-dynamical Structure of the Reaction involving four Spin-½ Particles. Annals of Physics, 84, 535 (1974). A general discussion, using the full set of tools of the spin-structure studies, of the structure of the four-spin ½ reaction. Detailed tables, lists of symmetry relations, etc. A little on amplitude determination.

In addition, of course, many other individuals and groups have also contributed to this subject. Perhaps the best source of information in that respect is provided by the proceedings of the various polarization conferences. See for example, F.Halzen, Polarization Experiments - A Theoretical Review, in Deepter Pathways in High Energy Physics, B.Kursunoglu, A.Perlmutter and L.F.Scott, Eds, Plenum Publishing Company, New York (1977), p. 33-69.

FUNDAMENTAL CONSTITUENTS OF MATTER AND UNIFICATION OF WEAK AND ELECTROMAGNETIC INTERACTIONS

Riazuddin

Department of Physics

Quaid-i-Azam University, Islamabad, Pakistan

1. INTRODUCTION

In this lecture I wish to review some recent developments in particle physics, in particular the prediction and subsequent experimental discovery of a new type of weak interaction, the so-called neutral weak interaction. This interaction resulted as a consequence of unification of weak and electromagnetic interaction by Salam and Weinberg. Also I would discuss how the proliferation of strongly interacting particles, called hadrons, the well-known example of which is proton, has lead us to regard them as composite entities made up of quarks.

Physicists have been concerned with ultimate constituents of matter and the basic interactions which control the physical phenomena. They have also been concerned with the question whether we can give a unified picture of mutual interactions of fundamental constituents of matter. As is well-known an atom, which was once considered to be indivisible, consists of electrons and nucleus. Further we know that complex nuclei are composite of two particles, namely protons and neutrons. The question arises whether the proton or neutron has a substructure. This question particularly became important in 1950's when a great number of hadrons were discovered. Although so many hadrons were found, nevertheless they showed some regularities in their mass spectrum, in their production characteristics as well as in their decays. It was therefore felt that some simplified picture should be introduced to provide unity in diversity. It was realised that indeed it is possible and one could describe all these hadrons as composite of three fundamental constituents which were given the name of quarks[1,2]. In this picture one could understand in a simple way the regularities which the mass spectrum of these hadrons shows as well as certain other symmetries which were noticed in their production and decay characteristics. More importantly many predictions made on this simple picture were also tested.

The hadron family consists of two main branches, the baryons and mesons. The former are Fermi particles (fermins) and carry half integral spin and baryon number 1, the conservation of which is responsible for the stability or near stability of proton. The latter are Bose particles (bosons) and carry zero or integral spin and baryon number zero. According to the above picture the lowest states of the baryon spectrum are interpreted as consequences of three quark structure, while meson spectrum is interpreted as quark and antiquark structure. Thus the quarks play the same role in hadron spectroscopy as electrons play in atomic spectroscopy and nucleons (neutrons and protons) in nuclear spectroscopy. The quarks themselves carry spin 1/2 and possess fractional charges. To be more specific let us denote the three quarks by u, d, s or collectively by q_a (a = 1, 2, 3). It is assumed that u, d, s quarks carry 2/3, -1/3 and -1/3 of the protonic charge respectively. Since d and s quarks have the same charge, therefore we must intorduce some physical attribute or quantum number which should make a distinction between d and s quarks This additional quantum number is called strangeness and it distinguishes a s quark from u and d quarks. The u and d quarks are identical in all aspects except in their charges.

Thus in analogy with ordinary spin, they may be considered as a doublet in some internal spin space, called isopin space. In Table 1 below I give the charge Q, isospin I, strangeness S and baryon number B of the 3 quarks. More often instead of strangeness, hypercharge Y = S + B is used to distinguish s quark from d quark.

Table 1

Quantum Numbers of Quarks

Quark Type	u	d	s
Q	2/3	-1/3	-1/3
I	1/2	1/2	0
I_3	1/2	-1/2	0
S	0	0	-1
B	1/3	1/3	1/3
Y	1/3	1/3	-2/3

The u, d, s quarks are said to carry three flavors, called up, down and strange. In group theoretic language, the three quarks

FUNDAMENTAL CONSTITUENTS OF MATTER

can be regarded as belonging to 3 representation of group SU(3) (unitary group in some internal three dimensional space defined by three flavors) while antiquarks $\bar{u}$, $\bar{d}$, $\bar{s}$ belong to 3* representation of SU(3). In what follows we shall often write this group as $S^fU(3)$ (and call it SU(3) flavor group)[3] to distinguish it from another SU(3) group to be introduced later. Thus we can write

Mesons $\sim q\bar{q}$,

$3 \otimes 3^* = 8 \oplus 1$

Baryons $\sim qqq$,

$3 \otimes 3 \otimes 3 = 10 \oplus 8 \oplus 8 \oplus 1$

i.e. to say mesons belong to octet or singlet representation of $S^fU(3)$ while baryons belong to decuplet, octet and singlet representations. This is what is seen in nature. Further q and $\bar{q}$ spins may be combined to form a total spin S, which is either 0 or 1. Similarly total quark spin for qqq system is 1/2 or 3/2. When coupled with relative internal orbital angular momentum L, we can form a total angular momentum $J = L + S$. The ground state of $q\bar{q}$ or qqq system is assumed to have $L = 0$. Further as q and $\bar{q}$ have opposite intrinsic parity, therefore parity of $q\bar{q}$ system is $P = (-1)(-1)^L = -1$ for the ground state. Thus we have for ground states

Mesons

$(q\bar{q})_{L=0}$

$S=0,1$

$J^P = 0^-, 1^-$

$S^fU(3)$ content : $8 \oplus 1, 8 \oplus 1$

Baryons

$(qqq)_{L=0}$

$S=1/2, 3/2$.

$J^P = 1/2^+, 3/2^+$

$S^fU(3)$ content : 8, 10.

Below we give the quark flavor content for the lowest lying meson and baryon states found in nature (we suppress the quark spin indices)

1^- (or $3S_1$) nonet (8+1) of Vector Mesons:

(I=1,Y=0) : $\rho^+ \sim u\bar{d}$, $\rho^- \sim d\bar{u}$, $\rho^0 \sim (u\bar{u} - d\bar{d})/\sqrt{2}$

(I=½,Y=1) : $K^{*+} \sim u\bar{s}$, $K^{*0} \sim d\bar{s}$

(I=½,Y=-1): $K^{*-} \sim s\bar{u}$, $\bar{K}^{*0} \sim s\bar{d}$

(I=0,Y=0) : $\omega_8 \sim (u\bar{u} + d\bar{d} - 2s\bar{s})/\sqrt{6}$

(I=0,Y=0) : $\omega_1 \sim (u\bar{u} + d\bar{d} + s\bar{s})/\sqrt{3}$

The 0^- (or $1S_0$) nonet of pseudoscalar mesons $\pi^\pm$, π^0, K^+, K^0, K^-, $\bar{K}^0$, η_8 and η_1 have also exactly the above quark content.

$1/2^+$ octet of Baryons

$(I=\tfrac{1}{2}, Y=1)$: $p \sim uud$, $n \sim udd$

$(I=0, Y=0)$: $\Lambda \sim uds$

$(I=1, Y=0)$: $\Sigma^+ \sim uus$, $\Sigma^\circ \sim uds$, $\Sigma^- \sim dds$

$(I=\tfrac{1}{2}, Y=-1)$: $\Xi^\circ \sim uss$, $\Xi^- \sim ds\bar{s}$

$3/2^+$ decuplet of Baryons

$(I=0, Y=-2)$: $\Omega^- \sim sss$

$(I=\tfrac{1}{2}, Y=-1)$: $\Xi^{*-} \sim ssd$, $\Xi^{*\circ} \sim ssu$

$(I=1, Y=0)$: $\Sigma^{*+} \sim suu$, $\Sigma^{*\circ} \sim sdu$, $\Sigma^{*+} \sim suu$

$(I=3, Y=1)$: $\Delta^{++} \sim uuu$, $\Delta^+ \sim duu$, $\Delta^\circ \sim ddu$, $\Delta^- \sim ddd$

There is a difficulty with the above picture. Consider $\Delta^{++}(S_z = +3/2)$ which according to the above picture can be written as

$$\Delta^{++} \sim |u\!\uparrow u\!\uparrow u\!\uparrow\rangle$$

where $\uparrow$ denotes that u quark has S_z (the component of spin) $= +1/2$. This state is symmetric in quark flavor and spin indices. The space part of the wave function for this state is also symmetric, being $L = 0$ state. Thus the wave function for this state is totally symmetric and thus violates Pauli's exclusion principle. To remedy it, concept of color[4] has been introduced i.e. to say quarks in addition to carrying flavor also carry color. Each quark flavor comes in three different colors, say, red, blue and yellow. Thus

$$q \longrightarrow q_i$$

where $i = 1,2,3$ refer to color indices. Thus we can form the following wave function for 3 quarks

$$|qqq\rangle \longrightarrow \frac{1}{3!} \epsilon_{ijk} |q_i q_j q_k\rangle \qquad (1)$$

which is symmetric in space, flavor and spin but is antisymmetric in color and thus satisfy Pauli's Exclusion Principle. Thus q's belonging to triplet representation of $S^f U(3)$ also belong to triplet representation of color SU(3) which we write as $S^c U(3)$. With respect of $S^c U(3)$,

$$3 \otimes 3 \otimes 3 = 10 \oplus 8 \oplus 8 \oplus 1, \quad 3 \otimes 3^* = 8 \oplus 1$$

Note the important fact that for 3 quark system only singlet (1 in $3 \otimes 3 \otimes 3$ decomposition) is totally antisymmetric with respect to $S^c U(3)$ while for $(q\bar{q})$ system the singlet is totally symmetric as it

FUNDAMENTAL CONSTITUENTS OF MATTER

should be. Thus the baryons formed above in (1) and mesons represented by

$$|q\bar{q}\rangle = \frac{1}{\sqrt{3}} |q_1 \bar{q}_1 + q_2 \bar{q}_2 + q_3 \bar{q}_3\rangle \qquad (2)$$

are color singlets. It is thus postulated (or it is a prejudice, an alternative point of view will be mentioned later) that all known hadrons are color singlets. Thus color is hidden and this is the postulate of color confinement[5]. As seen above in (1) and (2) the simplest color singlet configurations are mesons ($q\bar{q}$) and baryons (qqq). The principle of color confinement forbids the existence of states like single quark (q) or antiquark ($\bar{q}$), diquarks (qq) etc.

The above situation remained for 10 years (1964-1974). However, we now know that there is a fourth quark flavor, called charm. We shall discuss how the need for this fourth flavor arose and its subsequent experimental verification. We shall discuss its need in the context of unification of weak and electromagnetic interactions.

2.   UNIFICATION OF WEAK AND ELECTROMAGNETIC INTERACTIONS

Basic forces in nature can be divided into four classes: (i) Gravitational: Its strength is characterised by Newton's Gravitational constant G_N which when made dimensionless by using mass of the proton m_p (we use the units $\hbar = c = 1$, where $\hbar$ is Planck's constant divided by 2π and c is velocity of light) has the value $G_N m_p^2 \simeq 10^{-38}$. (ii) Weak: This is responsible, for example, for radioactivity, the well-known example of which is neutron beta decay, namely

$$n \rightarrow p + e^- + \bar{\nu}_e$$

or the decay of muon

$$\mu^- \rightarrow e^- + \bar{\nu}_e + \nu_\mu$$

Here e^- denotes electron, while ν_e and ν_μ denote the neutrinos associated with electron and muon respectively. $\bar{\nu}_e$ and $\bar{\nu}_\mu$ denote the corresponding antineutrinos. The weak interaction is characterised by the Fermi constant G_F, which when made dimensionless again by using mass of the proton m_p, has the value $G_F m_p^2 \simeq 10^{-5}$. (iii) Electromagnetic: This is a very well understood force and acts between electrically charged particles. It is possible, for example, for the attraction between electrons and protons and therefore for the binding of the atoms. Its strength is characterised by the fine structure constant $\alpha = e^2/4\pi = 1/137$. Here e denotes the electric charge of the proton. (iv) Strong: Strong interaction is responsible for binding of the nuclei or for confinement of quarks in a hadron. Its strength is much stronger than the electromagnetic interaction, at least by a factor of 30-100.

The gravitational interaction on microscopic scale is too weak compared to the other three forces that it is ignored. This assumption may very well be wrong; it is conceivable that this force may play some role in particle physics. In any case we will not consider this force any further. We shall take up the strong interaction later and consider first weak and electromagnetic interactions which show some similarities (there are some differences also) as indicated in Table II below.

Table II

Similarities and Differences of Electromagnetic and Weak Interactions

Electromagnetic	Weak
Electromagnetic current $J_\mu^{e.m.}$ ($\mu = 1,2,3,4$) is a vector; it cannot distinguish between left and right	Weak current J_μ^W consists of vector and axial vector parts since weak interaction can distinguish between left and right or does not conserve parity.
The interaction Lagrangian is $$L_{int} = e\, J_\mu^{e.m.}\, A_\mu$$ A_μ is a vector field whose quanta photons are the carrier of electromagnetic interaction.	Interaction Lagrangian is $$L_{int} = g_W\, J_\mu^W\, W_\mu^- + h.c.$$ (h.c. denotes Hermitian conjugate) W_μ is a vector field whose quanta, called the weak vector bosons, are carrier of weak interaction.
Photons are electrically neutral	Weak vector bosons must be able to carry charges ± since in the well-known manifestations of weak interaction, charge is exchanged (e.g. in neutron beta decay, charge is exchanged between neutron and proton).
Universality of coupling strength $e^2/4\pi = 1/137$	Universality of coupling strength $g_W^2/m_W^2 = G_F = 10^{-5}\, m_p^{-2}$
Photons are massless	Weak vector W bosons must be massive since weak interaction is of very short range.
Gauge Invariant	Gauge invariant in the limit of zero W masses
Abelian Gauge Theory as gauge particle photon does not carry charge.	Non-Abelian Gauge Theory (Yang-Mills type[6]) as gauge particles W-bosons carry charge.

FUNDAMENTAL CONSTITUENTS OF MATTER

As we have seen above both electromagnetic and weak currents are vector in character and both individually possess universality of the coupling strength for their respective couplings with photon and weak boson fields A_μ and W_μ. Since the universality of the coupling and vector character are attributes of gauge invariance, their unification was studied by Salam[7] and independently by Weinberg[8] in a gauge invariant theory. Gauge invariance is a symmetry in which not only does the theory have a symmetry with respect to some kind of internal rotation performed everywhere all at once, but it also is gauge invariant even if one performs independent internal transformations at every point in space-time. Thus if, for example, we demand the physical law expressed by Schrödinger equation or Dirac equation to be invariant under the space-time dependent phase transformation ($x \equiv \underline{x}$, it):

$$\psi(x) \to \psi'(x) = e^{ie\chi(x)} \psi(x), \qquad (3)$$

one can easily see that free particle Schrödinger or Dirac equation is not invariant under the above transformation. In order to restore gauge invariance, it is necessary to introduce a new field $A_\mu = (\underline{A}, i\phi)$ which is vector in character and which under the above gauge transformation transforms according to the following law ($\partial_\mu \equiv \partial/\partial x_\mu$)

$$A_\mu \to A'_\mu = A_\mu + \partial_\mu \chi(x)$$
$$\partial_\mu \partial_\mu \chi(x) = 0 \qquad (4)$$

A_μ here is interpreted as the electromagnetic field. The invariance of Schrödinger or Dirac equation under the gauge transformation is assured if A_μ field is coupled to the ψ field only through the replacement

$$\partial_\mu \psi(x) \to \left(\partial_\mu - \text{i.e. } A_\mu(x)\right) \psi(x) \qquad (5)$$

This would lead to the following gauge invariant Schrödinger and Dirac equations respectively

$$-\frac{\hbar^2}{2m} (\underline{\nabla} - \text{i.e. } \underline{A})^2 \psi = i(\frac{\partial}{\partial t} + \text{i.e. } \phi)\psi$$

$$\left(\gamma_\mu (\partial_\mu - \text{i.e. } A_\mu) + m\right) \psi(x) = 0$$

The above equations give correctly the form of the interaction of a particle of charge e with electromagnetic field in non-relativistic and relativistic case respectively. Moreover in the relativistic Dirac theory of spin 1/2 particle in the presence of electromagnetism we see that the gauge invariant Lagrangian as obtained by making the replacement (5) in the free particle-lagrangian density

$$- \bar{\psi}(x) \gamma_\mu \partial_\mu \psi(x)$$

is
$$L = -\bar{\psi}\gamma_\mu(\partial_\mu - ie A_\mu)\psi - \frac{1}{4}(\partial_\mu A_\nu - \partial_\nu A_\mu)(\partial_\mu A_\nu - \partial_\nu A_\mu) \quad (6)$$

where the second term in (6) denotes the Lagrangian density for the electromagnetic field A_μ. Note the important fact that the gauge invariance of electromagnetism is itself connected with the fact that photon is massless. For example the mass term $m^2 A_\mu A_\mu$ in the Lagrangian (6) would not be invariant under the gauge transformation given in Eqs. (3) and (4).

Now the vector bosons W mediating the weak interaction must be massive as remarked earlier. On the other hand we have seen that the gauge invariance, which plays a crucial role in the success of quantum electrodynamics, requires gauge vector bosons to be massless. In analogy with quantum electrodynamics we want weak interaction theory to be gauge invariant; but then we have to provide for mass of the vector bosons W. Rather than by introducing the explicit mass terms in the Lagrangian, masses of W bosons come from the interaction of these fields with a new kind of scalar field, the so called Higgs field[9]. The gauge symmetry is then broken spontaneously by allowing some of the scalar fields to have non-zero vacuum expectation values (for details the reader is referred to reference 10). Such a breaking of gauge symmetry introduces massive vector bosons in the theory except photon for which the gauge symmetry is exact.

We note from (6) that the interaction of an electrically charged Dirac particle of charge e represented by ψ with the electromagnetic field is given by

$$L_{int} = i.e.\ \bar{\psi}\gamma_\mu \psi A_\mu$$

so that
$$J_\mu^{e.m.} = i\ \bar{\psi}\gamma_\mu \psi$$

Thus in Table II we can easily see that for particles e^-, ν_e, $\bar{\mu}, \nu_\mu$ (note e^- and μ^- have charge -1 in units of proton charge, while ν_e and ν_μ are electrically neutral)

$$J_\mu^{e.m.}(\ell) = i[-\bar{e}\gamma_\mu e - \bar{\mu}\gamma_\mu \mu] \quad (7)$$

$$J_\mu^W(\ell) = i[\bar{\nu}_e \gamma_\mu(1+\gamma_5)e + \bar{\nu}_\mu \gamma_\mu(1+\gamma_5)\mu] \quad (8)$$

Here γ_μ ($\mu = 1,2,3,4$) are Dirac matrices; $\gamma_5 = \gamma_1 \gamma_2 \gamma_3 \gamma_4$ and its appearance ensures that parity is not conserved by weak interaction. The four particles e, ν_e, μ, ν_μ together with their antiparticles are known as leptons (i.e. why we have put ℓ in parenthesis in Eqs. (7) and (8)). Their main characteristic is that they do not have strong interaction. They have only weak interaction responsible, for example, for the muon decay $\mu \to e + \bar{\nu}_e + \nu_\mu$ and charged members of this family have also electromagnetic interaction. For hadrons the corresponding currents are (note that quarks u, d, s have charges 2/3, -1/3, -1/3)

$$J_\mu^{e.m.}(h) = i\left(\frac{2}{3}\bar{u}\gamma_\mu u - \frac{1}{3}\bar{d}\gamma_\mu d - \frac{1}{3}\bar{s}\gamma_\mu s\right) \quad (9)$$

$$J_\mu^W(h) = i\bar{u}\gamma_\mu(1+\gamma_5)d' \quad (10)$$

where

$$d' = \cos\theta_c\, d + \sin\theta_c\, s \quad (11)$$

Here θ_c is known as Cabibbo angle and has experimental value $\approx 15°$. This is to satisfy the experimental requirement that the decay rates for strangeness changing $|\Delta S| = 1$ (note that quark s has strangeness S = -1) decay processes (e.g. $\Lambda \to p + e^- + \bar{\nu}_e$) are suppressed by a factor of about 1/16 compared to those for strangeness conserving $\Delta S = 0$ decay processes (e.g. $n \to p + e^- + \bar{\nu}_e$) involving charged weak currents $J_\mu^W(h)$ and $J_\mu^W(\ell)$. Note, the important fact that weak interaction recognises the quarks d and s only in the combination d´ and not d and s separately. What about the other orthogonal combination

$$s' = -\sin\theta_c\, d + \cos\theta_c\, s \quad ? \quad (12)$$

We shall come to this later.

2.1   Gauge group $SU(2) \otimes U(1)$ of Salam and Weinberg

In order to unify weak and electromagnetic interactions we have to consider the charge carrying weak vector bosons W^+, W^- as well as the electrically neutral photon γ together in such a way that they are coupled to the weak and electromagnetic currents with the same coupling strength. In other words the weak vector fields W_μ^+, W_μ^- and the electromagnetic field A_μ must be regarded as different components of one and the same field. Let us denote by Q_W, $Q_{\bar{W}}$ and $Q_{e.m.}$ the charges associated with the weak currents J_μ^W, $\bar{J}_\mu^W$ ($\bar{J}_\mu^W$ is hermitian conjugate of J_μ^W) and the electromagnetic current $J_\mu^{e.m.}$ respectively. Now we see that the charges Q_W and $Q_{\bar{W}}$ do not commute with each other nor they commute with $Q_{e.m.}$ and nor they in general form a closed algebra. Thus

$$[Q_W, Q_{\bar{W}}] = Q_{W^0} \neq Q_{e.m.}$$

In order to incorporate $Q_{e.m.}$, we introduce another neutral current J_μ^B with associated charge Q_B and coupled to a neutral vector boson field B_μ. Thus we consider the group whose generators are Q_W, $Q_{\bar{W}}$, Q_{W^0}.

and Q_B i.e. a group $SU(2) \otimes U(1)$, the former three generating $SU(2)$ and the latter generating a $U(1)$ group. This is the model considered by Salam[7] and independently by Weinberg[8]. Then $Q_{e.m.}$ is a linear combination of Q_{W^0} and Q_B and is coupled to the electromagnetic field A_μ, which is now a linear combination of B_μ and W_μ fields:

$$A_\mu = \cos\theta_W B_\mu + \sin\theta_W W^0_\mu \qquad (13a)$$

while the other orthogonal combination

$$Z_\mu = -\sin\theta_W B_\mu + \cos\theta_W W^0_\mu \qquad (13b)$$

is coupled to a neutral current J^Z_μ, which is a linear combination of J^B_μ and $J^{W^0}_\mu$ or equivalently of $J^{W^0}_\mu$ and $J^{e.m.}_\mu$, the structure of which we shall consider below.

When the gauge symmetry corresponding to the group $SU(2) \otimes U(1)$ is spontaneously broken by introducing scalar Higgs fields and assuming for the scalar fields the simplest representation under the group, the vector bosons W^+, W^-, Z^0 acquire masses while the photon γ still remains massless. The angle θ_W tells us that we take the actual particles to be $W^\pm$, Z^0 and γ and not $W^\pm$, W^0 and B. The boson Z^0 mediates the neutral current J^Z_μ.

2.2   Assignment of Leptons and Hadrons in $SU(2) \otimes U(1)$ Group; Form of Neutral Weak Interaction and Fourth Flavor Charm

We have two pairs of leptons (ν_e, e^-), (ν_μ, μ^-) out of which we know that neutrinos are left handed (i.e. their spins always point opposite to their direction of motion) while electron e and muon µ have both left handed and right handed components. Accordingly if we write

$$e_{L,R} = \frac{1 \pm \gamma_5}{2} e, \quad \mu_{L,R} = \frac{1 \pm \gamma_5}{2} \mu$$

we put

$$L_e = \begin{pmatrix} \nu_e \\ e_L \end{pmatrix}, \quad L_\mu = \begin{pmatrix} \nu_\mu \\ \mu_L \end{pmatrix} \qquad (14)$$

as doublets under $SU(2)$ while e_R, μ_R are singlets under this group.

The group $SU(2)$ may be regarded as generated, by a 'weak' isospin $\underline{I}_L$ while $U(1)$ may be considered as generated by 'weak' hypercharge Y_L

$$\tfrac{1}{2} Y_L = Q - T_{3L} \qquad (15)$$

where I_{3L} is the third component of isospin $\underline{I}_L$ and Q is electric

FUNDAMENTAL CONSTITUENTS OF MATTER

charge. The gauge vector bosons corresponding to SU(2) are denoted by $A_{i\mu}$ (i = 1,2,3) while that corresponding to U(1) is denoted by B_μ. The relevant gauge transformations (analogues to Eq.(5)) which make the Lagrangian invariant under them corresponds to replacements

$$\partial_\mu \to \partial_\mu - i Q_i g' B_\mu \qquad (16a)$$

acting on the singlets (Q_i being charges of the singlets)

$$\partial_\mu \to \partial_\mu - ig\, \underline{t}\cdot\underline{A} - ig'(Q - t_3) B_\mu \qquad (16b)$$

acting on the doublets, where $t_i = \tfrac{1}{2}\tau_i$ are isospin matrices (τ_i are Pauli matrices acting in isospin space of $\underline{L}_L$) and Q is the charge matrix for the doublets.

The relevant Lagrangian for weak and electromagnetic interactions which is invariant under the above gauge transformations (according to the rule discussed previously e.g. Eq.(6)) as far as leptons are concerned is

$$L_{int} = -\bar{L}_e \gamma_\mu (\partial_\mu - ig\underline{t}\cdot\underline{A} - ig'(Q-t_3) B_\mu) L_e$$
$$- \bar{e}_R \gamma_\mu (\partial_\mu - Q_e g' B_\mu) e_R + (e \leftrightarrow \mu) \qquad (17)$$

Note that $t^\pm = \tfrac{1}{2}(\tau_1 \pm i\tau_2)$ so that

$$t^+ = \begin{pmatrix} 0 & 1 \\ 0 & 0 \end{pmatrix}, \quad t^- = \begin{pmatrix} 0 & 0 \\ 1 & 0 \end{pmatrix}$$

$$t_3 = \tfrac{1}{2}\tau_3 = \tfrac{1}{2}\begin{pmatrix} 1 & 0 \\ 0 & -1 \end{pmatrix}, \quad Q = \begin{pmatrix} 0 & 0 \\ 0 & -1 \end{pmatrix}$$

while the weak vector fields W_μ, $\bar{W}_\mu$, Z_μ and the electromagnetic field A_μ introduced previously are given by

$$\bar{W}_\mu = 1/\sqrt{2}\,(A_{1\mu} - iA_{2\mu}), \quad W_\mu = 1/\sqrt{2}\,(A_{1\mu} + iA_{2\mu})$$

$$W_\mu^\circ = A_{3\mu}$$

$$Z_\mu = (g^2 + g'^2)^{-\tfrac{1}{2}} (g A_{3\mu} - g' B_\mu) \qquad (18)$$

$$A_\mu = (g^2 + g'^2)^{-\tfrac{1}{2}} (g' A_{3\mu} + g B_\mu) \qquad (19)$$

with

$$g'/g = \tan\theta_W \qquad (20)$$

Then it is easy to see that L_{int} (17) can be written as

$$\mathcal{L}_{int} = \frac{gg'}{\sqrt{g^2+g'^2}} J_\mu^{e.m.} A_\mu + \frac{\sqrt{g^2+g'^2}}{2} J_\mu^Z Z_\mu +$$
$$+ \frac{g}{2\sqrt{2}} \left[J_\mu^W \bar{W}_\mu + \bar{J}_\mu^W W_\mu \right] \quad (21)$$

where for the lepton sector $J_\mu^{e.m.}$ and J_μ^W are given in Eqs.(7) and (8) while

$$J_\mu^Z(\ell) = J_{3\mu}(\ell) - 2\sin^2\theta_W J_\mu^{e.m.}(\ell) \quad (22a)$$

with

$$J_{3\mu}(\ell) = i\bar{L}_e \frac{1}{2}\tau_3 L_e + i\bar{L}_\mu \frac{1}{2}\tau_3 L_\mu$$
$$= \frac{i}{2} [\bar{\nu}_e \gamma_\mu(1+\gamma_5)\nu_e - \bar{e}\gamma_\mu(1+\gamma_5)e + \bar{\nu}_\mu \gamma_\mu(1+\gamma_5)\nu_\mu$$
$$- \bar{\mu}\gamma_\mu(1+\gamma_5)\mu] \quad (22b)$$

We have above dealt with four well known leptons. If heavy leptons exist, we can assign them to doublets or singlets under the group $SU(2) \otimes U(1)$. In fact there is already firm evidence for a heavy lepton[11] called τ with mass $m_\tau = 1.807 \pm 0.02$ GeV. If its associated neutrino ν_τ is also found, τ_L and ν_τ can form a doublet similar to L_e or L_μ in Eq.(14).

We now come to the question of assignment of hadrons. If we confine ourselves to three quark flavors u, d, s, then as already remarked the weak interactions see only the linear combinations u,d' where d' is given in Eq.(11). Thus the possible assignment under the group $SU(2) \otimes U(1)$ is

$$L_q = \begin{pmatrix} u_L \\ d'_L \end{pmatrix}, \quad I_L = \tfrac{1}{2}, \quad Y_L = -1$$

u_R, d_R, s_R are singlets with $I_L = 0$, $Y_L = -2$

Then it is easy to see that, as in case of leptons, in the Lagrangian (21), $J_\mu^{e.m.}$ and J_μ^W for hadrons are given in Eqs.(9) and (10) while

$$J_\mu^Z = J_{3\mu} - 2\sin^2\theta_W J_\mu^{e.m.} \quad (23a)$$

with

$$J_{3\mu} = i/2[\bar{u}\gamma_\mu(1+\gamma_5)u - \bar{d}'\gamma_\mu(1+\gamma_5)d'] \quad (23b)$$

The second term in Eq.(23b) contains factors like $\sin\theta_c \cos\theta_c$ $\bar{d}\gamma_\mu(1+\gamma_5)s$ and hence we see from Eqs.(22) and (23) that the neutral current J_μ^Z weak interaction through Z boson can give rise to proce-

FUNDAMENTAL CONSTITUENTS OF MATTER

sses of the form

$$s \to d + \ell + \bar{\ell} \quad , \quad \ell = \nu_e, \nu_\mu, e \text{ or } \mu$$

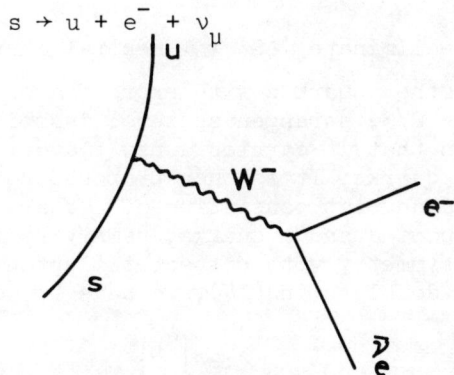

This occurs with the same strength as the process

$$s \to u + e^- + \nu_\mu$$

The latter process occurs through the interaction of charged vector boson with charged current J_μ^W. Thus we see that

$$K^- \to \pi^- + \ell + \bar{\ell}$$

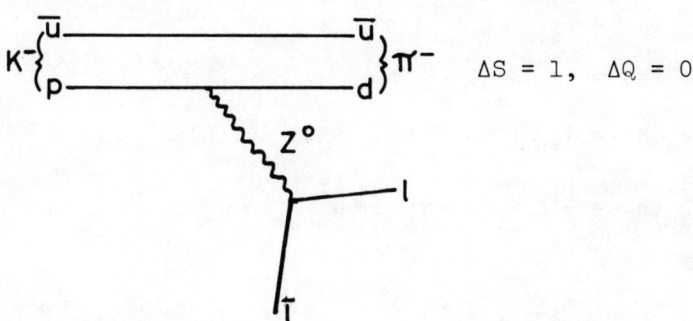

$\Delta S = 1, \quad \Delta Q = 0$

occurs with the same strength as

$$K^- \to \pi^0 + e^- + \bar{\nu}_e$$

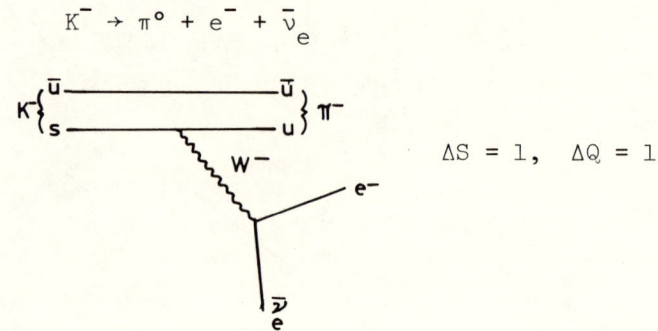

$\Delta S = 1, \quad \Delta Q = 1$

Experimentally $K^- \to \pi^- \ell\bar{\ell}$ and other $|\Delta S| = 1$ decay processes through $|\Delta S| = 1$ neutral current weak interaction are very very much suppressed compared to those through the $|\Delta S| = 1$ charged current weak interaction.

In order to eliminate $|\Delta S| = 1$ neutral current, it is necessary to introduce another quark c with extra flavor called charm[12]. The quark c has charge 2/3, strangeness zero, isospin zero and differs from u, d, s quarks in that it carries a new flavor, charm, equal to 1 relative to u, d, s quarks. It is then proposed to introduce an additional left handed doublet consisting of c quark and the other orthogonal combination of d and s quarks, namely s´ given in Eq.(12). We have now complete symmetry with respect to lepton doublets; corresponding to two lepton doublets in (14), we have two quark doublets

$$L_q = \begin{pmatrix} u_L \\ d'_L \end{pmatrix} \qquad L'_q = \begin{pmatrix} c_L \\ s'_L \end{pmatrix} \qquad (24)$$

while c_R, u_R, d_R and s_R are singlets. Then it is easy to see (e.g. Eq.17) that with respect to the Lagrangian density (21),

$$J_\mu^{e.m.}(h) = i(2/3\,\bar{c}\,\gamma_\mu\,c + 2/3\,\bar{u}\,\gamma_\mu\,u - 1/3\,\bar{d}\,\gamma_\mu\,d - 1/3\,\bar{s}\,\gamma_\mu\,s) \tag{25a}$$

$$J_\mu^W(h) = i\,\bar{u}\,\gamma_\mu(1+\gamma_5)\,d' + i\bar{c}\,\gamma_\mu(1+\gamma_5)\,s' \tag{25b}$$

$$J_\mu^Z(h) = J_{3\mu}(h) - 2\sin^2\theta_W\,J_\mu^{e.m.}(h) \tag{25c}$$

with

$$J_{3\mu}(h) = \tfrac{i}{2}[\bar{c}\,\gamma_\mu(1+\gamma_5)c + \bar{u}\,\gamma_\mu(1+\gamma_5)u - \bar{d}'\,\gamma_\mu(1+\gamma_5)d'$$
$$- \bar{s}'\,\gamma_\mu(1+\gamma_5)s']$$
$$= \tfrac{i}{2}[\bar{c}\,\gamma_\mu(1+\gamma_5)c + \bar{u}\,\gamma_\mu(1+\gamma_5)u - \bar{d}\,\gamma_\mu(1+\gamma_5)d$$
$$- \bar{s}\,\gamma_\mu(1+\gamma_5)s] \tag{25d}$$

FUNDAMENTAL CONSTITUENTS OF MATTER

The current $J_{3\mu}$ and hence J_μ^Z is now diagonal in quark fields and does not contain unwanted $|\Delta S| = 1$ piece. Hence one has $\Delta S = 0$ neutral current weak interaction only.

We note from the Lagrangian (21) that

$$e = gg'/\sqrt{g^2 + g'^2} \;, \quad g^2/8m_W^2 = G_F/\sqrt{2}$$

$$= g \sin \theta_W \tag{26}$$

Further it can be shown that when the gauge symmetry is spontaneously broken and gauge bosons $W^\pm$, Z^0 acquire masses while photon remains massless, then

$$\frac{m_Z}{m_W} = \frac{1}{\cos \theta_W} \tag{27}$$

The Eqs. (26) and (27) lead to

$$m_W^2 = \frac{e^2}{4\sqrt{2}\, G_F \sin^2 \theta_W} \;, \quad m_Z^2 = \frac{e^2}{\sqrt{2}\, G_F \sin^2 2\theta_W} \tag{28}$$

Then

$$m_W \geq (\sqrt{2}\, e^2/8\, G_F\, m_p^2)^{\frac{1}{2}}\, m_p = 37.3 \text{ GeV} \tag{29a}$$

$$m_Z \geq (\sqrt{2}\, e^2/2\, G_F\, m_p^2)^{\frac{1}{2}}\, m_p = 74.6 \text{ GeV} \tag{29b}$$

These are the predictions of the model for the physical masses of the vector bosons W and Z.

Finally we note that parameters of the theory outlined above are electric charge e, m_W, m_Z and θ_W which are related by Eqs. (26) and (27). Since e^2 and G_F are known, there are only two parameters which we may choose to be θ_W and m_Z. Furthermore, for low momentum transfer phenomena m_W and m_Z will always occur in the combination

$$\frac{g^2}{8m_W^2} = \frac{e^2}{8m_W^2 \sin^2 \theta_W} = \frac{G_F}{\sqrt{2}}$$

$$\frac{g^2 + g'^2}{16m_Z^2} = \frac{g^2}{16m_Z^2 \cos^2 \theta_W} = \frac{g^2}{16m_W^2} = \frac{G_F}{2\sqrt{2}}$$

Hence the theory must describe all low momentum transfer $(q^2 \ll m_W^2)$ experiments on neutral current weak interaction in terms of one parameter only, namely θ_W.

To sum up:

(i) It has been possible to obtain a gauge unification of weak and electromagnetic interactions. The weakness of the weak interaction relative to electromagnetic interaction is a low energy phenomena. For very high energies ($q^2 \gg M_W^2$) it would exhibit the same strength as electromagnetic interaction and same elegant simplicity which in principle, allows to calculate phenomena to all orders of approximation. In other words, it is renormalizable like quantum electrodynamics.

(ii) The direct test of the above ideas would be the experimental discoveries of the vector bosons W and Z which, as we have seen, the theory predicts to be very massive. This would require very high energies of the order of 100 GeV in centre of mass and such a test is not possible at present.

(iii) The main prediction of the gauge unification of weak and electromagnetic interaction in Salam-Weinberg model is the existence of neutral current weak interaction and that all low energy ($q^2 \ll m_W^2, m_Z^2$) phenomena involving this interaction is described in terms of just one parameter, θ_W, introduced above. At low energies neutral weak interaction effects are extremely difficult to isolate since the resulting transitions are completely swamped by much stronger electromagnetic interaction. Thus one way to isolate neutral current weak interaction effects is to consider processes involving neutrinos which are the only ones having no other interaction except weak. Another possibility is the following: Since weak interaction can distinguish between left and right or does not conserve parity, therefore interference effects between neutral current weak and electromagnetic interactions would give rise to minute parity violating effects in the resulting transitions which may be observable particularly in those phenomena where the main electromagnetic effect either cancels out (e.g. in the difference of scattering cross-sections of right and left polarised electrons on unpolarised targets or is suppressed by some selection rule (e.g. in certain atomic transitions).

(iv) The absence of strangeness changing $|\Delta S| = 1$ neutral current weak interaction necessitate the introduction of a new quark with fourth flavor called charm.

Next we discuss experimental evidence for the predictions of Salam-Weinberg theory referred to para (iii) above. Then we shall also discuss the evidence for the fourth flavor, charm.

2.3 Experimental evidence for neutral current weak interaction

We first consider the processes involving neutrinos. For purely leptonic case, consider the process

$$\nu_\mu(\bar{\nu}_\mu) + e \to \nu_\mu(\bar{\nu}_\mu) + e \qquad (31)$$

which cannot arise due to the third term in the Lagrangian (21) with J_μ^W given in (8) i.e. through charged current weak interaction. It can arise only through the second term of (21) with J_μ^Z given in Eq.(22) i.e. from the coupling of the neutral lepton current with itself and we have for the effective Hamiltonian

$$H_{eff} = \frac{g^2 + g'^2}{16 m_Z^2} [\{-\bar{e}\,\gamma_\mu(1+\gamma_5)e + 4\sin^2\theta_W\,\bar{e}\,\gamma_\mu\,e) \times$$
$$\times (\bar{\nu}_\mu\,\gamma_\mu(1+\gamma_5)\nu_\mu]$$
$$= \frac{G_F}{\sqrt{2}} [\bar{\nu}_\mu\,\gamma_\mu(1+\gamma_5)\nu_\mu][\bar{e}\,\gamma_\mu(C_V + C_A\,\gamma_5)e] \qquad (32a)$$

where
$$C_V = -\frac{1}{2} + 2\sin^2\theta_W$$
$$C_A = -\frac{1}{2} \qquad (32b)$$

and use has been made of Eq.(30). For charged current weak interaction only, $C_V = 0 = C_A$. The process (31) has now been seen with the cross-sections σ, $\bar{\sigma}$ for $\nu_\mu e$ and $\bar{\nu}_\mu e$ scattering given by[13]

$$\sigma = (1.7 \pm 0.5) \times 10^{-42} cm^2\ E_\nu/GeV$$

$$\bar{\sigma} = (1.8 \pm 0.9) \times 10^{-42} cm^2\ E_{\bar{\nu}}/GeV$$

These values are compatible with Salam-Weinberg model for

$$\sin^2\theta_W = 0.21 \begin{array}{c}+\ 0.09\\ -\ 0.06\end{array}$$

and
$$\sin^2\theta_W = 0.30 \begin{array}{c}+\ 0.10\\ -\ 0.30\end{array} \qquad (33)$$

respectively.

Let us now consider the semi-leptonic processes (i.e. which involve hadrons and leptons) and in particular inclusive processes in the deep inelastic region of the type

$$\nu N \to \nu X$$
$$\bar{\nu} N \to \bar{\nu} X, \qquad (34)$$

which arise from neutral current weak interaction and compare them with

$$\nu N \to \mu X$$
$$\bar{\nu} N \to \mu X \qquad (35)$$

which arise from charged current weak interaction. In (34) and (35) N denotes a nucleon and X denotes hadrons produced. Let us define the ratios:

$$R^{\nu N} = \frac{\sigma^{nc}(\nu)}{\sigma^{cc}(\nu)} \tag{36a}$$

$$R^{\bar{\nu} N} = \frac{\sigma^{nc}(\bar{\nu})}{\sigma^{cc}(\bar{\nu})} \tag{36b}$$

where nc and cc respectively denote neutral and charged current weak interaction. These ratios are functions of $\sin^2 \theta_W$. The experimental numbers[14] for these ratios for an isoscalar target, namely,

$$R^{\bar{\nu}} = 0.34 \pm 0.03, \quad R^{\nu} = 0.295 \pm 0.01 \tag{37}$$

are consistent with the value of

$$\sin^2 \theta_W \approx 0.25 \tag{38}$$

consistent with the value (33) of $\sin^2 \theta_W$.

The above tests of the theory involve neutrinos only. It would be interesting to have independent other tests which involve, say, the neutral weak coupling of electrons for which Salam-Weinberg model predicts a definite form. Below we shall discuss two such tests

The neutral current weak interaction can cause minute parity violating effects in transitions between atomic levels. The Z^0 coupling to electrons and nucleons contains both vector and axial vector parts so that Z^0 exchange between electrons and nucleons can cause parity admixture in atomic levels. Such effects have been studied with a heavy atom, Bismuth, for which the term

$$\langle e|J_A|e\rangle \langle B_i|J_V|B_i\rangle \tag{39}$$

dominates (where A and V refer to axial vector and vector parts of the neutral current). As can be seen from the second term of Lagrangian (21) with $J^Z(\ell)$ given in Eq. (22) and $J^Z(h)$ given in Eq. (25), and using Eq.(30), the effective Lagrangian for (39) is

$$\mathcal{L}_{eff} = \frac{G_F}{2\sqrt{2}} \left[\bar{e} \gamma_\mu \gamma_5 e\right] \left[(1 - \frac{8}{3} \sin^2 \theta_W)\bar{u} \gamma_\mu u \right.$$
$$\left. + (-1 + \frac{4}{3} \sin^2 \theta_W)\bar{d} \gamma_\mu d\right] \tag{40}$$

From (39) and (40), one sees that the parity violating charge apart from $G_F/2\sqrt{2}$ is

$$Q_{pv} = c_A^e \left[(2Z + N) c_V^u + (2N + Z)c_V^d\right] \tag{41a}$$

where Z denotes the number of protons and N the number of neutrons in the atom (for Bismuth Z = 83, N = 126) and from (40)

$$c_A^e = 1, \quad c_V^u = (1 - \frac{8}{3} \sin^2 \theta_W), \quad c_V^d = -1 + \frac{4}{3} \sin^2 \theta_W \qquad (41b)$$

Three independent experiments have measured the optical rotations of polarised light in Bismuth which give the ratio $R_\lambda (\equiv I_m \, E_1/M_1)$ of electric and magnetic dipole transition amplitudes between atomic states with <u>same parity</u>. The electric dipole transition due to electromagnetic interaction is forbidden (but not the magnetic one) and it arises only due to parity violating charge Q_{pv} multiplied by $G_F/2\sqrt{2}$. With $\sin^2 \theta_W \approx 0.24$ the parity violating charge $Q_{pv} = -123$. According to the theoretical calculations[15] using the above value of the Q_{pv}, the optical rotations for the two transitions which have been measured are given below together with their experimental values[16]:

	6476A°Line	8757A°Line
Salam-Weinberg Model with $\sin^2 \theta_W = 0.24$	-13×10^{-8} radians	-10×10^{-8} radians
Experiment	$(2.7 \pm 4.7) \times 10^{-8}$ Oxford (old)	$(-0.5 \pm 1.7) \times 10^{-8}$ (Seattle)
	$(-5 \pm 1.6) \times 10^{-8}$ Oxford (new)	
	$(-21 \pm 6) \times 10^{-8}$ Novosibirsk	

Thus Oxford and Seattle experiments yield results for the ratio R_λ, consistent with zero although new Oxford results now seem to be seeing non zero value of R_λ but still smaller than the theoretical estimate with $\sin^2 \theta_W \approx 0.24$. By contrast, the Novosibirsk experiment supports the theoretical estimate based on Salam-Weinberg model with $\sin^2 \theta_W = 0.24$. In arriving at numbers for optical rotations from the theory to be compared with the experimental results, a detailed knowledge of Bismuth atomic wave functions is necessary which is a non-trivial matter. This apparent contradiction between Salam-Weinberg SU(2) ⊗ U(1) model and some of the experimental results on Bismuth may not be so serious because of difficult nature of experiments, large errors in the experimental results and great uncertainty in the atomic theory calculations for Bismuth atomic wave functions. Below we describe a recent experiment which involves neutral weak coupling of electrons but is free from atomic theory calculation complications and decisively confirm Salam-Weinberg SU(2) ⊗ U(1) model.

The experiment mentioned above involves the scattering of polarised electrons from deuteron in the deep inelastic region

$$e^- D \to e^- X \qquad (42)$$

and measures the asymmetry

$$A = \frac{\sigma_R - \sigma_L}{\sigma_R + \sigma_L} \qquad (43)$$

where R and L denote right and left polarised electrons. The parity conserving electromagnetic interaction cannot contribute to the above asymmetry. This can occur only due to parity violating effects which in Salam-Weinberg model can arise due to the interference of $Z^°$ exchange (which has both vector and axial vector interactions) and photon exchange as shown below diagramatically:

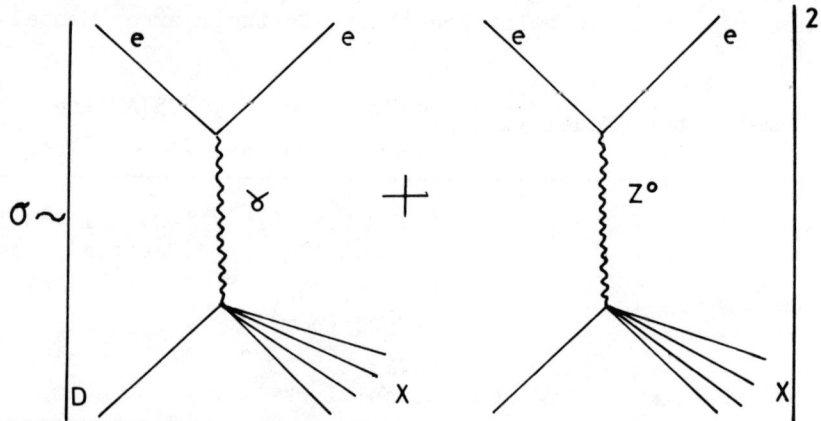

Only interference of the two diagrams contribute to $[\sigma_R - \sigma_L]$ while the unpolarised cross-section $[\sigma_R + \sigma_L]$ is dominated by the γ-exchange diagram. The asymmetry (43) has been measured[17] to be

$$\frac{A}{q^2} = (-9.5 \pm 1.6) \times 10^{-5} \text{ GeV}^{-2} \qquad (44)$$

for $q^2 = 1.4$ GeV^2 where q^2 denotes the momentum transfer. Thus it shows that parity violation exists and the negative sign indicates that left polarised electrons are scattered more than right polarised ones as predicted by Salam-Weinberg Model. Moreover, using quark-parton model[18] for the deep inealstic region, Salam-Weinberg model predicts the above asymmetry to be

$$\frac{A}{q^2} \approx -10^{-4} \; f(\sin^2 \theta_{W,y})$$
$$= -\frac{G_F}{2\sqrt{2} \, \pi\alpha} \frac{9}{10} \left\{ c_A^e \, c_V^q + h(y) \, c_V^e \, c_A^q \right\} \qquad (45)$$

where
$$h(y) = [1 - (1-y)^2]/[1 + (1-y)^2],$$

y being the fractional energy loss of the electron. The result (44) is for $y = 0.21$. The factor 10^{-4} is essentially the factor $9/10(G_F\, m_p^2/2\sqrt{2}\pi\alpha)$. $C_{A,V}^e$ and $C_{A,V}^q$ which arise from neutral current term are specified by $\sin^2\theta_W$ and in fact are given by

$$C_A^e\, C_V^q = (1 - \tfrac{20}{9}\sin^2\theta_W),\quad C_V^e\, C_A^q = (1 - 4\sin^2\theta_W).$$

Thus the experimental value (44) of the asymmetry determines $\sin^2\theta_W$. The value (44) is in good agreement with Salam-Weinberg model with

$$\sin^2\theta_W = 0.2 \pm 0.03 \tag{46}$$

which is consistent with its value (38) found from neutrino data.

To sum up there is now little doubt that Salam-Weinberg idea of unification of weak and electromagnetic interactions is correct. Thus the weak and electromagnetic interactions are manifestations of the same force and that the four basic forces of nature are reduced to three only.

2.4   Experimental evidence for charm

We now discuss the evidence for the new quark flavor charm (the introduction of which was necessitated to suppress the unwanted strangeness changing $|\Delta S| = 1$ neutral current weak interaction) and also for the charm changing current which appears in $J_\mu^W(h)$ (c.f. Eq.(25b))

$$J_\mu^W = i\bar{u}\gamma_\mu(1+\gamma_5)d' + i\bar{c}\gamma_\mu(1+\gamma_5)s',$$
$$\quad\quad \Delta C = 0 \quad\quad\quad\quad \Delta C = 1 \tag{47}$$

namely

$$J_\mu^{\Delta C=1} = i\bar{c}\gamma_\mu(1+\gamma_5)s'$$
$$= i\bar{c}\gamma_\mu(1+\gamma_5)(s\cos\theta_c - d\sin\theta_c) \tag{48}$$

This is left handed V-A current, where first term satisfies $\Delta C = \Delta S = 1$ selection rule while the second term satisfies $\Delta C = 1$, $\Delta S = 0$ selection rule. Since $\theta_c \approx 15°$, it follows that $\Delta C = \Delta S = 1$ piece dominates over $\Delta C = 1$, $\Delta S = 0$ piece.

One indirect evidence for the charm is the following: A particle called J/ψ at 3.1 GeV and a family of states-connected to this by radiative transitions have been discovered[19] in 1974. The most important characteristic of these states is their narrow decay widths in spite of large phase space available for their decays into ordinary hadrons. Their narrow widths can be qualitatively understood if we regard this family of particles as s-wave $c\bar{c}$ bound states. Thus since ordinary hadrons are made up of predominantly uncharmed quarks, the decay of ψ-family of particles into hadrons will be possible only to the extent of extremely small contamination of c and $\bar{c}$ quarks in ordinary hadrons or of uncharmed quarks in ψ-family of particles. In this way one can understand the narrow widths of these particles. The ψ-family of particles do not carry charm quantum number. Charm carrying particles have also been discovered[20]. First we note that charm quantum number should be conserved by strong and electromagnetic interactions and thus these particles can be produced via these interactions only in pairs. Thus charm carrying D, $\bar{D}$ meson states ($D^\pm \sim c\bar{d}, d\bar{c}, D^\circ \sim c\bar{u}, \bar{D}^\circ \sim u\bar{c}$) have been discovered in reaction

$$e^+ e^- \longrightarrow \gamma \longrightarrow D\bar{D} X,$$

where D and $\bar{D}$ are identified through their decay products (via weak interaction) consisting of ordinary mesons and one can argue that these decay products cannot arise from the decay of any uncharmed resonance. $D^\pm$ occur at 1.876 ± 0.015 GeV while D° occurs at 1.865 ± 0.015 GeV.

The charm quantum number is violated by weak interaction; thus a charmed state can be produced singly via weak interaction and it can also decay via this interaction. In fact $\Delta C = 1$ component of charged current (48) can excite a charmed state singly in neutrino reactions. Thus, for example, the charmed baryon states $C_1^+ \sim cuu$ and $C_0^+ \sim cud$ at 2.4 GeV and 2.25 GeV have been discovered[21] in the following reaction

$$\nu + p \longrightarrow \mu^- + \Lambda + \pi^+ + \pi^+ + \pi^+ + \pi^-$$

which can be interpreted as the production and subsequent decay of a charmed baryon:

$$\nu + p \longrightarrow \mu^- + C_1^{++}$$
$$\hookrightarrow C_0^+ + \pi^+ \quad \text{(strongly as charm is conserved)}$$
$$\hookrightarrow \Lambda + \pi^+ + \pi^- + \pi^+ + \pi^- \quad \text{(weakly)}$$

There have also been considerable evidence for semi-leptonic and parity violating non-leptonic decays of these objects. One is the $\nu(\bar{\nu})$ induced dimuon events interpreted as

FUNDAMENTAL CONSTITUENTS OF MATTER

$$\nu + N \longrightarrow \mu^- + C + X$$
$$\longrightarrow \mu^+ + \nu$$
$$\longrightarrow \bar{K}^0 + \mu^+ + \nu$$
$$(\text{or } e^+)$$

For details reader is referred to reference 22.

3. STRONG INTERACTION AND COLOR GAUGE GROUP

We have seen in Section 1 that for strong interaction we introduced two SU(3) groups, $S^fU(3)$ flavor group and SU(3) color group. The flavor group $S^fU(3)$ (or $S^fU(4)$ if we include charm) is used for the classification of particles and all known low lying hadrons belong to singlet, octet, decuplet (for non-charm carrying hadrons) and singlet, triplet and sextet (for charm carrying hadrons) representations of $S^fU(3)$ while they and other physical observables are singlets with respect to $S^cU(3)$. The $S^cU(3)$ group is regarded as a gauge group[5] and exact color gauge symmetry is assumed. Thus there are eight massless (non-flavor carrying) color gauge vector bosons associated with this group. These are called color carrying gluons. The idea here is that the gluons would provide binding of the quarks in a hadron and since free quarks have not so far been seen, they would also provide the confining mechanism for the quarks. Here it would be useful to summarise in Table 3 below the differences (and similarities) between the well known electromagnetic gauge group and the color gauge group.

Table 3

Differences and similarities between electromagnetic gauge group and color gauge group

Electromagnetic Gauge Group	Color Gauge Group
U(1) symmetry $\Rightarrow$ single conserved charge, namely, electric. The electromagnetic interaction is connected with the exchange between charged particles of a single electrically neutral massless gauge boson, photon, the field quantum of electromagnetic field; gauge group Abelian.	$S^cU(3)$ symmetry $\Rightarrow$ 8 conserved color charges. The strong interaction is connected with the exchange of an octet of color carrying massless gauge vector bosons called gluons interacting with color degrees of freedom of quarks.
Deal with electrically neutral systems, e.g. atoms.	Deal with color singlet systems i.e. hadrons.

Exchange of photon give force of repulsion between like charges and attraction between unlike charges; the latter between electrons and protons is responsible for the binding of atoms.	Exchange of gluons give force of attraction between color singlet hadrons and thus can provide binding of quarks in a hadron.
In the non-relativistic limit, the potential between unlike charges is $$-\alpha/r,$$ where $\alpha = 1/137$.	In the non-relativistic limit, the two body potential for color singlet hadrons is $$-k\,\alpha_s/r$$ where $\alpha_s = g_s^2/4\pi$; g_s being the coupling of gluons to quarks while $k = 2/3$ or $4/3$ for s wave baryon or meson state respectively. α_s corresponds to fine structure constant for strong interaction. Its value has been estimated to be ≈ 0.2–0.5.
The theory here is called Quantum Electrodynamics (QED)	The theory here is called Quantum Chromodynamics (QCD)
The effective 'charge' or coupling increases for large momentum transfers q^2 (or short distances) compared to its normal value at small momentum transfers (or large distances)	The effective coupling is reduced for large momentum transfers q^2 or short distances. Ultimately for $q^2 \to \infty$, it may go to zero and so quarks become free at very short distances. This is known as 'asymptotic freedom'. Thus in the deep inelastic region, an electron, for example, would interact with quarks inside a hadron, where distances are short, as if quarks were free particles in spite of strong interaction between the quarks (quark-parton model)[18].

The binding energy provided by one gluon exchange potential of the form mentioned in Table 3 cannot be sufficient to confine the quarks in a hadron since as one can ionise an atom to knock out an electron, similarly a quark could be separated from a hadron if sufficient energy is supplied. Thus this potential can at best provide binding for quarks at short distances and cannot explain their confinement i.e. impossibility of separating a quark from a hadron. The hope here is that because of non-Abelian nature of QCD the self interaction of gluons may give rise to long distance behavior of potential in QCD completely different from that in QED which is Abelian. One hopes that long range potential in QCD would increase

FUNDAMENTAL CONSTITUENTS OF MATTER

with the distance so that the quarks would be confined in a hadron. If it turns out to be so then the postulate of color confinement (i.e. all hadrons to be color singlets) would be a necessary condition for the confinement of quarks. We must emphasise that while QCD is a promising approach to quark confinement, this has so far not been demonstrated and it is simply a hope. Phenomonologically, a potential of the form[23]

$$V(r) = -k \frac{\alpha_s}{r} + br,$$

where the first term is single gluon exchange potential in the non-relativistic limit (when the mass of the quark is large, which would be particularly true for ψ system of particles) while br is the long range confining potential, has been qualitatively successful in explaining the spectrum of ψ-family of particles. Moreover, the E1 radiative transitions possible between the ψ-family of particles also come to be in qualitative agreement with experimental data where such data are available. There seems, however, to be trouble with M1 transitions.

The single gluon exchange part of the above potential which dominates at short distances, when written in full turns out to contain quark spin - dependent terms. Such terms give rise to hyperfine mass splittings[24] e.g. between ρ (which is a member of spin triplet L = 0 states of $q\bar{q}$) and π (which is a member of spin singlet L = 0 states of $q\bar{q}$) and these splittings are in the right direction. Thus we have

$$m_\rho > m_\pi, \quad m_{K^*} > m_K, \quad m_\psi > m_{\eta_c}$$

$$m_\Delta > m_N, \quad m_\Sigma > m_\Lambda.$$

It should be emphasised that it is by no means certain that QCD is correct theory of quark interactions. Some of its qualitative features have met with success[25] but there is as yet no firm experimental evidence for it.

To sum up quarks carrying flavors and color together with their associated flavorless but color carrying gauge vector gluons are regarded to form the fundamental constituents of matter. Since a fractionally charged quark has never so far been seen in isolation, color gauge theory is hoped to provide the complete or near complete confinement of quarks in hadrons. There is, however, an alternative point of view, particularly emphasised by Pati and Salam[26], where quarks could be integrally charged and need not be confined[27] i.e. they can be liberated. The mass of a free quark could be large, although inside a hadron its effective mass may be quite small due to

Archimedes Principle. Moreover, the color could also be excited and one might hope to see color non-singlet hadrons in future (they are expected to be very heavy, however). The color gauge symmetry is not exact in such an approach and is spontaneously broken so that the color gauge bosons acquire large masses, even larger than W masses and they could be produced in nature; only they would require much larger energies than presently available for their production.

There have also been attempts to merge the $G_{EW} \equiv [SU(2) \otimes U(1)]$ gauge group of electromagnetic and weak interactions and the color gauge group G_S of strong interactions into a bigger gauge group G, i.e. G contains

$$G_{EW} \otimes G_S$$

in an attempt to unify strong and electromagnetic and weak interactions (the latter two being already unified). Let us denote the gauge bosons associated with the gauge group G which requires masses when G is spontaneously broken to $G_{EW} \otimes G_S$ by X. Now as already seen in Section 2 when G_{EW} is spontaneously broken down to $[U(1)]_{E.M.}$, the associated flavor carrying gauge bosons W and Z (except photon γ) acquire masses (of order of 100-200 GeV). Thus the mass scale for the direct manifestation of unification of electromagnetic and weak interactions is of the order of 100-200 GeV. Similarly when the group G is broken down to $G_{EW} \otimes G_S$, the gauge bosons X would acquire masses much larger than m_W, m_Z. Thus

$$G \xrightarrow[m_X \gg m_W]{\text{broken down}} G_{EW} \otimes G_S$$
$$\Big\Downarrow \text{broken down}$$
$$[U(1)]_{E.M.}$$

That m_X must be $\gg m_W$ is to account for large difference between α ($\approx 1/137$) and α_s ($\approx .2$), the couplings associated with G_{EW} and G_S respectively. There have been two main approaches to the above unification. In one approach[28] the gauge group G is simple (e.g. SU(5)) and G_S is simply $S^cU(3)$ introduced earlier, the quarks are fractionally charged and color is confined. Here one has only one free coupling constant (the gauge coupling constant, associated with the group G) and angle θ_W introduced in Section 2 is determined and quantization of charge can be explained. However, the unifying mass $m_X \geq 10^{15}$ GeV. In the other approach[29] the gauge group G is semi-simple and is of the form

FUNDAMENTAL CONSTITUENTS OF MATTER

$$[SU_L(4) \otimes SU_R(4)]_F \times [SU_L(4) \otimes SU_R(4)]_C$$

$$\xrightarrow[M_X \gg M_W]{\text{broken down}} G_{EW} \otimes G_S$$

$$\Big\downarrow \text{broken down}$$

$$[U(1)]_{E.M.}$$

Above F denotes flavor and C color while L and R signify left and right handed groups i.e. we deal with chiral groups. Here G_S is chiral color group

$$G_S \equiv S^cU_L(3) \otimes S^cU_R(3)$$

In this approach the unifying mass could be low, $m_X \approx 10^4 - 10^6$ GeV; the quarks are integerally charged. In fact they together with leptons could form just one fermion super-multiplet. Moreover color could be liberated and in that case, the axial vector color gluon could have small mass, $m_A \leq 100$ GeV. Thus in the former approach, there is no hope of seeing direct unification of strong interaction with electromagnetic and weak interactions in any foreseeable future while in the latter approach, it may be possible. Our final remark is that in the grand unification of gravity with electromagnetic-weak and strong interaction, the unifying mass would of the order of [G_N is gravitational constant]

$$(G_N)^{-\frac{1}{2}} = 10^{19} \text{ GeV}$$

For attempts in this connection, the reader is referred to reference 30.

4. CONCLUDING REMARKS

We have seen that there is now decisive evidence that two of nature's four forces have been unified i.e. weak and electromagnetic forces are different manifestations of the same force. Thus four fundamental forces of nature have been reduced to three. This may make easier the unification of all the forces.

Let us now go back to the fundamental constituents of matter, namely, quarks postulated to understand the proliferation of hadrons. What happens if quarks themselves proliferate? Already there is evidence for fifth quark flavor, given the name bottom. This evidence comes from the existence of a new state[31] T at 9.5 GeV.

which again has narrow width in spite of very large phase space available to decay to ordinary or charmed hadrons. To understand narrowness of its width fifth flavor has been introduced just as charm was introduced to understand the narrowness of ψ particle width. Already people are talking of sixth flavor. If quarks do proliferate, should then one talk about substructure of quarks themselves? This question might have a meaning in Pati-Salam approach where the quarks are not permanently confined and can be liberated. But in QCD, where quarks are confined, what is the sense of talking of quarks' substrcuture when they have not themselves been found experimentally? For the last two centuries, physicists have insisted on finding the fundamental constituents of matter in nature; now we are trying to invent reasons for not finding them. Is, then the question 'What are fundamental constituents of matter' meaningful at least in the conventional sense? Or should we talk in terms of something else which are directly observable. Future will shed light on these questions and future, we hope, would at least be as interesting and full of excitement as this field has been since man began to ask the question 'What are we made of'?

References

1. M.Gell-Mann, Phys. Lett., $\underline{8}$, 214 (1964);
   G.Zweig, CERN 8182/TH.401 and 8419/TH.412, (1964);
   For other references on quark model, see, for example,
   J.J.J.Kokkedee, The Quark Model, Benjamin, New York, 1969.
   For more recent reviews, see reference 2.

2. O.W.Greenberg, Annual Reviews of Nuclear and Particle Science, Vol. $\underline{28}$, 1978.
   J.D.Jackson, in 'Proceedings of Summer Institute on Particle Physics' SLAC Report No.198, Edited by Martha C.Zipf (1976), p. 147. F.G.Gilman, ibid (1977), p. 121.

3. See for example, M.Gell-Mann and Y.Ne'eman, The Eightfold Way, Benjamin, New York (1964).

4. O.W.Greenberg, Phys. Rev. Lett., $\underline{13}$, 598 (1964);
   M.Y.Han and Y.Nambu, Phys. Rev., $\underline{139}$ B, 1006 (1965).
   For a review, see for example, O.W.Greenberg and C.A.Nelson Phys. Reports, $\underline{32}$ C, 69 (1977).

5. H.D.Politzer, Phys. Rev. Lett., $\underline{30}$, 1346 (1973);
   D.J.Gross and F.Wilczek, ibid, $\underline{30}$, 1343 (1973); Phys. Rev. $\underline{D8}$, 3633 (1973);
   S.Weinberg, Phys. Rev. Lett., $\underline{31}$, 494 (1973).
   See also H.D.Politzer, Physics Report $\underline{4C}$, 129 (1974).

6. C.N.Yang and R.L.Mills, Phys. Rev., $\underline{96}$, 191 (1954).
   For a review, see for example, E.S.Abers and B.W.Lee, Phys. Reports $\underline{9C}$, 1 (1973);
   J.C.Taylor, Gauge Theories of Weak Interactions, Cambridge (1976).

7. A.Salam in 'Elementary Particle Physics', Edited by N.Svartholm (Almqvist and Wiksell, Stockholm) 1968, p.367.

8. S.Weinberg, Phys. Rev. Lett., $\underline{19}$, 1264 (1967).

9. P.W.Higgs, Phys. Lett., $\underline{12}$, 132 (1964);
Phys. Rev. Lett., $\underline{13}$, 508 (1964).

10. See for example, J.Bernstein, Rev. of Modern Physics, $\underline{46}$, 7 (1974);
B.W.Lee in 'Physics and Contemporary Needs' Vol. 1, Edited by Riazuddin, (Plenum Press, New York) 1976, p.321.

11. M.L.Perl et al., Phys. Rev. Lett., $\underline{35}$, 1489 (1975);
See also M.L.Perl, Proceedings of the International Symposium on Lepton and Photon Interactions at High Energies', Edited by F.Gutbrod, (DESY, Hamburg, Germany) 1977.

12. S.L.Glashow, J.Iliopoulas and L.Maiani, Phys. Rev. $\underline{D2}$, 1285 (1970).

13. F.J.Hasert et al., Phys. Lett., $\underline{46}$ B, 121 (1973);
J.Blietschau et al., Nucl. Phys., B $\underline{114}$, 189 (1976);
H.Faissner et al., Phys. Rev. Lett., $\underline{41}$, 213 (1978);
P.Alibran et al., Phys. Lett., $\underline{74}$ B, 422 (1978);
C.Baltay et al., Fermi Lab Report (1978).

14. For a review, see for example:
L.F.Abbott and R.M.Barnett, SLAC-Pub-2136 (1978)(To be published in Phys. Rev. D).

15. M.A.Bauchiat and C.C.Bauchiat, Phys. Lett., $\underline{4}$B, 111 (1974);
J.Bernabeu, T.E.O.Ericson and C.Jarlskog, Phys. Lett., $\underline{20}$, 315 (1974);
I.B.Kriplovich, JETP Lett., $\underline{20}$, 315 (1974);
E.M.Henley and L.Wilets, Phys. Rev., A $\underline{19}$, 1911 (1976).

16. P.E.Baird et al., Nature, $\underline{264}$, 528 (1976);
P.G.H.Sandars, in Proc. 1977 Int. Sym. on Lepton and Photon Interactions at High Energies, Edited by F.Gutbrod (DESY, Hamburg) 1977, p. 599;
N.Fortson, in Proc. of Neutrinos - 78 Conference, (Purdue University) 1978;
L.M.Barkov and M.S.Zalotorev, JETP Lett., $\underline{26}$, 379 (1978).

17. C.Y.Prescott et al., SLAC-Pub-2148 (1978).

18. See for example, R.P.Feynmann, Photon-Hadron Interactions, Benjamin (1972).

19. J.E.Aubert et al., Phys. Rev. Lett., $\underline{33}$, 1404 (1974);
J.E.Augustin et al., ibid, 1406 (1974);
G.S.Abrums et al., ibid, 1453 (1974);
For review see for example, G.Feldman and M.L.Perl, Phys. Reports, $\underline{33}$C, 285 (1977).

20. For a review see for example, G.Feldman and M.L.Perl, Reference 19.

21. E.G.Cazzoli et al., Phys. Rev. Lett., $\underline{34}$, 1125 (1975).

22. See for example, B.W.Lee in Physics and Contemporary Needs, Vol. 1, Edited by Riazuddin (Plenum Press) 1977, p.321; B.W.Lee, C.Quigg and J.L.Rosner, Phys. Rev., D$\underline{15}$, 157 (1977)

23. E.Eichten et al., Phys. Rev. Lett., $\underline{34}$, 369 (1975); Phys. Rev., D$\underline{17}$, 3090 (1978).

24. A.De Rujula, G.Georgi and S.L.Glashow, Phys. Rev., D$\underline{12}$, 147 (1975).

25. For a review, see for example, J.Ellis, SLAC-PUB-2121 (1978).

26. J.C.Pati and A.Salam, Phys. Rev., D$\underline{10}$, 275 (1974).

27. See for example, A.Salam in Physics and Contemporary Needs, Vol. 2, Edited by Riazuddin (Plenum Press) 1978, p.419.

28. H.Georgi and S.L.Glashow, Phys. Rev. Lett., $\underline{32}$, 438 (1974); H.Georgi, H.R.Quinn and S.Weinberg, Phys. Rev. Lett., $\underline{33}$, 451 (1974).

29. V.Elias, J.C.Pati and A.Salam, Phys. Rev. Lett., $\underline{40}$, 920 (1978).

30. See for example, A.Salam, reference 27.

31. S.W.Herb et al., Phys. Rev., $\underline{39}$, 252 (1977); W.R.Innes et al., ibid, 1240 (1977); Ch. Berger et al., Phys. Lett., $\underline{76B}$, 243 (1978); C.W.Darden et al., ibid, 246 (1978).

PART V

SCIENCE AND DEVELOPMENT

SCIENCE AND DEVELOPMENT

Michael J. Moravcsik

Institute of Theoretical Science

University of Oregon, Eugene, Oregon 97403, USA

In Section 1 I will analyze the concepts of science and development and their relationship on a very general and abstract but still realistic plane. In Section 2 I plan to discuss a very specific subject, namely the upcoming United Nations Conference on Science and Technology for Development.

1. SCIENCE AND DEVELOPMENT

1.1 Science and Technology

Let me start with science. Science is a human activity aiming at the production of knowledge about the world around us. Note that science is an activity and that its product is knowledge. This is important to keep in mind when we want to discuss the relationship between science and technology. Technology is another human activity the product of which are prototypes of gadgets, procedures, patents, ways of making things.

Note that the best way to make the distinction between science and technology is in terms of their respective products. Science produces knowledge, while technology produces ways of doing things or actual first-of-the-kind objects.

The distinction between science and technology, however important, is often greatly obscured in public discussion in any country. As a result, the different needs, requirements, and expectations of science and technology are confused, and the resulting misunderstandings harm science, technology, as well as the country as a whole.

In the remainder of this talk I will constrain myself to a discussion of science. Within science, some further distinctions are useful. People talk about basic (or 'pure') science and applied science. What is the distinction between these two?

It is more difficult to make, and less sharply drawn, as will be evident from the discussion below.

The best way to make the distinction is in terms of the intention of the person performing scientific research or the person providing the support for such research. In basic research the primary motivation is to enlarge our knowledge about the world because we value knowledge <u>per se</u>. The motivation for applied research, on the other hand, is to find new knowledge so that it can be made use of in some technological or other human activity.

This sounds clear and distinct enough, but in practice difficulties arise. For example, the scientists performing the research may have one motivation in mind and the supporter of research another. Then, even a given person (whether the researcher or the supporter of research) may have <u>both</u> of these motivations in mind in some particular proportion.

There is also the time element. There has been no discovery in the basic sciences which was significant by the <u>internal</u> criteria of basic science itself, which was not turned into momentously important technological applications later. The only examples which are yet undecided from this point of view are the most recent discoveries (for example some of elementary particle physics) which have not had time to mature to applications. Thus every good piece of basic research is also applied research in the long run.

Nevertheless, when dealing with decisions in science policy, it is useful to make some distinction between research that appears more basic at the time and research that is more applied. This is necessary in order to maintain a balanced program of scientific activity. So one might say, for example, that applied research is an activity the main motivation of which is application in a definite area of technology within the next 5-10 years.

There is also another distinction made within science, one that has nothing to do with the one we just discussed, but which is often confused with it. This is the classification into theoretical and experimental science. The difference here is in the method used in the investigation. Theoretical research aims at finding relationships and connections among observations already made and at making predictions for experiments yet to be performed. In contrast, experimental research concerns itself with devising ways of making observations and then with performing those experiments.

In the public mind, the two dichotomies of basic vs. applied, and theoretical vs. experimental are often confused and thought

SCIENCE AND DEVELOPMENT

to be the same. Such people would use theoretical and basic interchangably, and experimental and applied also interchangably. It is important to clear up this misunderstanding and emphasize that both basic and applied research needs both theorists and experimentalists, otherwise those in charge of science policy will end up depriving basic research of experimental facilities and deny theoretical support for the applied work.

1.2   Development

So far I discussed science. Since the title of this talk consists of three words, I will now turn to the third of these and analyze what 'development' is.

This is also a word that is used most carelessly and ambiguously, and so let me offer what I consider the most appropriate description of this concept. Development is an action aiming at the realization of human aspirations.

Again I want to stress that development is action. National development plans are only plans. Development will actually take place only if some action results from these plans.

I also want to elaborate on what we mean by human aspirations. All of us as individuals, and as groups of individuals, have certain desires, certain goals, certain purposes in our lives. These are our aspirations.

Some of these aspirations are material: We strive for better food, better shelter, better consumer goods, better health facilities, etc. A much larger class of aspirations, however, are non-material. One need not be a learned scholar of the details of human history to note that virtually all the great forces in history came from non-material aspirations of persons and people. Civilizations rise and fall as a result of them, and individuals perform far beyond expectations or fall short, depending on whether they are strengthened by internally felt personal non-material aspirations.

For example, a count of the wars presently in progress around the world (and there are always quite a few small ones, and even more that are not quite in the open but smoldering) would reveal that virtually all of these are fueled by non-material aspirations: Ideological, religious, or otherwise.

Development being action, it entails certain beliefs and attitudes without which development cannot take place. The belief must exist that by human action the aims can in fact be realized, that the world is in fact subject to change, and that knowledge about the world aids in finding the type of action that brings us closer to our aspirations.

So much for the word 'development'. Now we must turn to the third word in the title, namely 'and', and analyze the relationship between science and development.

1.3 Relationship Between Science and Development

What are in fact our reasons, or motivations to engage in scientific research? This is a typical multidimensional question. Depending on whom you ask, you may get a multitude of different answers. To the engineer science provides knowledge to use, to the philosopher science opens up new questions and delights our esthetic sense, to the military man science eventually results in better weapons, to the doctor science may provide a cure for cancer, to the politician science will contribute to national prestige, to the economist science contributes to the rise in the per capita gross national product, etc. One should delight in this great variety of answers, because most causes that strive for broad public support must enjoy a broad assortment of justifications to make them attractive to many people

One can, however, take all these motivations for pursuing science and group them under three general headings:

(a) Science is the basis of technology.

(b) Science is a human aspiration.

(c) Science has an influence on man's view of the world.

The first of these we have discussed. This is the justification of science that is most obvious to most people, though this does not mean necessarily that it is also the predominant one.

That science in the 20th century is a human aspiration is almost a truism. The universal fascination with science fiction, the widespread interest in popular presentations of scientific discoveries, the image of a scientist as a very brainy (though possible 'evil') specimen of rare talent probing into the unknown all attest to this. Scientific achievements are among the primary areas that countries compete in and national prestige is linked with, and great leaders of developed and developing countries alike have stressed in their political statements the strength of this human aspiration.

The main handicap in being a less developed country is not so much in having a small per capita gross national product, but in being completely dependent on other countries in everything that is not only economic but also cultural, intellectual, and determinants of life styles. The lack of choice, the absence of control over one's own future and fate is indeed a debilitating feeling. It leads to an erratic political behavior, to curious placebos in domestic matters in the place of creative measures to alleviate problems.

The area of scientific achievement is one in which countries can therefore gain confidence and self-assurance. Many small countries have raised their morale through the knowledge of an outstanding countryman who has been acclaimed the world over in one of the conspicuous human activities. A Niels Bohr in Denmark, a Sibelius in Finland, or a Tagore in India might be examples.

Such a high morale is in fact a crucial key to further achievement and hence for development. Success in any of the highly valued human activities, in turn, is very effective in raising morale. Scientific excellence is a mode of success that can be achieved relatively easier by developing countries than many other modes.

Finally, to turn toward the third motivation for pursuing science, one of the striking differences between the 'developed' and the 'developing' parts of the world is that the scientific revolution as an intellectual force and as a world view has come to the former several centuries ago while it is just reaching the latter. Even Japan, the country that most recently joined the 'developed' camp, started its scientific revolution over 100 years ago.

This difference has an enormous effect not only on the attitudes of the leading segment of the country, but also on the attitudes of the population as a whole. Science has many implicit assumptions which appear so natural in the developing countries that they are forgotten. Yet these assumptions are revolutionary and at odds with many traditional beliefs held around the world. To give a few examples, science considers the body of knowledge an open domain which can be enlarged by experimentation; science believes that through the acquisition and utilization of knowledge man can change his surroundings to his advantage; science believes that all (scientific) problems have a solution; science believes that the natural state of the world is change (this is why time is such an essential parameter in all descriptions of nature); etc.

It is not difficult to see how these characteristics of science are very similar to the prerequisites of development, and hence how indigenous scientific activity in a country, combined with an energetic transmission of the basic attitudes of science to the population as a whole, will have a major impact on development in the sense development was defined earlier.

I would like to close with a practical suggestion. The analysis of science and development given here is considerably broader than what one customarily sees elsewhere. Scientists, in particular, are often very timid to argue the full spectrum of movivations of science and of its connection with development. Instead, they are feverishly trying to prove that more science in the country will result in more rice or more cotton next year. Not only is this argument of dubious validity if used in such a narrow sense, but it is also erroneous in a phychological sense, since it assumes that politicians and other non-scientists are mainly influenced by material arguments. As discussed earlier, one might wish that they were indeed, since in that case the world would be a much more managable place to live in. In reality, however, the primary forces shaping the action of people are non-material, and hence science and scientists should also fight for their existence with such weapons.

This has been a very sketchy discussion of this involved subject. For further references, I want to refer you to some writings of mine where these ideas are more fully expounded[1,2,3].

2. <ins>THE 1979 UNITED NATIONS CONFERENCE ON SCIENCE AND TECHNOLOGY IN DEVELOPMENT</ins>

At the end of August 1979 in Vienna, the United Nations will hold a Conference on Science and Technology for Development. In this talk I want to summarize the features of this conference, the preparations for it, and suggest ways in which the international scientific community can contribute to the meeting.

This is not the first such conference arranged by the UN. The previous ones were complete failures. So have been most of the meetings and conferences arranged over the years by the various special agencies of the UN. By failure one means a lack of resulting action bringing about concrete improvements in the state of science and technology in the developing countries. Considering those failures, why would one expect this conference be different?

Indeed, the probability that this conference will accomplish anything is very small. But it is not quite zero. There are certain circumstances surrounding this conference that are somewhat different from the previous ones, and there are some opportunities which, if utilized, could represent some progress.

By August 1 this year, each country is supposed to submit to the conference secretariat a national paper of about 15,000 words long, outlining the state of S & T in the country and suggesting improved modes and programs for international cooperation in S & T. There is a chance that these preparations themselves might contribute to action in international S & T Cooporation.

For example, in the United States, a special office was established in the State Department to coordinate the preparations. Many governmental agencies have commisioned position papers on various aspects of international S & T, and workshops and other discussions are held. All this, of course, so far amounts only to organizational manipulations and talk, and there is yet no guarantee that action will be forthcoming also. Yet, the atmosphere is more favorable.

Countries which would like to have external cooperation in formulating their national papers can request such 'technical help' from the conference secretariat, preferable by naming specific persons they would have as advisors. A fairly large number of countries have availed themselves with this opportunity, though the number is not very impressive as a percentage of all UN member countries.

As far as I can ascertain from various direct and indirect sources, the seriousness and comprehensiveness of the preparations in various countries varies greatly from one country to another. At one extreme, some third rank ministerial bureaucrat is delegated to compose the paper in the privacy of his office, insulated from any other input. At the other end, some countries have rallied their S & T community, held national meetings and workshops, have written drafts and circulated them for critique, and thus arrived at a fairly extensive consensus. I would like to urge everybody in the audience to make sure that his country is closer to the second than to the first of these examples. Even after the country paper is submitted, there is a year before the Vienna conference during which preparations and deliberations must continue.

The national papers will be received by the secretariat of the conference. This body was established especially for this conference, and is called UNCSTD office. It is housed in New York, and has a professional staff of perhaps 10 people. It is headed by a Brasilian diplomat, Joao F. da Costa, not particularly versed in matters of science and technology but good in public relations, and he has so far spent most of his time establishing an international acceptance of UNCSTD at the highest levels.

The remaining staff of the UNCSTD office is a combination of older and younger people, people with some S & T background and people with mostly bureaucratic experience, people with energy and purposefulness as well as people with the more traditional demeanor of a UN employee. The office is already in dire financial straits, and it is doubtful whether it will be able to carry out its task at a sufficiently expert level with only a very small in-house staff and no funds to acquire outside expertise.

The task is a gigantic one, indeed. At the moment, the office tries to carry out a mandate to prepare an 'action plan' even before the national papers have arrived. This task is meaningless and hence the resulting document is not expected to have any functional significance. After the national documents arrive, the office will have to read through the several million words and distill from them some kind of a consensus and common denominator, agenda, and action program for the Vienna meeting itself.

The meeting itself will probably be attended by some 2,000 delegates and their entourage, followed by TV cameras, newspapermen, and other circumstances conducive to encourage political oratory and grandstanding instead of serious, low-keyed, and substantive consultations. Hence, if anything concrete is to come <u>out</u> of Vienna, it will have to be fed <u>into</u> Vienna in an equally concrete form, since the meeting itself will, at best, be able to say 'yes' or 'no' to prearranged resolutions. The effectiveness of the UNCSTD office is therefore crucial, since it must do all the homework and preparation in advance of the meeting in order to be able to propose specific resolutions to the conference.

For this reason, it is absolutely necessary that national papers be as specific in their suggestions and proposals as possible, so that the UNCSTD office can merge these specific proposals into equally specific resolutions for the conference.

There are, of course, many areas in which such resolutions must be drafted. In the area of assistance for the strengthening of scientific and technological infrastructures in the developing countries, I have prepared a set of such resolutions which are listed in the appendix of this talk. I would like to urge everybody in the audience to study them, and if they appear meritorious, to campaign to include them in the various national papers. Apart from the specific content of these resolutions, however, I would like to present them also as an illustration of the degree of specificity that, I believe, is needed if anything concrete is to come out of Vienna.

Each country will send a delegation to Vienna. On the average, I would expect these delegations to consist of 5 - 15 people. It is essential that these delegations consist of a balanced set of scientists, technologists, administrators, and politicians. Previous such conferences would suggest that one need not fear about the politicians being underrepresented, but that strenuous efforts will have to be made to assure adequate representation by scientists and technologists. The definition of a scientist or technologist is a person who, within the last two years, has been directly involved in scientific research or technological development work.

I would, therefore, like to urge everybody in this audience to rally the scientific and technological community in his country for a campaign to get adequate representation on his country's national delegation to Vienna.

So far I suggested that scientists and technologists in every country become active in (a) making a concrete input into the national papers, including the specific consideration of the set of resolutions I appended, (b) participating vigorously in the post-national-paper, pre-conference preparations for UNCSTD, (c) campaigning for an adequate representation on the national delegations. In addition, I would now like to call attention to another need, transcending the specific occasion of the Vienna meeting.

Problems of S & T in developing countries, and specific action to solve these problems have come in the past, whether in writing or orally, predominantly from individuals in the scientifically advanced countries. The few outstanding exceptions in the persons of eloquent, prolific, extraverted, and articulate scientists and technologists from the developing countries themselves only strengthen the rule.

This state of affairs is regrettable, because to many people, authenticity and credibility go only with material emerging from the developing countries themselves, and others claiming to speak 'for' the developing countries can be accused of mingling in other people's business without knowing much about it. Personal accounts of scientists from developing countries, writing impactfully about their own difficulties, also create an emotional effect that cannot be matched by the analyses of 'outsiders', no matter how skillfully they are composed.

Therefore my last appeal to everybody in the audience is to participate, to the best of his ability, in the articulation, in writing and orally at scientific meetings and other occasions, of the problems of science and technology in developing countries and of the concrete measures that can be taken to relieve those difficulties.

References

(1) M.J.Moravcsik, "Science Development - The Building of Science in Less Developed Countries", PASITAM, Bloomington Indiana 47401, USA.

(2) M.J.Moravcsik, Scientists and Development, Journ. of the Sci. Soc. of Thailand, $\underline{1}$, 89 (1975).

(3) M.J.Moravcsik, "The Context of Creative Science" Intersiencia, $\underline{1}$, 71 (1976).

APPENDIX

RESOLUTION

WHEREAS
it is generally recognized that the existence of a functional scientific and technological infrastructure in each country is an indispensible ingredient of the New World Economic Order, and

RECALLING THAT

this country paper repeatedly stressed the crucial importance of international cooperation in the creation of such an infrastructure, particularly in those countries where such an infrastructure is at the present not developed,

BE IT RESOLVED THAT

a NEW STRUCTURE FOR WORLD SCIENCE AND TECHNOLOGY be formed, aiming specifically to assist each country in establishing its own infrastructure.

As an important first step in the implementation of this NEW STRUCTURE FOR WORLD SCIENCE AND TECHNOLOGY.

BE IT RESOLVED THAT

the countries with an already developed scientific and technological infrastructure (henceforth designated as countries A) and the countries where such infrastructures are still in a rudimentary state (henceforth designated as countries B) jointly commit themselves to take the following concrete steps to accelerate the creation of such infrastructures in countries B:

1. Each country A assures that by 1990 key libraries in all countries B are supplied with back issues of those major scientific and technological journals (from 1960 on) which were published in that country A. The number of key libraries in countries B should be one per country B or one per 10 million population of country B, whichever higher. Criteria for selecting major journals should be such that the worldwide total number of journals thus reaching each country B is approximately 1,000. Correspondingly, each country B benefitting from this program commits itself to providing orderly space for the incoming journals, including cataloging and shelfing, and to assuring easy access to such journals by the scientific and technological community.

2. Each country A supplies each key library in each country B with current issues of those major scientific and technological journals which are published in that country A. The transaction of journals should be in exchange for 1/4 of the regular individual subscription rate of that journal and the payment should be made in the currency of country B. The key libraries and the number of journals are to be defined as in Section 1 above. Correspondingly, participating countries B assume obligations similar to those listed in Section 1 above.

3. Each country A assures that for any international scientific and technological conference held in that country, resources are made available to make possible the participation of scientists and/or technologists from countries B, in numbers up to one such participant from all countries B together for each 30 overall participants at that conference, provided that such participants from countries B are available and measure up to the usual scientific

and/or technological criteria used to select participants for that conference. Correspondingly, participating countries B pledge to facilitate the participation of their scientists and/or technologists through all administrative means.

4. Each country A assures that resources are available to support scientists and technologists from that country on short term visits to counterpart institutions or individuals in countries B. These short term visits, with durations up to a month or two, might include several countries B and the number of such visits per year should be one for every 1,000 research scientists and technologists in country A, provided that a sufficient amount of mutual interest exists between the participants of such a visit. The participating countries B correspondingly pledge, during each such visit agreed on, a sum equivalent to US $10 per day, in the currency of the country B in question, toward a partial defrayal of the local living expenses of the visitor.

5. Each country A assures that resources are available to support longer visits by scientists and technologists from countries B to counterpart institutions in country A. These longer term visits are to be up to one year in duration, with occasional renewals for a second year but never beyond that, and the number of these long term visits is to be one for every 300 research scientists and technologists in country A, provided that a sufficient number of qualified applicants are available as judged by the customary standards prevailing in the worldwide scientific and technological community. Correspondingly, participating countries B pledge that, (a) the customary local salary of the scientist or technologist continues to be paid in the currency of that country B to locally maintain the family of the scientist or technologist while he is abroad, and (b) the transportation of the scientist or technologist will be provided by country B on its national airline, if there is such an airline, to a point closest to the institution to be visited that is a regular stop of that national airline.

6. Each country A assures that resources are made available for the placing into scientifically or technologically meaningful practical positions, during the longer academic recesses, students of science or technology from countries B who are being educated toward a degree at an institution of higher learning in country A. The number of such positions is to be up to one for every 500 research scientists and technologists in country A, provided that qualified applicants for such positions are available. The participating countries B, correspondingly, pledge that the time spent in such practical positions will be counted as time spent in the service of country B when matters of seniority and promotion arise during the career of this student.

7. Each country A assures that resources are available for the formation of bilateral links between a group of scientists

or technologists in a country B on the one hand, and a counterpart group in country A, on the other. The nature of the link is to be determined by the two parties involved, but the resources should also allow for the possibility of some collaborative research between the two groups. The number of such bilateral links is to be one for every 2,000 research scientists and technologists in country A, provided that qualified groups are available, as judged by the usual standards of the worldwide scientific and technological community. Correspondingly, the participating country B pledges to contribute, in its own currency, to each such linked group in country B, a sum amounting to 10% of the amount contributed by country A to that bilateral link.

8. Each country A assures that resources are available for the support of technicians from country A stationed in a country B or covering a group of countries B. Such a technician is to be experienced in the repair and maintenance of scientific and technological research equipment, have some ability to offer on-the-job training to local technicians, and he should also have access to minor spare parts from a stockroom which is to be replenished through the eventual arrival of parts ordered through the usual purchase procedures of country B. The number of such technicians is to be one for every 10,000 research scientists and technologists in country A, and the distribution of such technicians among the participating countries B should be such that no country should have less than one-half such technicians from among all the technicians contributed by all countries A together. Correspondingly, each participating country B pledges that (a) it expedites to the greatest possible extent the order and delivery of the spare parts borrowed from the revolving stock room, free of custom duties, and (b) it contributes, in its local currency, an equivalent of US $10 for each day the technician spends in country B, in order to partially defray his living expense.

9. Each country A assures that it will make a contribution annually into an overall fund for home fellowships, to be administered by a designated agency of the United Nations or some other international entity so designated. Home fellowships are to be used for the salary and research expenses of outstanding and promising scientists and technologists from countries B for work by them to be carried out in countries B. The support of such researchers should be comparable to the level customary in most countries A. Each country A should provide funds sufficient for the support of one scientist or technologist for every 10,000 research scientists and technologists in that country A. The fellowships should be awarded on the recommendation of a committee of scientists and technologists representing the participating countries A, the selection being made from among individual applications received by this commitee from candidates from countries B.

The provisions of this resolution are to be in effect until 1990, at which time a review is to be undertaken to ascertain the effectiveness of these provisions and the advisability of continuing them in the original or a modified form.

APPENDIX I

LIST OF INVITED SEMINAR SPEAKERS

Ahmad Ali   Der Universitat Hamburg,
                            Hamburg

M.A.K. Lodhi   Texas Technical University,
                            Lubbock, USA

APPENDIX II

LIST OF SEMINARS

Ahmad, M	Perspectives on Pakistan's Energy Problems
Ahmed, Shafique	Relativistic Two Fermion Equation
Ali, A	i) Weak Interactions at High Energies
	ii) Weak Mixing and CP Violation
Aslam, J.	Powder Neutron Diffraction Studies of UF_4
Butt, N.M.	Mossbaüer Diffraction Methods for Solid State Physics Studies
Faude, D.	Development of Nuclear Energy in Federal Republic of Germany
Fayyazuddin	Weak Decays of Heavy Mesons
Haroon, M.R.	Preliminary Reactor Physics Calculation for a Fluidised Bed Nuclear Reactor Concept
Hofstadter, R.	My Life as a Physicist
Ishaq, A.H.M.	PDP 11/45 at PINSTECH
Khakpour, A.	Difuseness and Energy of the Nuclear Matter Surface
Lodhi, M.A.K.	i) Short Range Correlation with Reference to Structure of Light Nuclei

	ii) Fission and Alpha Decay Anomalies with reference to Superheavy Elements
Magd, A.Y.Abdul	i) Interaction of Relativistic Particles with Nuclei
	ii) Angular Momentum Coherence in Heavy Ion Reactions
Mahmud, Bashiruddin	Problems of Energy with Respect to Transfer of Technology in Developing Countries
Malik, G.M.	Computing Facilities at Engineering University, Lahore
Moravcsik, M.J.	i) Energy Supplies and Space Colonization
	ii) An Overview of Quantum Mechanical Few Body Problems
Murray, R.L.	Nuclear Proliferation and Safeguards
Nayyar, A.H.	Solition Propagation in Magnetic Transitions
Panchapakesan, N.	i) Particle Creation in a De Sitter Universe
	ii) Changing Superconducting Transition Temperatures in Materials
Paul, W.	Defects in Amorphous Semiconductors
Qadir, A.	Black Holes and Neutron Stars
Qazi, M.N.	Problems of Research in Reactor Physics Around Reactor
Riazuddin	Electromagnetic Decays of Mesons
Shaukat, M.A.	Backward Pion-Nucleon Scattering
Sheikh, P.	Inertial Confinement Fusion Reactors
Shrestha, V.M.	Optimum Tilt Angle for Solar Collector
Srinivasan, M.	i) A Brief Introduction to Indian Nuclear Power Programme
	ii) Criticality Experiments at Trombay
Tounsi, A.	Statistical Bootstrap
Yaqub, M.	Thermal Energy Transfer Across Liquid Helium

APPENDIX III

LIST OF PARTICIPANTS

Name	Institution	Country
Ahmad, Iqbal	PINSTECH, Rawalpindi	Pakistan
Ahmad, S.Khurshid	PINSTECH, Rawalpindi	Pakistan
Ahmad, Masud	Pakistan Atomic Energy Commission, Islamabad	Pakistan
Ahmad, Nasir,	PINSTECH, Rawalpindi	Pakistan
Ahmad, Nisar	PINSTECH, Rawalpindi	Pakistan
Ahmad, Shujaat	PINSTECH, Rawalpindi	Pakistan
Ahmed, Aminuddin	PINSTECH, Rawalpindi	Pakistan
Ahmed, Arif	Pakistan Atomic Energy Commission, Islamabad	Pakistan
Ahmed, Farooq	Pakistan Atomic Energy Commission, Islamabad	Pakistan
Ahmed, Firoz	University of Karachi, Karachi	Pakistan
Ahmed, Irshad	PINSTECH, Rawalpindi	Pakistan
Ahmed, M.Jamil	Pakistan Atomic Energy Commission, Islamabad	Pakistan
Ahmed, Mesbahuddin	University of Dacca, Dacca	Bangladesh
Ahmed, Mohammad	Pakistan Atomic Energy Commission, Islamabad	Pakistan
Ahmed, S.Mohammad	University of Multan, Multan	Pakistan
Ahmed, Shafique	'G.Marconi' Universita degli Studi-Roma Plazzale delle Scienze Rome	Italy
Aitquad, Nasreen	Aligarh Muslim University, Aligarh	India
Ajayi, O.B.	University of Ife, Ile-Ife	Nigeria
Akhtar, Javed	PINSTECH, Rawalpindi	Pakistan
Akhtar, K.M.	PINSTECH, Rawalpindi	Pakistan
Akhtar, Nasim	University of Karachi, Karachi	Pakistan
Akhtar, Riaz A.	PINSTECH, Rawalpindi	Pakistan
Al-Badri, M.B.H.	University of Baghdad, Baghdad	Iraq
Al-Hassoun, A.	University of Baghdad, Baghdad	Iraq
Ali, Ahmed	der Universitat Hamburg	Hamburg
Ali, Waqar	Quaid-i-Azam University, Islamabad	Pakistan
Ansari, Salim	PINSTECH, Rawalpindi	Pakistan
Anwar, Mohammad	Pakistan Atomic Energy Commission, Islamabad	Pakistan

Name	Institution	Country
Asif, G.M.	PINSTECH, Rawalpindi	Pakistan
Aslam, Javed	Pakistan Atomic Energy Commission, Islamabad	Pakistan
Aslam, Mohammad	Pakistan Atomic Energy Commission, Islamabad	Pakistan
Atta, Ashraf	PINSTECH, Rawalpindi	Pakistan
Atta, Manzoor A.	Pakistan Atomic Energy Commission, Islamabad	Pakistan
Awan, Saleem	PINSTECH, Rawalpindi	Pakistan
Azhar, Iqbal Ali	PINSTECH, Rawalpindi	Pakistan
Aziz, Abdul	PINSTECH, Rawalpindi	Pakistan
Bashir, Zia	Pakistan Atomic Energy Commission, Islamabad	Pakistan
Beg, Altaf A.	Ministry of Science & Technology Islamabad	Pakistan
Beg, M.M.	Pakistan Atomic Energy Commission Islamabad	Pakistan
Ben, M.A.	University of Niawey, Niaway	Niger
Bernacca, P.L.	Asiago Observatory Asiago Viencenia,	Italy
Bhatti, Miss Nasim	Pakistan Atomic Energy Commission Islamabad	Pakistan
Bukhari, Khalid M.	PINSTECH, Rawalpindi	
Bukhari, S.J.H.	Pakistan Atomic Energy Commission Islamabad	Pakistan
Butt, N.M.	PINSTECH, Rawalpindi	Pakistan
Chaghtai, M.S.	Aligarh Muslim University, Aligarh	India
Chaudhary, M.Saeed	PINSTECH, Rawalpindi	Pakistan
Cheema, S.U.	Pakistan Atomic Energy Commission Islamabad	Pakistan
Dabiri, A.E.	Arya-Mehr University of Tech.Tehran	Iran
Dahanayke, C.	University of Sari Lanka, Kelaniya	Sri Lanka
Dastgir, Ghulam	Quaid-i-Azam University, Islamabad	Pakistan
Dayday, N.	Cekmece Nuclear Research & Training Centre, Istanbul	Turkey
Durrani, Ijaz-ur Rehman	Pakistan Atomic Energy Commission, Islamabad	Pakistan
Elbek, E.	Neils Bohr Institute, Copenhagen	Denmark
Farinelli, Ugo	CNEN-CSN, Cassacia, Rome	Italy
Faruq, M.Umar	PINSTECH, Rawalpindi	Pakistan
Faude, D.	Institut fur Angewandte, Karlsruhe	Germany
Fayyazuddin	Quaid-i-Azam University, Islamabad	Islamabad

APPENDIX

Name	Institution	Country
Gaafar, M.A.	Helemiyt El-zaitum, Cairo	Egypt
Ghani, Abdul	Pakistan Council of Scientific and Industrial Research, Karachi	Pakistan
Gondal, M.A.	PINSTECH, Rawalpindi	Pakistan
Gul, Subhan	PINSTECH, Rawalpindi	Pakistan
Hashmi, Jamshed	Kanupp, Karachi	Pakistan
Haq, Ikramul	Pakistan Atomic Energy Commission Islamabad	Pakistan
Haque, Ziaul	Land Reclamation Punjab, Lahore	Pakistan
Haroon, M.R.	Arya-Mehr University of Technology Tehran	Iran
Hasnain, S.A.	PINSTECH, Rawalpindi	Pakistan
Hasni, S.M.	PINSTECH, Rawalpindi	Pakistan
Hofstadter, R.	Stanford University, Stanford	U.S.A.
Hofstadter, R.Mrs.	Stanford University, Stanford	U.S.A.
Hussain, A.Z.	Nuclear Research Centre, Cairo	Egypt
Hussain, Ather	PINSTECH, Rawalpindi	Pakistan
Hussain, Gulzar	PINSTECH, Rawalpindi	Pakistan
Hussain, Manzoor	Engineering University, Lahore	Pakistan
Hussain, S.D.	Pakistan Atomic Energy Commission Islamabad	Pakistan
Husseni, S.D.	Pakistan Atomic Energy Commission Islamabad	Pakistan
Iqbal, M.Saleem	Quaid-i-Azam University, Islamabad	Pakistan
Irfan, M.	Aligarh Muslim University, Aligarh	India
Ishaq, A.F.M.	Pakistan Atomic Energy Commission Islamabad	Pakistan
Jameel, M.	Pakistan Atomic Energy Commission Islamabad	Pakistan
Javed, N.A.	Pakistan Atomic Energy Commission Islamabad	Pakistan
Jawad, S.M.Abdul	Royal Scientific Society, Aman	Jordan
Kadis, M.S.Mohd	University of Malysia, Kuala Lampur	Malaysia
Karim, M.Abdel	Faculty of Science, Tripoli	Libya
Kayani, Salim Ahmed	University of Engineering and Technology, Lahore	Pakistan
Keita, B.	Faculty des Sciences Dakar-Senegal	Senegal
Khakpour, Asha	17th St.Bokharest Ave, Tehran	Iran
Khalid, Z.M.	PINSTECH, Rawalpindi	Pakistan
Khan, Abdul Rashid	Islamia University, Bahawalpur	Pakistan

Name	Institution	Country
Khan, Asghar Ali	Pakistan Atomic Energy Commission, Islamabad	Pakistan
Khan, Hayatullah	Peshawar University, Peshawar	Pakistan
Khan, Iqbal Ahmed	University of Khartoum, Khartoum	Sudan
Khan, M.Aslam	DESTO, Chaklala	Pakistan
Khan, M.Aslam	PINSTECH, Rawalpindi	Pakistan
Khan, Mustafa Yar	Government College, Lahore	Pakistan
Khan, Q.H.	Pakistan Atomic Energy Commission, Islamabad	Pakistan
Khan, Sohail Aziz	University of Baluchistan, Quetta	Pakistan
Khawaja, E.E.	PINSTECH, Rawalpindi	Pakistan
Khazal, K.A.R.	University of Basrah, Basrah	Iraq
Khurshid, S.	PINSTECH, Rawalpindi	Pakistan
Lane, J.A.	International Atomic Energy Agency, Vienna	U.S.A.
Lodhi, M.A.K.	Texas Tech. University, Lubbock	U.S.A.
Magd, A.Y.Abdul	Cario University, Gize	Egypt
Magasu, L.M.	University of Technology, LAE P.N.G.	New Guinea
Mahmood, Bashir-uddin	Pakistan Atomic Energy Commission, Islamabad	Pakistan
Mahmood, Sajjad	Quaid-i-Azam University, Islamabad	Pakistan
Mahmood, S.Tahir	PINSTECH, Rawalpindi	Pakistan
Majeed, C.A.	Pakistan Atomic Energy Commission, Islamabad	Pakistan
Malik, Ghulam Mohammad	University of Engieering and Technology, Lahore	Pakistan
Mir, Nazir Ahmed	University of Multan, Multan	Pakistan
Mirza, M.Ilyas	Quaid-i-Azam University, Islamabad	Pakistan
Moravscik, M.J.	University of Oregon, Eugene	U.S.A.
Murray, R.L.	North Carolina State University, Raleigh	U.S.A.
Murtaza, G.	Quaid-i-Azam University, Islamabad	Pakistan
Mustafa, Noor-ul	PINSTECH, Rawalpindi	Pakistan
Naqvi, S.Munir Mehdi Raza	University of Karachi, Karachi	Pakistan

APPENDIX

Name	Institution	Country
Naseem, M.	Pakistan Atomic Energy Commission, Islamabad	Pakistan
Naseem, Tufail	Pakistan Atomic Energy Commission, Islamabad	Pakistan
Nayyar, A.H.	Quaid-i-Azam University, Islamabad	Pakistan
Nazar, Fateh Mohammad	Punjab University, Lahore	Pakistan
Niazi, Ahsan Khan	PINSTECH, Rawalpindi	Pakistan
Panchapakesan, N.	University of Delhi, Delhi	India
Paul, W.	Harvard University, Cambridge	U.S.A.
Pervez, Masud	Quaid-i-Azam University, Islamabad	Pakistan
Qadir, Asghar	Quaid-i-Azam University, Islamabad	Pakistan
Qazi, M.N.	PINSTECH, Rawalpindi	Pakistan
Qureshi, Farhat Saeeda (Miss)	Agriculture University, Faisalabad	Pakistan
Qureshi, I.H.	PINSTECH, Rawalpindi	Pakistan
Qureshi, K.Q.	Pakistan Atomic Energy Commission, Islamabad	Pakistan
Rashid, Khalid	PINSTECH, Rawalpindi	Pakistan
Razmi, M.S.K.	Quaid-i-Azam University, Islamabad	Pakistan
Rehman, Inamur	PINSTECH, Rawalpindi	Pakistan
Rehman, Fazal	Peshawar University, Peshawar	Pakistan
Riazuddin	Quaid-i-Azam University, Islamabad	Pakistan
Roberts, K.V.	Chlham Lab, Abingdon	U.K.
Sadiq, Abdullah	PINSTECH, Rawalpindi	Pakistan
Sadiq, Nasim A.	PINSTECH, Rawalpindi	Pakistan
Saeed, M.	University of Khartoum, Khartoum	Sudan
Saghir, Ahmad	Ministry of Science & Technology, Islamabad	Pakistan
Samdani, G.	Chasnupp, Islamabad	Pakistan
Sani, L.	ENEL-DCO VIA, G-B MARONI	Italy
Sehgal, M.L.	Aligarh Muslim University, Aligarh	India
Shafee, Ahmed	86 Indire Road, Dacca	Bangladesh
Shafique, M.	Pakistan Atomic Energy Commission, Islamabad	Pakistan

Name	Institution	Country
Shahid, K.A.	Pakistan Atomic Energy Commission, Islamabad	Pakistan
Shaikh, M.Afzal	Pakistan Atomic Energy Commission,	Pakistan
Shaikh, P.A.	PINSTECH, Rawalpindi	Pakistan
Shaikh, M.Saleem	Pakistan Atomic Energy Commission, Islamabad	Pakistan
Shaukat, Mumtaz A.	Punjab University, Lahore	Pakistan
Shrestna, V.M.	Kirtipur University, Kathmandu	Nepal
Siddique, K.M.	DESTO, Chaklala	Pakistan
Slack, L.	Virginia Polytechnical Institute & State University, Blacksburg	U.S.A.
Srinivasan, M.	Bhabha Atomic Research Centre, Bombay	India
Syed, H.M.	PINSTECH, Rawalpindi	Pakistan
Tan, B.Ch.	36 SS 2/41, Subang	Malaysia
Tariq, M.M.	Pakistan Atomic Energy Commission, Islamabad	Pakistan
Tennakone, K.	Vidyodaya University, Nugede	Sri Lanka
Tounsi, Ahmed	Theoreque Et Hautes Energies Tour Paris	Algeria
Runcel, R.	Cekmece Nuc. Res. & Trg. Centre, P.K.I. Havaalani, Istanbul	Turkey
Ullah, Hasib	Pakistan Atomic Energy Commission, Islamabad	Pakistan
Waheed, Abdul	Pakistan Atomic Energy Commission, Islamabad	Pakistan
Yaldram, Khawaja	Pakistan Atomic Energy Commission, Islamabad	Pakistan
Yaqub, M.	The Ohio State University, Columbus	U.S.A.
Yasin, Mehboob	Pakistan Atomic Energy Commission, Islamabad	Pakistan
Zafar, M.S.	Punjab University, Lahore	Pakistan
Zaidi, H.A.	PINSTECH, Rawalpindi	Pakistan
Zaidi, S.M.H.	PINSTECH, Rawalpindi	Pakistan
Zaidi, S.M.N.	PINSTECH, Rawalpindi	Pakistan
Zaidi, S.M.S.	PINSTECH, Rawalpindi	Pakistan

APPENDIX

Name	Institution	Country
Zaidi, Z.H.	210 West End Road, Merrut	India
Zauq, M.Hanif	PINSTECH, Rawalpindi	Pakistan

INDEX

Accretion,
  critical rate, 482-483
  disks, 483-486
  observational evidence of
    Wakes, 491-492
Advective equation, 419-420
Amorphous semiconductors, 295-348
Amplification factor, 422
Anderson's Theory, 339-441
  modification of, 341-345
Area weighting, 443-444
Atomic arrangement,
  amorphous cluster models, 304
  micro crystallite models, 304
  random network models, 304

Band structure, 338
Black Hole
  accretion disk around, 483-486
  in binary systems, 476-492
  in globular clusters, 492-495
  isolated, 495-498
Boltzmann's equation, 234
Breeder
  see fast breeder
Breeding
  principle of, 77

Chalcogenide glasses, 335-347

Chalcogenide materials, 320-332
Charm, 524-530
  experimental evidence for 535-537
Coal, 57, 70, 71
Coding systems, 252-263
Computational physics, 406-433
  evolutionary calculations in, 406-407
Computer,
  advantages of, 389-392
  complexity of, 392
  disadvantages of, 389-392
  environment, 260
  hardware, 397-398
  powers and limitations of, 385-406
  relevance to physics, 386-389
  reliability of, 392-395
  software, 402
  and theoretical physics, 405-406
Computer-Man analogy, 403-404
Conservation
  laws, 426
  parity, 507
Crank-Nicholson schemes, 424, 430-431

Data pool, 254

Density of states,
  intrinsic, 308-315
Development, 547, 549-550
  and science, 547-555
Difference
  2-level scheme, 424-426
  eulerian method, 436-437
Differential equation,
  method of modified, 420-421
Diffraction, 301-303
  and modelling studies of a-Ge,
    305-308
Diffusion equation, 238
  explicit solution of, 430
Digital information, 389
Disorder,
  nature of, 298
Disordered materials
  determination of structure,
    298-301
  preparation of, 296-298

Eigenvalue problem, 250-252
Electromagnetic interactions,
  515, 519-537
Electron beam fusion, 113
Electronic band structure and
  properties, 308
Electrostatic systems
  NGP method for, 441-443
Energy
  additional supply possibilities,
    105-107
  aspects of economy, 63-66
  conservation, 36-46, 50
  demand and economic development, 23-36
  demand and GNP, 29-36
  demand in a growing world, 23-26
  and food production, 46-49
  gap, 51-60
  geothermal, 104-105
  global picture, 3-4
  growth of consumption, 66-69
  nuclear, 73-74
  ocean thermal, 106
  principal supply options, 69-70

Energy (cont'd)
  requirements for Western Europe,
    200-205
  situation in Western Europe,
    193-200
  solar, 89-93
  tidal, 106
  traditional sources, 70-73
  wave, 107
  see also Pakistan
Eulerian difference method, 436-437
Eulerian mesh,
  uniform, 417-418
Evolutionary calculations, 406-407

Fast breeder reactor, 73, 78, 79, 226
  international development, 80-83
  market introduction, 88-89
  safety considerations, 84-86
  technical concept of, 83-84
Fields and Particles, 407-408
Fission breeder, 111
Food production
  and energy, 46-49
Fortran IV
  olympus programming system,
    450-469
Fourier modes, 418-419
Fuel
  reprocessing, 223-226
  resources, 75-76, 79, 80
Fuelcash computer code, 159
Fuel costs, 154-155, 160
Fuel cycle, 155-157
  conventional, 277
  economic considerations, 215-216
  enrichment services, 221-223
  once through, 278
  problems relating to, 215-226
Fusion, 111, 112, 113

Gas, 70
Gauge group,
  $SU(2) \otimes U(1)$, 523-537
  color, 537-541

INDEX

Geothermal energy, 104-105
Gross National Product
  and energy demand, 29-31
  material content of, 26-29

Hydropower, 57, 105

Information engineering, 402-403
Interface, 253
Iterative solution, 246-250

Lagrangian
  particle method, 438
  plane isentropic scheme, 431-433
Laser,
  fusion, 111, 112, 118-120
  plasma coupling, 120-142
Leap frog,
  stability analysis, 428
  2D scheme, 428
Linearization, 418-419
Luminosity,
  critical, 482

Man-Computer analogy, 403-404
Maxwell equations, 429-430
Mesh,
  techniques, 417-433
  uniform eulerian, 417-418
Microcrystallite models
  magnetic properties of, 336
Mirror machines, 112
Modular coding systems, 252-263
Module, 252-253
Montecarlo methods, 239-240

Natural gas, 56, 70-71
Neutron transport theory, 285-291
Noise suppression
  techniques, 447-448
Nuclear desalination
  plants, 186-187
Nuclear energy,
  long term option, 73
  present state of, 73-74
  in Western Europe, 200, 205-210
Nuclear Fuel cycle
  costs, 155
Nuclear Fusion, 101-104

Nuclear non-proliferation,
  academic aspects of, 275-284
  political aspects of, 271-275
Nuclear Power,
  capital costs, 163
  cost trends in developing
    countries, 172-183
  cost trends in U.S., 149-154
  industrial aspects of plants, 210-212
  long term economics of, 183-187
  problems relating to plant
    construction, 210-215
  safety problems in plants, 213-215
  station siting for plants, 212-213

Ocean thermal energy, 106
Oil,
  concept of gap, 55
  substitute of, 55
Olympus,
  Fortran IV programming system, 450-469
Optical absorption, 386
Optical excitation, 322
Optimization techniques,
  for particle codes, 445-447
Optimum representation, 410-414
Orcost computer code, 170

Pakistan,
  agricultural energy needs, 13-15
  energy consumption in, 4-8
  energy deficits, 20-21
  energy projections, 15
  energy reserves, 8, 11
  future energy consumption, 12
  growth of electric power, 17-20
Parity,
  conservation, 503
Particle
  simulation, 440-441
Photo luminiscence, 337
Photostructural changes, 345-348
Photosynthesis, 101
Photovoltaic
  materials, 369-379
  solar energy conversion, 100-101

Plasma,
  simulation, 433-450
Plutonium recycle, 279
  spiked, 281
  coprocessed uranium and, 282
Poisson's equation, 444-445
Polarisation, 503
Power,
  hydro, 57-105
  nuclear, 163-213
Power generation,
  cost in the United States, 170
Power cost comparison,
  the system analysis approach 176
Proliferation,
  some remarks on, 86-88

Reaction matrix, 504
Reactor physics,
  computational methods in, 234-241
  requirements in a developing country, 263-270
  short review of, 230-234
Recombination, 322
Representation,
  choice of, 415-417
Reprocessing,
  irradiated fuel, 223-226
Roche Lobe overflow, 487

Science
  and development, 547-555
  and technology, 547-549
Silicon,
  amorphous, 323-335
  hydrogen alloys, 323-335
Simulation,
  development of a model, 399-402
  techniques, 444-445
Solar energy, 58, 89-93, 111
Solar materials, 353-359
Solar reversible window, 359-369
Solar thermal conversion,
  centralised, 97-100
  decentralised, 93-97
States,
  intrinsic density of, 308-315
Stellar wind, 487-491

Strong interactions, 537-541
Structural modeling, 303-305
Strucutre,
  determination of, 298-301

Target designs, 148
Thermal materials, 353-359
Tidal energy, 106
Transport, 49-50, 322, 335
Transport theory,
  unified neutron, 285-291
Tokamak, 103, 104, 112, 113

Unification,
  of weak and electromagnetic interactions, 515-519
United Nations,
  conference on science and technology in development 1979, 552-585
Uranium,
  cost trends, 157, 183
  procurement of, 216-221

Vlasov equations, 433

WASP code, 176
Water power, 105
Waterbag method, 437-438
Wave energy, 107
Weak inteaction, 515-519
  neutral, 515, 524-535

X-ray,
  heating of normal stars, 491
  optical identification of sources, 478
  Types of accretion in binaries, 486-491
  yield by mass accretion, 478

RAYMOND H. FOGLER LIBRARY
DATE DUE